Cell Biology of Bacteria

A subject collection from *Cold Spring Harbor Perspectives in Biology*

ALSO FROM COLD SPRING HARBOR LABORATORY PRESS

OTHER SUBJECT COLLECTIONS FROM COLD SPRING HARBOR PERSPECTIVES IN BIOLOGY

Cell–Cell Junctions
Generation and Interpretation of Morphogen Gradients
Immunoreceptor Signaling
NF-κB: A Network Hub Controlling Immunity, Inflammation, and Cancer
Symmetry Breaking in Biology
The Origins of Life
The p53 Family

RELATED LABORATORY MANUALS

Molecular Cloning: A Laboratory Manual, Third Edition
Single-Molecule Techniques: A Laboratory Manual

OTHER RELATED TITLES

A Genetic Switch, Third Edition, Phage Lambda Revisited

Cell Biology of Bacteria

A subject collection from *Cold Spring Harbor Perspectives in Biology*

EDITED BY

Lucy Shapiro
Stanford University

Richard M. Losick
Harvard University

COLD SPRING HARBOR LABORATORY PRESS
Cold Spring Harbor, New York • www.cshlpress.com

Cell Biology of Bacteria

A Subject Collection from *Cold Spring Harbor Perspectives in Biology*
Articles online at www.cshperspectives.org

Printed in the United States of America

Publisher	John Inglis
Acquisition Editor	Richard Sever
Project Manager	Joan Ebert
Production Editor	Kaaren Kockenmeister
Cover Designer	Denise Weiss

Front cover artwork: This image shows segmented features from an ECT reconstruction of an *H. neapolitanus* cell. The inner and outer membranes are semi-transparent. Invaginations of the inner membrane are visible near the cell poles. Polyphosphate granules are depicted in purple and carboxysomes are depicted in yellow. A partially-formed carboxysome is shown with some internal Rubiscos visible in red. Comb-like structures connecting polyphosphate granules to neighboring carboxysomes are shown in blue. Two filamentous structures are depicted in red. The small yellow structures in the bottom-right region of the cell are probable poly-ribosomes. These features highlight the complex organization of this bacterial cell. (Courtesy of Cristina V. Iancu, Dylan M. Morris, H. Jane Ding, and Grant J. Jensen.)

Library of Congress Cataloging-in-Publication Data

Cell biology of bacteria : a subject collection from Cold Spring Harbor Perspectives in Biology / edited by Lucy Shapiro, Richard M. Losick.
p. cm.
Includes bibliographical references and index.
ISBN 978-0-87969-907-9 (hardcover : alk. paper)
1. Bacteria. 2. Cytology. I. Shapiro, Lucy. II. Losick, Richard III. Title

QR74.8.C447 2010
572.8′293--dc22

2010038471

10 9 8 7 6 5 4 3 2 1

All World Wide Web addresses are accurate to the best of our knowledge at the time of printing.

All Cold Spring Harbor Laboratory Press publications may be ordered directly from Cold Spring Harbor Laboratory Press, 500 Sunnyside Blvd., Woodbury, New York 11797-2924. Phone: 1-800-843-4388 in Continental U.S. and Canada. All other locations: (516) 422-4100. FAX: (516) 422-4097. E-mail: cshpress@cshl.edu. For a complete catalog of Cold Spring Harbor Laboratory Press publications, visit our website at http://www.cshlpress.com/.

Contents

Preface

Arthur Kornberg admonished us to "never do clean experiments on dirty enzymes." There is a formidable history of the bacterial cell as an exquisite system to analyze the inner workings of the smallest and simplest cell. Biochemists broke open the bacterial cell, collected the soup, and carried out chemical analyses of its constituent parts and biosynthetic pathways. The biochemical clarity provided by this approach provided much of our knowledge about the reaction mechanisms that form the basis of molecular biology. The combination of in vitro biochemistry and the elegant logic of genetic manipulation led to the general perception that we understood the operation of the bacterial cell. But this was a misleading perception because it did not take into account something crucial: the inner and dynamic organization of the bacterial cell.

Little more than two decades ago, the leading microbiology textbooks of the day depicted the bacterial cell as a relatively amorphous vessel in which proteins and DNA were distributed in a relatively haphazard fashion. The bacterial cell was sometimes considered to be a "bag of enzymes" and its content a "cell sap." Electron microscopy revealed a layered cell envelope and various appendages, such as flagella and pili, that protruded from the cells. Ribosomes were seen as broadly distributed puncta across the cytoplasm, with the nuclear material occupying a broad central zone. It was not until the 1990s with the application of immunoelectron and immunofluorescence microscopy and the introduction of the Green Fluorescent Protein that a new era of bacterial cell biology commenced. As we now know, and despite its small size, the bacterial cell has an intricate and dynamic three-dimensional organization that is inseparable from its biology. It is this revolution in our understanding of the bacterial cell that is the subject of this volume.

The early chapters by Toro and Shapiro and Thanbichler in this book reveal the extraordinary three-dimensional organization of the bacterial chromosome, the nucleoid, its dynamic behavior during the cell cycle, and the coupling mechanisms that coordinate chromosome dynamics with cell division. A further, major advance in the cytology of bacteria was the discovery that like eukaryotic cells, these most primitive of cells have a dynamic cytoskeleton that guides cytokinesis, growth of the cell envelope, and the movement of chromosomes. The chapter by Shaevitz and Gitai tells us about the working of tubulin-like and actin-like cytoskeletal proteins in the bacterial cell. The next two chapters, by Rudner and Losick and by Kirkpatrick and Viollier, focus on the localization of specific proteins within the cell, the underlying mechanisms that govern protein positioning, and how this positioning gives rise to polar structures.

We don't usually think of bacteria as having organelles, but classic experiments by Leon Heppel revealed that Gram-negative bacteria have an outer compartment, the periplasm, that is a focus of great interest, as presented by Silhavy, Kahne, and Walker. Other bacteria also have specialized membrane-bounded compartments, such as the magnetosome, whose function and assembly are described in the chapter by Murat, Byrne, and Komeili. In addition, bacteria have macromolecular machines embedded in their membranes that carry out specialized functions. Erhardt, Namba, and Hughes describe the exquisite ultrastructure of the flagellum and type III injectisome and how these motors pump proteins out of cells. Another category of macromolecular machine mediates the transport of highly charged DNA molecules across membranes. Burton and Dubnau tell us how these DNA translocases carry out this feat.

The next series of chapters focus on multicellularity. Bacteria were traditionally viewed as solitary creatures that go about their business individually. Now we know that many, indeed likely most, bacteria form complex assemblies of cells. Two classic examples are fruiting body formation in myxobacteria, as presented by Kaiser, Robinson, and Kroos, and heterocyst formation, as presented by Kumar, Mella-Herrera, and Golden. In both cases, multicellular arrays—three-dimensional in the case of the former and one-dimensional in the case of the latter—of cells communicate with each other to govern the differentiation of specialized cell types. Another more prevalent example of multicellularity is the formation of architecturally complex communities of cells on surfaces. These biofilms, which are important in medicine, agriculture, and industry, are the subject of the review by López, Vlamakis, and Kolter.

This volume would not be complete without a peek into the future. A second revolution is now unfolding, driven by electron cryotomography, as explained by Tocheva, Li, and Jensen, single-molecule and super-resolution imaging as presented by Biteen and Moerner, and automated quantitative live-cell fluorescence microscopy, as summarized by Fero and Pogliano. These explosive advances in combination with the availability of full genome sequences of hundreds of bacterial species and innovative bioinformatics are providing a new and deeper understanding of biochemical processes in the living cell. Indeed, we now know that the dynamic, three-dimensional organization of the bacterial cell plays an inextricable role in the regulatory mechanisms that control life.

We are grateful to the outstanding investigators who contributed to this volume and to Richard Sever, Kaaren Kockenmeister, and Joan Ebert of Cold Spring Harbor Laboratory Press for not letting us drop the ball.

Lucy Shapiro
Stanford University

Richard M. Losick
Harvard University

Bacterial Chromosome Organization and Segregation

Esteban Toro and Lucy Shapiro

Department of Developmental Biology, Beckman Center, Stanford University School of Medicine, Stanford, California 94305

Correspondence: etoro@stanford.edu

Bacterial chromosomes are generally ~1000 times longer than the cells in which they reside, and concurrent replication, segregation, and transcription/translation of this crowded mass of DNA poses a challenging organizational problem. Recent advances in cell-imaging technology with subdiffraction resolution have revealed that the bacterial nucleoid is reliably oriented and highly organized within the cell. Such organization is transmitted from one generation to the next by progressive segregation of daughter chromosomes and anchoring of DNA to the cell envelope. Active segregation by a mitotic machinery appears to be common; however, the mode of chromosome segregation varies significantly from species to species.

The DNA molecule is a remarkably simple and elegant storage medium for genetic information. However, linear encoding of this kind demands an inordinately long molecule, and so, bacterial chromosomes are much longer than the cells in which they reside. For example, *Caulobacter crescentus* packages a 1.3 mm (4.0 Mbp) genome in a 2 micron cell. This spatial constraint creates a daunting organizational problem that is exacerbated on DNA replication, an act that not only doubles the amount of DNA in the cell, but also creates topologically linked molecules. Watson and Crick published a companion paper to their 1953 DNA structure in which they pointed out that the elegant simplicity of the copying mechanism they proposed was complicated by the fact that the two strands wind around each other. Hence, to create two separate chromosomes, it is necessary to unlink the original two strands, which, even in the case of a bacterial genome, amounts to unwinding several hundreds of thousands of turns. They wrote: "Although it is difficult at the moment to see how these processes occur without everything getting tangled, we do not feel that this objection will be insuperable" (Watson and Crick 1953). Indeed, in 1971, James Wang purified the first of a class of proteins we now know as DNA topoisomerases (Wang 1971), which solve this particular problem by catalyzing the passage of single strands or duplexes of DNA through each other (Schoeffler and Berger 2008).

Topoisomerases thus provide a solution to the problem of DNA tangling within the cell.

 doi: 10.1101/cshperspect.a000349
Cite this article as *Cold Spring Harb Perspect Biol* 2010;2:a000349

Other, less biochemically tractable aspects of DNA spatial organization, however, received little attention for a considerable time, partly for technical reasons, but also because bacteria were seen as the proverbial "bag of enzymes," i.e., small enough for diffusive processes to dominate and thus not requiring any spatial organization. We now know that the bacterial cell is in fact highly organized (Thanbichler and Shapiro 2008), and the chromosome is no exception.

Here, we concentrate on recent findings about the dynamic high-order organization of the bacterial chromosome; that is, how is DNA organized within the cell and how is it segregated faithfully to each daughter cell on division? Our understanding of these two questions has progressed greatly in the past few years, aided in large part by advances in imaging technologies. For the sake of clarity, we explore each question independently. However, as will be clear throughout, DNA segregation and organization are intimately linked.

DNA ORGANIZATION WITHIN THE CELL

The Early Years and the Rosette Model

The very first investigations into the structure of the bacterial nucleoid came in the form of light microscopy of stained specimens (e.g., Feulgen and Giemsa staining), and electron microscopy of thin sections (reviewed in Robinow and Kellenberger 1994). These studies showed that although the nucleoid is mostly separate from the rest of the cytoplasm, it is not bound by a nuclear membrane (Robinow 1956). Some DNA strands extrude from the nucleoid into the cytoplasm like loose pieces in an otherwise tight ball of yarn (Kellenberger 1991), and autoradiography of transcribing genes suggested that these excrescences contain most of the transcriptionally active DNA (Ryter and Chang 1975).

As important as they were, these first studies were based on only limited resolution. In 1971, Stonington and Pettijohn opened a new window by developing a method to gently lyse *Escherichia coli* cells in solutions of high ionic strength to isolate "folded" chromosomes, i.e., free DNA that does not increase the viscosity of the solution. These isolated chromosomes contained nearly all of the DNA present in the lysate, some protein, and all nascent RNA (Stonington and Pettijohn 1971). Depending on the conditions of isolation, chromosomes were obtained in either membrane-associated or membrane-free form (Kavenoff and Bowen 1976; Kavenoff and Ryder 1976).

When imaged by electron microscopy, isolated chromosomes appeared as unbroken rosettes with a central core from which several tens of plectonemic loops radiate out (Fig. 1) (Delius and Worcel 1974; Kavenoff and Ryder 1976). This organization was sensitive to RNAse, suggesting a role for RNA in maintaining the integrity of the core (Kavenoff and Bowen 1976). Isolated chromosomes also showed a biphasic response to increased concentrations of ethidium bromide, indicating that DNA

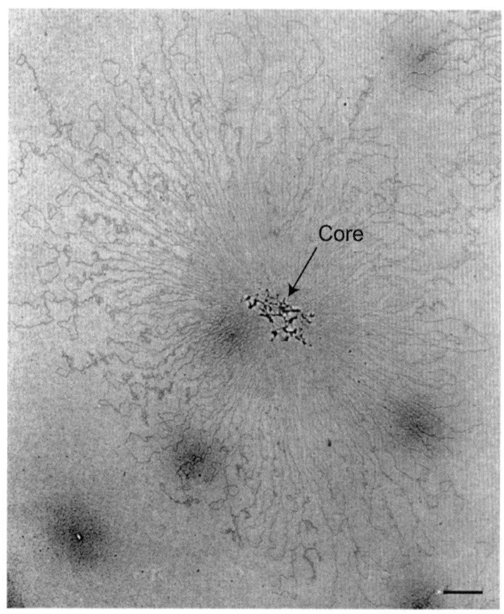

Figure 1. The "rosette" model of DNA organization. Electron micrograph of isolated membrane-free chromosomes from *E. coli*. The central core, from which several tens of loops radiate, is sensitive to RNAse. Single strand cuts to any one loop will only affect the supercoiling of that particular domain, leaving the rest of the chromosome unaffected. Bar = 1 μm. (reprinted from Kavenoff and Bowen 1976, with permission).

supercoiling was maintained. Consistent with the electron microscopy studies, a few tens of DNAse catalyzed nicks are required to obtain a fully relaxed complex (Worcel and Burgi 1972).

The rosette model thus provided some of the first insights into the higher-order structure of the bacterial chromosome. The picture that emerged was one in which the chromosome consists of a central core held together by RNA–DNA interactions. Between 12–80 topologically independent loops—or "domains of supercoiling"—radiate out from this core (Fig. 1). Although all domains have the same amount of superhelicity, a single strand nick in one of the loops eliminates only the superhelicity in that particular domain, leaving the rest of the chromosome unaffected (Worcel and Burgi 1972). Importantly, the existence of domains of supercoiling has been confirmed by in vivo studies (Sinden and Pettijohn 1981; Scheirer and Higgins 2001; Postow et al. 2004). The presence of a stabilizing RNA "core," however, has been subject to certain skepticism because of the possibility of isolation-induced artifacts (Pettijohn 1982) and, to this day, the specific role of RNA in maintaining nucleoid structure is not understood.

Localization of DNA in *Bacillus subtilis* and *Caulobacter*: Specific Orientation

Because chromosomes isolated using the Stonington and Pettijohn method are separated from the rest of the cell, the rosette model does not make any predictions about the subcellular localization of any given DNA sequence. The first indication that the bacterial chromosome has a specific orientation within the cell came from genetic studies in *Bacillus subtilis* (Wu and Errington 1994). Under starvation conditions, *B. subtilis* undergoes a sporulation program that divides the cell into a large mother cell compartment and a small forespore compartment (Barák and Wilkinson 2005). This asymmetric division traps ~30% of the chromosome in the forespore, and segregation of the rest of the chromosome is driven by SpoIIIE (Burton et al. 2007; Ptacin et al. 2008; see also Dubnau 2010). In SpoIIIE mutant cells, DNA translocation failed to occur and 70% of the chromosome was left stranded in the mother cell. Wu and Errington (1994) discovered that the identity of the stranded DNA was reproducible from cell to cell, suggesting that the chromosome has a specific orientation with respect to the division septum, at least during sporulation.

A few years later, direct visualization of the position of chromosomal loci was achieved. Mohl and Gober (1997) showed that the *Caulobacter* ParB protein, which binds specifically to a site near the origin of replication (termed *Cori* in *Caulobacter* and *oriC* in other bacteria), was localized consistently to the cell poles (Fig. 2A). Simultaneously, Webb and coworkers (Webb et al. 1997) adapted a method originally devised for visualizing DNA loci in eukaryotic cells (Robinett et al. 1996; Straight et al. 1996) to look at the position of *oriC* and the terminus of replication (*ter*) in *B. subtilis*. In this method, a large number (~250) of *lacO* sites, arranged in tandem, are inserted into the chromosome at a site of interest. Subsequent expression of a LacI-GFP fusion results in binding of LacI-GFP to the *lacO* arrays, and creates a localized fluorescent spot that can be visualized with a modern CCD camera. Using this method, Webb and coworkers confirmed the genetic prediction that the *B. subtilis oriC* localizes to the cell poles during sporulation. Furthermore, they showed that this localization pattern held true even in vegetatively growing cells, and that *ter* localized to the opposite pole in newborn cells (Fig. 2C,D) (Webb et al. 1997).

It is interesting to note that the total length of these *lacO* arrays is on the order of 10 kb, which, if stretched out, would measure ~3.4 μm; more than enough to span the length of the cell. Yet, in all cases examined, the arrays appeared as tight diffraction-limited foci. This observation is unlikely to be an artifact of the labeling procedure for two reasons: First, the version of LacI used in these experiments had a truncated carboxyl terminus that prevents dimers from associating with each other and making higher-order structures (Lau et al. 2003); and second, the same result was obtained in fixed cells with fluorescence in situ hybridization (FISH) probes of a similar length (Jensen

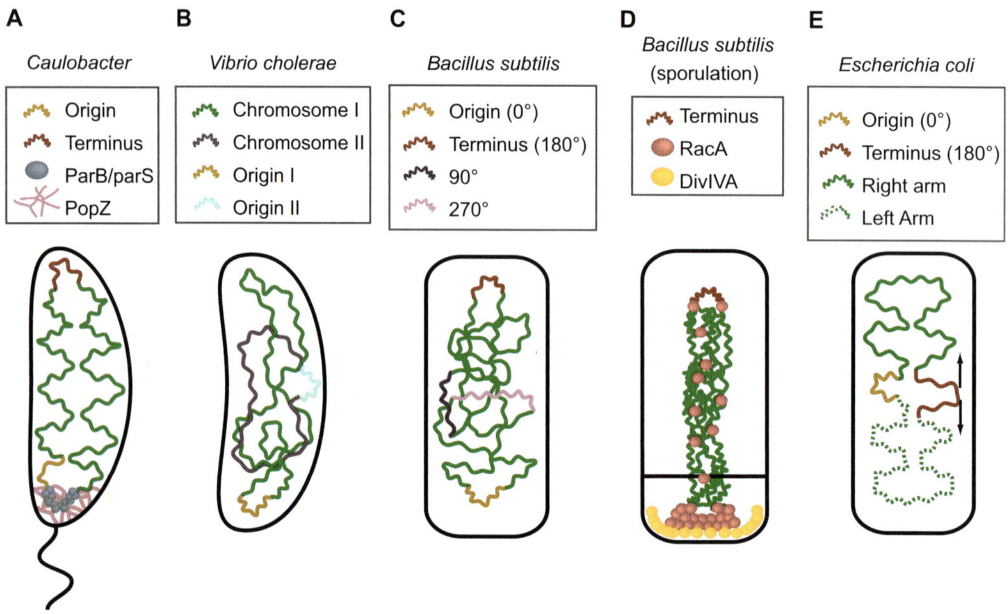

Figure 2. Chromosome organization in model bacteria. (*A*) The *Caulobacter* chromosome is linearly organized, and anchored to the flagellated pole via *parS*/ParB/PopZ. (*B*) In *Vibrio cholerae*, the origin region of the larger chromosome (chromosome I) is localized to the cell pole, whereas the origin of the smaller chromosome is localized to the cell center. The organization of the bulk of the chromosomes, as well as their separation or intermingling, are currently unknown. (*C*) Four loci have been localized in vegetative cells of *Bacillus subtilis*, and their organization is reminiscent of the linear order seen in *Caulobacter*. Although the origin region is localized near to one pole, it appears not to be anchored to the cell membrane. (*D*) Sporulating cells of *B. subtilis*, however, do anchor the origin region, through RacA/DivIVA, to the negatively curved membrane at the pole. RacA also binds all along the chromosome, compacting it into a long "axial filament" before sporulation. (*E*) The *E. coli* origin localizes to mid-cell, and the two replichores are separated into opposite cell halves. The terminus is broadly localized (arrows), and may be found on either side of the cell center.

and Shapiro 1999). Hence, it appears that in vivo DNA is highly and locally compacted.

The ability to synchronize *Caulobacter* cells played an important role in the next discovery, as it allowed investigators to easily follow the localization pattern of DNA through the cell cycle. Jensen and Shapiro (1999) used FISH to follow the positioning of *Cori* and *ter* in synchronized cells, and saw that *Cori* began the cell cycle at the old pole, whereas *ter* was localized to the new pole. After *Cori* duplicated, it quickly redistributed to a bipolar localization. Meanwhile, *ter* moved away from the pole and relocated to midcell. Near the end of the cell cycle, *ter* itself was duplicated, leaving a mirror image conformation of the chromosome (Jensen and Shapiro 1999).

Jensen and coworkers (2001) also looked at the localization pattern of members of the replication machinery. They found that the replisome formed near the old pole at the beginning of the cell cycle, close to *Cori*. As replication proceeded, the replisome migrated toward midcell, presumably being displaced from the pole by the accumulation of newly replicated DNA there.

Once it was established that both *Cori* and *ter* have rigorously defined cell-cycle-dependent localization, the question of how the rest of the chromosome is organized became more pressing (Breier and Cozzarelli 2004). Teleman and coworkers (Teleman et al. 1998) labeled four chromosomal positions (359°, 90°, 181°, and 270° on the circular chromosome; *oriC* corresponds to 0°) in fixed *B. subtilis* cells, and concluded that loci at intermediate positions on the chromosome lie between *oriC* and *ter* inside the cell (Fig. 2C). A similar model was proposed for *Caulobacter* based on pulse labeling of the chromosome in synchronized cells. Early-replicated DNA was shown to localize near the cell poles, whereas late-replicated DNA localized closer to midcell, and intermediate-replicated DNA localized in between the two (Jensen et al. 2001).

These results suggested that in bacterial cells, the entire chromosome was organized, but definitive proof only came in 2004, when Viollier and coworkers (Viollier et al. 2004) developed computational methods to measure the position of fluorescent foci with subdiffraction precision and in a high-throughput manner. Using a transposon-based strategy, they created a set of 112 strains each carrying the *lacO* arrays at a different position, and measured their location in >500 cells. They found a linear correspondence between the position of any given DNA locus on the chromosomal map and its position inside the cell. Thus, a locus that is 1/3 of the distance between *Cori* and *ter* on the chromosome map will be found 1/3 of the way between the old and new poles (Fig. 2A) (Viollier et al. 2004). It is important to remember that bacterial chromosomes are generally circular molecules, with the origin and terminus located at opposite ends of the circle. The two arms of the chromosome, or replichores, run parallel to each other from one cell pole to the other (Fig. 2A). Here, "linear organization" is used to refer to the linear relationship between a DNA locus's position in base pairs, and its subcellular location.

This striking organization raises the important question of how the chromosome is kept in place. Is there a site—or perhaps several sites—of attachment between the DNA and the cell envelope that ensures proper localization? Indeed, there are. During sporulation in *B. subtilis*, the RacA protein is required to bind the origin region of the chromosome to the cell pole (Ben-Yehuda et al. 2003; Wu and Errington 2003). RacA binds preferentially to 25 sites in a large region around the origin of replication and interacts directly with DivIVA (Ben-Yehuda et al. 2005; Lenarcic et al. 2009). DivIVA, in turn, binds to negatively curved membranes, such as those that occur on the cytoplasmic side of cell poles (Lenarcic et al. 2009; Ramamurthi and Losick 2009). Hence, polar localization of the origin region of *B. subtilis* results from a pole-DivIVA-RacA-DNA (RacA binding sites) connection (Fig. 2D).

However, RacA is not expressed in vegetatively growing *B. subtilis* and the origin of replication does not appear to be anchored at the cell pole under these conditions (Fig. 2C). Ectopic expression of RacA during vegetative growth caused the chromosome to shift position so that it came in close contact with the pole (Ben-Yehuda et al. 2003). Therefore, the question of how the chromosome is kept oriented during vegetative growth in *B. subtilis* is still unsolved.

The case appears to be more straightforward in *Caulobacter*. Two groups recently showed that the *Caulobacter* chromosome is anchored at the cell pole by the PopZ protein (Bowman et al. 2008; Ebersbach et al. 2008). PopZ is a small, proline-rich protein that self-associates to form a network at the *Caulobacter* cell pole. It interacts directly with ParB, a sequence-specific DNA binding protein that binds to the *parS* sequence (about which, more later), which is only 8 kb away from *Cori*. Thus, in *Caulobacter*, the connection is: pole-PopZ-ParB-DNA (*parS*) (Fig. 2A). Importantly, it was later shown that the polar localization of the origin of replication is not mediated by anchoring of *Cori* itself to the pole. When *Cori* was moved 100kb away from *parS*, its position in the cell shifted away from the pole (Toro et al. 2008). The polar position of *parS*, however, remained unchanged. Therefore, in contrast to previous belief, the determinants of chromosome orientation are not contained within the origin of replication.

Localization of DNA in *E. coli*: Replichore Separation and Nonrandom Segregation

Chromosome organization in *E. coli* appeared, at first, to be similar to that of *B. subtilis* and *Caulobacter*, with *oriC* at one pole and *ter* at the other (Gordon et al. 1997; Niki and Hiraga 1998; Bates and Kleckner 2005). Later work, however, showed a different, more complex picture. At the beginning of the cell cycle, slowly growing *E. coli* cells place *oriC* at midcell (Fig. 2E) (Nielsen et al. 2006b; Wang et al. 2006). FISH analysis suggested that the chromosome is linearly ordered in the cell, as in *Caulobacter*, but not along the *ori-ter* axis (Niki et al. 2000). Double-labeling experiments showed that each chromosomal arm, or replichore, localizes to a separate cell half (Nielsen et al. 2006b; Wang et al. 2006). The terminus region is broadly distributed, colocalizing with the left and right replichores in equal proportion (Fig. 2E, arrows) (Wang et al. 2006).

After duplication, the two *E. coli oriC* move to quarter cell positions and the bulk of the chromosome assumes a <L-R-L-R> configuration (L and R refer to the Left and Right replichores, respectively) (Fig. 3) (Nielsen et al. 2006b; Wang et al. 2006). Counter-intuitively, this translationally symmetric (as opposed to mirror-symmetric) organization of the replichores predicts mirror symmetry with regards to segregation of the leading versus lagging strands of replication (Wang et al. 2006). To achieve the <L-R-L-R> configuration observed, both leading strands have to be segregated to the distal edges of the cell while the lagging strands stay proximal, or vice-versa (White et al. 2008). Otherwise, the chromosome would toggle between configurations (see Fig. 4 for why this is so). Indeed, White and coworkers showed that leading strands are segregated to the edges of the cell. Using an ingenious method, they induced the specific degradation, in vivo, of the lagging-strand-replicated copy of one locus in the *E. coli* chromosome. By examining the localization of the remaining, leading-strand-replicated copy, they showed that in ~90% of cells, the leading strand is segregated to the outside edges of the cell (Fig. 4B, right) (White et al. 2008).

It should be noted that this leading strand preference for the outer edges of the cell does not necessarily lead to "immortal strand" segregation, which postulates nonrandom segregation of "old" and "new" DNA strands (Lansdorp 2007; Lew et al. 2008). Because of the semiconservative nature of DNA replication, every round of synthesis produces two chromosomes composed each of a newly synthesized strand annealed to an old strand. Similarly, every round of cell division produces two cells each with a newly synthesized cell pole as well as an old pole. The immortal strand hypothesis postulates that old DNA will be reliably partitioned to the old pole, thus generating a stable and potentially infinite association of one particular pole with one particular strand of DNA. Thus, the origin of the phrase—immortal strand segregation.

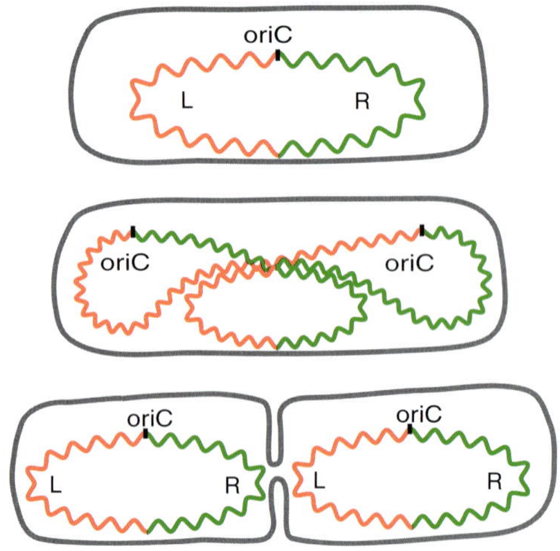

Figure 3. Chromosome organization in *E. coli*. The origin of replication of the *E. coli* chromosome (*oriC*) is located at midcell, and each arm is kept in a separate cell half (*top*). The terminus region is broadly distributed along the long axis of the cell (not shown). As replication proceeds (*middle*), each daughter *oriC* is segregated to the cell quarters and, when replication is complete, the daughter chromosomes adopt a translationally symmetric <L-R-L-R> configuration (*bottom*).

The leading versus lagging strand segregation discovered by White and coworkers, then, is slightly different from immortal strand segregation (see Fig. 5 for a visual explanation of this complicated point). Under leading strand segregation, it is possible for a given "old" pole to receive either daughter chromosome. As long as the leading strand of that chromosome is positioned closer to the old pole, White's law is not broken, even though the immortal strand hypothesis may not be kept. This situation, however, will change the chromosomal configuration in successive generations from

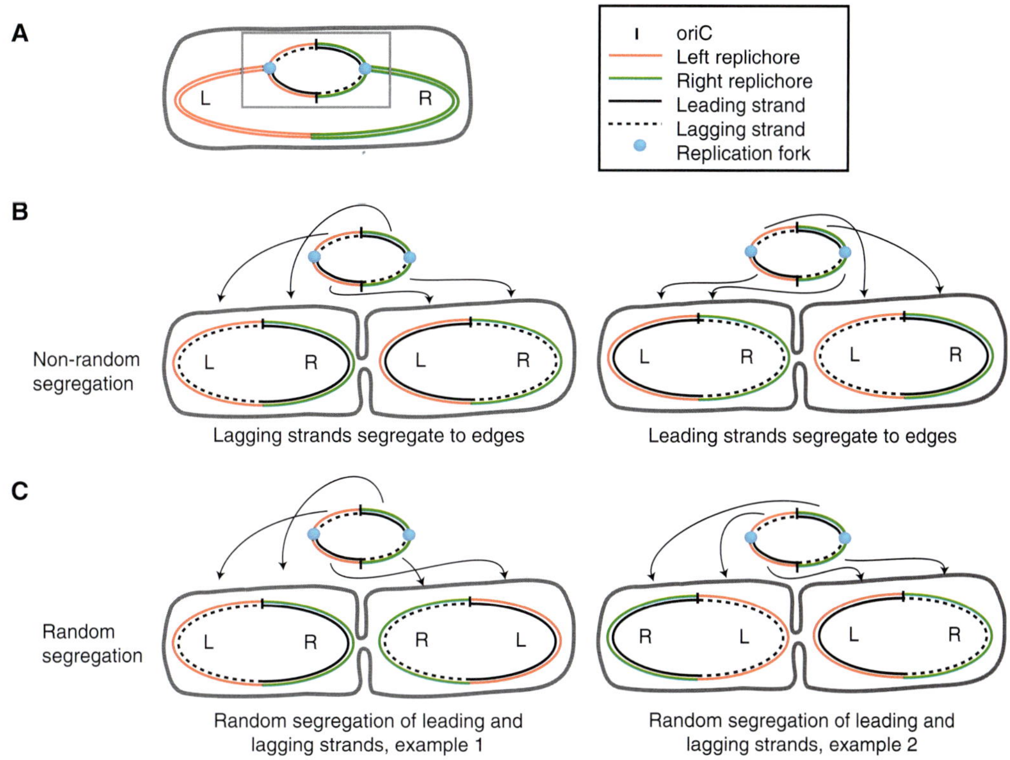

Figure 4. Nonrandom segregation in *E. coli*. (*A*) A replicating mother cell is shown, highlighting the Left and Right replichores as well as the difference between leading-strand-replicated (solid black lines) and lagging-strand-replicated DNA (dashed black lines). The replication bubble (gray box) is also shown, and repeated above each example in *B* and *C* to illustrate the origin of each configuration. (*B*) To achieve the <L-R-L-R> configuration seen, it is necessary for both lagging strands or both leading strands to be segregated to the distal edges of the cell. Note that in ~90% of cases, leading strand segregation to the distal edges (*right*) is observed. (*C*) If random segregation of *leading* and lagging strands is imposed, the <L-R-L-R> configuration cannot be achieved, and mirror symmetry appears (e.g., <L-R-R-L>).

<L-R-L-R> to <R-L-R-L> (Fig. 5, bottom). Thus, by following two markers in successive generations, it is possible to measure the degree to which *E. coli* undergoes immortal strand segregation, something that Wang and coworkers (Wang et al. 2006) measured: Of 17 divisions observed, 12 kept the <L-R-L-R> configuration, four changed to mirror symmetry (<L-R-R-L> or <R-L-L-R>), and one reversed to <R-L-R-L>. Hence, although ~90% of cells showed nonrandom leading-strand segregation, immortal strand segregation occurred in only ~70% (12/17) of cell divisions.

The immortal strand hypothesis has also been tested in *Caulobacter* (Osley and Newton 1974; Marczynski et al. 1990) and *B. subtilis* (Errington and Wake 1991), and in both cases it was found that chromosomes segregate randomly to daughter cells. Because neither of these species distinguishes between the Left and Right replichores, nonrandom segregation may be related to the orientation of the chromosome within the cell.

DNA SEGREGATION

The Replicon Model: Cell Growth as the Driver of Segregation

By far, the most influential hypothesis about bacterial segregation was the replicon model proposed by François Jacob, Sydney Brenner,

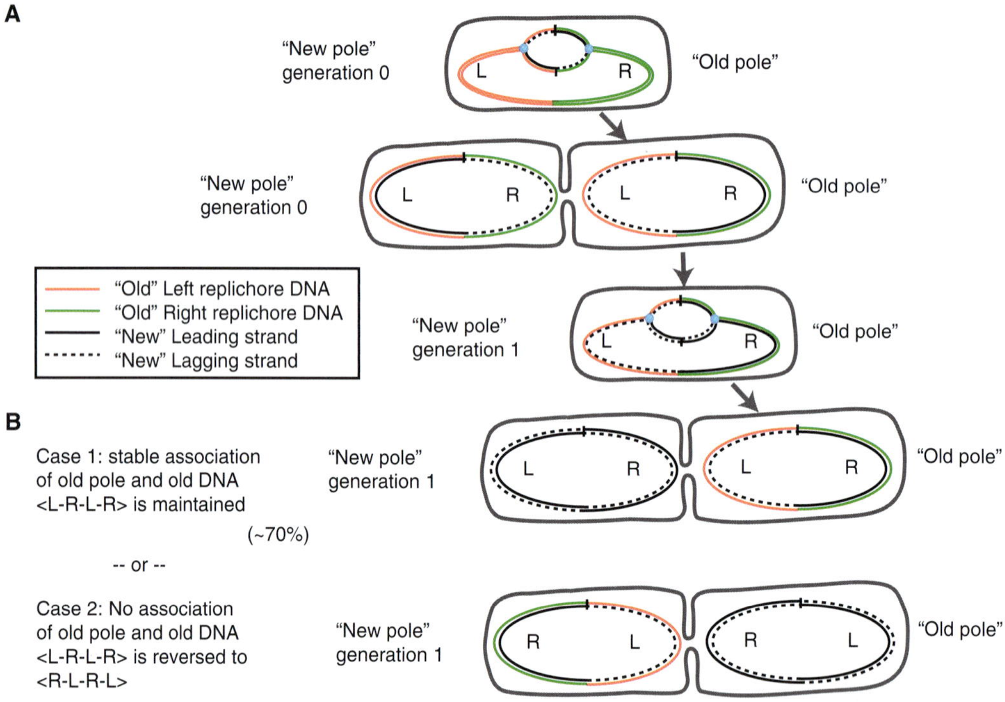

Figure 5. "Immortal strand" inheritance versus Leading strand segregation. Two generations of DNA replication and segregation are shown, to illustrate the association of an old DNA strand (colored) with the old cell pole. Note that in all cases, leading strand segregation to the distal cell edges is maintained. (*A*) After the first generation, both daughter cells carry one "new" and one "old" strand of DNA (black and colored, respectively). (*B*) During the second round of segregation, two scenarios are possible: (1) immortal strand segregation is kept, and the "old" pole stays associated with the "old" (colored) strand of DNA (*top*). Note that this is the case observed for *E. coli* cells in ~70% of cases; (2) Immortal strand segregation is not kept, and the "old" pole received two "new" strands of DNA. In this case, the configuration of the chromosome changes from <L-R-L-R> to <R-L-R-L>.

and François Cuzin in 1963 (Jacob et al. 1963). In a small section at the end of a paper dealing mostly with replication control, Jacob and coworkers noted that puzzling observations about the efficiency of conjugation as well as thin section electron micrographs were consistent with the anchoring of specific sections of DNA to the cell envelope. Hence, if sister chromosomes are attached to the cell membrane and growth takes place in between the two attachment sites, DNA segregation could be achieved passively as a byproduct of cell elongation.

In order for the Jacob model to be possible, two things must be true: First, the chromosome must be physically attached to the cell wall at a specific site; and second, the cell wall must grow in a nonuniform manner. These predictions led investigators to look for a specific association of the origin of replication with the membrane, which was found to be present in several species (Sueoka and Quinn 1968; Fielding and Fox 1970; Bowman et al. 2008; Ebersbach et al. 2008), as well as zonal cell wall growth, which was found to be present or absent, depending on the organism (Cole and Hahn 1962; Mobley et al. 1984; Aaron et al. 2007).

Many of the observed associations between DNA and membrane were sensitive to rifampin, a transcription inhibitor (Dworsky and Schaechter 1973), suggesting that they were mediated by transertion (i.e., coupled transcription, translation, and insertion of the nascent polypeptide into the membrane). However, some of the

DNA-membrane interactions were impervious to rifampin, implying that at least some sites on the chromosome, including *oriC*, were specifically anchored to the cell envelope by DNA binding proteins (Jacq and Kohiyama 1980). As mentioned previously, the chromosome-anchoring proteins for *Caulobacter* (ParB-PopZ) and *B. subtilis* (RacA-DivIVA) are now known (Fig. 2).

Two other observations seemed to support the replicon model: First, bacterial nucleoids stained with nonspecific DNA binding dyes appeared to separate gradually, in contrast to what happens in eukaryotes (Van Helvoort and Woldringh 1994); and second, early genetic screens for partitioning mutants in *E. coli* yielded hits involved in resolving topological problems associated with segregation (Hiraga et al. 1989; Adams et al. 1992; Drlica 1992), but no proteins that could provide the force to drive movement (Wheeler and Shapiro 1997). In the absence of force-generating proteins, passive segregation seemed the logical alternative.

It should be noted that the *E. coli* MukB protein, identified in a genetic screen for DNA partitioning mutants, was briefly hailed as the missing molecular motor based on the fact that it existed as a DNA-binding homodimer that could bind ATP/GTP and had a coiled-coil domain, a characteristic of motors like myosin and kinesin (Niki et al. 1992; Wake and Errington 1995). MukB turned out not to be a motor, however. The characterization of the *B. subtilis* SMC condensation complex (Britton et al. 1998; Moriya et al. 1998) later led to the realization that MukB is in fact part of the *E. coli* analogue of SMC (Bartosik and Jagura-Burdzy 2005).

Nondedicated Models of Segregation: Fast, Progressive Segregation in the Absence of a Mitotic Apparatus

Despite the popularity of the replicon model, evidence against it began mounting. Not only did zonal growth appear to be limited to a subset of species, as soon as direct labeling of specific DNA loci in living cells was achieved, it became clear that their speed of movement was too fast to be accounted for by cell growth (Gordon et al. 1997; Mohl and Gober 1997; Webb et al. 1997; Teleman et al. 1998; Jensen and Shapiro 1999; Viollier et al. 2004). Live tracking experiments also showed that chromosome segregation is progressive; that is, DNA loci were separated while replication was still ongoing (Viollier et al. 2004; Nielsen et al. 2006a). Interestingly, this is not an absolute rule. In *E. coli*, there is evidence that the origin region undergoes a short period of "sister chromatid cohesion," after which the entire region is segregated as a single unit (Bates and Kleckner 2005; Espeli et al. 2008). Other regions of the *E. coli* chromosome, however, segregate progressively (Nielsen et al. 2006a).

Abrupt DNA separation, coupled to the fact that no force-generating segregation proteins had been found, paved the way for a new wave of "nondedicated" models whose aim was to explain fast segregation in the absence of a mitotic apparatus. Several interesting proposals were put forth. It was hypothesized that the force produced by DNA polymerase (Lemon and Grossman 2001) or RNA polymerase (Dworkin and Losick 2002) could help separate daughter chromosomes. Intriguingly, it was also suggested that DNA segregation may be a self-organizing process, driven by separation of transertion domains (Woldringh 2002), or by entropic exclusion of sister chromosomes (Jun and Mulder 2006). Each of these models is supported by a subset of observations, and it is possible that these mechanisms contribute to bulk DNA segregation (Toro et al. 2008). However, none can fully explain the diverse behaviors observed in vivo.

Dedicated Models of Segregation: A Mitotic-like Apparatus

Around the same time that nondedicated models were being proposed, new observations began to suggest that perhaps a mitotic-like apparatus did exist in bacteria. We have already mentioned the transient suggestion that MukB could be a molecular motor in charge of driving chromosome segregation. A second, more promising candidate for force generation is the ParABS system.

Low copy number plasmids like the *E. coli* P1 prophage are inherited with high efficiency despite the fact that as few as two copies of the plasmid may be present before division (Prentki et al. 1977). These observations suggested that such plasmids contained a partitioning system, and deletion of P1 fragments showed that the ParABS locus was necessary and sufficient to confer stability to low copy number replicons (Abeles et al. 1985). This locus contains three components: ParB, a sequence-specific DNA binding protein that binds to the *parS* sequence; and ParA, a Walker type ATPase that has been shown to polymerize and/or oscillate over the nucleoid in several partition systems (Surtees and Funnell 2003; also see Mullins 2010).

Hence, by the early 2000s, it was clear that plasmid partitioning was actively mediated by the ParABS system, and also that the chromosomes of most bacteria (but not *E. coli*) contained homologs of the *par* locus. Based on sequence comparisons of their constituent ParA ATPases, chromosomal ParABS systems are distinct from plasmid ones (Gerdes et al. 2000). Nevertheless, the hypothesis that ParABS systems could indeed drive the segregation of bacterial chromosomes was supported by experiments showing that: (1) deletion/depletion of ParABS components resulted in higher proportions of anucleate cells (Ireton et al. 1994; Mohl and Gober 1997); and (2) addition of various chromosomal ParABS systems to an unstable plasmid increased its efficiency of transmission (Godfrin-Estevenon et al. 2002; Dubarry et al. 2006).

Work in *Vibrio cholerae* has been instrumental in demonstrating the importance of the ParABS system for segregation. *V. cholerae* carries two chromosomes (Fig. 2B), each with their own distinct segregation dynamics and ParABS system, and the larger of which (chromosome I) shows a segregation pattern reminiscent of that observed in *Caulobacter* (Fogel and Waldor 2005). Work by Fogel and Waldor (2006) showed that $parS_I$ segregates ahead of $oriC_I$, suggesting that it may act as the chromosomal centromere (i.e., the site of force exertion during segregation). Toro and coworkers (2008) expanded this observation in *Caulobacter* and found that *parS* invariably segregated ahead of all other loci, regardless of the order of replication. Furthermore, segregation was delayed when *parS* was moved to a site farther away from the origin of replication, indicating that segregation cannot begin before *parS* is duplicated. This showed that, at least in *Caulobacter*, nondedicated models of segregation are unable to initiate segregation.

How might the ParABS system perform its segregation function? Based on time-lapse observations of $ParA_I$ and $ParB_I$ dynamics, Fogel and Waldor proposed a pulling mechanism. In *V. cholerae*, $ParA_I$ forms a fluorescence "cloud" that grows from the pole opposite to $parS_I$. When it reaches $parS_I$, the cloud switches from growth to contraction, and one of the newly duplicated copies of $ParB_I$ follows the receding edge of the cloud until it reaches the opposite pole (Fogel and Waldor 2006). This mechanism is by no means proven, but it is certainly compelling. Work is currently under way to try to characterize the biochemical and structural characteristics of this system (Hester and Lutkenhaus 2007; Castaing et al. 2008).

Despite the promising results outlined above, some important aspects of the ParABS system remain to be explored. New work in *B. subtilis* has shown a link between ParA (Soj) and chromosome replication (Murray and Errington 2008), as well as a direct interaction between ParB (Spo0J) and the SMC complex (Gruber and Errington 2009; Sullivan et al. 2009). Thus, the perverse and often baffling results of *soj* and *spo0J* deletion in *B. subtilis* may be explained by a shift in function.

Compounding the issue, several species (with the exception of *Caulobacter* and *V. cholerae* chromosome II) still segregate their chromosome normally in most cells even when ParABS is inactivated (Ireton et al. 1994; Lewis et al. 2002; Fogel and Waldor 2006; Jakimowicz et al. 2007; Yamaichi et al. 2007). This suggests that there are redundant mechanisms that ensure chromosome segregation in the absence of ParABS. It has been proposed that the actin-like MreB protein contributes to DNA segregation (Gitai et al. 2005; Kruse et al. 2006; Hu et al. 2007; Karczmarek et al. 2007; Shebelut et al. 2009; see also Gitai 2010). Another candidate is the *migS* sequence found in *E. coli* (recall that *E. coli* does not contain a ParABS system on its chromosome).

migS was found by Yamaichi and Niki by cleverly forcing the cells to break off pieces of the chromosome and using them to create mini-chromosomes (Yamaichi and Niki 2004). Thus, the genome within the cell remains unchanged except for the fact that now a small piece of the chromosome is on a separate molecule. Using this strategy, Yamaichi and Niki found that the presence, in *cis*, of a 25 bp sequence (*migS*) was required for proper separation of *oriC*-proximal DNA, as measured by FISH. *migS* did not stabilize plasmids nor cause them to localize to a particular subcellular location (Yamaichi and Niki 2004). Despite this, there is good evidence that *migS* plays an important role in the segregation of the *E. coli* chromosome (Fekete and Chattoraj 2005).

CONCLUDING REMARKS

What We Know Thus Far

Taking into account all of the above, we can make the following generalizations about chromosome organization and segregation in bacteria: The circular chromosome exists within the cell as a compact structure that is organized into separate superhelical domains. At least one locus, and perhaps several, are specifically positioned within the cell, and the rest of the chromosome is kept linearly organized with respect to those landmarks. DNA is kept highly compacted at the local level by a balance of forces, including supercoiling, compaction by proteins, transcription, transertion, and perhaps even RNA–DNA interactions.

Segregation is mediated by an active and dedicated mechanism of which we know very little. In the case of bacteria like *E. coli*, which maintain the origin region at midcell and left and right replichores in separate cell halves, segregation is nonrandom and the leading strand-replicated arms are shuttled to the outer edges of the cell. Such distinction is not made in *Caulobacter* or *B. subtilis*, which keep the origin and terminus regions at opposite poles and do not appear to differentiate between the left and right replichores. Progressive segregation, coupled to fast and directed movement of the origin region at the beginning of replication, collaborate to duplicate the organization of the mother chromosome in each daughter cell.

Pressing Matters for the Future

Several critical questions remain to be explored with regards to chromosomal organization in all bacteria. We lack a mechanistic understanding of the requirements for maintaining chromosome orientation inside cells. Is one anchoring point enough to orient the entire chromosome, or do other sites need to be specifically positioned as well? In species like *E. coli* and *B. subtilis*, which apparently do not anchor the origin region to the pole during vegetative growth, are there compensating anchoring sites elsewhere? Indeed, does proper chromosome orientation *always* require an anchoring point to the cell envelope? It is known, for instance, that plasmids are positioned to specific subcellular locations in a self-organizing manner that does not seem to make use of membrane anchors (Ho et al. 2002; Garner et al. 2007). Hence, membrane anchors may be optional. On the other hand, early biochemical experiments in several bacteria—including *E. coli*—showed that multiple chromosomal sites, including not only the region around *oriC* but also the terminus and the site of ongoing replication, are found to associate with the membrane (Ryter 1968; Leibowitz and Schaechter 1975; Hendrickson et al. 1982). Perhaps, then, nonpolar anchoring sites exist in these organisms.

It is also still unclear why the chromosome is kept in a linear order within the cell, and not in the "bowl of spaghetti" configuration assumed in the past. This is a place in which physical modeling of polymers in confined spaces (e.g., Jun and Mulder 2006) may help guide our intuition. For example, linearity may be a simple consequence of progressive segregation and DNA compaction, i.e., as the DNA is segregated, it takes up space near the pole, forcing subsequent segments of the chromosome to reside farther away, with a simple linear dependence.

As for segregation, two avenues of investigation are likely to yield exciting results. On the one hand, understanding the mechanism of action of the ParABS and *migS* systems is of obvious importance. On the other hand, the fact that most bacteria seem to live surprisingly happy lives in the absence of ParABS or *migS* implies that there are other mechanisms at work. It is still unclear whether these are dedicated fail-safes, serendipitous consequences of other molecular machines (e.g., polymerases), or self-organizing processes.

Answering all of these questions will require new technologies and ingenious experiments, as well as an understanding of the similarities and differences between model systems. We expect the field of chromosome organization and segregation to be active and exciting in the coming years.

ACKNOWLEDGMENTS

The authors wish to thank A. Raphael and S. Goodwin for critical reading of the manuscript. Work in the Shapiro laboratory is supported by National Institutes of Health grants R01 GM5 1426 R24 and GM073011-04.

REFERENCES

Aaron M, Charbon G, Lam H, Schwarz H, Vollmer W, Jacobs-Wagner C. 2007. The tubulin homologue FtsZ contributes to cell elongation by guiding cell wall precursor synthesis in Caulobacter crescentus. *Mol Microbiol* **64:** 938–952.

Abeles AL, Friedman SA, Austin SJ. 1985. Partition of unit-copy miniplasmids to daughter cells. III. The DNA sequence and functional organization of the P1 partition region. *J Mol Biol* **185:** 261–272.

Adams DE, Shekhtman EM, Zechiedrich EL, Schmid MB, Cozzarelli NR. 1992. The role of topoisomerase IV in partitioning bacterial replicons and the structure of catenated intermediates in DNA replication. *Cell* **71:** 277–288.

Barák I, Wilkinson AJ. 2005. Where asymmetry in gene expression originates. *Mol Microbiol* **57:** 611–620.

Bartosik AA, Jagura-Burdzy G. 2005. Bacterial chromosome segregation. *Acta Biochim Pol* **52:** 1–34.

Bates D, Kleckner N. 2005. Chromosome and replisome dynamics in *E. coli*: Loss of sister cohesion triggers global chromosome movement and mediates chromosome segregation. *Cell* **121:** 899–911.

Ben-Yehuda S, Rudner DZ, Losick R. 2003. RacA, a bacterial protein that anchors chromosomes to the cell poles. *Science* **299:** 532–536.

Ben-Yehuda S, Fujita M, Liu XS, Gorbatyuk B, Skoko D, Yan J, Marko JF, Liu JS, Eichenberger P, Rudner DZ, et al. 2005. Defining a centromere-like element in *Bacillus subtilis* by identifying the binding sites for the chromosome-anchoring protein RacA. *Mol Cell* **17:** 773–782.

Bowman GR, Comolli LR, Zhu J, Eckart M, Koenig M, Downing KH, Moerner WE, Earnest T, Shapiro L. 2008. A polymeric protein anchors the chromosomal origin/ParB complex at a bacterial cell pole. *Cell* **134:** 945–955.

Breier AM, Cozzarelli NR. 2004. Linear ordering and dynamic segregation of the bacterial chromosome. *Proc Natl Acad Sci* **101:** 9175–9176.

Britton RA, Lin DC, Grossman AD. 1998. Characterization of a prokaryotic SMC protein involved in chromosome partitioning. *Genes Dev* **12:** 1254–1259.

Burton BM, Marquis KA, Sullivan NL, Rapoport TA, Rudner DZ. 2007. The ATPase SpoIIIE transports DNA across fused septal membranes during sporulation in *Bacillus subtilis*. *Cell* **131:** 1301–1312.

Castaing JP, Bouet JY, Lane D. 2008. F plasmid partition depends on interaction of SopA with non-specific DNA. *Mol Microbiol* **70:** 1000–1011.

Cole RM, Hahn JJ. 1962. Cell wall replication in *Streptococcus pyogenes*. *Science* **135:** 722–724.

Delius H, Worcel A. 1974. Electron microscopic studies on the folded chromosome of *Escherichia coli*. *Cold Spring Harb Symp Quant Biol* **38:** 53–58.

Drlica K. 1992. Control of bacterial DNA supercoiling. *Mol Microbiol* **6:** 425–433.

Dubarry N, Pasta F, Lane D. 2006. ParABS systems of the four replicons of *Burkholderia cenocepacia*: New chromosome centromeres confer partition specificity. *J Bacteriol* **188:** 1489–1496.

Dubnau D. 2010. DNA pumps. *Cold Spring Harb Perspect Biol* **2:** a000406.

Dworkin J, Losick R. 2002. Does RNA polymerase help drive chromosome segregation in bacteria? *Proc Natl Acad Sci* **99:** 14089–14094.

Dworsky P, Schaechter M. 1973. Effect of rifampin on the structure and membrane attachment of the nucleoid of *Escherichia coli*. *J Bacteriol* **116:** 1364–1374.

Ebersbach G, Briegel A, Jensen GJ, Jacobs-Wagner C. 2008. A self-associating protein critical for chromosome attachment, division, and polar organization in caulobacter. *Cell* **134:** 956–968.

Errington J, Wake RG. 1991. Chromosome strand segregation during sporulation in *Bacillus subtilis*. *Mol Microbiol* **5:** 1145–1149.

Espeli O, Mercier R, Boccard F. 2008. DNA dynamics vary according to macrodomain topography in the *E. coli* chromosome. *Mol Microbiol* **68:** 1418–1427.

Fekete RA, Chattoraj DK. 2005. A cis-acting sequence involved in chromosome segregation in *Escherichia coli*. *Mol Microbiol* **55:** 175–183.

Fielding P, Fox CF. 1970. Evidence for stable attachment of DNA to membrane at the replication origin of *Escherichia coli*. *Biochem Biophys Res Commun* **41:** 157–162.

Fogel MA, Waldor MK. 2005. Distinct segregation dynamics of the two *Vibrio cholerae* chromosomes. *Mol Microbiol* **55:** 125–136.

Fogel MA, Waldor MK. 2006. A dynamic, mitotic-like mechanism for bacterial chromosome segregation. *Genes Dev* **20:** 3269–3282.

Garner EC, Campbell CS, Weibel DB, Mullins RD. 2007. Reconstitution of DNA segregation driven by assembly of a prokaryotic actin homolog. *Science* **315:** 1270–1274.

Gerdes K, Møller-Jensen J, Bugge Jensen R. 2000. Plasmid and chromosome partitioning: Surprises from phylogeny. *Mol Microbiol* **37:** 455–466.

Gitai Z. 2010. The structure and function of bacterial actin and microfilaments. *Cold Spring Harb Perspect Biol* **2:** a000364.

Gitai Z, Dye NA, Reisenauer A, Wachi M, Shapiro L. 2005. MreB actin-mediated segregation of a specific region of a bacterial chromosome. *Cell* **120:** 329–341.

Godfrin-Estevenon AM, Pasta F, Lane D. 2002. The parAB gene products of *Pseudomonas putida* exhibit partition activity in both *P. putida* and *Escherichia coli*. *Mol Microbiol* **43:** 39–49.

Gordon GS, Sitnikov D, Webb CD, Teleman A, Straight A, Losick R, Murray AW, Wright A. 1997. Chromosome and low copy plasmid segregation in *E. coli*: Visual evidence for distinct mechanisms. *Cell* **90:** 1113–1121.

Gruber S, Errington J. 2009. Recruitment of condensin to replication origin regions by ParB/SpoOJ promotes chromosome segregation in *B. subtilis*. *Cell* **137:** 685–696.

Hendrickson WG, Kusano T, Yamaki H, Balakrishnan R, King M, Murchie J, Schaechter M. 1982. Binding of the origin of replication of *Escherichia coli* to the outer membrane. *Cell* **30:** 915–923.

Hester CM, Lutkenhaus J. 2007. Soj (ParA) DNA binding is mediated by conserved arginines and is essential for plasmid segregation. *Proc Natl Acad Sci* **104:** 20326–20331.

Hiraga S, Niki H, Ogura T, Ichinose C, Mori H, Ezaki B, Jaffé A. 1989. Chromosome partitioning in *Escherichia coli*: Novel mutants producing anucleate cells. *J Bacteriol* **171:** 1496–1505.

Ho TQ, Zhong Z, Aung S, Pogliano J. 2002. Compatible bacterial plasmids are targeted to independent cellular locations in *Escherichia coli*. *EMBO J* **21:** 1864–1872.

Hu B, Yang G, Zhao W, Zhang Y, Zhao J. 2007. MreB is important for cell shape but not for chromosome segregation of the filamentous cyanobacterium *Anabaena* sp. PCC 7120. *Mol Microbiol* **63:** 1640–1652.

Ireton K, Gunther NW, Grossman AD. 1994. spo0J is required for normal chromosome segregation as well as the initiation of sporulation in *Bacillus subtilis*. *J Bacteriol* **176:** 5320–5329.

Jacob F, brenner S, Cuzin F. 1963. On the regulation of DNA replication in bacteria. *Cold Spring Harbor Symp Quant Biol* **28:** 329–348.

Jacq A, Kohiyama M. 1980. A DNA-binding protein specific for the early replicated region of the chromosome obtained from Escherichia coli membrane fractions. *Eur J Biochem* **105:** 25–31.

Jakimowicz D, Brzostek A, Rumijowska-Galewicz A, Zydek P, Dołzbłasz A, Smulczyk-Krawczyszyn A, Zimniak T, Wojtasz L, Zawilak-Pawlik A, et al. 2007. Characterization of the mycobacterial chromosome segregation protein ParB and identification of its target in *Mycobacterium smegmatis*. *Microbiology* **153:** 4050–4060.

Jensen RB, Shapiro L. 1999. The *Caulobacter crescentus* smc gene is required for cell cycle progression and chromosome segregation. *Proc Natl Acad Sci* **96:** 10661–10666.

Jensen RB, Wang SC, Shapiro L. 2001. A moving DNA replication factory in *Caulobacter crescentus*. *EMBO J* **20:** 4952–4963.

Jun S, Mulder B. 2006. Entropy-driven spatial organization of highly confined polymers: Lessons for the bacterial chromosome. *Proc Natl Acad Sci* **103:** 12388–12393.

Karczmarek A, Martínez-Arteaga R, Baselga RM, Alexeeva S, Hansen FG, Vicente M, Nanninga N, den Blaauwen T. 2007. DNA and origin region segregation are not affected by the transition from rod to sphere after inhibition of *Escherichia coli* MreB by A22. *Mol Microbiol* **65:** 51–63.

Kavenoff R, Bowen BC. 1976. Electron microscopy of membrane-free folded chromosomes from *Escherichia coli*. *Chromosoma* **59:** 89–101.

Kavenoff R, Ryder OA. 1976. Electron microscopy of membrane-associated folded chromosomes of *Escherichia coli*. *Chromosoma* **55:** 13–25.

Kellenberger E. 1991. Functional consequences of improved structural information on bacterial nucleoids. *Res Microbiol* **142:** 229–238.

Kruse T, Blagoev B, Løbner-Olesen A, Wachi M, Sasaki K, Iwai N, Mann M, Gerdes K. 2006. Actin homolog MreB and RNA polymerase interact and are both required for chromosome segregation in *Escherichia coli*. *Genes Dev* **20:** 113–124.

Lansdorp PM. 2007. Immortal strands? Give me a break. *Cell* **129:** 1244–1247.

Lau IF, Filipe SR, Søballe B, Økstad OA, Barre FX, Sherratt DJ. 2003. Spatial and temporal organization of replicating *Escherichia coli* chromosomes. *Mol Microbiol* **49:** 731–743.

Leibowitz PJ, Schaechter M. 1975. The attachment of the bacterial chromosome to the cell membrane. *Int Rev Cytol* **41:** 1–28.

Lemon KP, Grossman AD. 2001. The extrusion-capture model for chromosome partitioning in bacteria. *Genes Dev* **15:** 2031–2041.

Lenarcic R, Halbedel S, Visser L, Shaw M, Wu LJ, Errington J, Marenduzzo D, Hamoen LW. 2009. Localisation of DivIVA by targeting to negatively curved membranes. *EMBO J* **28:** 2272–2282.

Lew DJ, Burke DJ, Dutta A. 2008. The immortal strand hypothesis: How could it work? *Cell* **133:** 21–23.

Lewis RA, Bignell CR, Zeng W, Jones AC, Thomas CM. 2002. Chromosome loss from par mutants of *Pseudomonas putida* depends on growth medium and phase of growth. *Microbiology* **148:** 537–548.

Marczynski GT, Dingwall A, Shapiro L. 1990. Plasmid and chromosomal DNA replication and partitioning during the *Caulobacter crescentus* cell cycle. *J Mol Biol* **212:** 709–722.

Mobley HL, Koch AL, Doyle RJ, Streips UN. 1984. Insertion and fate of the cell wall in *Bacillus subtilis*. *J Bacteriol* **158:** 169–179.

Mohl DA, Gober JW. 1997. Cell cycle-dependent polar localization of chromosome partitioning proteins in *Caulobacter crescentus*. *Cell* **88:** 675–684.

Moriya S, Tsujikawa E, Hassan AK, Asai K, Kodama T, Ogasawara N. 1998. A *Bacillus subtilis* gene-encoding protein homologous to eukaryotic SMC motor protein is necessary for chromosome partition. *Mol Microbiol* **29:** 179–187.

Mullins D. 2010. Bacterial cytoskeleton: The function of FtsZ and ParA polymers. *Cold Spring Harb Perspect Biol* **2:** a002550.

Murray H, Errington J. 2008. Dynamic control of the DNA replication initiation protein DnaA by Soj/ParA. *Cell* **135:** 74–84.

Nielsen HJ, Li Y, Youngren B, Hansen FG, Austin S. 2006a. Progressive segregation of the *Escherichia coli* chromosome. *Mol Microbiol* **61:** 383–393.

Nielsen HJ, Ottesen JR, Youngren B, Austin SJ, Hansen FG. 2006b. The *Escherichia coli* chromosome is organized with the left and right chromosome arms in separate cell halves. *Mol Microbiol* **62:** 331–338.

Niki H, Hiraga S. 1998. Polar localization of the replication origin and terminus in *Escherichia coli* nucleoids during chromosome partitioning. *Genes Dev* **12:** 1036–1045.

Niki H, Imamura R, Kitaoka M, Yamanaka K, Ogura T, Hiraga S. 1992. *E.coli* MukB protein involved in chromosome partition forms a homodimer with a rod-and-hinge structure having DNA binding and ATP/GTP binding activities. *EMBO J* **11:** 5101–5109.

Niki H, Yamaichi Y, Hiraga S. 2000. Dynamic organization of chromosomal DNA in *Escherichia coli*. *Genes Dev* **14:** 212–223.

Osley MA, Newton A. 1974. Chromosomes segregration and development in *Caulobacter crescentus*. *J Mol Biol* **90:** 359–370.

Pettijohn DE. 1982. Structure and properties of the bacterial nucleoid. *Cell* **30:** 667–669.

Postow L, Hardy CD, Arsuaga J, Cozzarelli NR. 2004. Topological domain structure of the *Escherichia coli* chromosome. *Genes Dev* **18:** 1766–1779.

Prentki P, Chandler M, Caro L. 1977. Replication of prophage P1 during the cell cycle of *Escherichia coli*. *Mol Gen Genet* **152:** 71–76.

Ptacin JL, Nollmann M, Becker EC, Cozzarelli NR, Pogliano K, Bustamante C. 2008. Sequence-directed DNA export guides chromosome translocation during sporulation in *Bacillus subtilis*. *Nat Struct Mol Biol* **15:** 485–493.

Ramamurthi KS, Losick R. 2009. Negative membrane curvature as a cue for subcellular localization of a bacterial protein. *Proc Natl Acad Sci* **106:** 13541–13545.

Robinett CC, Straight A, Li G, Willhelm C, Sudlow G, Murray A, Belmont AS. 1996. In vivo localization of DNA sequences and visualization of large-scale chromatin organization using lac operator/repressor recognition. *J Cell Biol* **135:** 1685–1700.

Robinow C, Kellenberger E. 1994. The bacterial nucleoid revisited. *Microbiol Rev* **58:** 211–232.

Robinow CF. 1956. The chromatin bodies of bacteria. *Bacteriol Rev* **20:** 207–242.

Ryter A. 1968. Association of the nucleus and the membrane of bacteria: A morphological study. *Bacteriol Rev* **32:** 39–54.

Ryter A, Chang A. 1975. Localization of transcribing genes in the bacterial cell by means of high resolution autoradiography. *J Mol Biol* **98:** 797–810.

Scheirer KE, Higgins NP. 2001. Transcription induces a supercoil domain barrier in bacteriophage Mu. *Biochimie* **83:** 155–159.

Schoeffler AJ, Berger JM. 2008. DNA topoisomerases: Harnessing and constraining energy to govern chromosome topology. *Q Rev Biophys* **41:** 41–101.

Shebelut CW, Jensen RB, Gitai Z. 2009. Growth conditions regulate the requirements for *Caulobacter* chromosome segregation. *J Bacteriol* **191:** 1097–1100.

Sinden RR, Pettijohn DE. 1981. Chromosomes in living *Escherichia coli* cells are segregated into domains of supercoiling. *Proc Natl Acad Sci* **78:** 224–228.

Stonington OG, Pettijohn DE. 1971. The folded genome of *Escherichia coli* isolated in a protein-DNA-RNA complex. *Proc Natl Acad Sci* **68:** 6–9.

Straight AF, Belmont AS, Robinett CC, Murray AW. 1996. GFP tagging of budding yeast chromosomes reveals that protein-protein interactions can mediate sister chromatid cohesion. *Curr Biol* **6:** 1599–1608.

Sueoka N, Quinn WG. 1968. Membrane attachment of the chromosome replication origin in *Bacillus subtilis*. *Cold Spring Harb Symp Quant Biol* **33:** 695–705.

Sullivan NL, Marquis KA, Rudner DZ. 2009. Recruitment of SMC by ParB-parS organizes the origin region and promotes efficient chromosome segregation. *Cell* **137:** 697–707.

Surtees JA, Funnell BE. 2003. Plasmid and chromosome traffic control: How ParA and ParB drive partition. *Curr Top Dev Biol* **56:** 145–180.

Teleman AA, Graumann PL, Lin DC, Grossman AD, Losick R. 1998. Chromosome arrangement within a bacterium. *Curr Biol* **8:** 1102–1109.

Thanbichler M, Shapiro L. 2008. Getting organized–how bacterial cells move proteins and DNA. *Nat Rev Microbiol* **6:** 28–40.

Toro E, Hong SH, McAdams HH, Shapiro L. 2008. *Caulobacter* requires a dedicated mechanism to initiate chromosome segregation. *Proc Natl Acad Sci* **105:** 15435–15440.

Van Helvoort JM, Woldringh CL. 1994. Nucleoid partitioning in *Escherichia coli* during steady-state growth and upon recovery from chloramphenicol treatment. *Mol Microbiol* **13:** 577–583.

Viollier PH, Thanbichler M, McGrath PT, West L, Meewan M, McAdams HH, Shapiro L. 2004. Rapid and sequential movement of individual chromosomal loci to specific subcellular locations during bacterial DNA replication. *Proc Natl Acad Sci* **101:** 9257–9262.

Wake RG, Errington J. 1995. Chromosome partitioning in bacteria. *Annu Rev Genet* **29:** 41–67.

Wang JC. 1971. Interaction between DNA and an *Escherichia coli* protein ω. *J Mol Biol* **55:** 523–533.

Wang X, Liu X, Possoz C, Sherratt DJ. 2006. The two *Escherichia coli* chromosome arms locate to separate cell halves. *Genes Dev* **20:** 1727–1731.

Watson JD, Crick FH. 1953. Genetical implications of the structure of deoxyribonucleic acid. *Nature* **171:** 964–967.

Webb CD, Teleman A, Gordon S, Straight A, Belmont A, Lin DC, Grossman AD, Wright A, Losick R. 1997. Bipolar localization of the replication origin regions of chromosomes in vegetative and sporulating cells of *B. subtilis. Cell* **88:** 667–674.

Wheeler RT, Shapiro L. 1997. Bacterial chromosome segregation: Is there a mitotic apparatus? *Cell* **88:** 577–579.

White MA, Eykelenboom JK, Lopez-Vernaza MA, Wilson E, Leach DR. 2008. Non-random segregation of sister chromosomes in *Escherichia coli. Nature* **455:** 1248–1250.

Woldringh CL. 2002. The role of co-transcriptional translation and protein translocation (transertion) in bacterial chromosome segregation. *Mol Microbiol* **45:** 17–29.

Worcel A, Burgi E. 1972. On the structure of the folded chromosome of *Escherichia coli. J Mol Biol* **71:** 127–147.

Wu LJ, Errington J. 1994. *Bacillus subtilis* spoIIIE protein required for DNA segregation during asymmetric cell division. *Science* **264:** 572–575.

Wu LJ, Errington J. 2003. RacA and the Soj-Spo0J system combine to effect polar chromosome segregation in sporulating *Bacillus subtilis. Mol Microbiol* **49:** 1463–1475.

Yamaichi Y, Niki H. 2004. migS, a cis-acting site that affects bipolar positioning of oriC on the *Escherichia coli* chromosome. *EMBO J* **23:** 221–233.

Yamaichi Y, Fogel MA, Waldor MK. 2007. par genes and the pathology of chromosome loss in *Vibrio cholerae. Proc Natl Acad Sci* **104:** 630–635.

Synchronization of Chromosome Dynamics and Cell Division in Bacteria

Martin Thanbichler[1,2]

[1]Max Planck Institute for Terrestrial Microbiology, Karl-von-Frisch-Straβe, D-35043 Marburg, Germany
[2]Laboratory for Microbiology, Department of Biology, Philipps University, Karl-von-Frisch-Straβe 8, D-35043 Marburg, Germany

Correspondence: thanbichler@mpi-marburg.mpg.de

Bacterial cells have evolved a variety of regulatory circuits that tightly synchronize their chromosome replication and cell division cycles, thereby ensuring faithful transmission of genetic information to their offspring. Complex multicomponent signaling cascades are used to monitor the progress of cytokinesis and couple replication initiation to the separation of the two daughter cells. Moreover, the cell-division apparatus actively participates in chromosome partitioning and, particularly, in the resolution of topological problems that impede the segregation process, thus coordinating chromosome dynamics with cell constriction. Finally, bacteria have developed mechanisms that harness the cell-cycle-dependent positioning of individual chromosomal loci or the nucleoid to define the cell-division site and control the timing of divisome assembly. Each of these systems manages to integrate a complex set of spatial and temporal cues to regulate and execute critical steps in the bacterial cell cycle.

In recent years, considerable progress has been made in understanding the cell biology of bacteria and, in particular, the organization and dynamics of their chromosomes (Shih and Rothfield 2006; Graumann 2007; Morris and Jensen 2008; Reyes-Lamothe et al. 2008b; Thanbichler and Shapiro 2008). It has emerged that bacterial cells have evolved a variety of mechanisms to closely coordinate replication and segregation of chromosomal DNA with cell division, thus ensuring that genetic information is passed on faithfully. In the absence of these regulatory circuits, premature formation of a division septum can lead to dissection of the nucleoid and generation of anucleate cells. Conversely, untimely origin firing before the end of cell division may result in the accumulation of supernumerary chromosomes, thereby interfering with proper cell cycle and gene regulation, and loading the cell with a substantial metabolic burden. This article first summarizes our current knowledge on checkpoints that couple replication initiation and the last stages of chromosome segregation to the progression

 doi: 10.1101/cshperspect.a000331
Cite this article as *Cold Spring Harb Perspect Biol* 2010;2:a000331

of cell constriction. Subsequently, it discusses effects of chromosome dynamics on the temporal and spatial control of divisome assembly.

ORGANIZATION AND DYNAMICS OF CHROMOSOMAL DNA

Bacteria usually contain a single, circular chromosome that floats freely in the cytoplasm. Although not encased in a specific membrane compartment, it frequently occupies a distinct region within the cell, termed the nucleoid. With an average size of approximately four megabases, the contour length of a chromosome measures approximately 1 mm and thus exceeds the length of a typical bacterial cell by more than 1000-fold. Apparently, this huge molecule needs to be compacted significantly to fit into the confined space of the cell body. Biochemical and electron microscopic analyses on bacterial chromatin have failed to detect the distinct hierarchical organization seen in eukaryotes. As a consequence, the nucleoid has long been envisioned as a compact tangle of DNA lacking higher-order structure. However, recent studies investigating the localization of individual chromosomal loci within cells from a variety of different species have revealed that the nucleoid in fact has a defined architecture. In all organisms analyzed, chromosomal DNA appears to be arranged in a circular superstructure, in which the subcellular position of each locus is directly reflected by its position on the circular chromosomal map (Teleman et al. 1998; Viollier et al. 2004; Wang et al. 2006b). The underlying principles are still unclear, but current data suggest that the chromosome is organized into numerous compact, supercoiled loops that are lined up like pearls on a necklace (Postow et al. 2004; Thanbichler et al. 2005). Proper temporal and spatial organization of bacterial chromatin is critically dependent on the action of various topological regulators. Bacterial structural maintenance of chromosome (SMC) complexes, for instance, act as molecular clamps that interconnect different DNA regions, thereby maintaining the chromosome in a state compatible with efficient DNA replication and segregation (Britton et al. 1998; Hirano 2006). In addition, every bacterium contains a varying set of small nucleoid-associated proteins, which bend or cross-link DNA, thereby adapting DNA topology to the needs of transcription and other cellular processes (Luijsterburg et al. 2006).

The conserved structure of bacterial chromosomes is a direct consequence of coreplicational DNA segregation. Chromosome replication is usually initiated at a single origin and then proceeds bidirectionally until the two replication forks meet in the terminus region (O'Donnell 2006). Immediately after duplication of the origin region, its two copies are rapidly moved apart by an active mechanism (Gordon et al. 1997; Webb et al. 1998; Viollier et al. 2004), setting the corner stones for the incipient sister nucleoids. As replication proceeds, newly synthesized sister duplexes are instantly separated once they emerge from the replication apparatus (replisome) and stacked on top of the preceding stretches of DNA that have already been deposited in the growing nucleoids, thus maintaining the original ordering of loci (Niki et al. 2000; Viollier et al. 2004; Nielsen et al. 2006; Wang et al. 2006b). Other than in the case of the origin regions, this process is likely to be driven by the pulling forces resulting from DNA recondensation, rather than by a dedicated segregation machinery. During the course of replication, the replisome migrates slowly along the two arms of the compacted chromosome, ending up near the cell-division plane when it reaches the terminus region (Bates 2008; Reyes-Lamothe et al. 2008a). After removal of DNA from the site of constriction, decatenation of the two sister chromosomes, and resolution of chromosome dimers, the cell divides in between the two fully established nucleoids, giving rise to siblings that each carry the full complement of genes present in the mother cell.

COORDINATION OF REPLICATION INITIATION WITH CELL DIVISION

Although bacterial replication origins differ considerably with respect to their precise architecture, most of them share several common features (Zakrzewska-Czerwinska et al. 2007). Importantly, all bacteria investigated require the AAA+ ATPase DnaA to initiate replisome assembly. On ATP binding, DnaA interacts with conserved motifs at the replication origin, forming an oligomeric nucleoprotein complex that modifies the topology of the origin such as to promote local unwinding of the DNA duplex at a nearby AT-rich sequence (Mott and Berger 2007). Generation of single DNA strands then facilitates loading of the replicative DNA helicase, which in turn recruits the other components of the replication apparatus. The activity of DnaA is controlled by a variety of mechanisms, affecting the nucleotide state, the free concentration, and the cellular abundance of the protein, as well as the accessibility of its binding sites (Kaguni 2006; Leonard and Grimwade 2009). However, the exact contribution of these different pathways to the inactivation of DnaA after replication initiation and to its reactivation at the start of a new cell cycle still remains to be elucidated. Apart from DnaA-binding sites, replication origins usually contain additional functional elements, such as promoters and recognition motifs for the DNA-bending protein IHF (integration host factor), which support the action of DnaA by inducing changes in the superhelicity and architecture of the origin region.

Once the replisome is assembled, each of the two replication forks proceeds at a speed of about 1000 base pairs/second, which means that duplication of a four-megabase chromosome takes about 33 minutes. The generation time of many bacteria, such as *Escherichia coli*, is considerably shorter than the duration of S-phase. In these cases, replication is initiated more than once in a single cell, giving rise to siblings that inherit already partially duplicated chromosomes (Cooper and Helmstetter 1968; Niki and Hiraga 1998; Nielsen et al. 2007). The regulatory pathways that ensure the proper number of initiation events in organisms with such overlapping cell cycles are largely obscure.

The situation is different in slow-growing bacteria that only perform a single round of replication in the mother cell. A representative of this group is *Caulobacter crescentus*, an organism that divides asymmetrically into a sessile stalked cell and a motile, flagellated swarmer cell. Whereas the stalked cell initiates chromosome replication immediately after birth, the swarmer cell rests in a replicationally quiescent, G1-like state until it differentiates into a stalked cell and thus continues its cell cycle (Degnen and Newton 1972). *C. crescentus* has developed an elaborate mechanism to synchronize replication initiation with the physical separation of the two daughter cells (Fig. 1). One of the key players in the underlying control circuit is the two-component response regulator CtrA, a central component of the regulatory network that drives the *C. crescentus* cell cycle (Quon et al. 1996; Laub et al. 2007). On phosphorylation, CtrA interacts with the promoters of about 50 operons, governing the expression of more than 95 genes (Laub et al. 2002). In addition, it binds to five sites in the replication origin, which overlap with the DnaA- and IHF-binding sites, and with a conspicuously AT-rich region that includes a strong promoter implicated in replication control. Mutations in these sites indeed lead to increased initiation frequencies, indicating that CtrA~P acts as an inhibitor of chromosome replication (Quon et al. 1998; Siam and Marczynski 2000; Marczynski and Shapiro 2002; Siam et al. 2003; Bastedo and Marczynski 2009). Consistently, the levels of CtrA are high throughout most of the cell cycle, whereas they fall sharply during a short interval around the onset of S-phase, concomitant with a peak in the cellular abundance and activity of the replication initiator DnaA (Domian et al. 1997; Gorbatyuk and Marczynski 2005; Collier and Shapiro 2009).

Aside from cell-cycle-regulated transcription of the *ctrA* gene, phosphorylation and targeted proteolysis are the key factors determining the abundance of active CtrA within the cell (Quon et al. 1996; Domian et al. 1997). Both processes are mediated by complex regulatory cascades, which converge at the single-domain response regulator DivK (Hecht et al. 1995) (Fig. 1A). The phosphorylation state of DivK is controlled by two proteins, the histidine kinase DivJ and the bifunctional histidine kinase/phosphatase PleC (Ohta et al. 1992;

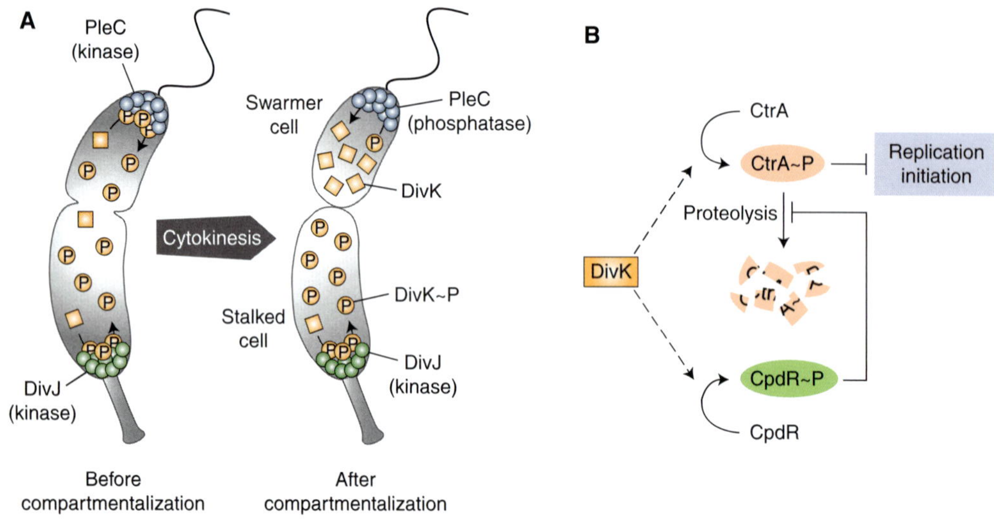

Figure 1. Synchronization of replication initiation with cell division in *C. crescentus*. (*A*) Monitoring of cell division via the phosphorylation state of DivK. In dividing cells, the histidine kinase DivJ and the bifunctional histidine kinase/phosphatase PleC form complexes that are localized to the stalked and the flagellated pole, respectively. At this stage, both proteins function as kinases, ensuring a high concentration of DivK∼P within the cell. After cell division, PleC switches from the kinase to the phosphatase mode. The DivK molecules captured in the swarmer sibling are thus dephosphorylated, whereas those remaining in the stalked sibling are still retained in the phosphorylated state. (*B*) Role of DivK in the regulation of replication initiation. Dephosphorylated DivK activates a signaling cascade that leads to the phosphorylation of the response regulators CtrA and CpdR. CtrA∼P binds to five sites within the origin of replication, thereby inhibiting replication initiation. Formation of CpdR∼P prevents proteolysis of CtrA∼P and thus reinforces the block of chromosome replication.

Hecht et al. 1995; Wu et al. 1998; Lam et al. 2003; Paul et al. 2008). In stalked cells, DivJ and PleC form complexes that are localized to the stalked and flagellated cell pole, respectively (Wheeler and Shapiro 1999; Jacobs et al. 2001; Viollier et al. 2002; Radhakrishnan et al. 2008). At this stage, both proteins function as kinases, ensuring a high level of DivK∼P within the cell (Paul et al. 2008). However, once cytokinesis physically separates the incipient daughter cell compartments, PleC is switched to the phosphatase mode. As a consequence, the DivK molecules captured in the swarmer sibling are rapidly dephosphorylated, whereas those segregated into the stalked offspring are still retained in the phosphorylated state (Jacobs et al. 2001; Lam et al. 2003; Matroule et al. 2004; Paul et al. 2008).

Nonphosphorylated DivK activates a phosphosignaling cascade that results in phosphoryl transfer to CtrA (Wu et al. 1998; Biondi et al. 2006) (Fig. 1B). The same pathway phosphorylates and thus inactivates the single-domain response regulator CpdR, a key regulator of targeted CtrA proteolysis (Biondi et al. 2006; Iniesta et al. 2006; Iniesta and Shapiro 2008). Consequently, two complementary pathways ensure accumulation of CtrA∼P in swarmer cells, allowing it to bind to the chromosomal origin and prevent replication initiation. Once the swarmer cell differentiates into a stalked cell, DivJ replaces PleC at the newly formed stalked pole (Wheeler and Shapiro 1999). The resulting increase in the level of DivK∼P abolishes CtrA and CpdR phosphorylation, thereby inactivating CtrA and triggering its degradation by the ClpXP protease (Iniesta et al. 2006; McGrath et al. 2006). Accordingly, CtrA is cleared from the replication origin, allowing DnaA to initiate replisome assembly. At the same time, elimination of CtrA∼P induces transcription of several essential cell-division genes, thereby marking the beginning of the next division cycle (Laub et al. 2002).

Later in the cell cycle, a new PleC cluster is formed at the pole opposite the stalk, again switched to kinase mode and thus stabilizing the DivK~P pool (Wheeler and Shapiro 1999; Paul et al. 2008). Because, as a consequence, CtrA~P levels remain low until the cell reaches the predivisional stage, CtrA-independent mechanisms must be in place to prevent supernumerary initiation events later in S-phase. Consistently, the replisome-associated protein Hda was found to contribute to proper replication control in *C. crescentus* by inhibiting the activity of DnaA once the first round of DNA synthesis has started (Collier and Shapiro 2009). Given that the replication origins of many α-proteobacteria contain CtrA binding sites that overlap elements critical for replication initiation, *C. crescentus* might not be the only bacterium using CtrA to synchronize its replication and division cycles (Brassinga et al. 2002; Ioannidis et al. 2007; Shaheen et al. 2009).

ROLE OF THE CELL DIVISION PROTEIN FtsK IN CHROMOSOME SEGREGATION

Sister chromosomes are moved apart in a multistep process, involving active segregation of the newly synthesized origin regions, condensation-driven partitioning of the bulk of the chromosomes, and, finally, separation of the terminus regions. The last step in this cascade can be complicated by several problems. Toward the end of the replication cycle, the terminus regions may become trapped in the closing septum, leading to a block in cell division (Lau et al. 2003). Clearance of the division site is impeded by catenation of the two sister chromosomes, an effect that is routinely observed during replication of circular DNA molecules (Schvartzman and Stasiak 2004). A similar obstacle is formed by chromosome dimers, arising from an odd number of homologous recombination events between two newly synthesized chromosomal regions (Lesterlin et al. 2004). The cell-division apparatus, and in particular its constituent FtsK, play an important role in resolving these issues, thereby coupling cell separation and the final steps of chromosome segregation.

FtsK is a hybrid protein, composed of various functional domains. Its membrane-integral, amino-terminal region ($FtsK_N$) is part of the cell division apparatus and responsible for localization of FtsK to the division site. In *E. coli* and *C. crescentus*, $FtsK_N$ is essential for cytokinesis, probably serving a structural function in divisome assembly (Begg et al. 1995; Liu et al. 1998; Wang and Lutkenhaus 1998; Yu et al. 1998a; Wang et al. 2006a). The soluble, carboxy-terminal part of FtsK ($FtsK_C$), by contrast, is generally dispensable for constriction. It is connected to $FtsK_N$ via a variable linker and constitutes a molecular motor, belonging to the AAA+ ATPase family (Liu et al. 1998; Yu et al. 1998b; Aussel et al. 2002). The motor domains of different FtsK molecules assemble into hexameric rings whose central opening is large enough to accommodate a DNA duplex (Massey et al. 2006). These complexes function as pumps that actively move chromosomal DNA from the constriction site into the daughter cell compartments, reaching translocation rates of up to 7 kb/s (Saleh et al. 2004; Pease et al. 2005). To ensure proper sorting of the chromosomes, the cell uses molecular signposts called KOPS (FtsK-orienting polar sequences) (Fig. 2A). KOPS are short, conserved sequence motifs that are highly overrepresented in the genome, with their orientation being skewed toward a defined site (*dif*) in the terminus region (Bigot et al. 2005; Levy et al. 2005). They are specifically recognized by the γ-subdomain of $FtsK_C$ and thus serve as the preferred loading zones of the translocation complex (Massey et al. 2006; Ptacin et al. 2006; Sivanathan et al. 2006). Their orientation dictates the positioning of $FtsK_C$ on the DNA molecule and, consequently, the direction of the translocation process, ensuring net movement of $FtsK_C$ toward the terminus region (Bigot et al. 2006). Although KOPS have initially been identified in *E. coli*, polarized sequences also appear to guide DNA translocation in other, related systems. The *Bacillus subtilis* chromosome, for instance, is interspersed with so-called SRS (SpoIIIE recognition sequences), which are required to direct

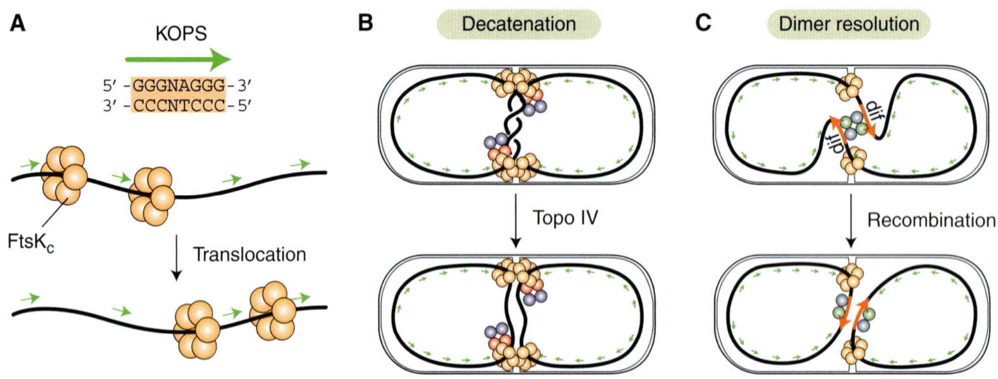

Figure 2. Role of KOPS-regulated DNA translocation by FtsK in the final steps of chromosome segregation. (*A*) Effect of KOPS (FtsK-orienting polar sequences) on the direction of FtsK movement. The boxed sequence indicates the *E. coli* KOPS consensus motif. Hexameric rings, assembled from the carboxy-terminal portions of different FtsK molecules ($FtsK_C$), load onto DNA in a KOPS-dependent manner, and then translocate in the direction determined by the polarity of the KOPS elements (green arrows). (*B*) Stimulation of chromosome decatenation by FtsK. Translocation of $FtsK_C$ toward the terminal *dif* site positions catenanes at the cell-division plane. Unlinking of the two chromosomes is catalyzed by topoisomerase IV, a tetrameric enzyme composed of the proteins ParC (red spheres) and ParE (blue spheres). FtsK directly interacts with ParC, thereby concentrating the activity of topoisomerase IV to the vicinity of the cell-division site. (*C*) Role of FtsK in chromosome dimer resolution. The translocase activity of $FtsK_C$ moves the two *dif* sites of a chromosome dimer to the cell-division plane, thereby promoting formation of a productive recombination synapse. In addition, $FtsK_C$ directly interacts with the recombinase XerD (green spheres) and thus induces the first pair of strand exchanges. The recombinase XerC (blue spheres) then completes the recombination reaction, restoring the two original chromosomes.

the SpoIIIE-driven import of chromosomal DNA into the nascent forespore at the onset of the sporulation process (Wu and Errington 1994; Wu and Errington 1997; Bath et al. 2000; Ptacin et al. 2008).

The pumping activity of FtsK has several implications for chromosome segregation. Sorting of the terminus regions not only clears the division site of DNA but also prevents excessive entanglement of the two sister chromosomes, which may facilitate their distribution to the daughter compartments and their decatenation by topoisomerase (Topo) IV (Adams et al. 1992; Peng and Marians 1993; Zechiedrich and Cozzarelli 1995; Ullsperger and Cozzarelli 1996). In addition to establishing a favorable DNA arrangement, FtsK also affects chromosome decatenation in a direct manner. Topo IV is a heterotetramer composed of two different subunits, named ParC and ParE (Kato et al. 1990). In *E. coli* and *C. crescentus*, ParC colocalizes with the replisome, whereas ParE is dispersed within the cell (Espeli et al. 2003b; Wang and Shapiro 2004). Formation of the active enzyme occurs mostly during the late stages of the cell cycle, in a short interval between replication termination and cell separation (Espeli et al. 2003b). $FtsK_C$ was shown to interact with ParC and stimulate Topo IV activity in vitro (Espeli et al. 2003a; Espeli et al. 2003b). It might, therefore, capture ParC after its release from the disassembling replisome and promote interaction between the ParC and ParE subunits, thereby constituting a functional Topo IV complex (Espeli et al. 2003b) (Fig. 2B). This mechanism could allow close temporal and spatial coordination of the processes involved in the final steps of chromosome and cell separation. However, given that decatenation can still occur, albeit inefficiently, in *E. coli* cells producing a carboxy-terminally truncated form of FtsK, the stimulatory role of $FtsK_C$ does not appear to be essential for Topo IV activity (Liu et al. 1998; Yu et al. 1998b).

Aside from catenation, dimer formation is a severe impediment to the completion of chromosome segregation (Steiner and Kuempel 1998). Cells have evolved a specialized machinery to cope with this problem, consisting of the two tyrosine recombinases XerC and XerD. These proteins cooperate to catalyze a site-specific recombination event between the two terminal *dif* sites, thereby restoring the original two chromosomes (Blakely et al. 1993). The DNA translocase activity of $FtsK_C$ helps align the two XerCD · *dif* complexes at the division site and thus promotes the formation of a productive recombination synapse (Fig. 2C). Moreover, it may induce topological changes in the vicinity of *dif* that are required for proper recombination (Perals et al. 2000; Aussel et al. 2002; Capiaux et al. 2002; Ip et al. 2003; Massey et al. 2004). Apart from its role in synapse formation, $FtsK_C$ was shown to interact directly with XerD and thereby stimulate XerD to perform a first pair of strand exchanges, resulting in the generation of a Holliday junction. This intermediate is subsequently converted to a crossover by a second pair of strand exchanges, catalyzed by XerC in an $FtsK_C$-independent manner (Grainge and Sherratt 1999; Barre et al. 2000; Aussel et al. 2002; Massey et al. 2004; Yates et al. 2006). Given that successive rounds of XerCD-mediated recombination can unlink catenated DNA molecules in vitro, FtsK may also have a direct role in chromosome decatenation that is independent of its interaction with Topo IV (Ip et al. 2003; Grainge et al. 2007). Interestingly, *dif* recombination requires a closing septum and only occurs shortly before cell division (Steiner and Kuempel 1998; Kennedy et al. 2008). Thus, an increase in the local concentration of FtsK, resulting from constriction of the divisome, might be required for DNA translocation and efficient stimulation of XerCD activity.

ROLE OF NUCLEOID OCCLUSION IN THE REGULATION OF CELL DIVISION

In most bacteria, the cell division apparatus comprises more than 15 different proteins, which assemble into an annular structure at the future division site. The fundament of this complex machinery is a ring-shaped polymer formed by the tubulin homolog FtsZ (Bi and Lutkenhaus 1991; Lowe and Amos 1998; Mukherjee and Lutkenhaus 1998; Li et al. 2007). The FtsZ ring recruits, directly and indirectly, all other components of the divisome, and its constriction is thought to provide a major driving force for the subsequent division process (Goehring and Beckwith 2005; Osawa et al. 2008). Reorganization of the FtsZ ring is facilitated by its rapid turnover kinetics, with subunits being exchanged at a half-time of only a few seconds (Anderson et al. 2004; Thanedar and Margolin 2004; Peters et al. 2007). Owing to its central role in cytokinesis, FtsZ is the primary target of pathways regulating cell division in bacteria.

Many organisms use a dual mechanism to control divisome positioning and assembly, involving both the Min system and nucleoid occlusion (Fig. 3). In *E. coli*, the Min system is composed of three proteins, encoded by the *minCDE* operon (de Boer et al. 1989). It establishes an autonomous oscillatory system that confines the FtsZ polymerization inhibitor MinC to the cell poles, thus limiting assembly of the divisome to the midcell region (Hu and Lutkenhaus 1999; Hu et al. 1999; Raskin and de Boer 1999a; Raskin and de Boer 1999b; Hale et al. 2001; Rothfield et al. 2005). *B. subtilis* and other Gram-positive bacteria, by contrast, use a different variant of the Min system, which does not produce pole-to-pole oscillations but stably tethers MinC at the late-division septum and the cell poles (Cha and Stewart 1997; Edwards and Errington 1997; Marston et al. 1998; Bramkamp et al. 2008; Gregory et al. 2008; Patrick and Kearns 2008). In mutants lacking the Min system, division septa are formed randomly within the DNA-free regions of the cell, but never on top of nucleoids (Marston et al. 1998; Yu and Margolin 1999). This phenomenon, termed *nucleoid occlusion*, indicates a negative effect of chromosomal DNA on FtsZ ring assembly (Mulder and Woldringh 1989; Woldringh et al. 1990; Margolin 2001). Its mechanistic basis lies in the activity of DNA-associated proteins, such as SlmA from *E. coli* and Noc from *B. subtilis*, that act as inhibitors of FtsZ polymerization (Wu and Errington 2004; Bernhardt and de

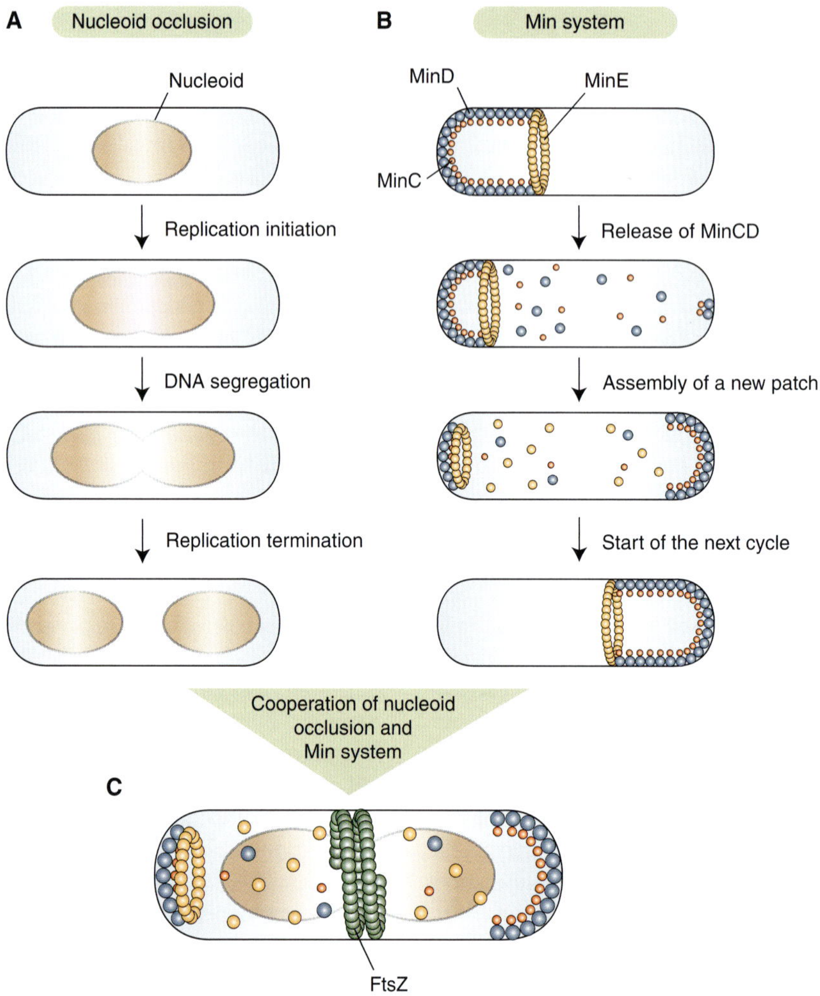

Figure 3. Model for the positioning of the FtsZ ring by the nucleoid occlusion and Min systems in *E. coli*. (*A*) Temporal and spatial regulation of cell division by nucleoid occlusion. The nucleoid occlusion protein SlmA preferentially associates with the pole-proximal regions of the nucleoid. At the beginning of the division cycle, the longitudinal dimensions of the nucleoid are small, thereby placing SlmA close to midcell and blocking FtsZ ring assembly. In the course of chromosome replication and segregation, the two nascent daughter nucleoids move apart. As a consequence, the midcell region is cleared of SlmA, allowing FtsZ polymerization to occur. (*B*) Inhibition of polar cell-division events by the Min system. MinD, bound to the cell division inhibitor MinC, assembles on the cytoplasmic membrane, forming a cap-like polymeric layer that prevents FtsZ ring formation in the polar region of the cell. MinE is organized into a ring-shaped structure that gradually displaces MinCD from the membrane. Free MinC and MinD subunits reassemble at the opposite cell pole, thus establishing a new polar cap and restarting the cycle. (*C*) Cooperation of the nucleoid occlusion and Min systems. The combined action of SlmA and the Min system targets the FtsZ ring to midcell and ensures that divisome formation is delayed to the final phase of the replication cycle.

Boer 2005). In the absence of these factors, Min-deficient cells accumulate clusters of FtsZ that overlap with the nucleoid and, under certain conditions, initiate cell-division events that lead to bisection of the chromosome. Thus, nucleoid occlusion may have evolved as a safeguard mechanism to protect DNA against guillotining during cytokinesis.

Although evolutionarily unrelated, SlmA and Noc both contain a helix-turn-helix DNA-binding domain. In addition, both proteins display similar localization patterns, as they concentrate predominantly in the pole-proximal regions of the nucleoid (Wu and Errington 2004; Bernhardt and de Boer 2005). In *B. subtilis*, this uneven subcellular distribution could be attributed to a skew in the distribution of Noc binding sites. Noc interacts specifically with a conserved 14-base pair sequence, from which it spreads laterally into the flanking chromosomal regions, thereby amplifying the amount of protein associated with the nucleoid (Wu et al. 2009). About 70 copies of this recognition motif are found in the genome, but none of them are located in the terminal quarter of the chromosome. Integration of ectopic binding sites in the terminus region causes a delay in cell division, suggesting that the asymmetric positioning of Noc is important for the timing of divisome formation (Wu et al. 2009). Noc might only be able to block the midcell region efficiently in early S-phase, when the nucleoid is small and its Noc-associated regions are in proximity to each other. However, once chromosome replication and segregation have started, expansion of the nucleoid likely displaces the majority of Noc from the cell center. As a consequence, divisome assembly may be allowed to initiate before the two daughter nucleoids are actually fully separated, facilitating closure of the septum immediately after completion of the replication cycle (Wu et al. 2009). The mechanism whereby SlmA and Noc affect divisome assembly is still unclear. SlmA can recruit FtsZ to the nucleoid when overproduced and promote bundling of FtsZ filaments in vitro. It could, therefore, act by out-competing other cell-division proteins that associate with and stabilize the FtsZ ring (Bernhardt and de Boer 2005).

CONTROL OF DIVISION SITE PLACEMENT BY THE SPATIAL REGULATOR MipZ

There are a number of bacteria that divide by medial fission, even though they lack the Min and nucleoid occlusion systems, suggesting the existence of alternative mechanisms for the control of divisome assembly. Work in *C. crescentus* has indeed identified a novel cell-division regulator, designated MipZ, which couples the positioning and assembly of the FtsZ ring to the initiation of chromosome replication and the bipolar positioning of the two sister origin regions (Thanbichler and Shapiro 2006). MipZ is a member of the Walker A cytoskeletal ATPase (WACA) family (Michie and Lowe 2006) and highly conserved among α-proteobacteria. Its function is critically dependent on the DNA-binding protein ParB, which acts as a central regulator of chromosome dynamics in bacteria (Thanbichler 2009). In *C. crescentus*, ParB recognizes four conserved sites (*parS*) in the vicinity of the chromosomal origin of replication (Mohl and Gober 1997; Figge et al. 2003). After initial site-specific binding, it spreads laterally into the neighboring chromosomal regions and forms a centromer-like nucleoprotein complex that cooperates with the DNA-partitioning protein ParA to actively separate the newly synthesized replication origins (Mohl and Gober 1997; Figge et al. 2003; Thanbichler and Shapiro 2006; Toro et al. 2008). Once segregation is finished, ParB additionally interacts with the scaffolding protein PopZ, thereby mediating attachment of the segregated origin regions at the cell poles (Bowman et al. 2008; Ebersbach et al. 2008).

Newborn *C. crescentus* cells contain a single chromosome, which extends throughout the whole cell, with its replication origin being attached to the old cell pole (Jensen and Shapiro 1999). On entry into S-phase, the *parS*-containing segment is among the first chromosomal regions to be duplicated. The two copies immediately reassociate with ParB and then move apart in a ParA-dependent manner (Thanbichler and Shapiro 2006; Toro et al. 2008; Shebelut et al. 2009). During this partitioning process, one of the sister origin regions remains at its original location, whereas the other one moves rapidly across the cell toward the opposite cell pole (Viollier et al. 2004). MipZ directly interacts with ParB and thus follows the movement of the origin regions (Thanbichler and Shapiro 2006) (Fig. 4). However, it is not stably tethered to the cell poles but rather distributed in a gradient, with its concentration progressively increasing toward the polar ParB · *parS* nucleoprotein complexes. In vitro studies showed that MipZ acts as an inhibitor of FtsZ polymerization (Thanbichler and Shapiro 2006). Accordingly, FtsZ is consistently localized to the subcellular region that exhibits the lowest concentration of MipZ. Before the start of chromosome replication, the MipZ · ParB complex resides at the old pole, resulting in the accumulation of FtsZ at the opposite end of the cell. However, on duplication and segregation of the origin regions, this polar FtsZ cluster is disassembled and a new one is formed at the cell center (Thanbichler and Shapiro 2006). Synthesis of other cell-division proteins, which occurs later in the *C. crescentus* cell cycle, may then lead to reorganization of these polymers into a septal ring that can establish a functional divisome and initiate cytokinesis (Sackett et al. 1998; Laub et al. 2000). MipZ thus serves as a molecular ruler that uses the two segregated origin regions as landmarks to determine the cell center, ensuring that the cell constricts in between the two nascent daughter nucleoids.

It is unknown why most α-proteobacteria use MipZ instead of the Min or nucleoid occlusion system to regulate division site placement. Conversely, however, a MipZ-based mechanism is clearly not applicable to organisms such as *E. coli* and *B. subtilis*, which allow overlapping replication cycles and thus accumulate multiple origin regions within a single cell. Thus, bacteria may have evolved a variety of regulatory mechanisms to control divisome assembly that are specifically tailored to their distinct structural and physiological needs.

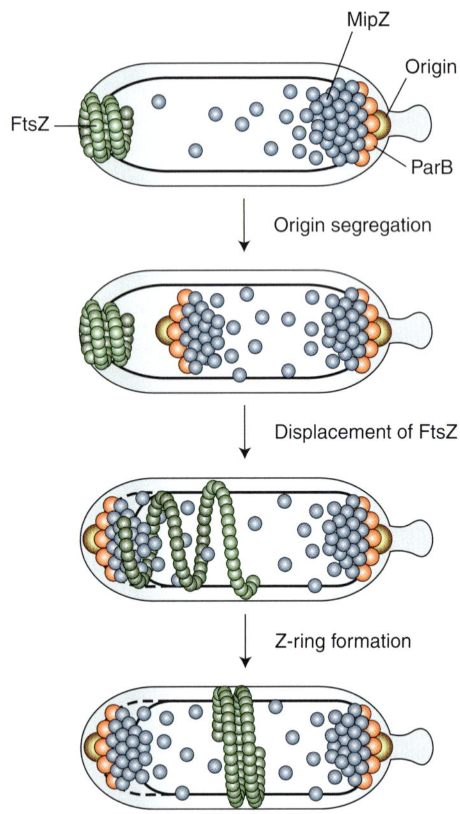

Figure 4. Division site placement by the MipZ · ParB system. The cell division regulator MipZ forms a complex with the DNA-binding protein ParB close to the chromosomal origin of replication, located at the stalked pole of the cell. FtsZ, by contrast, assembles into a polymer that is localized to the pole opposite the stalk. Chromosome replication generates two copies of the origin region, which immediately reassociate with the MipZ · ParB complex and then move apart toward the cell poles. As a consequence, a gradient of MipZ is established, with its concentration being highest in proximity of the two segregated origin regions and lowest at midcell. Owing to the inhibitory effect of MipZ on FtsZ polymerization, the polar FtsZ complex disintegrates and a new polymer is formed at the cell center (adapted from Thanbichler and Shapiro 2006).

FUTURE PERSPECTIVES

Although considerable advances have been made in understanding the mechanisms that interface chromosome dynamics and cell division in bacteria, many questions remain to be answered.

Importantly, the pathways that control the frequency of origin firing in *E. coli* and *B. subtilis* are still obscure. Furthermore, the biochemical basis for the inhibition of divisome assembly by nucleoid occlusion proteins and for the establishment of the MipZ gradient requires further investigation. It will be interesting to see how widespread the systems identified in the common model organisms are among other bacterial species and what kind of variations have evolved on these schemes. Given the enormous diversity of bacteria and the small number of phyla that are currently studied at the molecular biological level, our current knowledge on bacterial cell biology likely represents only a small piece of the overall picture. Thanks to the increasing ease of whole genome sequencing, transcriptome analyses, and high-throughput ORF cloning, the development of new model systems may become more straightforward in the near future. It will be fascinating to follow the progress of the field and see the surprises that still await discovery.

ACKNOWLEDGMENTS

This work was supported by the Max Planck Society and a Young Investigator Grant from the Human Frontier Science Program (HFSP).

REFERENCES

Adams DE, Shekhtman EM, Zechiedrich EL, Schmid MB, Cozzarelli NR. 1992. The role of topoisomerase IV in partitioning bacterial replicons and the structure of catenated intermediates in DNA replication. *Cell* **71:** 277–288.

Anderson DE, Gueiros-Filho FJ, Erickson HP. 2004. Assembly dynamics of FtsZ rings in *Bacillus subtilis* and *Escherichia coli* and effects of FtsZ-regulating proteins. *J Bacteriol* **186:** 5775–5781.

Aussel L, Barre FX, Aroyo M, Stasiak A, Stasiak AZ, Sherratt D. 2002. FtsK is a DNA motor protein that activates chromosome dimer resolution by switching the catalytic state of the XerC and XerD recombinases. *Cell* **108:** 195–205.

Barre FX, Aroyo M, Colloms SD, Helfrich A, Cornet F, Sherratt DJ. 2000. FtsK functions in the processing of a Holliday junction intermediate during bacterial chromosome segregation. *Genes Dev* **14:** 2976–2988.

Bastedo DP, Marczynski GT. 2009. CtrA response regulator binding to the *Caulobacter* chromosome replication origin is required during nutrient and antibiotic stress as well as during cell cycle progression. *Mol Microbiol* **72:** 139–154.

Bates D. 2008. The bacterial replisome: Back on track? *Mol Microbiol* **69:** 1341–1348.

Bath J, Wu LJ, Errington J, Wang JC. 2000. Role of *Bacillus subtilis* SpoIIIE in DNA transport across the mother cell-prespore division septum. *Science* **290:** 995–997.

Begg KJ, Dewar SJ, Donachie WD. 1995. A new *Escherichia coli* cell division gene, *ftsK*. *J Bacteriol* **177:** 6211–6222.

Bernhardt TG, de Boer PA. 2005. SlmA, a nucleoid-associated, FtsZ binding protein required for blocking septal ring assembly over chromosomes in *E. coli*. *Mol Cell* **18:** 555–564.

Bi EF, Lutkenhaus J. 1991. FtsZ ring structure associated with division in *Escherichia coli*. *Nature* **354:** 161–164.

Bigot S, Saleh OA, Cornet F, Allemand JF, Barre FX. 2006. Oriented loading of FtsK on KOPS. *Nat Struct Mol Biol* **13:** 1026–1028.

Bigot S, Saleh OA, Lesterlin C, Pages C, El Karoui M, Dennis C, Grigoriev M, Allemand JF, Barre FX, Cornet F. 2005. KOPS: DNA motifs that control *E. coli* chromosome segregation by orienting the FtsK translocase. *EMBO J* **24:** 3770–3780.

Biondi EG, Reisinger SJ, Skerker JM, Arif M, Perchuk BS, Ryan KR, Laub MT. 2006. Regulation of the bacterial cell cycle by an integrated genetic circuit. *Nature* **444:** 899–904.

Blakely G, May G, McCulloch R, Arciszewska LK, Burke M, Lovett ST, Sherratt DJ. 1993. Two related recombinases are required for site-specific recombination at *dif* and *cer* in *E. coli* K12. *Cell* **75:** 351–361.

Bowman GR, Comolli LR, Zhu J, Eckart M, Koenig M, Downing KH, Moerner WE, Earnest T, Shapiro L. 2008. A polymeric protein anchors the chromosomal origin/ParB complex at a bacterial cell pole. *Cell* **134:** 945–955.

Bramkamp M, Emmins R, Weston L, Donovan C, Daniel RA, Errington J. 2008. A novel component of the division-site selection system of *Bacillus subtilis* and a new mode of action for the division inhibitor MinCD. *Mol Microbiol* **70:** 1556–1569.

Brassinga AK, Siam R, McSween W, Winkler H, Wood D, Marczynski GT. 2002. Conserved response regulator CtrA and IHF binding sites in the α-proteobacteria *Caulobacter crescentus* and *Rickettsia prowazekii* chromosomal replication origins. *J Bacteriol* **184:** 5789–5799.

Britton RA, Lin DC, Grossman AD. 1998. Characterization of a prokaryotic SMC protein involved in chromosome partitioning. *Genes Dev* **12:** 1254–1259.

Capiaux H, Lesterlin C, Perals K, Louarn JM, Cornet F. 2002. A dual role for the FtsK protein in *Escherichia coli* chromosome segregation. *EMBO Rep* **3:** 532–536.

Cha JH, Stewart GC. 1997. The *divIVA* minicell locus of *Bacillus subtilis*. *J Bacteriol* **179:** 1671–1683.

Collier J, Shapiro L. 2009. Feedback control of DnaA-mediated replication initiation by replisome-associated HdaA protein in *Caulobacter*. *J Bacteriol* **191:** 5706–5716.

Cooper S, Helmstetter CE. 1968. Chromosome replication and the division cycle of *Escherichia coli* B/r. *J Mol Biol* **31:** 519–540.

De Boer PA, Crossley RE, Rothfield LI. 1989. A division inhibitor and a topological specificity factor coded for by the minicell locus determine proper placement of the division septum in *E. coli*. *Cell* **56:** 641–649.

Degnen ST, Newton A. 1972. Chromosome replication during development in *Caulobacter crescentus*. *J Mol Biol* **64:** 671–680.

Domian IJ, Quon KC, Shapiro L. 1997. Cell type-specific phosphorylation and proteolysis of a transcriptional regulator controls the G1-to-S transition in a bacterial cell cycle. *Cell* **90:** 415–424.

Ebersbach G, Briegel A, Jensen GJ, Jacobs-Wagner C. 2008. A self-associating protein critical for chromosome attachment, division, and polar organization in *Caulobacter*. *Cell* **134:** 956–968.

Edwards DH, Errington J. 1997. The *Bacillus subtilis* DivIVA protein targets to the division septum and controls the site specificity of cell division. *Mol Microbiol* **24:** 905–915.

Espeli O, Lee C, Marians KJ. 2003a. A physical and functional interaction between *Escherichia coli* FtsK and topoisomerase IV. *J Biol Chem* **278:** 44639–44644.

Espeli O, Levine C, Hassing H, Marians KJ. 2003b. Temporal regulation of topoisomerase IV activity in *E. coli*. *Mol Cell* **11:** 189–201.

Figge RM, Easter J, Gober JW. 2003. Productive interaction between the chromosome partitioning proteins, ParA and ParB, is required for the progression of the cell cycle in *Caulobacter crescentus*. *Mol Microbiol* **47:** 1225–1237.

Goehring NW, Beckwith J. 2005. Diverse paths to midcell: Assembly of the bacterial cell division machinery. *Curr Biol* **15:** R514–526.

Gorbatyuk B, Marczynski GT. 2005. Regulated degradation of chromosome replication proteins DnaA and CtrA in *Caulobacter crescentus*. *Mol Microbiol* **55:** 1233–1245.

Gordon GS, Sitnikov D, Webb CD, Teleman A, Straight A, Losick R, Murray AW, Wright A. 1997. Chromosome and low copy plasmid segregation in *E. coli*: Visual evidence for distinct mechanisms. *Cell* **90:** 1113–1121.

Grainge I, Bregu M, Vazquez M, Sivanathan V, Ip SC, Sherratt DJ. 2007. Unlinking chromosome catenanes *in vivo* by site-specific recombination. *EMBO J* **26:** 4228–4238.

Grainge I, Sherratt DJ. 1999. Xer site-specific recombination. DNA strand rejoining by recombinase XerC. *J Biol Chem* **274:** 6763–6769.

Graumann PL. 2007. Cytoskeletal elements in bacteria. *Annu Rev Microbiol* **61:** 589–618.

Gregory JA, Becker EC, Pogliano K. 2008. *Bacillus subtilis* MinC destabilizes FtsZ-rings at new cell poles and contributes to the timing of cell division. *Genes Dev* **22:** 3475–3488.

Hale CA, Meinhardt H, de Boer PA. 2001. Dynamic localization cycle of the cell division regulator MinE in *Escherichia coli*. *EMBO J* **20:** 1563–1572.

Hecht GB, Lane T, Ohta N, Sommer JM, Newton A. 1995. An essential single domain response regulator required for normal cell division and differentiation in *Caulobacter crescentus*. *EMBO J* **14:** 3915–3924.

Hirano T. 2006. At the heart of the chromosome: SMC proteins in action. *Nat Rev Mol Cell Biol* **7:** 311–322.

Hu Z, Lutkenhaus J. 1999. Topological regulation of cell division in *Escherichia coli* involves rapid pole to pole oscillation of the division inhibitor MinC under the control of MinD and MinE. *Mol Microbiol* **34:** 82–90.

Hu Z, Mukherjee A, Pichoff S, Lutkenhaus J. 1999. The MinC component of the division site selection system in *Escherichia coli* interacts with FtsZ to prevent polymerization. *Proc Natl Acad Sci* **96:** 14819–14824.

Iniesta AA, Shapiro L. 2008. A bacterial control circuit integrates polar localization and proteolysis of key regulatory proteins with a phospho-signaling cascade. *Proc Natl Acad Sci* **105:** 16602–16607.

Iniesta AA, McGrath PT, Reisenauer A, McAdams HH, Shapiro L. 2006. A phospho-signaling pathway controls the localization and activity of a protease complex critical for bacterial cell cycle progression. *Proc Natl Acad Sci* **103:** 10935–10940.

Ioannidis P, Hotopp JC, Sapountzis P, Siozios S, Tsiamis G, Bordenstein SR, Baldo L, Werren JH, Bourtzis K. 2007. New criteria for selecting the origin of DNA replication in *Wolbachia* and closely related bacteria. *BMC Genomics* **8:** 182.

Ip SC, Bregu M, Barre FX, Sherratt DJ. 2003. Decatenation of DNA circles by FtsK-dependent Xer site-specific recombination. *EMBO J* **22:** 6399–6407.

Jacobs C, Hung D, Shapiro L. 2001. Dynamic localization of a cytoplasmic signal transduction response regulator controls morphogenesis during the *Caulobacter* cell cycle. *Proc Natl Acad Sci* **98:** 4095–4100.

Jensen RB, Shapiro L. 1999. The *Caulobacter crescentus smc* gene is required for cell cycle progression and chromosome segregation. *Proc Natl Acad Sci* **96:** 10661–10666.

Kaguni JM. 2006. DnaA: Controlling the initiation of bacterial DNA replication and more. *Annu Rev Microbiol* **60:** 351–375.

Kato J, Nishimura Y, Imamura R, Niki H, Hiraga S, Suzuki H. 1990. New topoisomerase essential for chromosome segregation in *E. coli*. *Cell* **63:** 393–404.

Kennedy SP, Chevalier F, Barre FX. 2008. Delayed activation of Xer recombination at *dif* by FtsK during septum assembly in *Escherichia coli*. *Mol Microbiol* **68:** 1018–1028.

Lam H, Matroule JY, Jacobs-Wagner C. 2003. The asymmetric spatial distribution of bacterial signal transduction proteins coordinates cell cycle events. *Dev Cell* **5:** 149–159.

Lau IF, Filipe SR, Soballe B, Okstad OA, Barre FX, Sherratt DJ. 2003. Spatial and temporal organization of replicating *Escherichia coli* chromosomes. *Mol Microbiol* **49:** 731–743.

Laub MT, Shapiro L, McAdams HH. 2007. Systems biology of *Caulobacter*. *Annu Rev Genet* **41:** 429–441.

Laub MT, Chen SL, Shapiro L, McAdams HH. 2002. Genes directly controlled by CtrA, a master regulator of the *Caulobacter* cell cycle. *Proc Natl Acad Sci* **99:** 4632–4637.

Laub MT, McAdams HH, Feldblyum T, Fraser CM, Shapiro L. 2000. Global analysis of the genetic network controlling a bacterial cell cycle. *Science* **290:** 2144–2148.

Leonard AC, Grimwade JE. 2009. Initiating chromosome replication in *E. coli*: It makes sense to recycle. *Genes Dev* **23:** 1145–1150.

Lesterlin C, Barre FX, Cornet F. 2004. Genetic recombination and the cell cycle: What we have learned from chromosome dimers. *Mol Microbiol* **54:** 1151–1160.

Levy O, Ptacin JL, Pease PJ, Gore J, Eisen MB, Bustamante C, Cozzarelli NR. 2005. Identification of oligonucleotide sequences that direct the movement of the *Escherichia coli* FtsK translocase. *Proc Natl Acad Sci* **102:** 17618–17623.

Li Z, Trimble MJ, Brun YV, Jensen GJ. 2007. The structure of FtsZ filaments *in vivo* suggests a force-generating role in cell division. *EMBO J* **26:** 4694–4708.

Liu G, Draper GC, Donachie WD. 1998. FtsK is a bifunctional protein involved in cell division and chromosome localization in *Escherichia coli*. *Mol Microbiol* **29:** 893–903.

Lowe J, Amos LA. 1998. Crystal structure of the bacterial cell-division protein FtsZ. *Nature* **391:** 203–206.

Luijsterburg MS, Noom MC, Wuite GJ, Dame RT. 2006. The architectural role of nucleoid-associated proteins in the organization of bacterial chromatin: A molecular perspective. *J Struct Biol* **156:** 262–272.

Marczynski GT, Shapiro L. 2002. Control of chromosome replication in *Caulobacter crescentus*. *Annu Rev Microbiol* **56:** 625–656.

Margolin W. 2001. Spatial regulation of cytokinesis in bacteria. *Curr Opin Microbiol* **4:** 647–652.

Marston AL, Thomaides HB, Edwards DH, Sharpe ME, Errington J. 1998. Polar localization of the MinD protein of *Bacillus subtilis* and its role in selection of the mid-cell division site. *Genes Dev* **12:** 3419–3430.

Massey TH, Aussel L, Barre FX, Sherratt DJ. 2004. Asymmetric activation of Xer site-specific recombination by FtsK. *EMBO Rep* **5.** 399–404.

Massey TH, Mercogliano CP, Yates J, Sherratt DJ, Lowe J. 2006. Double-stranded DNA translocation: Structure and mechanism of hexameric FtsK. *Mol Cell* **23:** 457–469.

Matroule JY, Lam H, Burnette DT, Jacobs-Wagner C. 2004. Cytokinesis monitoring during development: Rapid pole-to-pole shuttling of a signaling protein by localized kinase and phosphatase in *Caulobacter*. *Cell* **118:** 579–590.

McGrath PT, Iniesta AA, Ryan KR, Shapiro L, McAdams HH. 2006. A dynamically localized protease complex and a polar specificity factor control a cell cycle master regulator. *Cell* **124:** 535–547.

Michie KA, Lowe J. 2006. Dynamic filaments of the bacterial cytoskeleton. *Annu Rev Biochem* **75:** 467–492.

Mohl DA, Gober JW. 1997. Cell cycle-dependent polar localization of chromosome partitioning proteins in *Caulobacter crescentus*. *Cell* **88:** 675–684.

Morris DM, Jensen GJ. 2008. Toward a biomechanical understanding of whole bacterial cells. *Annu Rev Biochem* **77:** 583–613.

Mott ML, Berger JM. 2007. DNA replication initiation: Mechanisms and regulation in bacteria. *Nat Rev Microbiol* **5:** 343–354.

Mukherjee A, Lutkenhaus J. 1998. Dynamic assembly of FtsZ regulated by GTP hydrolysis. *EMBO J* **17:** 462–469.

Mulder E, Woldringh CL. 1989. Actively replicating nucleoids influence positioning of division sites in *Escherichia coli* filaments forming cells lacking DNA. *J Bacteriol* **171:** 4303–4314.

Nielsen HJ, Li Y, Youngren B, Hansen FG, Austin S. 2006. Progressive segregation of the *Escherichia coli* chromosome. *Mol Microbiol* **61:** 383–393.

Nielsen HJ, Youngren B, Hansen FG, Austin S. 2007. Dynamics of *Escherichia coli* chromosome segregation during multifork replication. *J Bacteriol* **189:** 8660–8666.

Niki H, Hiraga S. 1998. Polar localization of the replication origin and terminus in *Escherichia coli* nucleoids during chromosome partitioning. *Genes Dev* **12:** 1036–1045.

Niki H, Yamaichi Y, Hiraga S. 2000. Dynamic organization of chromosomal DNA in *Escherichia coli*. *Genes Dev* **14:** 212–223.

O'Donnell M. 2006. Replisome architecture and dynamics in *Escherichia coli*. *J Biol Chem* **281:** 10653–10656.

Ohta N, Lane T, Ninfa EG, Sommer JM, Newton A. 1992. A histidine protein kinase homologue required for regulation of bacterial cell division and differentiation. *Proc Natl Acad Sci* **89:** 10297–10301.

Osawa M, Anderson DE, Erickson HP. 2008. Reconstitution of contractile FtsZ rings in liposomes. *Science* **320:** 792–794.

Patrick JE, Kearns DB. 2008. MinJ (YvjD) is a topological determinant of cell division in *Bacillus subtilis*. *Mol Microbiol* **70:** 1166–1179.

Paul R, Jaeger T, Abel S, Wiederkehr I, Folcher M, Biondi EG, Laub MT, Jenal U. 2008. Allosteric regulation of histidine kinases by their cognate response regulator determines cell fate. *Cell* **133:** 452–461.

Pease PJ, Levy O, Cost GJ, Gore J, Ptacin JL, Sherratt D, Bustamante C, Cozzarelli NR. 2005. Sequence-directed DNA translocation by purified FtsK. *Science* **307:** 586–590.

Peng H, Marians KJ. 1993. Decatenation activity of topoisomerase IV during *oriC* and pBR322 DNA replication *in vitro*. *Proc Natl Acad Sci* **90:** 8571–8575.

Perals K, Cornet F, Merlet Y, Delon I, Louarn JM. 2000. Functional polarization of the *Escherichia coli* chromosome terminus: The *dif* site acts in chromosome dimer resolution only when located between long stretches of opposite polarity. *Mol Microbiol* **36:** 33–43.

Peters PC, Migocki MD, Thoni C, Harry EJ. 2007. A new assembly pathway for the cytokinetic Z ring from a dynamic helical structure in vegetatively growing cells of *Bacillus subtilis*. *Mol Microbiol* **64:** 487–499.

Postow L, Hardy CD, Arsuaga J, Cozzarelli NR. 2004. Topological domain structure of the *Escherichia coli* chromosome. *Genes Dev* **18:** 1766–1779.

Ptacin JL, Nollmann M, Becker EC, Cozzarelli NR, Pogliano K, Bustamante C. 2008. Sequence-directed DNA export guides chromosome translocation during sporulation in *Bacillus subtilis*. *Nat Struct Mol Biol* **15:** 485–493.

Ptacin JL, Nollmann M, Bustamante C, Cozzarelli NR. 2006. Identification of the FtsK sequence-recognition domain. *Nat Struct Mol Biol* **13:** 1023–1025.

Quon KC, Marczynski GT, Shapiro L. 1996. Cell cycle control by an essential bacterial two-component signal transduction protein. *Cell* **84:** 83–93.

Quon KC, Yang B, Domian IJ, Shapiro L, Marczynski GT. 1998. Negative control of bacterial DNA replication by a cell cycle regulatory protein that binds at the chromosome origin. *Proc Natl Acad Sci* **95:** 120–125.

Radhakrishnan SK, Thanbichler M, Viollier PH. 2008. The dynamic interplay between a cell fate determinant and a lysozyme homolog drives the asymmetric division cycle of *Caulobacter crescentus*. *Genes Dev* **22:** 212–225.

Raskin DM, de Boer PA. 1999a. MinDE-dependent pole-to-pole oscillation of division inhibitor MinC in *Escherichia coli*. *J Bacteriol* **181:** 6419–6424.

Raskin DM, de Boer PA. 1999b. Rapid pole-to-pole oscillation of a protein required for directing division to the middle of *Escherichia coli*. *Proc Natl Acad Sci* **96:** 4971–4976.

Reyes-Lamothe R, Wang X, Sherratt D. 2008b. *Escherichia coli* and its chromosome. *Trends Microbiol* **16:** 238–245.

Reyes-Lamothe R, Possoz C, Danilova O, Sherratt DJ. 2008a. Independent positioning and action of *Escherichia coli* replisomes in live cells. *Cell* **133:** 90–102.

Rothfield L, Taghbalout A, Shih YL. 2005. Spatial control of bacterial division-site placement. *Nat Rev Microbiol* **3:** 959–968.

Sackett MJ, Kelly AJ, Brun YV. 1998. Ordered expression of *ftsQA* and *ftsZ* during the *Caulobacter crescentus* cell cycle. *Mol Microbiol* **28:** 421–434.

Saleh OA, Perals C, Barre FX, Allemand JF. 2004. Fast, DNA-sequence independent translocation by FtsK in a single-molecule experiment. *EMBO J* **23:** 2430–2439.

Schvartzman JB, Stasiak A. 2004. A topological view of the replicon. *EMBO Rep* **5:** 256–261.

Shaheen SM, Ouimet MC, Marczynski GT. 2009. Comparative analysis of *Caulobacter* chromosome replication origins. *Microbiology* **155:** 1215–1225.

Shebelut CW, Jensen RB, Gitai Z. 2009. Growth conditions regulate the requirements for *Caulobacter* chromosome segregation. *J Bacteriol* **191:** 1097–1100.

Shih YL, Rothfield L. 2006. The bacterial cytoskeleton. *Microbiol Mol Biol Rev* **70:** 729–754.

Siam R, Marczynski GT. 2000. Cell cycle regulator phosphorylation stimulates two distinct modes of binding at a chromosome replication origin. *EMBO J* **19:** 1138–1147.

Siam R, Brassinga AK, Marczynski GT. 2003. A dual binding site for integration host factor and the response regulator CtrA inside the *Caulobacter crescentus* replication origin. *J Bacteriol* **185:** 5563–5572.

Sivanathan V, Allen MD, de Bekker C, Baker R, Arciszewska LK, Freund SM, Bycroft M, Lowe J, Sherratt DJ. 2006. The FtsK γ domain directs oriented DNA translocation by interacting with KOPS. *Nat Struct Mol Biol* **13:** 965–972.

Steiner WW, Kuempel PL. 1998. Sister chromatid exchange frequencies in *Escherichia coli* analyzed by recombination at the *dif* resolvase site. *J Bacteriol* **180:** 6269–6275.

Teleman AA, Graumann PL, Lin DC, Grossman AD, Losick R. 1998. Chromosome arrangement within a bacterium. *Curr Biol* **8:** 1102–1109.

Thanbichler M. 2009. Closing the ring: A new twist to bacterial chromosome condensation. *Cell* **137:** 598–600.

Thanbichler M, Shapiro L. 2006. MipZ, a spatial regulator coordinating chromosome segregation with cell division in *Caulobacter*. *Cell* **126:** 147–162.

Thanbichler M, Shapiro L. 2008. Getting organized–how bacterial cells move proteins and DNA. *Nat Rev Microbiol* **6:** 28–40.

Thanbichler M, Wang SC, Shapiro L. 2005. The bacterial nucleoid: A highly organized and dynamic structure. *J Cell Biochem* **96:** 506–521.

Thanedar S, Margolin W. 2004. FtsZ exhibits rapid movement and oscillation waves in helix-like patterns in *Escherichia coli*. *Curr Biol* **14:** 1167–1173.

Toro E, Hong SH, McAdams HH, Shapiro L. 2008. *Caulobacter* requires a dedicated mechanism to initiate chromosome segregation. *Proc Natl Acad Sci* **105:** 15435–15440.

Ullsperger C, Cozzarelli NR. 1996. Contrasting enzymatic activities of topoisomerase IV and DNA gyrase from *Escherichia coli*. *J Biol Chem* **271:** 31549–31555.

Viollier PH, Sternheim N, Shapiro L. 2002. Identification of a localization factor for the polar positioning of bacterial structural and regulatory proteins. *Proc Natl Acad Sci* **99:** 13831–13836.

Viollier PH, Thanbichler M, McGrath PT, West L, Meewan M, McAdams HH, Shapiro L. 2004. Rapid and sequential movement of individual chromosomal loci to specific subcellular locations during bacterial DNA replication. *Proc Natl Acad Sci* **101:** 9257–9262.

Wang L, Lutkenhaus J. 1998. FtsK is an essential cell division protein that is localized to the septum and induced as part of the SOS response. *Mol Microbiol* **29:** 731–740.

Wang SC, Shapiro L. 2004. The topoisomerase IV ParC subunit colocalizes with the *Caulobacter* replisome and is required for polar localization of replication origins. *Proc Natl Acad Sci* **101:** 9251–9256.

Wang SC, West L, Shapiro L. 2006a. The bifunctional FtsK protein mediates chromosome partitioning and cell division in *Caulobacter*. *J Bacteriol* **188:** 1497–1508.

Wang X, Liu X, Possoz C, Sherratt DJ. 2006b. The two *Escherichia coli* chromosome arms locate to separate cell halves. *Genes Dev* **20:** 1727–1731.

Webb CD, Graumann PL, Kahana JA, Teleman AA, Silver PA, Losick R. 1998. Use of time-lapse microscopy to visualize rapid movement of the replication origin region of the chromosome during the cell cycle in *Bacillus subtilis*. *Mol Microbiol* **28:** 883–892.

Wheeler RT, Shapiro L. 1999. Differential localization of two histidine kinases controlling bacterial cell differentiation. *Mol Cell* **4:** 683–694.

Woldringh CL, Mulder E, Valkenburg JA, Wientjes FB, Zaritsky A, Nanninga N. 1990. Role of the nucleoid in the toporegulation of division. *Res Microbiol* **141:** 39–49.

Wu LJ, Errington J. 1994. *Bacillus subtilis* SpoIIIE protein required for DNA segregation during asymmetric cell division. *Science* **264:** 572–575.

Wu LJ, Errington J. 1997. Septal localization of the SpoIIIE chromosome partitioning protein in *Bacillus subtilis*. *EMBO J* **16:** 2161–2169.

Wu LJ, Errington J. 2004. Coordination of cell division and chromosome segregation by a nucleoid occlusion protein in *Bacillus subtilis*. *Cell* **117:** 915–925.

Wu J, Ohta N, Newton A. 1998. An essential, multicomponent signal transduction pathway required for cell cycle regulation in *Caulobacter. Proc Natl Acad Sci* **95:** 1443–1448.

Wu LJ, Ishikawa S, Kawai Y, Oshima T, Ogasawara N, Errington J. 2009. Noc protein binds to specific DNA sequences to coordinate cell division with chromosome segregation. *EMBO J* **28:** 1940–1952.

Yates J, Zhekov I, Baker R, Eklund B, Sherratt DJ, Arciszewska LK. 2006. Dissection of a functional interaction between the DNA translocase, FtsK, and the XerD recombinase. *Mol Microbiol* **59:** 1754–1766.

Yu XC, Margolin W. 1999. FtsZ ring clusters in *min* and partition mutants: Role of both the Min system and the nucleoid in regulating FtsZ ring localization. *Mol Microbiol* **32:** 315–326.

Yu XC, Weihe EK, Margolin W. 1998b. Role of the C terminus of FtsK in *Escherichia coli* chromosome segregation. *J Bacteriol* **180:** 6424–6428.

Yu XC, Tran AH, Sun Q, Margolin W. 1998a. Localization of cell division protein FtsK to the *Escherichia coli* septum and identification of a potential N-terminal targeting domain. *J Bacteriol* **180:** 1296–1304.

Zakrzewska-Czerwinska J, Jakimowicz D, Zawilak-Pawlik A, Messer W. 2007. Regulation of the initiation of chromosomal replication in bacteria. *FEMS Microbiol Rev* **31:** 378–387.

Zechiedrich EL, Cozzarelli NR. 1995. Roles of topoisomerase IV and DNA gyrase in DNA unlinking during replication in *Escherichia coli. Genes Dev* **9:** 2859–2869.

The Structure and Function of Bacterial Actin Homologs

Joshua W. Shaevitz[1] and Zemer Gitai[2]

[1]Department of Physics and the Lewis-Sigler Institute for Integrative Genomics, Princeton University, Princeton, New Jersey 08544

[2]Department of Molecular Biology, Princeton University, Princeton, New Jersey 08544

Correspondence: zgitai@princeton.edu

During the past decade, the appreciation and understanding of how bacterial cells can be organized in both space and time have been revolutionized by the identification and characterization of multiple bacterial homologs of the eukaryotic actin cytoskeleton. Some of these bacterial actins, such as the plasmid-borne ParM protein, have highly specialized functions, whereas other bacterial actins, such as the chromosomally encoded MreB protein, have been implicated in a wide array of cellular activities. In this review we cover our current understanding of the structure, assembly, function, and regulation of bacterial actins. We focus on ParM as a well-understood reductionist model and on MreB as a central organizer of multiple aspects of bacterial cell biology. We also discuss the outstanding puzzles in the field and possible directions where this fast-developing area may progress in the future.

The discovery of cytoskeletal proteins in bacteria has fundamentally altered our understanding of the organization and evolution of bacteria as cells. Homologs of eukaryotic actin represent the most molecularly and functionally diverse family of bacterial cytoskeletal elements. Recent phylogenetic studies have identified more than 20 subgroups of bacterial actin homologs (Derman et al. 2009) (Fig. 1). Many of these bacterial actins are encoded on extrachromosomal plasmids, but most bacterial species with nonspherical morphologies also encode chromosomal actin homologs (Daniel and Errington 2003). The two earliest proteins to be characterized as bacterial actins were the chromosomal protein MreB (Jones et al. 2001) and the plasmidic protein ParM (Jensen and Gerdes 1997). MreB and ParM remain the best-characterized of the bacterial actins and we will thus focus on these two proteins for most of this article.

The appreciation that bacteria possess actin homologs only occurred in the past decade. MreB was first identified as a protein involved in cell shape regulation in *Escherichia coli* in the late 1980s (Doi et al. 1988). In the early 1990s, pioneering bioinformatic studies identified similarities in a group of ATPases that have five conserved motifs (Bork et al. 1992), a feature dubbed the actin superfamily fold. Although

 doi: 10.1101/cshperspect.a000364
Cite this article as *Cold Spring Harb Perspect Biol* 2010;2:a000364

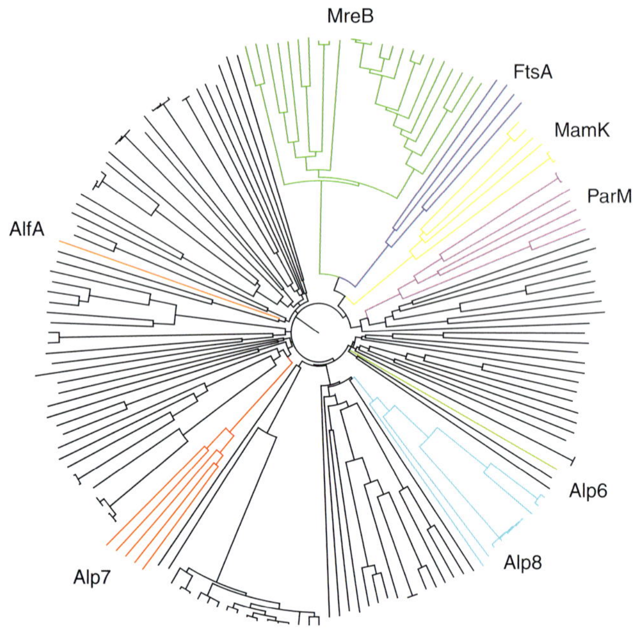

Figure 1. The superfamily of bacterial actin homologs. Shown is a phylogenetic tree of the bacterial actin subfamilies that have been identified to date based on sequence homology. The subfamilies that have been experimentally shown to polymerize are labeled and colored. (Courtesy of Joe Pogliano, based on Derman et al. 2009.)

this group includes actin and MreB, it also contains proteins that do not polymerize into filaments, such as sugar kinases like hexokinase and chaperones like Hsp70. A number of bacterial proteins are present in the actin superfamily, including the bacterial cell division protein FtsA which interacts with the tubulin homolog FtsZ and may or may not form filaments in different contexts (van den Ent and Lowe 2000). Because MreB did not appear significantly more related to actin than these nonfilamentous proteins, the weak sequence similarity with actin was largely ignored for the better part of a decade. This changed in 2001 when two seminal papers showed that *Bacillus subtilis* MreB forms cytoskeletal filaments in vivo (Jones et al. 2001) and that *Thermotoga maritima* MreB forms cytoskeletal filaments in vitro (van den Ent et al. 2001). Indeed, structural and biochemical studies of both MreB and ParM have convincingly showed that these proteins closely resemble actin and polymerize into linear filaments in a nucleotide-dependent manner (Fig. 2).

Research following the identification of bacterial cytoskeletal proteins has focused on understanding their assembly, regulation, and function. Here, we will summarize our current understanding of these issues and highlight the outstanding questions. We will begin with ParM, whose well-characterized assembly and dynamics represent a model for future studies of all cytoskeletal proteins. We will then focus on MreB, whose diverse activities appear to be central to the cell biology of many bacterial species.

PARM AS A MODEL FOR STUDYING BACTERIAL ACTINS

Perhaps the most completely understood bacterial actin is ParM, the plasmid segregating protein from the R1 plasmid. An interdisciplinary effort by a number of groups over the last decade has illuminated the details of ParM function from its fundamental molecular properties to a reconstitution of plasmid segregation in an

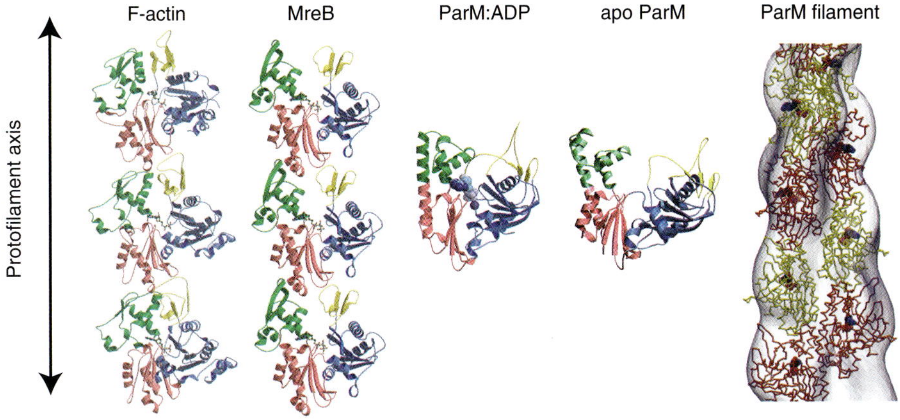

Figure 2. Structures of F-actin (Holmes et al. 1990), MreB (van den Ent et al. 2001), and ParM (van den Ent et al. 2002). (*Left*) Structures of F-actin filaments (PDB entry 1YAG). (Second from the *left*) MreB filaments from *T. maritima* (PDB entry 1JCE). (*Center*) ParM:ADP monomer in the "closed" conformation. (Second from the *right*) apo ParM monomer in the "open" conformation. (*Right*) ParM filament. Shown are the position of the nucleotide within the interdomain cleft, the conservation of fold, and the axis of the protofilament extension (arrow). Note that the conformational change shown for ParM from the "open" to "closed" state is predicted for all actin homologs. (Adapted, with permission from, Michie and Löwe 2006.)

artificial system. This multipronged approach brings together techniques from biochemistry, molecular biology, imaging, and materials science to provide an integrated solution to how ParM filaments form and how they lead to the movement of DNA. It is likely that this kind of approach will be useful in the study of the other, lesser-understood bacterial actins.

ParM Function and Mechanism

Plasmids are naturally occurring molecular parasites that often exploit their host cells' machinery for replication but provide their own segregation machinery. By 1997 it was known that the low-copy number *E. coli* plasmid R1 is actively partitioned during cell division so that each daughter cell retains the drug resistance conferred by the plasmid (Jensen and Gerdes 1997). This partitioning is achieved by the active positioning of two R1 sister plasmids to opposite ends of a cell during cell division (Jensen and Gerdes 1997). The machinery that directs the two plasmids to opposite cell poles is grouped in a locus termed *par*. The presence of this operon in a plasmid lowers the frequency of plasmid loss during division by several orders of magnitude (Gerdes et al. 1985). The Par machinery consists of three parts: a *cis*-acting region of DNA, *parC*, that acts as a centromere; the ParR protein that binds to 10 repeats within *parC* and has kinetichore-like activity; and the actin-homolog ParM, originally labeled as the partitioning motor, which drives segregation through polymerization dynamics (Dam and Gerdes 1994).

The details of plasmid segregation are elegant both in their simplicity and their efficiency. Throughout the cytoplasm, ParM filaments actively lengthen and shorten in a process known as dynamic instability that somewhat mimics the action of spindle microtubules (Garner et al. 2004). However, when the ParM filaments are bound to a plasmid via the ParR/*parC* complex they are stabilized against depolymerization (Garner et al. 2004). Consequently, plasmid bound ParM filaments proceed to elongate as much as they can, which in a rod-shaped cell pushes the plasmids to the cell poles at opposite ends of the cell axis (Garner et al. 2007) (Fig. 3).

This mechanism was discovered through an interdisciplinary approach to the study of plasmid segregation. Imaging work in vivo first

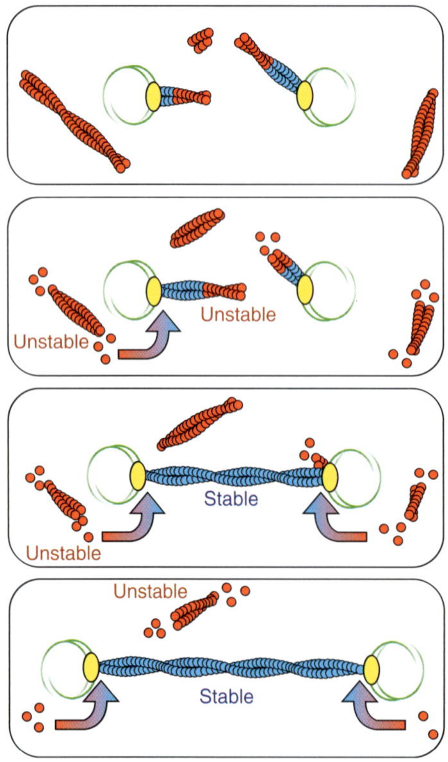

Figure 3. The assembly dynamics of ParM drive R1 plasmid segregation. Free ParM filaments ends (red) are dynamically unstable and undergo catastrophe. ParR (yellow) associates with the *parC* loci on the R1 plasmid (plasmids are green). Plasmid-bound ParR captures and stabilizes ParM filaments (blue). When both ends of a ParM filament are stabilized by ParR, a productive spindle is formed. Insertional polymerization at the filament ends serves to drive the two plasmids apart. (Adapted, with permission, from Garner 2008.)

defined the segregation pattern of plasmids and showed that the ParM protein forms long filaments in cells (Moller-Jensen et al. 2002; Moller-Jensen et al. 2003). Mutagenesis and other molecular techniques led to the mapping of the *par* locus (Dam and Gerdes 1994). In vitro work, including structural, biochemical, and imaging experiments confirmed the initial hypotheses from in vivo experiments and allowed for the reconstitution of the entire segregation process (Moller-Jensen et al. 2003; Garner et al. 2004; Garner et al. 2007). Richard Feynman once said that "What I cannot create, I do not understand." It is in this spirit that we think that ParM is well understood. In the paragraphs to follow, we discuss in greater detail what is currently known about ParM from the molecular to the cellular scales.

ParM Polymerization and Structure

ParM forms filaments both in vivo and in vitro. Initial immunofluorescence imaging showed that ParM formed long, curved filaments (Moller-Jensen et al. 2002). These authors observed that ParM filaments often spanned the length of rod-shaped cells and that the filaments were gently curved. Interestingly, the length of the filaments was observed to be dynamic. In vitro, purified ParM was shown to polymerize into long, straight filaments. Polymerization requires ATP, or the nonhydrolyzable ATP analogs ATPγS and AMPPNP (Moller-Jensen et al. 2002). In the presence of ADP, ParM does not polymerize. This observation, together with the dynamic nature of ParM filaments in cells, led to a dynamic instability model for plasmid segregation, similar to that for tubulin, where ATP-bound ParM monomers bind stably to the end of ParM filaments whereas posthydrolysis, ADP-bound monomers lead to depolymerization from the filament end.

The three-dimensional crystal structure of monomeric ParM shows remarkable similarity to that of actin (Fig. 2). ParM is an asymmetric protein with a barbed and a pointed end made up of four domains: IA, IB, IIA, and IIB, which respectively correspond to the actin domains 1, 2, 3, and 4 (van den Ent et al. 2002). ParM has been successfully crystallized in the nucleotide-free and ADP bound states. These structures show that the nucleotide binds in the interdomain cleft of the barbed end. On nucleotide binding, a rigid-body rotation of about 25° of domains I and II closes the cleft slightly. This conformational change is thought to cause the nucleotide-induced depolymerization of ParM filaments.

The structure of ParM filaments was solved by docking the monomeric crystal structures into three-dimensional electron microscopy images of filaments (van den Ent et al. 2002).

Filaments assemble with the pointed end of one monomer enclosed by the barbed-end of the next. Along the filament, adjacent monomers are rotated by about 166° and translated by 2.5 nm from each other. Even though this is a slightly tighter twist and shorter axial spacing than that of actin, ParM still forms a twisted, two-start helix. One significant difference between actin and ParM filaments is that the ParM helix is left-handed whereas actin is right-handed. The juxtaposition of a similar but chirally opposite geometry suggests that this helical shape might afford the filaments a high level of stability that has been arrived at independently through convergent evolution. Similar conclusions about the shape of ParM filaments in vivo have been made using high-resolution cryo-EM tomography (Salje et al. 2009). More recently, analysis of additional bacterial actin homologs supports the idea that though these proteins all polymerize, the details of their filament formation can be quite divergent (Polka et al. 2009). Although the sequence and 3D structure of the ATPase domains of ParM have significant homology with actin, the exposed residues of ParM are very different in their chemical properties from those of actin. Consequently, proteins that bind to actin and to ParM are not expected to share much similarity.

ParM Biochemistry and Biophysics

In vitro experiments have been very successful at verifying and enhancing the early models of plasmid segregation derived from in vivo experiments. Jensen and Gerdes showed that mutation of the aspartate at position 170 of ParM abolished both the protein's ATPase activity and its ability to segregate plasmids in vivo (Jensen and Gerdes 1997; Jensen and Gerdes 1999). Immunofluorescence imaging showed that these mutants showed hyperfilamentation, lending support to the idea that ParM dynamics are required for proper plasmid segregation.

Right-angle light scattering measurements have been used to measure the kinetics of ParM filament polymerization. Moller-Jensen and others showed that after an initial bout of polymerization, ParM slowly hydrolyzes bound ATP leading to filament depolymerization (Moller-Jensen et al. 2002). Additional ATP added after this depolymerization phase causes a new round of fast polymerization and slow depolymerization, indicating that the monomers are capable of going through multiple rounds of nucleotide stimulated polymerization and depolymerization. Most interestingly, filaments are stabilized by the addition of both ParR and *parC*. When these three components are mixed together, filaments do not depolymerize and the critical concentration for polymerization is reduced (Garner et al. 2007). These results were confirmed by subsequent FRET measurements that verified the nucleation condensation mechanism of ParM polymerization. Garner and coworkers found a critical nucleus size of three monomers that defines the two-stranded helical filament geometry (Garner et al. 2004).

Although light scattering is a useful tool to measure the typical size of an ensemble of oligomers in solution, it cannot resolve the nature of individual filament dynamics. For example, it can be difficult to distinguish whether some monomers are incapable of forming polymers or if the ensemble of polymers is very dynamic. Garner et al. used total-internal fluorescence microscopy to directly visualize single Alexa-dye labeled ParM filaments in vitro (Garner et al. 2004). Using a two-color assay in which red-labeled monomers were added to green-labeled filament seeds, they were able to distinguish growth from the two filament ends. Using the nonhydrolyzable substrate AMPPNP, they observed steady growth from both ends of long, stable filaments. In the presence of ATP, however, filaments grew symmetrically for a while but then rapidly depolymerized to completion from one end. The time scales associated with the filament growth and shrinking were consistent with a mechanism in which an ATP cap stabilizes the ParM filament. If the terminal monomers hydrolyze their bound nucleotide, the filament becomes unstable. These studies thus showed that in contrast to actin, which preferentially polymerizes at one filament end at a constant rate, ParM polymerizes symmetrically and experiences dynamic instability with bouts of steady polymerization interrupted by rapid depolymerization.

Putting all of these ideas together, Garner and colleagues showed that ParM, ParR, and *parC* are sufficient to reconstitute plasmid segregation in vitro (Garner et al. 2007). The authors coated small, 350-nm diameter beads with a piece of DNA that contained the *parC* sequence. When added to a solution of ParR and fluorescently labeled ParM, the beads formed aster-like clouds of highly dynamic ParM. When two beads were in proximity, their ParM structures stabilized one another to generate long spindle-like structures. EM images indicated that single filaments extended all the way from one bead to the other, suggesting that the long structures resulted from stabilization of both ends of the ParM filaments. These filaments were able to elongate by insertional polymerization between the tips of the filaments and the beads, thereby pushing the DNA-coated beads farther and farther apart (Fig. 3). The energetics of this elongation were proposed to be driven by a monomer excess generated by the dynamic instability of the nonstabilized ParM. Most excitingly, when the *parC*-coated beads, ParM, and ParR were placed in long microfabricated channels, the ParM spindle elongated along the long axis of the channel, thereby breaking symmetry to separate pairs of beads to opposite ends of the cylinder.

Although the ability to reconstitute the entire plasmid segregation process is nothing short of remarkable, it is worth noting that it shows sufficiency for how the system could work, but does not prove how it does work in vivo. Thus, several exciting questions about ParM remain to be answered. For example, does the rest of the plasmid DNA really just play a passive role in the whole process? Also, why is it that ParM and actin filaments are structurally similar yet one polymerizes symmetrically at both ends while the other polymerizes asymmetrically? Along similar lines, EM studies both in vitro and in vivo suggest that both ends of the filament associate with the ParR/*parC* complex (Garner et al. 2007; Salje et al. 2009), and work by Choi and colleagues used very small gold beads to show that in a ParM spindle, a single copy of *parC* is bound to the end of a single ParM filament (Choi et al. 2008). So how does the ParR/*parC* manage to interact with both ends of an asymmetric polymer? Other unresolved issues surround the nature of the dynamic instability and the ability of other nucleotides such as GTP to stabilize ParM filaments (Popp et al. 2008). Nevertheless, ParM-mediated plasmid segregation remains a prototype for how combining genetics, cell biology, biochemistry, and biophysics can lead to an emergent understanding of dynamic systems.

THE LOCALIZATION AND FUNCTION OF MREB

ParM is encoded by a subset of plasmids and carries out one highly specific function, plasmid segregation. In contrast, the most widely conserved bacterial actin homolog, MreB, is encoded in the chromosomes of many different species and can participate in many different cellular activities (Daniel and Errington 2003). MreB homologs have been implicated in nearly every spatially organized cellular process, including cell growth, morphogenesis, polarity, protein localization, organelle positioning, division, and differentiation, as well as chromosome segregation, replication, and decatenation (reviewed in Carballido-Lopez 2006). The current challenge is to understand the mechanism by which MreB impacts these cellular processes and distinguish its primary roles from their secondary consequences. MreB also assembles into an interesting, often helical, localization pattern that may help it execute its many functions (Jones et al. 2001). Finally, MreB clearly cannot be doing all of these things on its own; indeed, a growing number of MreB interactors are being identified. In this section we will focus on the best characterized of these MreB localizations, functions, and interactors.

MreB Proteins are Relatively Diverse

Discussion of MreB is inherently complicated because of the vast diversity of MreB homologs in the bacterial and archael kingdoms. At least one MreB homolog is found in most nonspherical bacteria (Daniel and Errington 2003). The exceptions to this rule include a number of plant and animal pathogens that are rod-shaped but lack a clear MreB homolog, including *Mycoplasmas*, *Mycobacteria*, and *Rhizobiae*. There are also several spherical bacteria such as *Cyanobacteria* and *Planctomycetes* that have MreB homologs. Many bacteria also encode multiple MreB homologs. The best-characterized example is *B. subtilis*, which has three MreB homologs: MreB, Mbl, and MreBH (Jones et al. 2001). The three *B. subtilis* MreB homologs each has similar homology to other MreBs (~50% identity to *T. maritima* MreB), such that even though only one of them bears the name "MreB", they should each be viewed equivalently. *Spiroplasmas* have even more MreB homologs, as many of these species have five MreB proteins with more divergent sequences (Kurner et al. 2005). Yet other species, like *Magnetospirillum magnetotactum*, have both a relatively well-conserved MreB homolog and a second more divergent actin superfamily member, MamK (Komeili et al. 2006). All MreB homologs share the same basic actin superfamily signature, and all MreB homologs characterized to date have been found to be able to polymerize. Nevertheless, the diversity in number and sequence of MreB homologs means that MreB cannot be treated as a single entity and that it is important to specify which MreB homolog one is discussing in a specific context. Most of the structural and biochemical work on MreB has been performed on the *T. maritima* homolog, whereas the in vivo properties of MreB have best been characterized in *B. subtilis*, *Caulobacter crescentus*, and *E. coli*. Our discussion here will primarily focus on these systems.

MreB Localization and Dynamics

MreB localization was first characterized in *B. subtilis* for both the *mreB* and *mbl* genes (Jones et al. 2001). These proteins were found to form right-handed helical structures by deconvolution microscopy. In some cells a double helix is observed, whereas others resemble a single helix. Assessing the exact dimensions and topologies of these helices remains an active area of investigation that is complicated by two issues. First, the small size of these structures is near the diffraction-limited resolution of light microscopy. Second, immunofluorescence (IF) microscopy requires fixation that can alter cell ultrastructure, whereas both amino- and carboxy-terminal fluorescent protein fusions to MreB can perturb both the function and localization of native MreB. Very recently, an internal fusion that placed mCherry in the middle of the *E. coli* MreB protein was found to be largely functional (Bendezu et al. 2009), suggesting that these issues, perhaps in combination with recent advances in subdiffraction-limited microscopy, might soon be resolved.

Nevertheless, in many cases the IF and the GFP-MreB images agree, leading to a consensus view that MreB prefers, at least locally, to form a helix with several turns per cell length. In *B. subtilis*, the three MreB homologs colocalize with an approximate pitch of 0.75 μm, suggesting that the three MreB isoforms may functionally interact or copolymerize into a single structure (Carballido-Lopez et al. 2006). As cells elongate, the pitch appears to remain constant with addition of new helical turns. In *Caulobacter*, both IF and GFP-MreB reporter studies indicate that the localization of MreB is regulated during the cell cycle (Figge et al. 2004; Gitai et al. 2004). Early in the cell cycle, *Caulobacter* MreB forms a patchy, potentially helical, pattern that extends from pole to pole. As the cell cycle progresses, this helix condenses into a ring at the presumptive division plane. Before cell division occurs, MreB expands from a ring back into a helical form such that each daughter cell inherits a similar polymer structure. The transition from helix to ring depends on the FtsZ tubulin homolog, the central organizer of the division machinery (Figge et al. 2004). Localization experiments in other species such as *E. coli*, *P. aeruginosa*, and *Rhodobacter* have also characterized both helical and medial MreB distributions (Slovak et al. 2005; Vats et al. 2009; Cowles and Gitai 2010), though the physiological consequences of this transition remain unclear.

In contrast to *B. subtilis* cells that colocalize three closely related MreB homologs, MreB does not appear to colocalize with more highly diverged actin homologs. For example, in *M. magnetotactum*, the divergent actin MamK forms a straight filament structure unlike any MreB localization pattern that has been characterized (Komeili et al. 2006). Similarly, plasmidic actin homologs like ParM, AlfA, and the Alp proteins form straight structures that do not colocalize with MreB (Moller-Jensen et al. 2002; Becker et al. 2006; Derman et al. 2009). Some of these proteins, including ParM and AlfA, have highly divergent polymer interaction surfaces and geometries (van den Ent et al. 2001; van den Ent et al. 2002; Polka et al. 2009), which may function to actively prevent their copolymerization with MreB.

While the entire MreB structure is dynamically rearranged in *Caulobacter*, it also appears that the subunits of the MreB helix are highly dynamic in all cells examined. These dynamics have been assessed by FRAP for Mbl (Carballido-Lopez and Errington 2003), by speckle tracking for all three MreB proteins in *B. subtilis* (Defeu Soufo and Graumann 2004), and by single-molecule imaging of MreB in *Caulobacter* (Kim et al. 2006). The *Caulobacter* MreB appears to have faster turnover kinetics than observed in *B. subtilis*, but it remains unclear whether this reflects species-specific differences or differences in the types of assays used (FRAP vs. single-molecule imaging). Although the *Caulobacter* studies suggest that individual MreB filaments may be polarly assembled (Kim et al. 2006), there does not appear to be an overall pole-to-pole polarity in the MreB helix, suggesting that the MreB helix includes individual filaments with mixed polarities.

Further evidence for the dynamic nature of MreB assembly comes from experiments with the small molecule A22 (S-(3,4-dichlorobenzyl) isothiourea). A22 was first found in a chemical genetic screen for compounds that increased the rate of chromosome loss in *E. coli* (Iwai et al. 2002). A22 was found to cause cells to become round, but its cellular target was unknown until A22-resistant mutations were mapped to the *mreB* gene, first in *Caulobacter* (Gitai et al. 2005), and subsequently in other species such as *E. coli* and *P. aeruginosa* (Kruse et al. 2006; Robertson et al. 2007; Cowles and Gitai 2010). In *Caulobacter*, A22 rapidly delocalizes MreB filaments in vivo, and the resistant mutants all map to the nucleotide-binding pocket (Gitai et al. 2005). Indeed, biochemical studies (detailed below) show that A22 functions by binding the MreB nucleotide-binding pocket and mimicking the ADP-bound low polymerization affinity monomer state (Bean et al. 2009). When A22 is applied to either purified MreB in vitro (Bean et al. 2009) or MreB that has been heterologously expressed in the eukaryote *S. pombe* (Srinivasan et al. 2007), A22 inhibits new MreB polymerization but does not stimulate depolymerization. Consequently, the rapid delocalization of MreB observed in bacterial cells likely results from dynamic MreB depolymerization that can no longer be balanced by A22-inhibited polymerization. The increased dynamics of MreB assembly in vivo also suggests that additional cellular factors exist to stimulate these dynamics. The identification and characterization of such factors will be an important area of future investigation.

Another important yet unresolved issue is the mechanism by which MreB adopts its localization pattern. MreB filaments are not obviously helical either in vitro or on heterologous expression in eukaryotes (van den Ent et al. 2001; Srinivasan et al. 2007). It is possible that MreB helices are the mechanical consequence of forcing a linear polymer into a cylindrical container. Alternatively, accessory factors may change the preferred conformation of MreB filaments in bacteria. The dynamic redistribution of MreB during the *Caulobacter* cell cycle supports the idea that MreB conformation and localization can be regulated. One candidate for such regulation is RodZ, which was recently characterized as a protein that binds MreB and potentially links it to the inner membrane (Shiomi et al. 2008; Alyahya et al. 2009; Bendezu et al. 2009; van den Ent et al. 2010). RodZ and MreB have an interdependent genetic relationship and similar loss-of-function phenotypes, making it difficult to dissect a linear localization hierarchy between these proteins. Yet other MreB-interacting proteins such as Ef-Tu (Defeu

Soufo et al. 2010), Pbp2 (Figge et al. 2004), or other as-yet-unidentified proteins could also influence MreB localization.

MreB is a Key Regulator of Cell Shape Determination

MreB was first characterized as an *E. coli* mutant that caused normally rod-shaped cells to become spherical (Doi et al. 1988). Indeed, loss of proper cell shape has emerged as the most common defect associated with MreB proteins across most species. The central importance of MreB for achieving rodlike elongation is supported by the phylogenetic observation that MreB proteins are primarily found in nonspherical cells (Daniel and Errington 2003). However, there are exceptions to this rule. For example, *Helicobacter pylori* MreB has been reported to influence chromosome dynamics and virulence factor secretion without affecting cell shape (Waidner et al. 2009), and in *Streptomyces coelicolor* MreB affects sporulation but does not affect the shape of vegetatively growing cells (Mazza et al. 2006). To more generally survey how rodlike bacteria grow, the Errington lab chemically labeled nascent cell wall synthesis and characterized two distinct cylindrical growth modes in different species (Daniel and Errington 2003). One mode is more common and uses MreB to direct the insertion of new cell wall material along the length of the cylinder. These species generally have inert cell poles. The second mode is MreB-independent and involves insertion of new cell wall material at the cell poles. A third, FtsZ-dependent elongation mechanism has also been more recently proposed (Aaron et al. 2007).

MreB regulates cell shape in three well-characterized model systems: *B. subtilis*, *E. coli*, and *Caulobacter*. In *B. subtilis*, mutants in the three MreB homologs have different morphologies (Jones et al. 2001). It is possible that these MreB proteins have specialized for distinct functions. Alternatively, a study demonstrating partial redundancy between these mutants suggests that they may all influence the same process to different extents, perhaps because of differential expression levels (Kawai et al. 2009). In *E. coli* grown under normal conditions, cells lacking MreB become spheres and eventually lyse (Bendezu and de Boer 2008). However, the lethality of loss of MreB can be suppressed either by overexpressing cell division proteins or reducing the rate of cell growth (Bendezu and de Boer 2008). In these conditions *mreB* mutants are viable but accumulate intracellular vesicles, suggesting that membrane production is disregulated (Bendezu and de Boer 2008). The lethality of a mutation in one of the *B. subtilis mreB* genes can also be suppressed by growth in increased levels of Mg^{++} (Formstone and Errington 2005).

Studies from both *B. subtilis* and *Caulobacter* suggest that at least one way in which MreB influences the cell wall is by directing the insertion of new cell wall material in a helical pattern. The bacterial cell wall is composed of stiff peptidoglycan strands that are polymerized by transglycosylases and crosslinked by transpeptidases to generate a meshlike superstructure (Holtje and Heidrich 2001). The cell wall has been thought to be the pressure-bearing cell shape determinant because cell wall lysis causes cells to round up and isolated cell walls retain their general morphology (Young 2006). Because both MreB and new cell wall insertion follow helical patterns, the basic model for how MreB directs cell shape determination is that MreB helically localizes proteins that in turn lead to helical peptidoglycan assembly (Daniel and Errington 2003). It is also possible that as in eukaryotes, the MreB cytoskeleton plays a mechanical role in directing proper morphogenesis.

Consistent with the cell wall patterning model, *Caulobacter* MreB biochemically associates with the cytoplasmic tail of a cell wall assembly enzyme, the Pbp2 peptidoglycan transpeptidase, and directs its localization (Figge et al. 2004; Dye et al. 2005). *Caulobacter* MreB has also been proposed to direct the localization of cytoplasmic proteins that direct peptidoglycan subunit synthesis (White et al. 2010) and inhibition of MreB with the small molecule A22 leads to shortened cell wall glycan strands (Takacs et al. 2010), indicating that MreB may have a general influence on peptidoglycan assembly. In *B. subtilis*, the Pbp proteins adopt a helical localization pattern (Scheffers et al. 2004). However, single *mreB* mutants do not disrupt this localization (Scheffers et al. 2004), possibly because of partial redundancy between the different MreB. Unlike *E. coli* and *Caulobacter,* which are gram-negative bacteria, *B. subtilis* are gram-positive and their cell wall is also composed of teichoic acids. The proteins that assemble teichoic acids are also helically localized, though it remains unclear if this helical distribution is related to that of the MreB proteins (Formstone et al. 2008). In addition, MreBH interacts with the cell wall hydrolase LytE (Carballido-Lopez et al. 2006), and theoretical studies suggest that cell shape could be dictated by patterning either peptidoglycan insertion or cleavage (Huang et al. 2008).

In many species, *mreB* is found in the same operon as two other genes, *mreC* and *mreD.* MreB, MreC, and MreD, along with the cell wall assembly proteins Pbp2 and RodA have been proposed to form a complex that collaborates to coordinate cell elongation (Kruse et al. 2005). This complex of predominantly transmembrane proteins may explain how the cytoplasmic MreB structure can have such an impact on the assembly of the cell wall, which occurs outside of the inner membrane. In *Caulobacter,* MreC is cleaved and released into the periplasm where it forms a helical structure whose assembly is independent of MreB (Divakaruni et al. 2005). The crystal structure of MreC suggests that it may form polymers and could thus act as a periplasmic cytoskeleton (van den Ent et al. 2006). Further genetic and colocalization studies in *Caulobacter* suggest that MreC acts to localize peptidoglycan assembly in the periplasm (Divakaruni et al. 2005; Dye et al. 2005), whereas MreB acts to localize peptidoglycan precursor synthesis in the cytoplasm (Figge et al. 2004; White et al. 2010). However, this model must either be oversimplified or species-specific because some species that have MreB lack MreC or vice-versa, and in yet other species, MreB localization depends on MreC (Kruse et al. 2005).

The growing body of evidence that MreB can influence nearly every aspect of cell wall assembly suggests either that all of these processes are highly interdependent or that MreB has multiple independent roles in regulating cell shape. Dissecting the specific roles of MreB in this process has been complicated by the fact that we do not have very good tools for studying peptidoglycan structure and dynamics. Fluorescent derivatives of antibiotics that target nascent peptidoglycan structures have proved useful (Daniel and Errington 2003; Tiyanont et al. 2006), but can also perturb cell wall assembly and occasionally produce conflicting results. Recent studies have used high-resolution atomic force microscopy to study the exposed cell walls of gram positives or isolated gram-negative cell walls (Yao et al. 1999; Touhami et al. 2004), and an elegant electron cryotomography study gave the first glimpses into the structure of Gram-negative cell walls (Gan et al. 2008). By combining these new approaches for studying peptidoglycan, the next few years promise to reveal a great deal about the assembly and regulation of bacterial cell walls.

MreB Regulates Protein Localization

MreB has also emerged as a key regulator of the subcellular organization of bacterial proteins. The role of MreB in polar protein localization was first characterized in *Caulobacter,* where MreB was found to be necessary for the localization of multiple polar protein markers (Gitai et al. 2004). Subsequently, MreB has been implicated in a wide range of protein localization processes in many experimental systems. In addition to the proteins involved in cell wall assembly discussed in the previous section on cell shape, these include chemotaxis receptors in *E. coli* (Shih et al. 2005), gliding motility proteins in *Myxococcus xanthus* (Mauriello et al. 2010), and pilus-associated proteins in *Pseudomonas aeruginosa* (Cowles and Gitai 2010). There is also evidence that bacterial actins are important for localizing larger protein complexes or organelles. For example, MamK is essential for magnetosome localization in *M. magnetotactum* (Komeili et al. 2006). Similarly, MreB influences stalk assembly and localization in *Caulobacter* (Wagner et al. 2005; Divakaruni et al. 2007), pilus assembly in *P. aeruginosa* and

M. Xanthus (Cowles and Gitai 2010; Mauriello et al. 2010), inclusion body localization in *E. coli* (Rokney et al. 2009), and carboxysome localization in *Synechoccus elongates* (Savage et al. 2010). These larger structures may particularly benefit from cytoskeletal-mediated localization because of their inherently reduced rates of diffusion.

One issue with studying MreB's functions is that MreB is pleiotropic, such that it can be difficult to untangle the direct effects of MreB on subcellular localization from indirect effects caused by decreased growth or disrupted morphology. One advance that has been of immense help in this area has been the discovery of small molecules, such as A22, that rapidly perturb MreB. The rapid action of these compounds helps temporally uncouple primary and secondary consequences of MreB disruption. For example, although MreB can be delocalized in as little as 30 s with these agents, the cell shape defects induced on MreB disruption require hours of new growth to manifest, producing a time window during which MreB is disrupted but cell shape is still unaffected (Gitai et al. 2005). One concern with using any small molecule inhibitor is the degree of specificity for the target protein of interest. Although A22 appears to primarily affect MreB, genetic studies from both *E. coli* and *P. aeruginosa* suggest that it may have additional off-target effects (Cowles and Gitai 2010; Takacs et al. 2010). Recently, two additional MreB antagonist compounds, CBR-4830 and MP265 have been characterized (Robertson et al. 2007; Takacs et al. 2010). By combining comparisons of multiple distinct MreB inhibitors and the proper use of drug-resistant control strains, small molecule antagonists remain the premier method for dissecting MreB function.

Although MreB is clearly important for protein localization, and this role may explain how MreB can affect so many cellular processes, the mechanism by which MreB directs protein localization remains mysterious. Increasing evidence suggests that MreB may often be involved in the initiation, but not the maintanence, of protein localization. For example, in *Caulobacter*, the polar marker PopZ localizes to one cell pole early in the cell cycle and later redistributes to the other cell pole (Bowman et al. 2008). Although MreB is not required for maintaining the localization of PopZ at the initial pole, it is required for relocalizing PopZ to the second pole (Bowman et al. 2008). By treating cells with A22 before or after the induction of a fluorescent fusion to a protein of interest, one can directly distinguish whether MreB is involved in maintaining the localization of old protein (which would be delocalized regardless of whether A22 is administered before or after induction), or whether MreB is involved in initiating new protein localization (which would only be delocalized when A22 is administered before induction). Consistent with the PopZ example, MreB is also required for the initiation of Pbp2 localization in *Caulobacter* (Dye et al. 2005). These functions can also be regulated, as seen in *P. aeruginosa*, where MreB is required for both the initiation and maintenance of polar PilT localization when cells are grown in liquid media but is only required for initiation of PilT localization when cells are grown on solid surfaces (Cowles and Gitai 2010). Thus, although several studies have taken a protein's persistent localization in the presence of A22 as indication of MreB-independent localization, such experiments only address the importance of MreB for the maintenance of protein localization. A better understanding of both the proteins that directly interact with MreB and the single-molecule motions of these target proteins should lead to a better understanding of the molecular mechanism by which MreB directs the dynamics of initiating protein localization.

MreB and Chromosome Dynamics

Bacterial actins have been implicated in regulating the organization of bacterial DNA. This function is clearest for the plasmidic actins such as ParM and AlfA, as discussed earlier. MreB proteins have also been implicated in chromosome dynamics, though their specific functions and mechanisms remain unclear. In *E. coli*, a dominant-negative MreB mutant was shown to perturb chromsome segregation (Kruse et al. 2003), and A22 treatment was found to both affect chromosome segregation and increase the rate of chromosome loss (Iwai et al. 2002; Kruse et al. 2006). A mechanism for this effect was suggested by the interaction of MreB with RNA polymerase, which could serve as a motor to push the two chromosomes apart (Kruse et al. 2006). In contrast, recent studies have shown that *E. coli* can survive and presumably segregate its chromosomes in the absence of MreB when cell growth is slowed (Bendezu and de Boer 2008), and can segregate its chromosomes when MreB localization is aberrant (Karczmarek et al. 2007). Similarly, MreB has been implicated in chromosome segregation in *B. subtilis* in some cases but not others (Soufo and Graumann 2003; Formstone and Errington 2005), though a careful analysis in the absence of all three MreB homologs has yet to be completed. Experiments in *Vibrio cholerae* and *Helicobacter pylori* support the role of MreB in chromosome segregation (Srivastava et al. 2007; Waidner et al. 2009), whereas the nonessential nature of MreB in other species such as *S. coelicolor* suggest that MreB is not necessary for segregation in these organisms (Mazza et al. 2006; Hu et al. 2007). Some insight into this apparent paradox might be gleaned from studies in *Caulobacter*. Under certain environmental conditions such as growth on agarose pads, the segregation of the region of the chromosome near the *origin* of replication is either blocked or significantly delayed by treatment with A22 (Gitai et al. 2005). However, under other conditions, such as growth in liquid media, A22 treatment only results in a mild delay in the onset of segregation, which can be largely attributed to a delay in the onset of DNA replication (Shebelut et al. 2009). Genetic studies suggest that there are multiple pathways that contribute to the process of segregation (Shebelut et al. 2009). It is possible that these pathways may act redundantly in some conditions, but that each may become essential under other, perhaps stressful, conditions. Although it remains unclear whether MreB directly or indirectly affects chromosome segregation, viewing segregation as a sequence of distinct processes may help define the specific mechanisms that collaborate to govern segregation in different contexts.

MreB has also been implicated in aspects of chromosome dynamics other than segregation. In *V. cholerae*, MreB perturbations affect chromosome condensation (Srivastava et al. 2007). Meanwhile, in *E. coli*, MreB was found to regulate chromosome decatenation (Madabhushi and Marians 2009). This effect appears to be direct, as in vitro studies showed that the activity of Topoisomerase IV was stimulated by purified MreB polymers and inhibited by purified MreB monomers. Finally, MreB may play a role in DNA replication, as A22 treatment slows replication in *Caulobacter* (Shebelut et al. 2009) and MreB can affect the positioning of DNA replication proteins in *B. subtilis* (Defeu Soufo and Graumann 2005). MreB may also be required for the replication of foreign DNA in bacteria. MreB is required for the replication of a number of phages that infect different bacterial species, and MreB specifically mediates the localization of phage replication proteins for the *B. subtilis* phage 29 (Munoz-Espin et al. 2009).

MreB May be Important for Bacterial Pathogenesis

The study of bacterial cell biology in general and bacterial actins in particular is rapidly advancing our understanding of the fundamentals of cellular organization and dynamics. Meanwhile, these studies also promise to yield exciting advances in our ability to combat infectious diseases. MreB is essential for the rapid growth of many bacteria, such that small molecule inhibitors of MreB could represent powerful broad-spectrum antibiotics. Recent studies have begun to suggest that MreB may also have been co-opted by pathogenic bacteria to organize their virulence mechanisms. In *P. aeruginosa*, MreB regulates the polar assembly of type IV pili, which are important for virulence (Cowles and Gitai 2010). In *H. pylori*, MreB regulates the secretion of virulence factors (Waidner et al. 2009), and *Vibrio parahaemalyticus* differentially regulates MreB expression on interaction with its host (Chiu et al. 2008). *Bdellovibrio bacteriovorus* is a bacterium that infects other bacteria, and perturbing *B. bacteriovorus*

MreB perturbs this ability (Fenton et al. 2010). Consequently, understanding bacterial actins may help understand and ultimately develop both broad and narrow spectrum approaches to combating bacterial pathogenesis.

MREB STRUCTURE AND ASSEMBLY

Like ParM, MreB shares significant sequence homology with actin. An early search of known bacterial sequences indicated that MreB was a likely member of the actin superfamily based on its catalytic core (Bork et al. 1992). By sequence, MreB is the most closely related to actin of all the actin family proteins although small differences between MreB and actin exist in both the nucleotide binding site and the residues that form the monomer–monomer interface within a protofilament. The differences at the monomer–monomer interface are interesting because they must have evolved concomitantly to retain the polymerization. In vitro polymerization of MreB and subsequent solution of the polymer crystal structure was performed in 2001 by van den Ent, Amos, and Lowe (van den Ent et al. 2001). Using purified MreB1 from *T. maritima*, they found that polymers formed in a variety of conditions and required ATP or GTP.

MreB was the first actin for which the polymeric crystal structure was solved (van den Ent et al. 2001), as opposed to conventional actin for which only EM-based polymer structures exist (Oda et al. 2009). This breakthrough revealed a two-domain, V-shaped configuration with the nucleotide bound in the interdomain cleft, in good agreement with models of actin filaments. The lateral spacing of monomers along an MreB protofilament is 5.1 nm and a high concentration of hydrophobic residues lies at the monomer–monomer interface, producing a strong binding interaction. Unlike actin and ParM, however, MreB appears to form straight protofilaments and does not assemble into two-filament twisted helices. The ability to express, purify, polymerize, and crystallize recombinant MreB polymers provides the potential to study structural properties of filamentous proteins in a way that has not been possible with actin or tubulin and should result in some major breakthroughs in the near future.

MreB Filament Structure and Biophysics

Although MreB appears to form a helical structure in cells, in vitro MreB usually forms very straight and stiff filaments, and sometimes tight, ringlike spirals depending on the polymerization conditions (van den Ent et al. 2001; Esue et al. 2005; Esue et al. 2006). The MreB protofilament is straight with a longitudinal spacing of 5.1 nm. These protofilaments associate laterally to build thick, crystalline bundles. Bean and Amann used fluorescent imaging to examine Alexa-fluor-labeled Cys332-MreB and observed thick bundles ~3 microns in length that appeared very rigid because of their lack of thermally driven bending fluctuations (Bean and Amann 2008). Similar features were observed when a GFP-labeled MreB was expressed in fission yeast (Srinivasan et al. 2007). Esue et al. (2006) measured the bulk rheological properties of MreB gels (Esue et al. 2006). At physiological concentrations, the MreB formed a very stiff gel (~10 dyn/cm^2) in about 2 min. However, these types of bulk measurements are difficult to interpret and the relationship between gel stiffness and MreB filament function remains to be explored. Very recent measurements have shown that MreB contributes significantly to the overall stiffness of *E. coli* cells (Wang et al. 2010).

MreB Biochemistry

A consensus view of the in vitro kinetics of MreB polymerization has been more difficult to measure than for ParM because different experimental groups have produced varying results using different MreB constructs. To date, the biochemical mechanisms of MreB polymerization and dynamics remain an active area of research.

Esue and coworkers published the first measurements of MreB polymerization kinetics in 2005 (Esue et al. 2005). Using a His-tagged form of *T. Maritima* MreB1, the authors found that polymerization was strongly dependant on temperature and the concentrations of different ions in solution. They measured a critical concentration of 3 nM, a hundred times smaller than that for actin. Further experiments showed that MreB can use ATP or GTP as a substrate equally well, unlike other proteins such as actin (Esue et al. 2006).

Work in 2008 from Bean and Amann focused on the same protein but measured a different behavior (Bean and Amann 2008). These authors purified the native form of *T. Maritima* MreB1 without a His-tag through ion exchange and gel filtration chromatography and differential centrifugation. They found two phases of polymerization, one that uses divalent cations and one that does not. These phases have been hypothesized to correspond with the nucleation and elongation phases of polymer growth. Their more "native" protein was also less temperature sensitive than those reported previously. ATP hydrolysis of the enzyme was fast, implying that in vivo most polymerized MreB is in an ADP-bound state. In addition, the authors measured the critical concentration for polymerization in the presence of ATP and ADP and found these two values to be close to each other—one requirement for treadmilling.

Very recent work from Mayer and Amman (2009) successfully purified and polymerized MreB from *B. subtilis* (Mayer and Amann 2009). This protein is 56% identical and 76% similar to MreB1 from *T. maritima*. The kinetic behavior of this enzyme was drastically different from that of *T. maritima* MreB1. Polymerization of *B. subtilis* MreB required millimolar concentrations of divalent cations, was favored by low pH, and was inhibited by monovalent salts and low temperatures. The authors found that *B. subtilis* MreB binds and hydrolyzes ATP and GTP, but surprisingly, does not require nucleotide to polymerize. Indeed, the critical concentration for polymerization was ~900 nM regardless of the presence or absence of nucleotide.

Bean and Amann also produced the first fluorescently labeled MreB in vitro by binding a dye to an engineered cysteine residue at position 332 (Bean and Amann 2008). Using a fluorescence-resonance energy transfer measurement between adjacent monomers in a filament, the authors found similar bulk polymerization kinetics to that measured with light scattering. Unlike the measurements from ParM, no one has yet measured the kinetics of MreB polymer elongation using fluorescence microscopy. These measurements are likely to be very important because of the combination of elongation and filament bundling seen in MreB.

Because of MreB's propensity to self-assemble laterally into bundles, many questions remain about translating in vitro kinetic data to an in vivo context. In vitro bundles can be very thick and quite crystalline. Is MreB bundled inside a cell, and if so, how thick are these bundles? Fast growing *E. coli* cells contain ~40,000 MreB monomers which, if fully polymerized into a single helical bundle that spans the cell, suggests a bundle thickness of greater than 50 protofilaments (Kruse et al. 2003). Assuming that these filaments are ~5 nm in width, a tightly packed bundle would have a width of at least 50 nm. The mechanisms by which cells control MreB filament size, geometry, and conformation remains to be discovered. Future experiments that take advantage of recent improvements in electron-microscopy or super-resolution fluorescence imaging will likely be required to address these issues.

CONCLUSIONS AND OUTLOOK

The ParM-based system of R1 plasmid segregation suggests that at least some bacterial cytoskeletons have relatively simple and specific functions. Here, regulated polymerization of ParM mechanically drives two plasmids to opposite cell poles. In contrast, the shear number of functions that have been associated with MreB suggests that the MreB situation may be far more complicated than that observed with other cytoskeletal proteins. The primary outstanding challenge for the field is to understand the mechanistic basis for these many functions and distinguish which effects are directly or indirectly mediated by MreB. It is possible that MreB carries out a small set of relatively simple functions that are used over and over for different purposes. For example, MreB's entire function could be to define a helical path along the long axis of rodlike cells, thereby directly or indirectly organizing the rest of the cellular components. Multiple helical structures have been identified in bacteria in the past decade. It will be important to determine whether these helices are independent such that the helix is perhaps simply an energetically favorable conformation for many structures, or whether all of these helices are ultimately patterned interdependently, either by MreB or another structure. Another outstanding question concerns the functional significance of the MreB helix. The fact that some cells transition from having helical to medial MreB suggests that different organizations of MreB filaments could play different functions, but how and why this may occur remains mysterious.

By analogy to eukaryotic actin, it has been largely assumed that MreB assembly is related to its function. However, it is possible that the eukaryotic analogy is stifling our perspective on MreB, which could be playing yet unconsidered functions. The properties of insertional polymerization, treadmilling, and structural mechanics could be combined and controlled by different proteins for different cellular purposes. Alternatively, assembly dynamics may only be important for forming the helical structure, and a multitude of MreB-binding proteins may have been adapted to tailor MreB's function for each individual activity. Indeed, eukaryotes use many regulators to control each step of actin assembly and use divergent myosins to traffic cargoes along actin. No such assembly regulators or motor proteins have thus far been identified for MreB, and it will be of interest to see if they exist and whether they are phylogenetically conserved. Advancing our understanding of MreB will thus require bringing to MreB the level of mechanistic detail that we currently enjoy for ParM. The analysis of ParM was largely driven by a bottom-up approach of reconstructing the system in vitro from its individual parts. It remains unclear whether this approach will work for more complex systems with many components and regulators. In light of the fact that a decade ago bacteria were not even thought to possess actin proteins, we now know a great deal about the bacterial actin superfamily. The prospect of combining new molecular, genetic, biophysical, and imaging approaches promises to reward the bacterial actin field with a far deeper understanding in the decade to come.

REFERENCES

Aaron M, Charbon G, Lam H, Schwarz H, Vollmer W, Jacobs-Wagner C. 2007. The tubulin homologue FtsZ contributes to cell elongation by guiding cell wall precursor synthesis in *Caulobacter crescentus*. *Mol Microbiol* **64:** 938–952.

Alyahya SA, Alexander R, Costa T, Henriques AO, Emonet T, Jacobs-Wagner C. 2009. RodZ, a component of the bacterial core morphogenic apparatus. *Proc Natl Acad Sci U S A* **106:** 1239–1244.

Bean GJ, Amann KJ. 2008. Polymerization properties of the *Thermotoga maritima* actin MreB: roles of temperature, nucleotides, and ions. *Biochemistry* **47:** 826–835.

Bean GJ, Flickinger ST, Westler WM, McCully ME, Sept D, Weibel DB, Amann KJ. 2009. A22 disrupts the bacterial actin cytoskeleton by directly binding and inducing a low-affinity state in MreB. *Biochemistry* **48:** 4852–4857.

Becker E, Herrera NC, Gunderson FQ, Derman AI, Dance AL, Sims J, Larsen RA, Pogliano J. 2006. DNA segregation by the bacterial actin AlfA during *Bacillus subtilis* growth and development. *Embo J* **25:** 5919–5931.

Bendezu FO, de Boer PA. 2008. Conditional lethality, division defects, membrane involution, and endocytosis in mre and mrd shape mutants of *Escherichia coli*. *J Bacteriol* **190:** 1792–1811.

Bendezu FO, Hale CA, Bernhardt TG, de Boer PA. 2009. RodZ (YfgA) is required for proper assembly of the MreB actin cytoskeleton and cell shape in *E. coli*. *EMBO J* **28:** 193–204.

Bork P, Sander C, Valencia A. 1992. An ATPase domain common to prokaryotic cell cycle proteins, sugar kinases, actin, and hsp70 heat shock proteins. *Proc Natl Acad Sci U S A* **89:** 7290–7294.

Bowman GR, Comolli LR, Zhu J, Eckart M, Koenig M, Downing KH, Moerner WE, Earnest T, Shapiro L. 2008. A polymeric protein anchors the chromosomal origin/ParB complex at a bacterial cell pole. *Cell* **134:** 945–955.

Carballido-Lopez R. 2006. The bacterial actin-like cytoskeleton. *Microbiol Mol Biol Rev* **70:** 888–909.

Carballido-Lopez R, Errington J. 2003. The bacterial cytoskeleton: in vivo dynamics of the actin-like protein Mbl of *Bacillus subtilis*. *Dev Cell* **4:** 19–28.

Carballido-Lopez R, Formstone A, Li Y, Ehrlich SD, Noirot P, Errington J. 2006. Actin homolog MreBH governs cell morphogenesis by localization of the cell wall hydrolase LytE. *Dev Cell* **11:** 399–409.

Chiu SW, Chen SY, Wong HC. 2008. Localization and expression of MreB in *Vibrio parahaemolyticus* under different stresses. *Appl Environ Microbiol* **74:** 7016–7022.

Choi CL, Claridge SA, Garner EC, Alivisatos AP, Mullins RD. 2008. Protein-nanocrystal conjugates support a single filament polymerization model in R1 plasmid segregation. *J Biol Chem* **283:** 28081–28086.

Cowles KN, Gitai Z. 2010. Surface association and the MreB cytoskeleton regulate pilus production, localization and function in *Pseudomonas aeruginosa*. *Mol Microbiol.*

Dam M, Gerdes K. 1994. Partitioning of plasmid R1. Ten direct repeats flanking the *parA* promoter constitute a centromere-like partition site *parC*, that expresses incompatibility. *J Mol Biol* **236:** 1289–1298.

Daniel RA, Errington J. 2003. Control of cell morphogenesis in bacteria: Two distinct ways to make a rod-shaped cell. *Cell* **113:** 767–776.

Defeu Soufo HJ, Graumann PL. 2004. Dynamic movement of actin-like proteins within bacterial cells. *EMBO Rep* **5:** 789–794.

Defeu Soufo HJ, Graumann PL. 2005. *Bacillus subtilis* actin-like protein MreB influences the positioning of the replication machinery and requires membrane proteins MreC/D and other actin-like proteins for proper localization. *BMC Cell Biol* **6:** 10.

Defeu Soufo HJ, Reimold C, Linne U, Knust T, Gescher J, Graumann PL. 2010. Bacterial translation elongation factor EF-Tu interacts and colocalizes with actin-like MreB protein. *Proc Natl Acad Sci U S A* **107:** 3163–3168.

Derman AI, Becker EC, Truong BD, Fujioka A, Tucey TM, Erb ML, Patterson PC, Pogliano J. 2009. Phylogenetic analysis identifies many uncharacterized actin-like proteins (Alps) in bacteria: Regulated polymerization, dynamic instability and treadmilling in Alp7A. *Mol Microbiol* **73:** 534–552.

Divakaruni AV, Baida C, White CL, Gober JW. 2007. The cell shape proteins MreB and MreC control cell morphogenesis by positioning cell wall synthetic complexes. *Mol Microbiol* **66:** 174–188.

Divakaruni AV, Loo RR, Xie Y, Loo JA, Gober JW. 2005. The cell-shape protein MreC interacts with extracytoplasmic proteins including cell wall assembly complexes in *Caulobacter crescentus*. *Proc Natl Acad Sci* **102:** 18602–18607.

Doi M, Wachi M, Ishino F, Tomioka S, Ito M, Sakagami Y, Suzuki A, Matsuhashi M. 1988. Determinations of the DNA sequence of the *mreB* gene and of the gene products of the *mre* region that function in formation of the rod shape of *Escherichia coli* cells. *J Bacteriol* **170:** 4619–4624.

Dye NA, Pincus Z, Theriot JA, Shapiro L, Gitai Z. 2005. Two independent spiral structures control cell shape in *Caulobacter*. *Proc Natl Acad Sci* **102:** 18608–18613.

Esue O, Wirtz D, Tseng Y. 2006. GTPase activity, structure, and mechanical properties of filaments assembled from bacterial cytoskeleton protein MreB. *J Bacteriol* **188:** 968–976.

Esue O, Cordero M, Wirtz D, Tseng Y. 2005. The assembly of MreB, a prokaryotic homolog of actin. *J Biol Chem* **280:** 2628–2635.

Fenton AK, Lambert C, Wagstaff PC, Sockett RE. 2010. Manipulating each MreB of *Bdellovibrio bacteriovorus* gives diverse morphological and predatory phenotypes. *J Bacteriol* **192:** 1299–1311.

Figge RM, Divakaruni AV, Gober JW. 2004. MreB, the cell shape-determining bacterial actin homologue, co-ordinates cell wall morphogenesis in *Caulobacter crescentus*. *Mol Microbiol* **51:** 1321–1332.

Formstone A, Errington J. 2005. A magnesium-dependent *mreB* null mutant: implications for the role of *mreB* in *Bacillus subtilis*. *Mol Microbiol* **55:** 1646–1657.

Formstone A, Carballido-Lopez R, Noirot P, Errington J, Scheffers DJ. 2008. Localization and interactions of teichoic acid synthetic enzymes in *Bacillus subtilis*. *J Bacteriol* **190:** 1812–1821.

Gan L, Chen S, Jensen GJ. 2008. Molecular organization of Gram-negative peptidoglycan. *Proc Natl Acad Sci* **105:** 18953–18957.

Garner EC. 2008. GE Prize essay. Understanding a minimal DNA-segregating machine. *Science* **322:** 1486–1487.

Garner EC, Campbell CS, Mullins RD. 2004. Dynamic instability in a DNA-segregating prokaryotic actin homolog. *Science* **306:** 1021–1025.

Garner EC, Campbell CS, Weibel DB, Mullins RD. 2007. Reconstitution of DNA segregation driven by assembly of a prokaryotic actin homolog. *Science* **315:** 1270–1274.

Gerdes K, Larsen JE, Molin S. 1985. Stable inheritance of plasmid R1 requires two different loci. *J Bacteriol* **161:** 292–298.

Gitai Z, Dye N, Shapiro L. 2004. An actin-like gene can determine cell polarity in bacteria. *Proc Natl Acad Sci* **101:** 8643–8648.

Gitai Z, Dye NA, Reisenauer A, Wachi M, Shapiro L. 2005. MreB actin-mediated segregation of a specific region of a bacterial chromosome. *Cell* **120:** 329–341.

Holmes KC, Popp D, Gebhard W, Kabsch W. 1990. Atomic model of the actin filament. *Nature* **347:** 44–49.

Holtje JV, Heidrich C. 2001. Enzymology of elongation and constriction of the murein sacculus of *Escherichia coli*. *Biochimie* **83:** 103–108.

Hu B, Yang G, Zhao W, Zhang Y, Zhao J. 2007. MreB is important for cell shape but not for chromosome segregation of the filamentous cyanobacterium *Anabaena* sp. PCC 7120. *Mol Microbiol* **63:** 1640–1652.

Huang KC, Mukhopadhyay R, Wen B, Gitai Z, Wingreen NS. 2008. Cell shape and cell-wall organization in Gram-negative bacteria. *Proc Natl Acad Sci* **105:** 19282–19287.

Iwai N, Nagai K, Wachi M. 2002. Novel S-benzylisothiourea compound that induces spherical cells in *Escherichia coli* probably by acting on a rod-shape-determining protein(s) other than penicillin-binding protein 2. *Biosci Biotechnol Biochem* **66:** 2658–2662.

Jensen RB, Gerdes K. 1997. Partitioning of plasmid R1. The ParM protein exhibits ATPase activity and interacts with the centromere-like ParR-parC complex. *J Mol Biol* **269:** 505–513.

Jensen RB, Gerdes K. 1999. Mechanism of DNA segregation in prokaryotes: ParM partitioning protein of plasmid R1 co-localizes with its replicon during the cell cycle. *EMBO J* **18:** 4076–4084.

Jones LJ, Carballido-Lopez R, Errington J. 2001. Control of cell shape in bacteria: helical, actin-like filaments in *Bacillus subtilis*. *Cell* **104:** 913–922.

Karczmarek A, Martinez-Arteaga R, Alexeeva S, Hansen FG, Vicente M, Nanninga N, den Blaauwen T. 2007. DNA and origin region segregation are not affected by the transition from rod to sphere after inhibition of *Escherichia coli* MreB by A22. *Mol Microbiol* **65:** 51–63.

Kawai Y, Asai K, Errington J. 2009. Partial functional redundancy of MreB isoforms, MreB, Mbl and MreBH, in cell morphogenesis of *Bacillus subtilis*. *Mol Microbiol* **73:** 719–731.

Kim SY, Gitai Z, Kinkhabwala A, Shapiro L, Moerner WE. 2006. Single molecules of the bacterial actin MreB undergo directed treadmilling motion in *Caulobacter crescentus*. *Proc Natl Acad Sci U S A* **103:** 10929–10934.

Komeili A, Li Z, Newman DK, Jensen GJ. 2006. Magnetosomes are cell membrane invaginations organized by the actin-like protein MamK. *Science* **311:** 242–245.

Kruse T, Bork-Jensen J, Gerdes K. 2005. The morphogenetic MreBCD proteins of *Escherichia coli* form an essential membrane-bound complex. *Mol Microbiol* **55:** 78–89.

Kruse T, Blagoev B, Lobner-Olesen A, Wachi M, Sasaki K, Iwai N, Mann M, Gerdes K. 2006. Actin homolog MreB and RNA polymerase interact and are both required for chromosome segregation in *Escherichia coli*. *Genes Dev* **20:** 113–124.

Kruse T, Moller-Jensen J, Lobner-Olesen A, Gerdes K. 2003. Dysfunctional MreB inhibits chromosome segregation in *Escherichia coli*. *EMBO J* **22:** 5283–5292.

Kurner J, Frangakis AS, Baumeister W. 2005. Cryo-electron tomography reveals the cytoskeletal structure of *Spiroplasma melliferum*. *Science* **307:** 436–438.

Madabhushi R, Marians KJ. 2009. Actin homolog MreB affects chromosome segregation by regulating topoisomerase IV in *Escherichia coli*. *Mol Cell* **33:** 171–180.

Mauriello EM, Mouhamar F, Nan B, Ducret A, Dai D, Zusman DR, Mignot T. 2010. Bacterial motility complexes require the actin-like protein, MreB and the Ras homologue, MglA. *EMBO J* **29:** 315–326.

Mayer JA, Amann KJ. 2009. Assembly properties of the *Bacillus subtilis* actin, MreB. *Cell Motil Cytoskeleton* **66:** 109–118.

Mazza P, Noens EE, Schirner K, Grantcharova N, Mommaas AM, Koerten HK, Muth G, Flardh K, van Wezel GP, Wohlleben W. 2006. MreB of *Streptomyces coelicolor* is not essential for vegetative growth but is required for the integrity of aerial hyphae and spores. *Mol Microbiol* **60:** 838–852.

Michie KA, Löwe J. 2006. Dynamic filaments of the bacterial cytoskeleton. *Annu Rev Biochem* **75:** 467–492.

Moller-Jensen J, Borch J, Dam M, Jensen RB, Roepstorff P, Gerdes K. 2003. Bacterial mitosis: ParM of plasmid R1 moves plasmid DNA by an actin-like insertional polymerization mechanism. *Mol Cell* **12:** 1477–1487.

Moller-Jensen J, Jensen RB, Lowe J, Gerdes K. 2002. Prokaryotic DNA segregation by an actin-like filament. *EMBO J* **21:** 3119–3127.

Munoz-Espin D, Daniel R, Kawai Y, Carballido-Lopez R, Castilla-Llorente V, Errington J, Meijer WJ, Salas M. 2009. The actin-like MreB cytoskeleton organizes viral DNA replication in bacteria. *Proc Natl Acad Sci* **106:** 13347–13352.

Oda T, Iwasa M, Aihara T, Maeda Y, Narita A. 2009. The nature of the globular- to fibrous-actin transition. *Nature* **457:** 441–445.

Polka JK, Kollman JM, Agard DA, Mullins RD. 2009. The structure and assembly dynamics of plasmid actin AlfA imply a novel mechanism of DNA segregation. *J Bacteriol* **191:** 6219–6230.

Popp D, Narita A, Oda T, Fujisawa T, Matsuo H, Nitanai Y, Iwasa M, Maeda K, Onishi H, Maeda Y. 2008. Molecular structure of the ParM polymer and the mechanism leading to its nucleotide-driven dynamic instability. *EMBO J* **27:** 570–579.

Robertson GT, Doyle TB, Du Q, Duncan L, Mdluli KE, Lynch AS. 2007. A novel indole compound that inhibits *Pseudomonas aeruginosa* growth by targeting MreB is a substrate for MexAB-OprM. *J Bacteriol* **189:** 6870–6881.

Rokney A, Shagan M, Kessel M, Smith Y, Rosenshine I, Oppenheim AB. 2009. *E. coli* transports aggregated proteins to the poles by a specific and energy-dependent process. *J Mol Biol* **392:** 589–601.

Salje J, Zuber B, Lowe J. 2009. Electron cryomicroscopy of *E. coli* reveals filament bundles involved in plasmid DNA segregation. *Science* **323:** 509–512.

Savage DF, Afonso B, Chen AH, Silver PA. 2010. Spatially ordered dynamics of the bacterial carbon fixation machinery. *Science* **327:** 1258–1261.

Scheffers DJ, Jones LJ, Errington J. 2004. Several distinct localization patterns for penicillin-binding proteins in *Bacillus subtilis*. *Mol Microbiol* **51:** 749–764.

Shebelut CW, Jensen RB, Gitai Z. 2009. Growth conditions regulate the requirements for *Caulobacter* chromosome segregation. *J Bacteriol* **191:** 1097–1100.

Shih YL, Kawagishi I, Rothfield L. 2005. The MreB and Min cytoskeletal-like systems play independent roles in prokaryotic polar differentiation. *Mol Microbiol* **58:** 917–928.

Shiomi D, Sakai M, Niki H. 2008. Determination of bacterial rod shape by a novel cytoskeletal membrane protein. *EMBO J* **27:** 3081–3091.

Slovak PM, Wadhams GH, Armitage JP. 2005. Localization of MreB in *Rhodobacter sphaeroides* under conditions causing changes in cell shape and membrane structure. *J Bacteriol* **187:** 54–64.

Soufo HJ, Graumann PL. 2003. Actin-like proteins MreB and Mbl from *Bacillus subtilis* are required for bipolar positioning of replication origins. *Curr Biol* **13:** 1916–1920.

Srinivasan R, Mishra M, Murata-Hori M, Balasubramanian MK. 2007. Filament formation of the *Escherichia coli* actin-related protein, MreB, in fission yeast. *Curr Biol* **17:** 266–272.

Srivastava P, Demarre G, Karpova TS, McNally J, Chattoraj DK. 2007. Changes in nucleoid morphology and origin localization upon inhibition or alteration of the actin-homolog, MreB, of *Vibrio cholerae*. *J Bacteriol*.

Takacs CN, Poggio S, Charbon G, Pucheault M, Vollmer W, Jacobs-Wagner C. 2010. MreB drives de novo rod morphogenesis in *Caulobacter crescentus* via remodeling of the cell wall. *J Bacteriol* **192:** 1671–1684.

Tiyanont K, Doan T, Lazarus MB, Fang X, Rudner DZ, Walker S. 2006. Imaging peptidoglycan biosynthesis in *Bacillus subtilis* with fluorescent antibiotics. *Proc Natl Acad Sci U S A* **103:** 11033–11038.

Touhami A, Jericho MH, Beveridge TJ. 2004. Atomic force microscopy of cell growth and division in *Staphylococcus aureus*. *J Bacteriol* **186:** 3286–3295.

van den Ent F, Amos LA, Lowe J. 2001. Prokaryotic origin of the actin cytoskeleton. *Nature* **413:** 39–44.

van den Ent F, Johnson CM, Persons L, de Boer P, Lowe J. 2010. Bacterial actin MreB assembles in complex with cell shape protein RodZ. *EMBO J* **29:** 1081–1090.

van den Ent F, Lowe J. 2000. Crystal structure of the cell division protein FtsA from *Thermotoga maritima*. *EMBO J* **19:** 5300–5307.

van den Ent F, Leaver M, Bendezu F, Errington J, de Boer P, Lowe J. 2006. Dimeric structure of the cell shape protein MreC and its functional implications. *Mol Microbiol* **62:** 1631–1642.

van den Ent F, Moller-Jensen J, Amos LA, Gerdes K, Lowe J. 2002. F-actin-like filaments formed by plasmid segregation protein ParM. *EMBO J* **21:** 6935–6943.

Vats P, Shih YL, Rothfield L. 2009. Assembly of the MreB-associated cytoskeletal ring of *Escherichia coli*. *Mol Microbiol* **72:** 170–182.

Wagner JK, Galvani CD, Brun YV. 2005. *Caulobacter crescentus* requires RodA and MreB for stalk synthesis and prevention of ectopic pole formation. *J Bacteriol* **187:** 544–553.

Waidner B, Specht M, Dempwolff F, Haeberer K, Schaetzle S, Speth V, Kist M, Graumann PL. 2009. A novel system of cytoskeletal elements in the human pathogen *Helicobacter pylori*. *PLoS Pathog* **5:** e1000669.

Wang S, Arellano-Santoyo H, Combs PA, Shaevitz JW. 2010. An actin-like cytoskeleton contributes to cell mechanics in bacteria. *Proc Natl Acad Sci* **107:** 9182–9185.

White CL, Kitich A, Gober JW. 2010. Positioning cell wall synthetic complexes by the bacterial morphogenetic proteins MreB and MreD. *Mol Microbiol* **76:** 616–633.

Yao X, Jericho M, Pink D, Beveridge T. 1999. Thickness and elasticity of gram-negative murein sacculi measured by atomic force microscopy. *J Bacteriol* **181:** 6865–6875.

Young KD. 2006. The selective value of bacterial shape. *Microbiol Mol Biol Rev* **70:** 660–703.

Protein Subcellular Localization in Bacteria

David Z. Rudner[1] and Richard Losick[2]

[1]Department of Microbiology and Molecular Genetics, Harvard Medical School, Boston, Massachusetts 02115
[2]The Biological Laboratories, Harvard University, Cambridge, Massachusetts 02138
Correspondence: rudner@hms.harvard.edu

Like their eukaryotic counterparts, bacterial cells have a highly organized internal architecture. Here, we address the question of how proteins localize to particular sites in the cell and how they do so in a dynamic manner. We consider the underlying mechanisms that govern the positioning of proteins and protein complexes in the examples of the divisome, polar assemblies, cytoplasmic clusters, cytoskeletal elements, and organelles. We argue that geometric cues, self-assembly, and restricted sites of assembly are all exploited by the cell to specifically localize particular proteins that we refer to as anchor proteins. These anchor proteins in turn govern the localization of a whole host of additional proteins. Looking ahead, we speculate on the existence of additional mechanisms that contribute to the organization of bacterial cells, such as the nucleoid, membrane microdomains enriched in specific lipids, and RNAs with positional information.

Our view of the organization of the bacterial cell has changed radically over the past two decades. Once seen as an amorphous vessel harboring a homogeneous solution of proteins, these primitive organisms are now known to have an intricate subcellular architecture in which individual proteins localize to particular sites in the cell, often in a dynamic manner. Of course, bacteria frequently show conspicuous morphological features, such as division septa, flagella, pili, and stalks, which implied a nonuniform, underlying distribution of proteins. But it was not until the early 1990s that it became clear that proteins can, and often do, have distinctive subcellular addresses. Among the earliest discoveries were: (1) the formation of a ringlike structure at the mid cell position by the cytokinetic protein FtsZ (Bi and Lutkenhaus 1991), (2) the clustering of chemotaxis proteins at the poles of cells (Alley et al. 1992), (3) the compartment-specific production of sporulation proteins and their assembly into shell-like structures (Driks and Losick 1991), and (4) the asymmetric distribution of proteins involved in actin polymerization along the cell surface (Goldberg et al. 1993; Kocks et al. 1993). These discoveries were initially made by immunoelectron and immunofluorescence microscopy with fixed cells, but the discovery of green fluorescent protein (GFP) and the demonstration that proteins could retain their proper subcellular localization as GFP fusions

 doi: 10.1101/cshperspect.a000307
Cite this article as *Cold Spring Harb Perspect Biol* 2010;2:a000307

opened the way to visualizing proteins and their dynamic behavior in living cells, including, importantly, in bacteria (Arigoni et al. 1995).

Knowing where proteins are in the cell is often critical to understanding their function. Thus, the position of the aforementioned FtsZ ring (the Z-ring) dictates where cytokinesis will take place (Margolin 2005). The clustering of chemotaxis proteins plays an important role in the extraordinary gain in the responsiveness of chemotatic behavior to small changes in attractants (Ames and Parkinson 2006). Where sporulation proteins are produced and the way in which they assemble governs spore morphogenesis (Stragier and Losick 1996; Errington 2003). The asymmetric distribution of actin-polymerization proteins on the cell surface explains how certain pathogens harness host cytoskeletal proteins for their own motility (Smith et al. 1995). From these and other examples emerge a view of the bacterial cell as a dynamic, three-dimensional system in which protein localization and changes in protein localization over time orchestrate growth, the cell cycle, behavior, and differentiation.

Here, after an initial discussion of general principles governing the positioning of proteins within the cell, we consider five broad categories of subcellular localization: the divisome, polar assemblies, cytoplasmic clusters, cytoskeletal elements, and organelles. We end by looking ahead to exciting new aspects of bacterial cytology just emerging from current research. Our goal is not to be comprehensive but rather to focus on examples that are illustrative of general principles of protein localization. Comprehensive treatment of individual topics can be found in other articles on this topic.

DIFFUSION-AND-CAPTURE AND SELF-ASSEMBLY

What principles govern subcellular protein localization? We distinguish between two mechanisms. Perhaps the most pervasive and best established mechanism is diffusion-and-capture, whereby a protein diffuses in one dimension along the DNA until it encounters its DNA binding site, in two dimensions in the membrane or in three-dimensions through the cytoplasm until it encounters, and is captured by, a target protein to which it adheres. A clear example in which diffusion-and-capture mediates the localization of integral membrane proteins is the three-protein complex that governs the activation of the sporulation transcription factor σ^K (Rudner and Losick 2002). The complex localizes to a particular membrane that will become the outer membrane around the spore. One member of the complex serves as an anchor that is responsible for capturing the other two proteins, which reach their target by diffusion in the cytoplasmic membrane. A key observation is that under circumstances in which the target membrane has become topologically isolated from the cytoplasmic membrane, proteins in the cytoplasmic membrane cannot reach the target protein (Rudner et al. 2002).

Complementary evidence in support of the diffusion-and-capture mechanism comes from studies on the localization of the histidine kinase PleC to the cell pole in *Caulobacter* (Deich et al. 2004). Individual fluorescently labeled molecules of PleC were visualized in living cells, revealing a focus of stationary molecules at the cell pole and other molecules that were diffusing without a directional bias in the membrane. Thus, PleC molecules appear to diffuse in a random manner until they are captured at the pole.

Another example that is best understood in the context of diffusion-and-capture is the localization of proteins to the divisome, which is composed of components that bind to the scaffold of the Z-ring in an ordered sequence (Margolin 2005). Thus, each division protein (or protein subcomplex) diffuses until it encounters, and then adheres to, the Z-ring or to another protein that has already assembled into the divisome. As we will argue, it seems likely that diffusion-and-capture is the principal mechanism by which protein localization takes place in bacteria in that most proteins find their position in the cell by interacting with another protein (or protein complex). Of course, and as we shall see, diffusion-and-capture begs the question of the ultimate cue(s) by which the location of the target protein itself is determined.

A related mechanism that does not depend on a positional cue is self-assembly. A striking example is the self-assembly of the MinD and MinE proteins of *Escherichia coli*, which show a dynamic pattern of pole-to-pole oscillation in the cell (Raskin and de Boer 1999; Fu et al. 2001; Hale et al. 2001). Recent work has shown that MinD and MinE alone are sufficient to form dynamic waves on a planar surface (Loose et al. 2008). The frequencies of these waves are consistent with those observed in vivo when models for the dynamic behavior of the proteins are adjusted to account for the expected lower diffusion constants of the cytoplasm. Thus, key features of the dynamic behavior of MinD and MinE can be explained without invoking cues from the cell other than the ends of the cell, which restrict how far the waves can travel. In a sense, self-assembly is a variation on diffusion-and-capture in that the proteins reach their targets by diffusion and are then captured by the self-assembled complexes. Self-assembly differs from other examples of diffusion-and-capture to the extent that it does not rely on a separate anchor protein or other cellular cue.

ARE PROTEINS TARGETED IN BACTERIA?

The word "targeted" is often used to describe protein localization in bacteria. That is, proteins are said to be targeted to one or another location. But the word targeted denotes an active process in which the protein in question is somehow aimed at its destination. For example, the secretion of some proteins are said to be "targeted" to a particular location either because the secretory apparatus is located there or because the signal sequence directs secretion to a particular location. It is likely, however, in the first scenario, that the secretory apparatus localizes to a particular location either by diffusion to, and capture at, some cellular cue or by a process of self-assembly, as described previously. Likewise, in the second scenario, the signal sequence undoubtedly finds its proper location by a passive process in which by diffusion it eventually encounters a landmark that restricts its secretion to a particular location. Perhaps the use of the word targeted is best justified for plasmid segregation in which an active mechanism driven by protein polymerization (and depolymerization) pushes (or in some cases pulls) newly duplicated plasmid molecules apart toward opposite poles of the cell (Ebersbach and Gerdes 2005). Likewise, in yeast, the mRNA for a protein involved in mating type switching (Ash1) is directed to the daughter cell by an energy-driven process involving motor proteins (Takizawa et al. 1997). Here, too, it seems justified to refer to *ASH1* mRNA as being targeted to the bud. It is our contention, however, that in general, protein localization in bacteria cannot properly be said to be targeted to a particular location.

ASSEMBLY AND PLACEMENT OF THE DIVISOME

The best studied and most conspicuous example of subcellular localization is the divisome, the cytokinetic structure that mediates, and dictates the site of, cell division (see deBoer 2010). The divisome is generally located at the mid cell position but in some circumstances it is near a pole (see later). The divisome poses two issues in protein subcellular localization: its assembly and its position. The divisome is composed of at least 10 proteins that show an elaborate web of interactions (Goehring and Beckwith 2005; Margolin 2005). The function of these proteins is to anchor the complex in the membrane and to remodel the cell envelope biosynthetic machinery so as to mediate constriction and cytokinesis. At the heart of the divisome is a ring-like structure composed of the tubulin-like protein FtsZ. This Z-ring is a scaffold on which the remaining divisome proteins assemble, although only some of them directly interact with the ring. This is therefore a simple diffusion-and-capture system (as far as we know) in which some proteins (e.g., FtsA, ZipA, ZapA) are captured by the Z-ring and they, in turn, capture additional members (e.g., FtsK) of the structure and so forth.

How does the Z-ring ring assemble? This is largely through self-assembly as was most vividly shown with a modified FtsZ that had been fused to a membrane anchor (Osawa et al. 2008). The membrane-anchored FtsZ was capable of assembling into ring-like structures in liposomes in the absence of any other proteins. However, in the cell, the dynamics of the FtsZ polymers are influenced by a variety of proteins that either stabilize or destabilize the ring, much as is in the case of microtubules in eukaryotes. How the division machinery once assembled mediates cytokinesis remains an outstanding and unsolved problem in bacterial cell biology.

Finally, what are the factors that dictate the site in the cell at which the Z-ring, and hence the entire divisome, assembles? Two principal processes normally limit Z-ring assembly to the mid cell position in *E. coli* and *Bacillus subtilis*. One is nucleoid occlusion, which restricts Z-ring formation to nucleoid-free regions of the cell, that is, space between newly duplicated and segregated chromosomes (Yu and Margolin 1999). Nucleoid occlusion in *E. coli* is mediated in part by SlmA, a protein that both binds to the nucleoid and interacts with FtsZ (Bernhardt and de Boer 2005). An unrelated but analogous DNA binding protein, Noc, functions in nucleoid occlusion in *B. subtilis* (Wu and Errington 2004). Interestingly, it was recently shown that Noc binds to sequence elements that are largely absent from the terminus region, suggesting that the organization of the chromosome plays an important role in nucleoid occlusion (Wu et al. 2009). The second process is destabilization of Z-rings near the poles, which in *E. coli* is mediated by MinC, MinD, and MinE (Margolin 2005). MinC, which is anchored to the membrane by MinD (Hu et al. 2003), destabilizes FtsZ polymers (Hu et al. 1999). MinC and MinD rapidly oscillate from pole to pole in a process that is mediated by MinE (Raskin and de Boer 1999; Fu et al. 2001; Hale et al. 2001). As a consequence of this oscillation, the time-averaged concentration of the Z-ring destabilizing protein MinC is highest near the poles and lowest at the mid cell position. As discussed in the introduction, the Min system is itself a remarkable example of dynamic self-assembly, which can be largely reconstructed in vitro in the absence of any other protein. Thus, at least in *E. coli*, positioning of the divisome is ultimately dictated by chromosome segregation and a self-assembly process.

Interestingly, *B. subtilis* lacks MinE, and MinC and MinD do not oscillate. Instead, they are found statically at the poles and at mature divisomes (Marston et al. 1998; Gregory et al. 2008). As a consequence, Z-ring formation is normally excluded from the poles and at zones near existing divisomes. Yet another mechanism restricts Z-ring formation to the mid cell in *Caulobacter*, a bacterium in which the replication origin regions of the chromosome segregate to the extreme opposite poles of the cell (Jensen and Shapiro 1999). *Caulobacter* does not have the cell division inhibitor MinC. Instead, it has a protein MipZ, which shares similiarity to MinD and the partitioning protein ParA (Thanbichler and Shapiro 2006). MipZ localizes to the replication origin region (by binding to the chromosome segregation protein ParB) and directly inhibits Z-ring formation (Fig. 1). Thus, and in analogy to MinC, MipZ is thought to form a gradient with its highest concentration at the cell pole. This gradient is likely to involve nucleoprotein filaments of MipZ molecules that are ultimately anchored to the origin region (and hence the pole) via ParB.

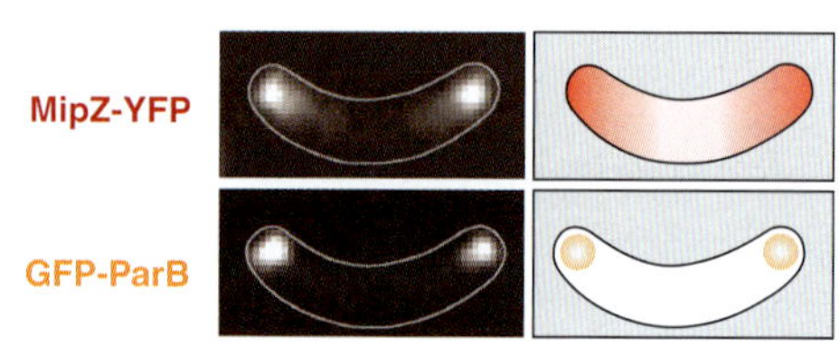

Figure 1. MipZ localization in *C. crescentus*. MipZ is anchored adjacent to the cell pole through its interaction with ParB (bound to the origin of replication) and appears to form a nucleoprotein gradient. The site of lowest concentration of the FtsZ inhibitor dictates future division site. Images show MipZ-YFP in a predivisional cell and GFP-ParB in a cell at a similar stage. Images kindly provided by M. Thanbichler.

In all of the examples considered earlier, the cell is, in effect, using inhibition (at the poles and atop the nucleoid in *E. coli* and *B. subtilis*) to dictate where the Z-ring can assemble. These negatively acting cues contrast with other cues we will consider that act positively to recruit proteins and promote localization.

Finally, we come to the unusual case of the asymmetrically positioned divisome of sporulating *B. subtilis*. On entry into sporulation, Z-ring formation switches from the midcell to sites near the poles, one of which mediates the formation of a division septum (Levin and Losick 1996). Polar Z-rings arise from a medial Z-ring in which FtsZ molecules appear to redistribute to the poles via a striking spiral intermediate (Ben-Yehuda and Losick 2002). Two factors govern this redistribution. One is an increase in FtsZ protein levels, which occurs as cells enter sporulation. The other is a sporulation protein called SpoIIE that interacts both with the membrane and with FtsZ (Lucet et al. 2000). That these two factors are sufficient to account for asymmetric placement of Z-rings is shown by experiments in which vegetative (nonsporulating) cells can be made to undergo asymmetric division by engineering the expression and overexpression of *spoIIE* and *ftsZ*, respectively, during growth (Ben-Yehuda and Losick 2002). How high levels of FtsZ decorated with SpoIIE overcomes nucleoid occlusion and the effects of MinC to favor polar assembly remains an outstanding question for the future.

POLAR ASSEMBLIES

Next, we turn our attention to the poles of the cell. What feature or features of the pole allow certain proteins to coalesce at the cell extremities? Three broad categories of models can be envisaged for how proteins localize to the extreme ends of cells, which are referred to as the geometric cue model, the division remnant model, and the self-assembly model. In what follows, we present examples of polar localization that are illustrative of these models.

Perhaps the simplest is the geometric cue model, which we illustrate with the example of DivIVA of *B. subtilis*, which localizes both to the poles and even more strongly to mature division septa (Edwards and Errington 1997). Multiple functions have been ascribed to this highly conserved protein (Fig. 2) (Flardh 2003), among which is the recruitment of the replication origin region to the poles during sporulation (Thomaides et al. 2001). In *B. subtilis*, entry into sporulation is associated with the movement of the replication origin regions of the two chromosomes of the early-stage sporangium to the extreme opposite poles of the cell (Webb et al. 1997). Polar localization is mediated by RacA, a sporulation protein that binds to multiple sites in the origin-proximal region of the chromosome and directly or indirectly adheres to DivIVA, which is located at the poles (Ben-Yehuda et al. 2003). Interestingly, something analogous takes place in *Caulobacter* in which origin regions are propelled to extreme opposite poles during the cell cycle. In this case, the chromosome segregation protein ParB (which as we saw also interacts with the Z-ring inhibiting protein MipZ) directly binds to a DivIVA-like protein, PopZ, which is localized to the poles (Bowman et al. 2008; Ebersbach et al. 2008). Both DivIVA and PopZ, which are not homologous to each other, are thought to coat the inside surface of the cell pole to create surfaces that capture origin-region-bound RacA (Lenarcic et al. 2009) or ParB.

What causes DivIVA (and PopZ) to localize to the cell poles and mature septa? Recent findings suggest that DivIVA recognizes the negative curvature of the membrane at the end of the cell as well as at the newly completed septum (Howard 2004; Lenarcic et al. 2009; Ramamurthi and Losick 2009). It is further suggested that DivIVA prefers the greater negative curvature of mature septa over the gentler curvature of the hemispherical poles. The chief evidence for these conclusions is: (1) when cytokinesis is blocked, DivIVA redistributes itself to the poles; (2) in aberrantly shaped cells created by use of a mutation in the gene for a cytoskeleton protein, DivIVA localizes preferentially to regions of negative curvature; and (3) in spherical cells (protoplasts) with uniform negative curvature, the protein shows a more or less uniform distribution (Lenarcic et al. 2009;

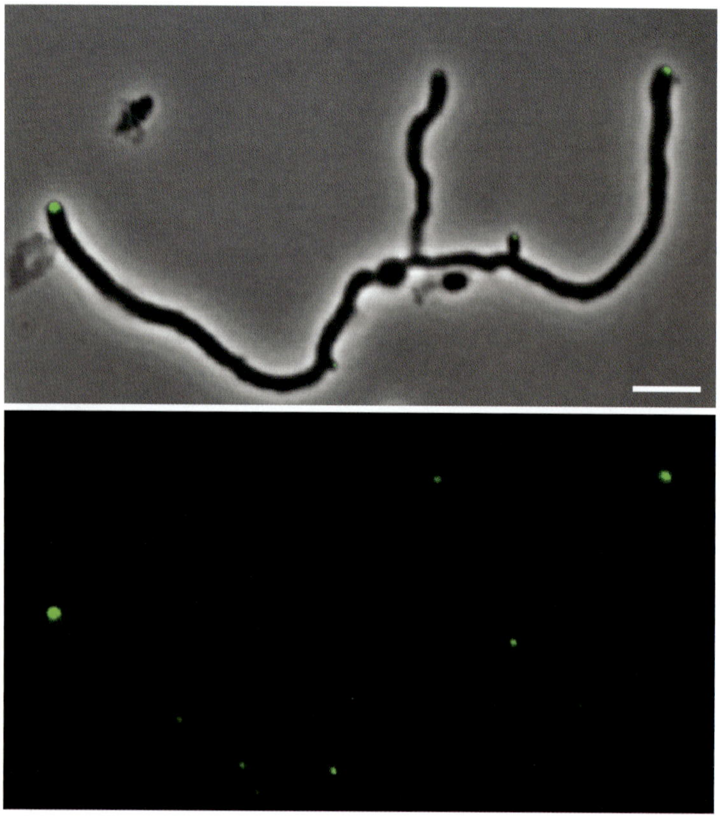

Figure 2. Polarly localized DivIVA in *Streptomyces coelicolor* A3(2). *S. coelicolor* grows as a branching hyphae similar to filamentous fungi. Growth occurs at the hyphal tips in a manner dependent on DivIVA (Flardh 2003). Images show a DivIVA-GFP fusion. Lower panel: DivIVA-GFP foci. Upper panel: a merged image of DivIVA-GFP and phase contrast. Scale bar indicates 4 μm. Image kindly provided by K. Flardh.

Ramamurthi and Losick 2009). This compelling model awaits biochemical confirmation using purified DivIVA and lipid tubules. Conceivably, negative curvature is also a cue for the polar localization of PopZ. Of course, for both DivIVA and PopZ, other as yet unidentified proteins could assist in proper localization.

Next, we turn to the division remnant model, an example of which is TipN in *Caulobacter crescentus* (see Viollier 2010). TipN localizes to the divisome after constriction has initiated and is retained at this newly created pole after division is complete (Huitema et al. 2006; Lam et al. 2006). TipN then remains at the pole to create a mark or "birth scar," which in turn recruits other proteins to the pole. TipN is required for the assembly of polar flagellum at the newly formed pole and is thought to serve as an organizational center for polar structures. In its absence, flagella are often assembled at the old cell pole instead of the new one. Once cytokinesis begins anew, TipN relocalizes from the soon to be old pole to the incipient new one.

It is not known what recruits TipN to the divisome or how it is retained at the new pole once division is completed. Conceivably, the geometry of the inside surface of the pole contributes both to the recruitment of TipN to the divisome and its retention at the pole. Alternatively, or additionally, TipN binds to the newly synthesized septal peptidoglycan. Once cytokinesis is complete and the divisome complex disassembles, it remains (a division remnant) anchored at the newly formed poles through interactions with the cell wall.

Finally, we illustrate polar localization by self-assembly with the case of the polar clustering of chemoreceptors at the cell pole of *E. coli*. When visualized by super-high resolution fluorescence microscopy (photoactivated localization microscopy or PALM), chemoreceptors fused to a photoactivatable fluorescent reporter protein are found to show an exponential distribution of cluster sizes, which are spread along the length of the cell with the largest clusters at the poles (Fig. 3) (Greenfield et al. 2009). The arrangement of these membrane clusters is consistent with a model in which a receptor is most likely to be captured by an existing cluster if one is nearby and to nucleate the formation of a new cluster when existing clusters are far (Wang et al. 2008). As a consequence, large clusters preferentially assemble as far as possible from each other, i.e., the two poles of the cell. If this model is correct, then self-assembly into clusters would be sufficient to explain the pattern of large polar clusters of chemoreceptors, without having to invoke the existence of an anchor or other cue at the pole. Of course, it is possible that a polar cue, such as negative curvature or a remnant protein from cell division, also contributes to the polar localization of chemoreceptors. Conversely, it is possible that self-assembly contributes to the localization of all of the proteins considered previously in this section.

One other case of polar localization merits comment because it does not seem to conform to any of the polar localization mechanisms we have considered so far. The virulence protein IcsA of *Shigella flexneri* localizes to the cell pole where it is incorporated into the outer membrane and mediates the assembly of an actin tail (Goldberg et al. 1993). The actin tail drives the movement of the bacterium through the cytoplasm of its mammalian host cell. Polar localization can occur in the absence of translocation across the inner membrane, a finding that reveals that IcsA localizes to the cytoplasmic face of the inner membrane at the pole before being secreted (Charles et al. 2001). Interestingly, IcsA localizes to the pole in *E. coli* as well as in *Shigella*, which suggests

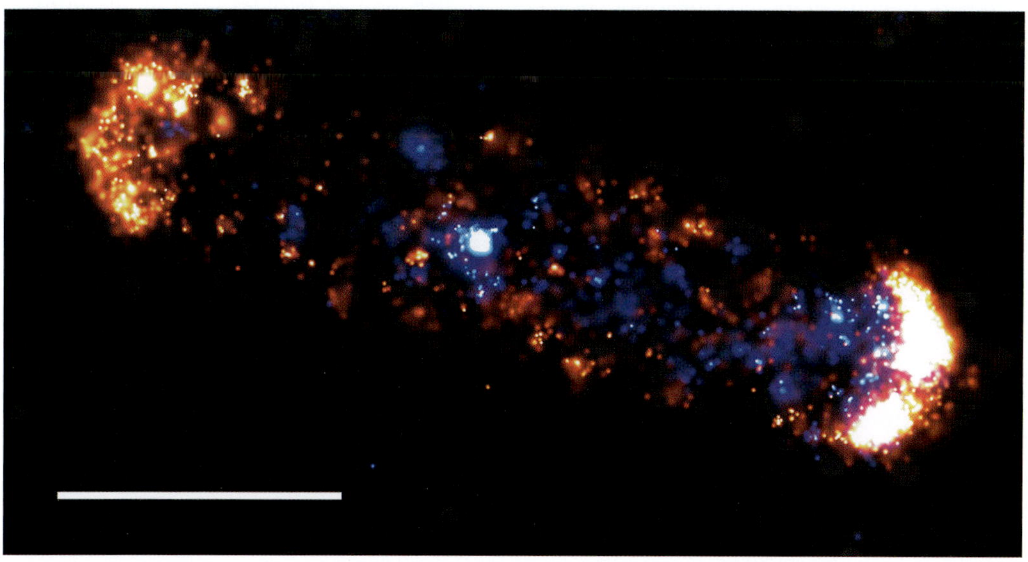

Figure 3. Chemoreceptors in *E. coli* show an exponential distribution of cluster sizes with the largest clusters at the poles. The image shows the Tar receptor fused to mEos visualized by super-high resolution fluorescence microscopy (photoactivated localization microscopy or PALM). Blue foci are from a PALM image in TIR-illumination. Red foci are from a PALM image in epi-fluourescence illumination (taken after Tar-mEos protein in the TIR region were bleached). Scale bar indicates 1 μm. Reprinted, with permission, from Greenfield et al. 2009.

that the protein recognizes conserved features in both species. Using *E. coli* as a host, Goldberg and coworkers have made the discovery that under conditions in which cell division is blocked, IcsA localizes to future division sites (which will become cell poles after cytokinesis) in the resulting filaments. Remarkably, this localization does not appear to depend on FtsZ or the factors (nucleoid occlusion and MinCDE) that are known to govern the position of the divisome in *E. coli* (Janakiraman and Goldberg 2004). Evidently, IcsA is able to recognize some other as yet unknown feature of the future division site. Just what this feature is and how it is recognized by IcsA remain exciting and as yet unanswered questions.

PROTEIN CLUSTERS AND MICROCOMPARTMENTS

So far, we have considered examples of proteins or protein assemblies that are localized to the membrane. But some bacteria also produce clusters of proteins within the cytoplasm that lack any obvious membrane attachment. A striking example of such a cytoplasmic cluster is that formed by certain chemotaxis proteins in *Rhodobacter sphaeroides* (Wadhams et al. 2002). (*R. sphaeroides* also produces a separate cluster of chemotaxis proteins at the cell pole similar to that of many other bacteria.) The cytoplasmic chemotaxis proteins are seen as one large cluster, which gives rise to two clusters early in the cell cycle. The daughter clusters are located at the one-fourth and three-fourths positions of the cell, a pattern reminiscent of that seen for certain low-copy number plasmids. Remarkably, separation of a parental cluster into two daughter clusters requires an ortholog of the plasmid partitioning protein ParA (Thompson et al. 2006) (see Mullins 2010). Evidently, the same kinds of mechanisms that mediate plasmid separation in bacteria have also been harnessed to drive the separation and positioning of protein clusters. It will be interesting to see if these unusual ParA proteins take advantage of the nucleoid to segregate these cytoplasmic clusters as is hypothesized for plasmid segregation (Ebersbach and Gerdes 2005). In a sense, the positioning of the cytoplasmic chemotaxis cluster of *R. sphaeroides* is an example of protein targeting in that a ParA-like protein drives the daughter clusters apart to a characteristic position in the cell. It will be interesting to see whether other proteins or protein complexes take advantage of partitioning proteins for proper localization.

If so, an attractive candidate would be the carboxysome, a protein microcompartment in cyanobacteria that contains the carboxylase RuBisCO (Yeates et al. 2008). The carboxysome, which mediates the fixation of carbon dioxide, is believed to isolate RuBisCO from the cytoplasmic environment and thereby facilitate the incorporation of carbon dioxide into 3-phosphoglycerate. Recent studies in which carboxysome proteins in *Synechococcus elongatus* were fused to the green fluorescent protein revealed fluorescent foci in a strikingly even spacing along the length of the cell (Fig. 4) (D. Savage and P. Silver, pers. comm.). In light of the discovery that the spacing of cytoplasmic chemotaxis protein clusters in *R. sphaeroides* is governed by a ParA homolog, it is appealing to imagine that plasmid partitioning proteins

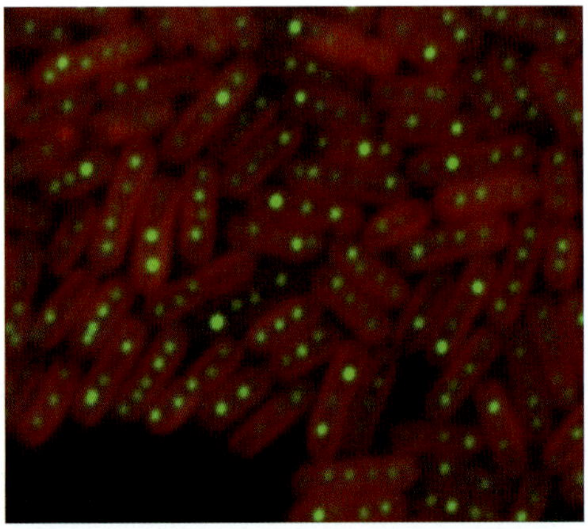

Figure 4. Carboxysomes in *S. elongatus* are found evenly spaced along the length of the cell. The image shows the carboxysome-localized enzyme RuBisCO fused to the green fluorescent protein. The native chlorophyll fluorescence (red) labels the peripheral membranes. Image kindly provided by D. Savage and P. Silver.

might also contribute to the stereotyped distribution of carboxysomes in cyanobacteria.

CYTOSKELETAL ELEMENTS

As alluded to earlier, proteins can also localize as polymeric assemblies inside the cell. These assemblies are called cytoskeletal elements, and the proteins that form them include the tubulin-like protein FtsZ, the actin-like protein MreB (Jones et al. 2001), and the *Caulobacter* protein Crescentin (Ausmees et al. 2003), which resembles an intermediate filament protein. As discussed earlier, FtsZ filaments assemble into a ring-like structure at mid-cell and the localization of this cytoskeletal element is dictated by the inhibitory activities of MinC/D and nucleoid. MreB forms helical filaments that extend the entire length of the cell and localizes adjacent to the cytoplasmic membrane. A linear polymer forming along the periphery of a cylinder will naturally trace out a spiral unless otherwise constrained. Accordingly, it is possible that MreB adopts this helical localization pattern without cellular cues. If this is the case, we can consider the localization of MreB as an example of self-assembly. Interestingly and importantly, in some bacteria, the MreB helical filament appears to coalesce into a ring-like structure at mid-cell during cytokinesis (in an FtsZ-dependent manner), suggesting that other proteins can influence its organization (Figge et al. 2004).

Many functions have been attributed to MreB filaments, the principal of which is cell shape determination. In its absence (or on depletion or inhibition), cells lose their rod-shaped appearance and become spherical before lysing (Jones et al. 2001; Gitai et al. 2005). Despite the word "skeleton" embedded within cytoskeleton and the helical localization pattern, MreB does not appear to serve as a bacterial skeleton. This job is the purview of the cell wall, the peptidoglycan, which acts as an exoskeleton, maintaining the cell's shape. The current thinking is that MreB and its homologs participate in cell shape by organizing the peptidoglycan biosynthetic enzymes into a helical arrangement that is oriented along the long axis of the cell. In support of this idea, fluorescent probes that label sites of active cell wall synthesis have a spiral-like pattern (Daniel and Errington 2003; Tiyanont et al. 2006). Whether the reorganization of MreB at the mid-cell is involved in redirecting these enzymes toward peptidoglycan synthesis at or adjacent to the septum remains unclear. The localization of these biosynthetic enzymes to the MreB filament is presumably achieved by diffusion and capture, much like the localization of the divisome components to the cytoskeletal FtsZ ring. The emerging view is that MreB and other cytoskeletal components serve as major organizing centers inside the bacterial cell. One clear example is the arrangement of magnetosomes inside *Magnetospirillum* (to follow).

Yet a third variety of cytoskeletal elements exists in *Caulobacter crescentus*. These polymers are composed of a protein called Crescentin that resembles intermediate filaments found in eukaryotes (Ausmees et al. 2003). Crescentin is responsible for the curved crescent, comma-shape of *Caulobacter*. Cells lacking Crescentin are rod-shaped and expression of Crescentin in *E. coli* results in a transition from a rod to a crescent (Cabeen et al. 2009). Crescentin forms an extended filament that localizes adjacent to the cell envelope along the concave side of *Caulobacter* cells. This filament appears to govern cell shape by a different mechanism than the one proposed for MreB. Current data suggest that physical strain imparted by the Crescentin structure can alter the kinetics of cell wall insertion, such that less growth occurs along the Crescentin-containing side the of cell, resulting in curvature (Cabeen et al. 2009). Interestingly, the MreB cystoskeleton appears to help anchor the Crescentin filament to the cell envelope (Charbon et al. 2009).

ORGANELLES

We do not usually think of bacteria as having organelles, but some bacteria do have organelle-like structures that pose fascinating issues of protein localization and assembly. Indeed, the cell envelope of Gram-negative bacteria can be thought of as an organelle that is bound on the inside by the cytoplasmic membrane and on the outside by the outer membrane. Because an entire article is devoted to the envelope (see Silhavy et al. 2010), we do not consider it further here except to point out that proteins in every layer of the envelope have been found to localize to specific subcellular sites. For example, some periplasmic proteins specifically localize to the cytokinetic ring during cell division (Uehara et al. 2009), whereas others involved in cell wall remodeling localize in spiral-like structures reminiscent of cytoplasmic MreB spirals (Divakaruni et al. 2007).

A paradigm for a bacterial organelle is the magnetosome of *Magnetospirillum* (see Fig. 1 in Murat et al. 2010). The magnetosome is a chain of magnetite crystals that acts as a compass to passively orient the bacterium in a geomagnetic field (Komeili 2007; Schuler 2008). Each crystal is surrounded by a membrane and long chains of membrane-encased crystals are contained within a tubular membrane compartment that grows out by invagination of the cytoplasmic membrane. The crystals are organized in a linear array within the bacterium by a dedicated cytosekeleton composed of an MreB homolog called MamK (Komeili et al. 2006). The magnetosome is specified by a genomic island that consists of about 100 genes that are responsible for biogenesis of the magnetosome membrane and cytoskeleton and the formation of the crystals. Importantly, the genomic island encodes membrane and cytoplasmic proteins that localize uniquely to the magnetosome (Schubbe et al. 2003; Fukuda et al. 2006). Thus, the magnetosome raises fascinating and as yet unanswered questions about its biogenesis and the mechanisms that direct proteins to a membrane compartment.

A second example of an organelle-like body in a bacterium is the developing spore or, more properly, the *endo*spore, of *B. subtilis*. The endospore is a special kind of dormant cell that is generated internally to its host cell (Stragier and Losick 1996; Errington 2003). Endospore formation takes place in two principal stages. First, a cell that has entered the pathway to sporulate undergoes asymmetric division to create a small compartment known as the forespore and an adjacent, large compartment, the mother cell. Next, in a process that resembles phagocytosis, the mother-cell membrane migrates around and engulfs the forespore, eventually pinching it off as a free protoplast in the mother-cell cytoplasm. In subsequent development, a shell composed of more than 70 proteins is produced in the mother cell and deposited around the forespore to create the spore coat (Kim et al. 2006).

Interestingly, engulfment resembles magnetosome formation in the sense that the engulfing membrane is initially continguous with the cytoplasmic membrane of the mother cell. (The magentosome membrane, as we saw, arises as an invagination of the cytoplasmic membrane.) Unlike the magnetosome, however, the forespore becomes topologically isolated from the cytoplasmic membrane when engulfment is complete (Sharp and Pogliano 1999).

How is this protein shell around the forespore assembled? As in the case of other multiprotein complexes we have considered, the spore coat poses two questions: What is the primary cue that dictates the location of coat protein deposition and how are the myriad proteins properly assembled? The first step in coat assembly is the localization of a small amphipathetic helix, SpoVM, to the outer membrane that surrounds the forespore (van Ooij and Losick 2003). How does this small protein discriminate between the outer forespore membrane and the cytoplasmic membrane of the mother cell? Evidence indicates that SpoVM preferentially binds to positively curved surfaces, thereby discriminating between the concave shape of the cytoplamic membrane and the convex shape of the outer forespore membrane (Ramamurthi et al. 2009). SpoVM, in turn, recruits a cytosolic protein called SpoIVA, which polymerizes into long filaments in a process that is driven by ATP hydrolysis (Ramamurthi et al. 2006; Ramamurthi and Losick 2008). These long filaments wrap around the forespore to create the basement layer of the coat. A complex web of still-poorly understood interactions involving scores of additional proteins drive the remaining steps in coat assembly.

Thus, in the case of the protein coat, localization to the prospective spore appears to be achieved by geometric cues followed by the recruitment of additional proteins by diffusion-and-capture.

A second case of protein localization to the membranes surrounding the forespore merits comment because it does not conform to any of the mechanisms we have considered previously. In this case, protein localization is governed by the interaction of membrane proteins in adjacent daughter cells. A membrane protein (called SpoIIIAH) produced in the mother cell becomes anchored in the membranes surrounding the spore through its interaction with a membrane protein (SpoIIQ) synthesized in the forespore (Blaylock et al. 2004; Doan et al. 2005). The extracellular domains of these two proteins interact in the space between the double-membrane that separates them. Recent evidence suggests that these proteins are, in fact, part of a larger complex that serves as a channel or secretion apparatus that links the mother cell and forespore (Camp and Losick 2008; Meisner et al. 2008; Camp and Losick 2009; Doan et al. 2009). The SpoIIIAH-SpoIIQ complex also plays a key role in anchoring several other proteins in the forespore membranes (Doan et al. 2005, Jiang et al. 2005; Campo et al. 2008). In this unusual case of protein localization, the ultimate anchor is the interaction of two proteins across adjacent membranes that separate daughter cells. This type of localization mechanism is reminiscent of the interactions between cadherins or claudins in neighboring cells at tight junctions in eukaryotes (Tsukita et al. 2001) and of the interaction between pathogens and host cells at sites of adherence (Kenny et al. 1997).

CONCLUSIONS AND FUTURE PROSPECTS

As stated from the outset, we have only touched on a few representative examples of protein localization to highlight general principles. Many, many proteins (and clearly more to come) have been found to show specific patterns of subcellular localization. We suspect that most of these proteins attain their proper localization by diffusion-and-capture but anticipate more examples like the clusters of cytoplasmic chemotaxis proteins in *R. sphaeroides*, where proteins or protein complexes are actively "targeted" to specific sites within the cell.

One of the principal challenges for the future is defining the mechanisms by which proteins are anchored at specific subcellular positions. We have argued that geometric cues, self-assembly, and restricted sites of assembly are all exploited by the cell to specifically localize a subset of proteins. These "founder" proteins, in turn, govern (directly or indirectly) the localization of a whole host of others. We anticipate that many other anchor proteins will be identified that either recognize geometric cues or self-assemble. Nonetheless, we suspect that other mechanisms exist that anchor proteins at specific sites. If, for example, we look at the rich literature of cell biology in eukaryotes, it is not hard to imagine that the phospholipid bilayer of bacteria contains rafts enriched in specific lipids and/or cholesterol-like molecules that recruit and anchor a subset of integral membrane proteins (Brown and London 1998). In fact, there is growing evidence in support of lipid microdomains in bacteria (D. López and R. Kolter, Membrane rafts in bacteria, submitted; reviewed in Matsumoto et al. 2006). Of course, lipid rafts are yet another example of self-assembly, although where these membrane patches localize is thought, at least in some cases, to be dictated by geometric cues. It is also not hard to imagine, by analogy to eukaryotic cell biology, that specific mRNAs in bacteria localize to the sites where the proteins they encode are destined (Shav-Tal and Singer 2005). If such messages exist, protein localization in these cases could be governed by mRNA localization. Understanding how these mRNAs are themselves anchored at specific sites will likely reveal new and interesting biology.

We only briefly touched on the role of the largest macromolecule in the cell, the chromosome, in orchestrating protein localization. The example we highlighted was division site selection and the role of the nucleoid in preventing FtsZ assembly. We hypothesize that this highly organized macromolecule (bound by numerous proteins) exerts its influence on the subcellular organization of many proteins and protein complexes, the extent of which will be revealed in the coming years. Along these lines, we eagerly await definitive proof that membrane protein insertion is indeed coupled to transcription and translation. Once established, a future challenge will be to assess the role of the "transertion" process in localizing membrane proteins at specific sites and conversely in organizing and sculpting the nucleoid itself.

Finally, we anticipate that the ability to more precisely define the localization of proteins in the cell using super-high resolution microscopy will shed light on the organization of the nucleoid and other macromolecules and macromolecular complexes that have remained refractory to diffraction-limited microscopy (Bates et al. 2008). Super-resolution single-molecule tracking holds similar promise for characterizing the dynamic behaviors of proteins and protein complexes (Xie et al. 2008). It is our expectation that these new cytological approaches will reveal ultrastructure and protein dynamics in bacteria that parallel eukaryotic cells.

REFERENCES

Alley MR, Maddock JR, Shapiro L. 1992. Polar localization of a bacterial chemoreceptor. *Genes Dev* **6:** 825–836.

Ames P, Parkinson JS. 2006. Conformational suppression of inter-receptor signaling defects. *Proc Natl Acad Sci* **103:** 9292–9297.

Arigoni F, Pogliano K, Webb CD, Stragier P, Losick R. 1995. Localization of protein implicated in establishment of cell type to sites of asymmetric division. *Science* **270:** 637–640.

Ausmees N, Kuhn JR, Jacobs-Wagner C. 2003. The bacterial cytoskeleton: An intermediate filament-like function in cell shape. *Cell* **115:** 705–713.

Bates M, Huang B, Zhuang X. 2008. Super-resolution microscopy by nanoscale localization of photo-switchable fluorescent probes. *Curr Opin Chem Biol* **12:** 505–514.

Ben-Yehuda S, Losick R. 2002. Asymmetric cell division in *B. subtilis* involves a spiral-like intermediate of the cytokinetic protein FtsZ. *Cell* **109:** 257–266.

Ben-Yehuda S, Rudner DZ, Losick R. 2003. RacA, a bacterial protein that anchors chromosomes to the cell poles. *Science* **299:** 532–536.

Bernhardt TG, de Boer PA. 2005. SlmA, a nucleoid-associated, FtsZ binding protein required for blocking septal ring assembly over Chromosomes in *E. coli. Mol Cell* **18:** 555–564.

Bi EF, Lutkenhaus J. 1991. FtsZ ring structure associated with division in *Escherichia coli. Nature* **354:** 161–164.

Blaylock B, Jiang X, Rubio A, Moran CP Jr, Pogliano K. 2004. Zipper-like interaction between proteins in adjacent daughter cells mediates protein localization. *Genes Dev* **18:** 2916–2928.

Bowman GR, Comolli LR, Zhu J, Eckart M, Koenig M, Downing KH, Moerner WE, Earnest T, Shapiro L. 2008. A polymeric protein anchors the chromosomal origin/ParB complex at a bacterial cell pole. *Cell* **134:** 945–955.

Brown DA, London E. 1998. Functions of lipid rafts in biological membranes. *Annu Rev Cell Dev Biol* **14:** 111–136.

Cabeen MT, Charbon G, Vollmer W, Born P, Ausmees N, Weibel DB, Jacobs-Wagner C. 2009. Bacterial cell curvature through mechanical control of cell growth. *Embo J* **28:** 1208–1219.

Camp AH, Losick R. 2008. A novel pathway of intercellular signalling in *Bacillus subtilis* involves a protein with similarity to a component of type III secretion channels. *Mol Microbiol* **69:** 402–417.

Camp AH, Losick R. 2009. A feeding tube model for activation of a cell-specific transcription factor during sporulation in *Bacillus subtilis. Genes Dev* **23:** 1014–1024.

Campo N, Marquis KA, Rudner DZ. 2008. SpoIIQ anchors membrane proteins on both sides of the sporulation septum in *Bacillus subtilis. J Biol Chem* **283:** 4975–4982.

Charbon G, Cabeen MT, Jacobs-Wagner C. 2009. Bacterial intermediate filaments: In vivo assembly, organization, and dynamics of crescentin. *Genes Dev* **23:** 1131–1144.

Charles M, Perez M, Kobil JH, Goldberg MB. 2001. Polar targeting of *Shigella* virulence factor IcsA in Enterobacteriacae and Vibrio. *Proc Natl Acad Sci* **98:** 9871–9876.

Daniel RA, Errington J. 2003. Control of cell morphogenesis in bacteria: Two distinct ways to make a rod-shaped cell. *Cell* **113:** 767–776.

deBoer PA. 2010. Cytokinesis and the min system. *Cold Spring Harb Perspect Biol* **2:** a000323.

Deich J, Judd EM, McAdams HH, Moerner WE. 2004. Visualization of the movement of single histidine kinase molecules in live *Caulobacter* cells. *Proc Natl Acad Sci* **101:** 15921–15926.

Divakaruni AV, Baida C, White CL, Gober JW. 2007. The cell shape proteins MreB and MreC control cell morphogenesis by positioning cell wall synthetic complexes. *Mol Microbiol* **66:** 174–188.

Doan T, Marquis KA, Rudner DZ. 2005. Subcellular localization of a sporulation membrane protein is achieved through a network of interactions along and across the septum. *Mol Microbiol* **55:** 1767–1781.

Doan T, Morlot C, Meisner J, Serrano M, Henriques AO, Moran CP Jr, Rudner DZ. 2009. Novel secretion apparatus maintains spore integrity and developmental gene expression in *Bacillus subtilis. PLoS Genet* **5:** e1000566.

Driks A, Losick R. 1991. Compartmentalized expression of a gene under the control of sporulation transcription factor σ E in *Bacillus subtilis. Proc Natl Acad Sci* **88:** 9934–9938.

Ebersbach G, Gerdes K. 2005. Plasmid segregation mechanisms. *Annu Rev Genet* **39:** 453–479.

Ebersbach G, Briegel A, Jensen GJ, Jacobs-Wagner C. 2008. A self-associating protein critical for chromosome attachment, division, and polar organization in caulobacter. *Cell* **134:** 956–968.

Edwards DH, Errington J. 1997. The *Bacillus subtilis* DivIVA protein targets to the division septum and controls the site specificity of cell division. *Mol Microbiol* **24:** 905–915.

Errington J. 2003. Regulation of endospore formation in *Bacillus subtilis*. *Nat Rev Microbiol* **1:** 117–126.

Figge RM, Divakaruni AV, Gober JW. 2004. MreB, the cell shape-determining bacterial actin homologue, co-ordinates cell wall morphogenesis in *Caulobacter crescentus*. *Mol Microbiol* **51:** 1321–1332.

Flardh K. 2003. Essential role of DivIVA in polar growth and morphogenesis in *Streptomyces coelicolor* A3(2). *Mol Microbiol* **49:** 1523–1536.

Fu X, Shih YL, Zhang Y, Rothfield LI. 2001. The MinE ring required for proper placement of the division site is a mobile structure that changes its cellular location during the *Escherichia coli* division cycle. *Proc Natl Acad Sci* **98:** 980–985.

Fukuda Y, Okamura Y, Takeyama H, Matsunaga T. 2006. Dynamic analysis of a genomic island in *Magnetospirillum* sp. strain AMB-1 reveals how magnetosome synthesis developed. *FEBS Lett* **580:** 801–812.

Gitai Z, Dye NA, Reisenauer A, Wachi M, Shapiro L. 2005. MreB actin-mediated segregation of a specific region of a bacterial chromosome. *Cell* **120:** 329–341.

Goehring NW, Beckwith J. 2005. Diverse paths to midcell: Assembly of the bacterial cell division machinery. *Curr Biol* **15:** R514–R526.

Goldberg MB, Barzu O, Parsot C, Sansonetti PJ. 1993. Unipolar localization and ATPase activity of IcsA, a *Shigella flexneri* protein involved in intracellular movement. *Infect Agents Dis* **2:** 210–211.

Greenfield D, McEvoy AL, Shroff H, Crooks GE, Wingreen NS, Betzig E, Liphardt J. 2009. Self-organization of the *Escherichia coli* chemotaxis network imaged with super-resolution light microscopy. *PLoS Biol* **7:** e1000137.

Gregory JA, Becker EC, Pogliano K. 2008. *Bacillus subtilis* MinC destabilizes FtsZ-rings at new cell poles and contributes to the timing of cell division. *Genes Dev* **22:** 3475–3488.

Hale CA, Meinhardt H, de Boer PA. 2001. Dynamic localization cycle of the cell division regulator MinE in *Escherichia coli*. *Embo J* **20:** 1563–1572.

Howard M. 2004. A mechanism for polar protein localization in bacteria. *J Mol Biol* **335:** 655–663.

Hu Z, Saez C, Lutkenhaus J. 2003. Recruitment of MinC, an inhibitor of Z-ring formation, to the membrane in *Escherichia coli*: Role of MinD and MinE. *J Bacteriol* **185:** 196–203.

Hu Z, Mukherjee A, Pichoff S, Lutkenhaus J. 1999. The MinC component of the division site selection system in *Escherichia coli* interacts with FtsZ to prevent polymerization. *Proc Natl Acad Sci* **96:** 14819–14824.

Huitema E, Pritchard S, Matteson D, Radhakrishnan SK, Viollier PH. 2006. Bacterial birth scar proteins mark future flagellum assembly site. *Cell* **124:** 1025–1037.

Janakiraman A, Goldberg MB. 2004. Evidence for polar positional information independent of cell division and nucleoid occlusion. *Proc Natl Acad Sci* **101:** 835–840.

Jensen RB, Shapiro L. 1999. The Caulobacter crescentus smc gene is required for cell cycle progression and chromosome segregation. *Proc Natl Acad Sci* **96:** 10661–10666.

Jiang X, Rubio A, Chiba S, Pogliano K. 2005. Engulfment-regulated proteolysis of SpoIIQ: Evidence that dual checkpoints control σ activity. *Mol Microbiol* **58:** 102–115.

Jones LJ, Carballido-Lopez R, Errington J. 2001. Control of cell shape in bacteria: Helical, actin-like filaments in *Bacillus subtilis*. *Cell* **104:** 913–922.

Kenny B, DeVinney R, Stein M, Reinscheid DJ, Frey EA, Finlay BB. 1997. Enteropathogenic *E. coli* (EPEC) transfers its receptor for intimate adherence into mammalian cells. *Cell* **91:** 511–520.

Kim H, Hahn M, Grabowski P, McPherson DC, Otte MM, Wang R, Ferguson CC, Eichenberger P, Driks A. 2006. The *Bacillus subtilis* spore coat protein interaction network. *Mol Microbiol* **59:** 487–502.

Kocks C, Hellio R, Gounon P, Ohayon H, Cossart P. 1993. Polarized distribution of *Listeria monocytogenes* surface protein ActA at the site of directional actin assembly. *J Cell Sci* **105:** 699–710.

Komeili A. 2007. Molecular mechanisms of magnetosome formation. *Annu Rev Biochem* **76:** 351–366.

Komeili A, Li Z, Newman DK, Jensen GJ. 2006. Magnetosomes are cell membrane invaginations organized by the actin-like protein MamK. *Science* **311:** 242–245.

Lam H, Schofield WB, Jacobs-Wagner C. 2006. A landmark protein essential for establishing and perpetuating the polarity of a bacterial cell. *Cell* **124:** 1011–1023.

Lenarcic R, Halbedel S, Visser L, Shaw M, Wu LJ, Errington J, Marenduzzo D, Hamoen LW. 2009. Localisation of DivIVA by targeting to negatively curved membranes. *Embo J* **28:** 2272–2282.

Levin PA, Losick R. 1996. Transcription factor Spo0A switches the localization of the cell division protein FtsZ from a medial to a bipolar pattern in *Bacillus subtilis*. *Genes Dev* **10:** 478–488.

Loose M, Fischer-Friedrich E, Ries J, Kruse K, Schwille P. 2008. Spatial regulators for bacterial cell division self-organize into surface waves in vitro. *Science* **320:** 789–792.

Lucet I, Feucht A, Yudkin MD, Errington J. 2000. Direct interaction between the cell division protein FtsZ and the cell differentiation protein SpoIIE. *Embo J* **19:** 1467–1475.

Margolin W. 2005. FtsZ and the division of prokaryotic cells and organelles. *Nat Rev Mol Cell Biol* **6:** 862–871.

Marston AL, Thomaides HB, Edwards DH, Sharpe ME, Errington J. 1998. Polar localization of the MinD protein of *Bacillus subtilis* and its role in selection of the mid-cell division site. *Genes Dev* **12:** 3419–3430.

Matsumoto K, Kusaka J, Nishibori A, Hara H. 2006. Lipid domains in bacterial membranes. *Mol Microbiol* **61:** 1110–1117.

Meisner J, Wang X, Serrano M, Henriques AO, Moran CP Jr. 2008. A channel connecting the mother cell and forespore during bacterial endospore formation. *Proc Natl Acad Sci* **105:** 15100–15105.

Mullins D. 2010. Bacterial cytoskeleton: the function of FtsZ and ParA polymers. *Cold Spring Harb Perspect Biol* **2:** a002550.

Murat D, Byrne M, Komeili A. 2010. Cell biology of prokaryote organelles. *Cold Spring Harb Perspect Biol* **2:** a000422.

Osawa M, Anderson DE, Erickson HP. 2008. Reconstitution of contractile FtsZ rings in liposomes. *Science* **320:** 792–794.

Ramamurthi KS, Losick R. 2008. ATP-driven self-assembly of a morphogenetic protein in *Bacillus subtilis. Mol Cell* **31:** 406–414.

Ramamurthi KS, Losick R. 2009. Negative membrane curvature as a cue for subcellular localization of a bacterial protein. *Proc Natl Acad Sci* **106:** 13541–13545.

Ramamurthi KS, Clapham KR, Losick R. 2006. Peptide anchoring spore coat assembly to the outer forespore membrane in *Bacillus subtilis. Mol Microbiol* **62:** 1547–1557.

Ramamurthi KS, Lecuyer S, Stone HA, Losick R. 2009. Geometric cue for protein localization in a bacterium. *Science* **323:** 1354–1357.

Raskin DM, de Boer PA. 1999. Rapid pole-to-pole oscillation of a protein required for directing division to the middle of *Escherichia coli. Proc Natl Acad Sci* **96:** 4971–4976.

Rudner DZ, Losick R. 2002. A sporulation membrane protein tethers the pro-σK processing enzyme to its inhibitor and dictates its subcellular localization. *Genes Dev* **16:** 1007–1018.

Rudner DZ, Pan Q, Losick RM. 2002. Evidence that subcellular localization of a bacterial membrane protein is achieved by diffusion and capture. *Proc Natl Acad Sci* **99:** 8701–8706.

Schubbe S, Kube M, Scheffel A, Wawer C, Heyen U, Meyerdierks A, Madkour MH, Mayer F, Reinhardt R, Schuler D. 2003. Characterization of a spontaneous nonmagnetic mutant of *Magnetospirillum gryphiswaldense* reveals a large deletion comprising a putative magnetosome island. *J Bacteriol* **185:** 5779–5790.

Schuler D. 2008. Genetics and cell biology of magnetosome formation in magnetotactic bacteria. *FEMS Microbiol Rev* **32:** 654–672.

Sharp MD, Pogliano K. 1999. An in vivo membrane fusion assay implicates SpoIIIE in the final stages of engulfment during *Bacillus subtilis* sporulation. *Proc Natl Acad Sci* **96:** 14553–14558.

Shav-Tal Y, Singer RH. 2005. RNA localization. *J Cell Sci* **118:** 4077–4081.

Silhavy TJ, Kahne D, Walker S. 2010. The bacterial cell envelope. *Cold Spring Harb Perspect Biol* **2:** a000414.

Smith GA, Portnoy DA, Theriot JA. 1995. Asymmetric distribution of the *Listeria monocytogenes* ActA protein is required and sufficient to direct actin-based motility. *Mol Microbiol* **17:** 945–951.

Stragier P, Losick R. 1996. Molecular genetics of sporulation in *Bacillus subtilis. Annu Rev Genet* **30:** 297–341.

Takizawa PA, Sil A, Swedlow JR, Herskowitz I, Vale RD. 1997. Actin-dependent localization of an RNA encoding a cell-fate determinant in yeast. *Nature* **389:** 90–93.

Thanbichler M, Shapiro L. 2006. MipZ, a spatial regulator coordinating chromosome segregation with cell division in *Caulobacter. Cell* **126:** 147–162.

Thomaides HB, Freeman M, El Karoui M, Errington J. 2001. Division site selection protein DivIVA of *Bacillus subtilis* has a second distinct function in chromosome segregation during sporulation. *Genes Dev* **15:** 1662–1673.

Thompson SR, Wadhams GH, Armitage JP. 2006. The positioning of cytoplasmic protein clusters in bacteria. *Proc Natl Acad Sci* **103:** 8209–8214.

Tiyanont K, Doan T, Lazarus MB, Fang X, Rudner DZ, Walker S. 2006. Imaging peptidoglycan biosynthesis in *Bacillus subtilis* with fluorescent antibiotics. *Proc Natl Acad Sci* **103:** 11033–11038.

Tsukita S, Furuse M, Itoh M. 2001. Multifunctional strands in tight junctions. *Nat Rev Mol Cell Biol* **2:** 285–293.

Uehara T, Dinh T, Bernhardt TG. 2009. LytM-domain factors are required for daughter cell separation and rapid ampicillin-induced lysis in *Escherichia coli. J Bacteriol* **191:** 5094–5107.

van Ooij C, Losick R. 2003. Subcellular localization of a small sporulation protein in *Bacillus subtilis. J Bacteriol* **185:** 1391–1398.

Viollier P. 2010. Polar structures and the proteins that regulate their biogenesis. *Cold Spring Harb Perspect Biol* **2:** a000372.

Wadhams GH, Martin AC, Porter SL, Maddock JR, Mantotta JC, King HM, Armitage JP. 2002. TlpC, a novel chemotaxis protein in *Rhodobacter sphaeroides*, localizes to a discrete region in the cytoplasm. *Mol Microbiol* **46:** 1211–1221.

Wang H, Wingreen NS, Mukhopadhyay R. 2008. Self-organized periodicity of protein clusters in growing bacteria. *Phys Rev Lett* **101:** 218101.

Webb CD, Teleman A, Gordon S, Straight A, Belmont A, Lin DC, Grossman AD, Wright A, Losick R. 1997. Bipolar localization of the replication origin regions of chromosomes in vegetative and sporulating cells of *B. subtilis. Cell* **88:** 667–674.

Wu LJ, Errington J. 2004. Coordination of cell division and chromosome segregation by a nucleoid occlusion protein in *Bacillus subtilis. Cell* **117:** 915–925.

Wu LJ, Ishikawa S, Kawai Y, Oshima T, Ogasawara N, Errington J. 2009. Noc protein binds to specific DNA sequences to coordinate cell division with chromosome segregation. *Embo J* **28:** 1940–1952.

Xie XS, Choi PJ, Li GW, Lee NK, Lia G. 2008. Single-molecule approach to molecular biology in living bacterial cells. *Annu Rev Biophys* **37:** 417–444.

Yeates TO, Kerfeld CA, Heinhorst S, Cannon GC, Shively JM. 2008. Protein-based organelles in bacteria: Carboxysomes and related microcompartments. *Nat Rev Microbiol* **6:** 681–691.

Yu XC, Margolin W. 1999. FtsZ ring clusters in min and partition mutants: Role of both the Min system and the nucleoid in regulating FtsZ ring localization. *Mol Microbiol* **32:** 315–326.

Poles Apart: Prokaryotic Polar Organelles and Their Spatial Regulation

Clare L. Kirkpatrick and Patrick H. Viollier

Department of Microbiology and Molecular Medicine, Centre Médicale Universitaire, Faculty of Medicine, University of Geneva, Rue Michel Servet 1, CH-1211 Geneva 4, Switzerland

Correspondence: patrick.viollier@unige.ch

While polar organelles hold the key to understanding the fundamentals of cell polarity and cell biological principles in general, they have served in the past merely for taxonomical purposes. Here, we highlight recent efforts in unraveling the molecular basis of polar organelle positioning in bacterial cells. Specifically, we detail the role of members of the Ras-like GTPase superfamily and coiled-coil-rich scaffolding proteins in modulating bacterial cell polarity and in recruiting effector proteins to polar sites. Such roles are well established for eukaryotic cells, but not for bacterial cells that are generally considered diffusion-limited. Studies on spatial regulation of protein positioning in bacterial cells, though still in their infancy, will undoubtedly experience a surge of interest, as comprehensive localization screens have yielded an extensive list of (polarly) localized proteins, potentially reflecting subcellular sites of functional specialization predicted for organelles.

Since the first electron micrographs that revealed flagella at the cell poles of bacteria, we have known that bacterial cells are polarized and that they are able to decode the underlying positional information to confine the assembly of an extracellular organelle to a polar cellular site (Fig. 1). Foraging into this unknown territory has been challenging, but recent efforts that exploit the power of bacterial genetics along with modern imaging methods to visualize proteins in the minute bacterial cells has yielded several enticing entry points to dissect polarity-based mechanisms and explore potentially contributing subdiffusive characteristics (Golding and Cox 2006).

While polar organelles are a visual manifestation of polarity, it is important to point out that polarity can also be inherent to cells, at least in molecular terms, even in the absence of discernible polar structures. In other words, molecular anatomy can reveal that a bacterial cell, such as an *Escherichia coli* cell, features specialized protein complexes at or near the poles, despite a perfectly symmetrical morphology (Maddock and Shapiro 1993; Lindner et al. 2008). Such systemic polarization in bacteria, likely stemming from the distinctive division history of each pole, has the potential to be widespread and to be exploited for positioning of polar organelles and protein complexes. As excellent reviews have been published detailing the interplay between cell polarity and protein localization (Dworkin 2009; Shapiro et al. 2009; Kaiser et al. 2010; Rudner and Losick

 doi: 10.1101/cshperspect.a006809
Cite this article as *Cold Spring Harb Perspect Biol* 2011;3:a006809

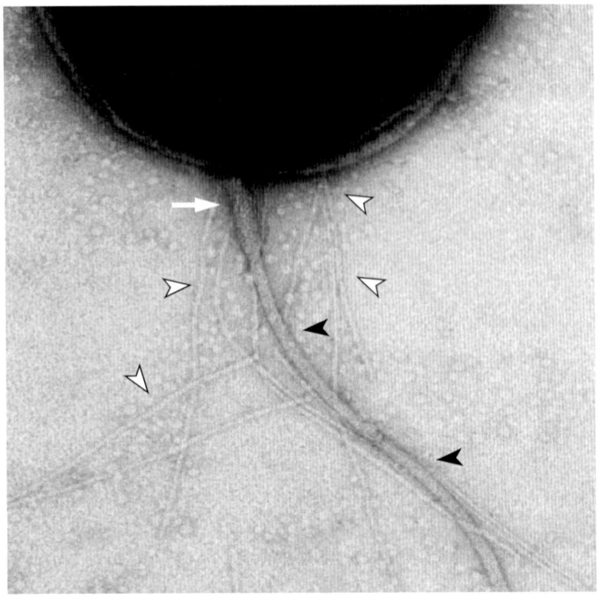

Figure 1. Transmission electron micrograph (taken by Jeff Skerker) of a *Caulobacter crescentus* swarmer cell showing the polar pili (empty arrowheads), the polar flagellum with the flagellar filament (filled arrowheads), and the hook (white arrow) (see Fig. 2A).

2010), here we focus on recent progress in understanding the function and localization of spatial regulators of polar organelles. Considering that the ever-growing list of polar protein complexes emerging from systematic and comprehensive localization studies (Kitagawa et al. 2005; Russell and Keiler 2008; Werner et al. 2009; Hughes et al. 2010) is suggestive of multiple polarly confined (organelle-like) functions, understanding their spatial regulation is also of critical relevance in the realm of medical bacteriology, as many virulence determinants also underlie polarity (Goldberg et al. 1993; Scott et al. 2001; Judd et al. 2005; Jain et al. 2006; Jaumouille et al. 2008; Carlsson et al. 2009). Below, we highlight a few prominent examples of overtly polar organelles and the proteins known to date that regulate their polar positioning.

THE POLARITY OF MOTILITY ORGANELLES

Flagellar Motility: FlhFG

The Ras-like GTPase FlhF is required for proper assembly and placement of the polar flagellum in several bacterial lineages, including *Vibrio*, *Pseudomonas*, *Helicobacter*, and *Campylobacter* species (Pandza et al. 2000; Niehus et al. 2004; Murray and Kazmierczak 2006; Kusumoto et al. 2008; Balaban et al. 2009; Green et al. 2009). Flagellum biogenesis occurs through an "inside-out" process by which components of the early structure first assemble a platform on the cytoplasmic membrane (the MS-ring) that will facilitate the subsequent polymerization and growth of the trans-envelope structures through the periplasm (the flagellar rod), the peptidoglycan (P-ring), and outer membrane (L-ring), ultimately to enable assembly of the extracellular hook and flagellar filament (Macnab 2003) (Fig. 2A). FlhF mutants often exhibit a reduced ability to assemble flagella and in the case of *Vibrio cholerae* this seems to be attributable to a requirement of FlhF for the polar recruitment of the flagellar baseplate (MS-ring) protein FliF (Green et al. 2009). In the absence of FlhF, FliF is dispersed from the poles and flagella are formed at a reduced frequency and at aberrant (nonpolar) locations. Conversely, functional FlhF-GFP is localized to the flagellated pole in *V. cholerae* and *Campylobacter*

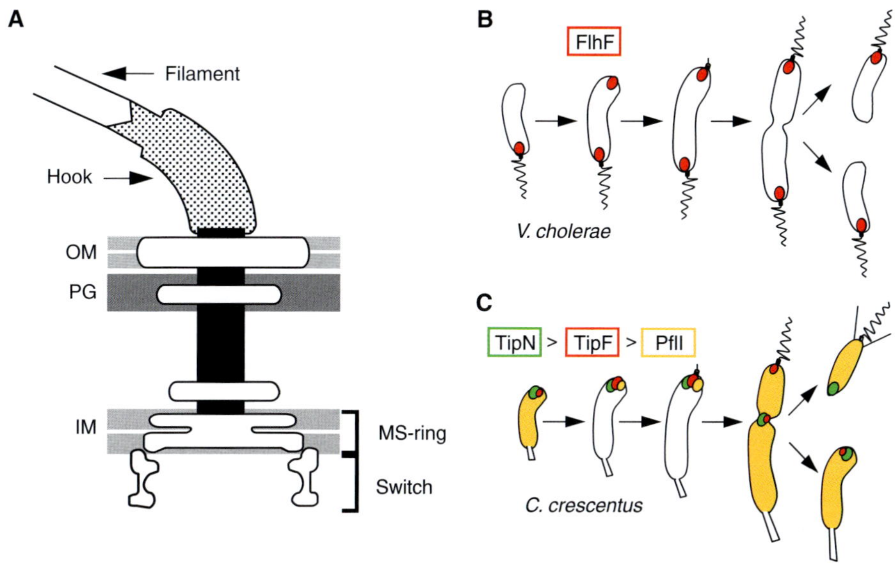

Figure 2. Spatial regulation of the polar flagellum (*A*) by the FlhF GTPase in *Vibrio cholerae* (*B*) and by the TipF c-di-GMP receptor protein, the TipN birth scar protein, and the PflI positioning factor in *Caulobacter crescentus* (*C*). In Panel A, the outer membrane, peptidoglycan layer, and inner membrane are abbreviated by OM, PG, and IM, respectively. In Panels B and C, the wavy and straight lines denote the flagellum and the pili, respectively.

jejuni (Murray and Kazmierczak 2006; Ewing et al. 2009; Green et al. 2009) independently of flagellar structural proteins (Ewing et al. 2009; Green et al. 2009) (Fig. 2B). Thus, FlhF plays a role in both flagellum positioning and assembly, and the latter function may in part be mediated at the level of gene expression (Niehus et al. 2004; Correa et al. 2005). A member of the (SIMIBI) Ras-like GTPase superfamily, FlhF associates peripherally with the cytoplasmic membrane (Green et al. 2009) and exhibits extensive sequence homology to the signal recognition particle (SRP) component Ffh, a GTPase that mediates the co-translational insertion of membrane proteins (Driessen and Nouwen 2008). This event requires a heteromeric interaction of Ffh with its receptor, the FtsY GTPase. A truncated fragment of *Bacillus subitilis* FlhF has been crystallized as a dimer, forming a composite active site in which the two GTP molecules are arranged in a head-to-tail configuration (Bange et al. 2007). While this same configuration in the composite active site of the Ffh-FtsY heterodimer induces GTP hydrolysis, this is not the case for *B. subtilis* FlhF. However, recent in vitro assays with recombinant full-length FlhF from *C. jejuni* harboring an N-terminal His6-tag exhibited GTPase and comparatively low ATPase activity (Balaban et al. 2009). Point mutations that cripple the GTPase activity of *C. jejuni* FlhF result in flagellar placement defects that resemble the phenotypes observed for the *flhF* deletion mutant of *V. cholerae*, *Pseudomonas putida*, and *Pseudomonas aeruginosa* (Pandza et al. 2000; Murray and Kazmierczak 2006; Balaban et al. 2009; Green et al. 2009). Moreover, FlhF with a mutation in the conserved nucleotide-binding P-loop (K295A) has a reduced ability to bind GTP and is unable to support motility in *V. cholerae* (Green et al. 2009). These findings indicate a link between flagellar positioning and the ability to bind and/or hydrolyze GTP. The FlhF polypetide is sequentially subdivided into the B-N-G regions. The conserved GTPase domain of FlhF resides in the C-terminal G-region. The G-region is preceded by the central N-region that is rich in α-helices, while the basic and unstructured (protease-sensitive) B-region is located at the N-terminus of FlhF (Bange et al. 2007). Unlike the G-region, the B- and N-regions are quite variable in length among FlhF orthologs, raising the possibility that these regions account for the structural and functional differences observed in vitro and in vivo. In fact, the N-region included in the crystal structure of *B. subtilis* FlhF is significantly shorter than that from FlhF encoded in the genomes of polarly flagellated bacteria. Deletion analysis in *V. cholerae* pointed to a key role for the N-region in polar localization of FlhF (Green et al. 2009). FlhF-GFP deleted for the N-region was not localized to the pole and unable to confer motility. The N-region could support polar localization, but not motility, when fused to either the B region (i.e., as BN-GFP fusion) or the G-region (i.e., as NG-GFP fusion). However, the N-region alone (i.e., as N-GFP construct) was unable to direct GFP to the pole (Green et al. 2009). It is conceivable that an association with the membrane is a prerequisite for FlhF to recognize localization determinant(s) that may reside in the envelope, as is the case for other polarly localized proteins (Jacobs et al. 1999; Boyd 2000). If so, the N-region might be required for this membrane association, while the B- and G-regions recognize the localization determinant(s). Perhaps one region is used for polar targeting, while the other region functions as polar retention domain. This would imply that the mechanism and targets for the polar localization for BN-GFP and NG-GFP is different. In support of the findings with truncated FlhF derivatives that are localized but not functional, full length FlhF-GFP with the GTP-binding K295A mutation did not support motility either, while still allowing polar recruitment, indicating that bound GTP is not a prerequisite for polar localization (Green et al. 2009). The identity of the polar determinant(s) recognized by FlhF is not known, but it is not a structural component of the flagellum as FlhF-GFP is still at the pole in *Vibrio* mutants that do not express any flagellar subunits (Kusumoto et al. 2008; Green et al. 2009). It would be interesting to examine whether the BN-GFP and NG-GFP constructs are still polar in the absence of the flagellar components to determine whether each of the FlhF subdomains relies on flagellar parts for polar recruitment and/or retention. The principal determinant for FlhF localization appears to be widely conserved across bacterial species, since FlhF-GFP from *V. cholerae* is also polar in *E. coli* (Green et al. 2009). Thus, even a peritrichously-flagellated bacterium such as *E. coli* encodes a positional signal that is recognized by the B- and/or G-region of FlhF. Interestingly, FlhF is localized to both poles in *E. coli*, but in the native host FlhF-GFP was also seen in a predominantly unipolar localization pattern (Murray and Kazmierczak 2006; Kusumoto et al. 2008; Green et al. 2009), suggesting that a specific mechanism exists to impart polar specificity on FlhF localization and, thus, the new flagellum assembly site.

Although FlhF may regulate the correct positioning of the core flagellum assembly platform in the cytoplasmic membrane through FliF localization, recent work in *C. jejuni* has unearthed flagellar glycosylation pathway components that are localized to the cell pole independently of FlhF (Ewing et al. 2009). In many polarly flagellated bacteria, the flagellar filament is polymerized from glycosylated flagellins. Glycosylation seems to precede the secretion of the flagellin (Ewing et al. 2009), yet its general function is not understood. Many components of the glycosylation pathway are dispensable for motility. This may be attributable to a certain redundancy in glycosylation components, as several polarly flagellated bacteria of the genera *Campylobacter*, *Helicobacter*, and *Caulobacter* are unable to polymerize a flagellar filament and unable to secrete unglycosylated flagellins when other glycosylation proteins are inactivated (Leclerc et al. 1998; Goon et al. 2003; Schirm et al. 2003). Thus, glycosylation seems to fulfill an important and mysterious role in the assembly of polar flagella. Pertinent to the tentative link between polar flagellation and glycosylation, the components of the glycosylation machinery PseC and PseE are localized to the pole in *C. jejuni* (Ewing et al. 2009). Although FlhF is also at the cell pole in *C. jejuni*, it is not required for the polar positioning of PseC and PseE. In support of this, PseC and PseE are also still at the poles in the absence of the flagellar export apparatus.

Thus, a FlhF-independent flagellar localization pathway exists that remains to be uncovered (Ewing et al. 2009).

FlhF is also found in peritrichously flagellated bacteria and an investigation on FlhF in *Bacillus cereus* determined that inactivation of FlhF reduced flagellation and resulted in a bias toward polar flagella from a normal peritrichous flagellar arrangement (Salvetti et al. 2007). While the function of FlhF in flagellum biogenesis does not appear to be specialized for polar flagellation, certain bacteria may have appropriated FlhF to recruit flagellar components to the pole and, therefore, to confine the assembly of the flagellar structure to a polar site, simply by directing FlhF to the pole. In other cases, such as *B. cereus*, FlhF may be localized to nonpolar sites to obtain a peritrichous flagellar disposition, implying an underlying default pathway for polar flagellation in *B. cereus* that FlhF acts upon to guide flagellar assembly factors to lateral sites.

The mechanism(s) of spatial regulation of FlhF likely holds the key to understanding the FlhF-dependent polar flagellation in *Vibrios*, *Pseudomads*, and others. Circumstantial evidence has implicated FlhG (also known as FleN in *Pseudomonas spp.*), which resembles the ParA/MinD superfamily of ATPases (cd01983) that feature a deviant Walker A-type nucleotide-binding pocket (the P-loop), in the regulation of the polar localization of FlhF (Kusumoto et al. 2008). MinD and ParA adopt key roles as part of positioning systems that direct the proper placement of the cytokinetic organelle or (chromosomal or plasmid) DNA, respectively, within the prokaryotic cell or the eukaryotic chloroplasts (de Boer et al. 1991; Margolin 2005; Gerdes et al. 2010; Thanbichler 2010). In the absence of FlhG, the fluorescent cluster of FlhF-GFP at the cell pole was brighter than when FlhG was present (Kusumoto et al. 2008). While this result suggested that the polar localization of FlhF is improved in FlhG-deficient cells, they also had higher steady-state levels of FlhF. Nevertheless, the diffuse background fluorescence of FlhF-GFP in the cytoplasm was also reduced in cells lacking FlhG, pointing toward a role of FlhG in interference with the polar sequestration of FlhF (Kusumoto et al. 2008). Interestingly, mutant forms of FlhF have been identified that show the opposite phenotype of the FlhG-deficient cells (i.e., strong dependency on the presence of FlhG for polar localization), and GFP-tagged FlhG is also polar (Kusumoto et al. 2009). These observations are consistent with the proposed role of the ParA/MinD ATPases acting as binary regulatory switches that control the localization of cellular constituents, a recurring theme throughout this article. Below we outline the role of other ParA/MinD-homologs in the localization of organellar components, including that of a chemosensory complex (Thompson et al. 2006), a pilus biogenesis apparatus (Viollier et al. 2002a), a DNA transfer machine (Atmakuri et al. 2007), and a polysaccharide export system (Le Quere and Ghigo 2009).

Flagellar Motility: TipNF-PflI

While FlhG may be a key element for the spatial regulation of FlhF, many polarly flagellated bacterial lineages, primarily from the α-proteobacteria, do not encode FlhF in their genomes. Thus, nature has devised another path to polar flagellation, one that seems to involve a conserved positioning protein, PflI. PflI was identified using a bioinformatic approach and then characterized in the vibroid-shaped bacterium *Caulobacter crescentus* (Obuchowski and Jacobs-Wagner 2007). In *C. crescentus*, the flagellum is always built at the new pole (i.e., the pole that emerged from the most recent division event). PflI, a bitopic membrane protein with a cytoplasmic coiled-coil domain followed by a proline-rich stretch of 102 residues, is localized to the future flagellated pole well before the flagellar proteins are expressed. Moreover, PflI is still localized to the new pole in the absence of the MS-ring protein FliF, indicating that, akin to FlhF, PflI does not depend on an intact flagellar structure for polar localization. Importantly, inactivation or overexpression of PflI results in a high proportion of cells with misplaced flagella, while not affecting the localization of polar marker proteins; for example, that of the pilus assembly machinery (see below) or signaling proteins that normally colocalize with the flagellum. Thus, PflI seems to function as a positioning factor specifically for the flagellum. How PflI influences the flagellum positioning is unknown, but it is likely that additional localization pathways work in concert with PflI. This conjecture is based on the finding that in PflI-deficient or -overproducing cells, the flagellum misplacement is not as pervasive, with 30–60% of the flagellated cells still having a polar flagellum (Obuchowski and Jacobs-Wagner 2007). Progress has recently been made in understanding how PflI localization is regulated (Davis et al. 2010, in prep). As hinted by the similar localization pattern during the cell cycle (Fig. 2C) and in flagellar mutants, the recruitment of PflI to the future flagellated pole depends on the EAL-domain protein TipF (Huitema et al. 2006; Obuchowski and Jacobs-Wagner 2007). A polytopic membrane protein previously known to regulate flagellar assembly and implicated in flagellar positioning, TipF is also localized to the new pole even in the absence of early flagellar structural components such as the FliF MS-ring protein (Huitema et al. 2006) (Davis et al. 2010, in prep). The cytoplasmic EAL-domain is responsible for TipF function in flagellum assembly and the recruitment of PflI to the pole (Huitema et al. 2006). In related proteins, the EAL-domain is responsible for the binding and subsequent hydrolysis of cyclic-di-GMP, a signaling molecule that drives the switch from motility to sessility in eubacteria. (Jenal and Malone 2006; Hengge 2009). The EAL domain can also act as c-di-GMP binding domain or protein–protein interaction domain in the absence of hydrolytic activity (Newell et al. 2009; Tschowri et al. 2009). The TipF EAL-domain also lacks c-di-GMP hydrolytic activity, but has retained the ability to bind c-di-GMP (Davis et al. 2010, in prep). When TipF is removed or when c-di-GMP-binding is prevented either by point mutation or by depletion of cellular c-di-GMP levels, flagella are not assembled and PflI is not localized (Davis et al. 2010, in prep). TipF is also required for the localization of the components of the flagellar switch complex (Fig. 2A), but conversely the switch complex is not necessary for the polar positioning of TipF. Intriguingly, in the absence of an association with c-di-GMP, TipF remains dispersed in the cell envelope and does not cluster at the new cell pole. Moreover, under these conditions, flagella are also not assembled and PflI also remains dispersed (Davis et al. 2010, in prep). Thus, binding of c-di-GMP induces the polar localization of TipF to enable the recruitment of a placement factor and structural proteins for flagellum assembly at the new pole. TipF localization was previously shown to be determined by the TipN birth scar protein, a polytopic membrane protein with an extensive coiled-coil-rich domain facing the cytoplasm, that is required for proper localization of the flagellum at the new pole (Huitema et al. 2006; Lam et al. 2006). When TipN was inactivated, flagella and TipF (but not pili) were frequently mispositioned at seemingly identical aberrant locations (Huitema et al. 2006), hinting that akin to FlhF, TipF regulates flagellar placement and assembly. In dividing cells, TipN is localized at the division plane in a manner that requires the FtsZ cytokinetic tubulin homolog and then retained at the newborn pole that emerges from division (Huitema et al. 2006; Lam et al. 2006), the site of flagellum assembly and TipF/PflI localization. Thus, the protein localization and ensuing flagellum positioning pathway seems to follow the order: FtsZ > TipN > TipF > PflI. The key element of this polarity pathway for flagellar assembly and positioning is that protein localization relies on a cytokinetic mark or remnant that is inherited by the newborn pole of each daughter cell, thereby preselecting these poles as sites of flagellum biogenesis that attract the downstream assembly factors (Huitema et al. 2006; Lam et al. 2006). Intriguingly, TipN was recently shown to serve as the spatial cue that provides the correct directionality to the new pole in bacterial chromosome segregation (Ptacin et al. 2010; Schofield et al. 2010), showing that other polarized cellular functions also rely on the TipN-dependent polarity axis.

In summary, two mechanisms are known, FlhFG and TipNF-PflI, that both use a guanine nucleotide derivative to regulate the polar placement of the flagellum. The physiological significance of the guanine nucleotide-dependent input is unclear, but may be related to the variation in cellular concentration of the nucleotides over time. C-di-GMP levels fluctuate during the cell cycle (Paul et al. 2008; Christen et al. 2010b) and GTP levels change as cells transition from logarithmic growth into stationary phase (Freese et al. 1979; Buckstein et al. 2008; Paul et al. 2008; Christen et al. 2010b). Exploiting such fluctuation(s) would allow cells to tie the assembly of flagellar structures with the cell cycle or with the growth phase when the energetic cost for the assembly of such a complex nanomachine can be afforded.

Gliding Motility: MglAB

In a striking parallel to the role of the FlhF GTPase in determining the polarity of the flagellum-dependent motility system, recent research illustrates the role of the Ras-like GTPase MglA in regulating the dynamic polarity switch of (flagellum-independent) gliding motility of *Myxcoccus xanthus* cells on surfaces (Leonardy et al. 2010; Patryn et al. 2010; Zhang et al. 2010). *M. xanthus* cells move by way of cooperative, but distinct, motility systems, termed S (conferred by pilus motor proteins) and A (unidentified motor proteins), whose activity are modulated by polarized effectors (see Kaiser et al. 2010). A hallmark of motile *M. xanthus* cells is that they exhibit periodic reversals in the direction of movement and in a fashion that, at the subcellular level, is exquisitely tuned with the dynamic and often anticyclical polar localization of the A-motility effectors RomR and AglZ (Leonardy et al. 2007; Mignot et al. 2007) and the S-motility effectors PilT and FrzS (Mignot et al. 2005; Bulyha et al. 2009; Mauriello et al. 2010). The dynamic pole-to-pole oscillation of several of these effectors shows a strong dependency on the active, presumably GTP-bound, form of MglA (Leonardy et al. 2010; Zhang et al. 2010). Mutations in MglA that impair hydrolysis but not GTP binding 1) alter the cellular reversal frequency, 2) block polar oscillations of RomR, AglZ, and PilT and 3) prevent motility. Moreover, these MglA point mutant strains phenocopy those with a deletion in the gene encoding MglB, which is encoded by the open reading frame immediately upstream of that for MglA. MglA is localized to the leading pole (Mauriello et al. 2010; Patryn et al. 2010) while MglB clusters at the lagging pole (Leonardy et al. 2010; Zhang et al. 2010). Recent biochemical studies demonstrated that MglB essentially functions as a GTPase activating protein (GAP) to MglA to regulate its polar localization dynamics and activity (Leonardy et al. 2010; Zhang et al. 2010). In moving cells, MglA and MglB oscillate from pole to pole with a seemingly identical period that is out-of-phase with one another and coordinated with cell reversals (Fig. 3). The anticyclical polar localization of MglA is ablated when GTP hydrolysis is crippled by inactivation of MglB or by GTPase mutations in MglA (Leonardy et al. 2010; Zhang et al. 2010). Moreover, MglA also modulates the dynamics of MglB localization (Leonardy et al. 2010). Based on the reciprocal regulation of MglAB localization dynamics and the dependency of downstream motility effectors on MglAB (Mauriello et al. 2010), it stands to reason that the MglAB system is a core constituent of the molecular oscillator that regulates the dynamic polarity of cell reversals in *M. xanthus*. Fine-tuning of the MglAB localization and activity is accomplished through the Frz chemosensory system and the actin-like MreB cytosleketon, albeit in an unknown way (Leonardy et al. 2010; Mauriello et al. 2010; Zhang et al. 2010). Additional mechanisms in polarity control remain to be identified, as the motility effectors are still recruited to the pole in the absence of MglA, albeit often to the erroneous one.

POLARITY CONTROL OF PILI AND OTHER ORGANELLAR PROTEINS BY ACTIN HOMOLOGS, PARA-LIKE ATPASES, AND A POLARITY FACTOR

Polar Pili

The pili assembled by *P. aeruginosa* are very similar with respect to structure, function, and localization to the pili that confer gliding

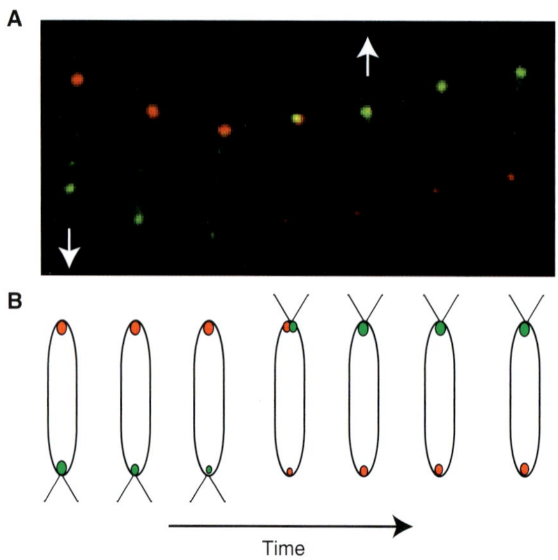

Figure 3. Polarity switching by the Ras-like GTPase MglA (green oval) and its GTPase-activating protein (GAP) MglB (red oval) in gliding *Myxococcus xanthus* cells. (*A*) A time series of gliding *M. xanthus* cells expressing MglA-YFP and MglB-mCherry that were imaged every 20 seconds by fluorescence microscopy (by Yong Zhang and Tâm Mignot). The yellow coloring indicates the colocalized MglA and MglB. The white arrow indicates the direction of movement that reverses during the course of the experiment. (*B*) A schematic of the cells and polarly localized proteins from panel A. The straight lines denote the pili that confer S-motility.

motility in *M. xanthus* (see the schematic of the pilus in Fig. 3 of Kaiser et al. 2010). They are built from homologous components into polar trans-envelope structures that confer a form of flagellar-independent motility. In *P. aeruginosa*, this second form of motility is called (pilus-mediated) twitching motility. Akin to the aforementioned gliding motility, twitching motility involves the regular and coordinated extension/retraction cycles of the pilus filament from the cell surface (Mattick 2002). Consistent with the role of the actin-like MreB cytoskeleton in maintaining the polar localization of the S-motility regulator FrzS in *M. xanthus* (Mauriello et al. 2010), MreB seems to play a crucial role in the localization of pili in *P. aeruginosa* (Cowles and Gitai 2010). When *P. aeruginosa* cells were pretreated with the small molecule MreB-inhibitor A22 before being placed on medium that induces pilus production, cells accumulated pili at lateral sites instead of the pole. Also, unlike untreated cells, those pre-exposed to A22 mislocalized the PilT retraction ATPase to nonpolar sites. Thus, MreB maintains the polar disposition of pili and pilus regulators in an unknown way. A likely model would hold that pilus structural and regulatory proteins such as PilT and FrzS travel to the pole along the helical tracks of MreB filaments lining the cytoplasmic face of the inner membrane (Cowles and Gitai 2010; Mauriello et al. 2010). Remarkably, the PilM pilus assembly protein that is encoded in the *pil* clusters of both *P. aeruginosa* and *M. xanthus* is a member of the actin superfamily of proteins (Martin et al. 1995). As MreB isoforms are thought to inter-digitate into heteromeric cytoskeletal filaments (Carballido-Lopez et al. 2006), it is conceivable that PilM is also integrated into a cytoskeletal structure to connect the localization of the pilus structure and its constituents with the cytoskeleton. It will be interesting to determine the subcellular distribution of PilM and whether it is affected by treatment with A22. MreB has also been shown, directly or indirectly, to influence the polar localization of the virulence determinant IcsA of *Shigella spp.*, the *E. coli* chemoreceptor Tar, several developmental kinases, and the origin of replication (*Cori*) in *C. crescentus* (Goldberg et al. 1993; Gitai et al. 2004; Shih et al. 2005). Actin-like proteins also determine the subcellular distribution of plasmids in *E. coli* (Gerdes et al. 2010), organelles like magnetosomes in *Magnetospirrillum magneticum* (Komeili et al. 2006; Murat et al. 2010), and cell wall biosynthetic enzymes (Dye et al. 2005).

Pilus machineries, evolutionarily less-related to that of *P. aeruginosa* and *M. xanthus*, are also polarized. The pili of *C. crescentus* are assembled from a prototypical pilus/protein secretion apparatus encoded by a widely-conserved locus (the *cpa* or *tad* locus) at the new (flagellated) pole (Skerker and Shapiro 2000; Tomich et al. 2007). The CpaE component of the *C. crescentus* pilus system (also known as TadZ in *Aggregatibacter actinomycetemcomitans*) has extensive sequence homology to the ParA/MinD-like ATPases and is localized to the piliated pole, where it induces the formation of the polar pilus secretion channel (CpaC) in the outer membrane (Skerker and Shapiro 2000; Viollier et al. 2002a). How this occurs is unknown, but it could involve the spatial regulation of the CpaF secretion ATPase or the CpaD lipopotein that is presumably anchored to the inner leaflet of the outer membrane. Both CpaF and CpaD are polarly localized in *C. crescentus* (Werner et al. 2009), and a likely functional analog to CpaD (Tgl) is required for the formation of the pilus channel at both *M. xanthus* poles (Nudleman et al. 2006). In support of the notion that the ParA/MinD-like CpaE protein regulates the localization of the CpaF secretion ATPase, the polarly localized TadZ protein (the CpaE ortholog) was recently found to recruit the hexameric TadA secretion ATPase (the CpaF ortholog) (Bhattacharjee et al. 2001) to the poles of *A. actinomycetemcomitans* (B. Perez and D. Figurski, pers. comm.; Perez 2007). Thus, the common theme that emerges from the CpaEF and FlhGF (or FleN-FlhF) regulatory pairs is that ParA/MinD-like regulator influences the correct localization of the adjacently encoded NTPase.

Polar Polysaccharide Clusters

The central function of ParA/MinD-homologs in coordinating subcellular organization, potentially from within a cytoskeletal structure (Shih et al. 2003), is evident from several studies investigating how DNA and/or proteins are precisely positioned at the cell center or the cell extremities (reviewed in Gerdes et al. 2010). RpfA and MipZ determine the medial localization of TlpT chemoreceptor complex in *Rhodobacter sphaeroides* and the cytokinetic FtsZ-ring in *C. crescentus*, respectively (Thanbichler and Shapiro 2006; Thompson et al. 2006). The ParA/MinD-homologs VirC1 and BcsQ are polarly localized and dictate the recruitment of T-DNA nucleoprotein complexes or cellulose fibers to the cell pole, respectively (Atmakuri et al. 2007; Le Quere and Ghigo 2009). The activity of VirC1 and BcsQ, but not their polar localization, is impeded by mutations in the deviant Walker A motif that are thought to be required for ATP binding and hydrolysis (Leonard et al. 2005; Atmakuri et al. 2007; Le Quere and Ghigo 2009). For example, VirC1 bearing a mutation (K15Q) of lysine at the corresponding position 6 in the Walker-A motif (Fig. 4) is still

Deviant Walker-A motif:	KGGxxKT S
MinD:	KGGVGKT
ParA:	KGGVSKT
MipZ:	KGGAGKS
PpfA:	KGGVGKT
FlhG:	KGGVGKS
VirC1:	KGGAGKT
CpaE:	KGGVGAS
TadZ:	KGGIGAS
BcsQ:	RGGVGTT
WssJ:	NGGVGRS

Figure 4. Alignment of conserved residues of deviant Walker-A boxes found in polarly localized members of the ParA/MinD family of ATPases. Shown are *Escherichia coli* MinD and BcsQ, *Caulobacter crescentus* ParA, MipZ and CpaE, *Rhodobacter sphaeroides* PpfA, *Vibrio cholerae* FlhG, *Agrobacterium tumefaciens* VirC1, *Aggregatibacter actinomycetemcomitans* TadZ, and *Pseudomonas fluorescens* WssJ. Residues that are not identical, but similar, to the consensus sequence in red are shown in blue. Black color codes for variable residues.

localized to the cell poles. Interestingly, the corresponding mutation is naturally present in CpaE and BcsQ, which are also polarly localized. It is unknown whether CpaE, BcsQ, VirC1, and the K15Q derivative can bind or hydrolyze ATP on their own in vitro. If this is not the case, CpaE and BcsQ might rely on ancillary factors to augment ATP binding/hydrolysis in vivo. Alternatively, the modified P-loop might accommodate a nucleotide (derivative) other than ATP, possibly even the signaling molecule c-di-GMP, which is known to regulate cellulose production at the post-translational level (Jenal and Malone 2006; Le Quere and Ghigo 2009). It has been hypothesized that BcsQ-orthologs of *Pseudomonas fluorescens*, WssA, and WssJ, control the localization of the cellulose synthase complex (Spiers et al. 2002), raising the intriguing possibility that this regulation is mediated by c-di-GMP via BcsQ/WssA/J.

The adhesive holdfast of *C. crescentus*, the UPP (unipolar polysaccharide) of its cousins *A. tumefaciens and Rhizobium leguminosarum*, reflect other polar sites where exopolysaccharides are concentrated (Merker and Smit 1988; Laus et al. 2006; Tomlinson and Fuqua 2009; Hardy et al. 2010). While studies on the localization of UPP are still in their infancy, a recent study revealed that the polar localization of the *C. crescentus* holdfast attachment proteins (Hfàs) requires the PodJ polarity factor that is localized to the new cell pole (Crymes et al. 1999; Viollier et al. 2002b; Hinz et al. 2003; Hardy et al. 2010). PodJ is a bitopic membrane protein that features N-terminal coiled-coil rich domain facing the cytoplasm and a C-terminal putative peptidoglycan-binding domain. The Hfa proteins are thought to tether the holdfast polysaccharide to the flagellated cell pole (Cole et al. 2003; Smith et al. 2003; Levi and Jenal 2006). PodJ is sequestered to the pole well before the Hfàs are expressed (Janakiraman and Brun 1999; Viollier et al. 2002b; Hinz et al. 2003). In the absence of PodJ, HfaA/B/D were no longer polar (Hardy et al. 2010), indicating that PodJ provides positional information to enable the polar sequestration of these holdfast proteins. Interestingly, PodJ is also required for the polar clustering of the ParA/MinD-paralog and pilus assembly regulator CpaE (see above) and the cell fate histidine kinase/phosphatase PleC (Viollier et al. 2002b; Hinz et al. 2003). These multifunctional activities are reflected in the pleiotropic phenotype of the PodJ mutant that includes a pilus and holdfast assembly defect and aberrant expression of developmental genes. It is currently unknown whether PodJ regulates CpaE, PleC, and/or the Hfàs directly or not. However, since the polar localization of the Hfàs also depends on the presence of the holdfast synthesis (*hfs*) genes, it is possible that the effect of PodJ on Hfàs is indirect and is mediated through *hfs*-encoded proteins (Hardy et al. 2010). Recent evidence also suggests that CpaE contributes to the polar recruitment of PleC (Christen et al. 2010a), arguing that at least one PodJ-dependent mechanism indirectly regulates the localization of PleC to the new (flagellated) pole. How PodJ itself is attracted to the new cell pole is currently a

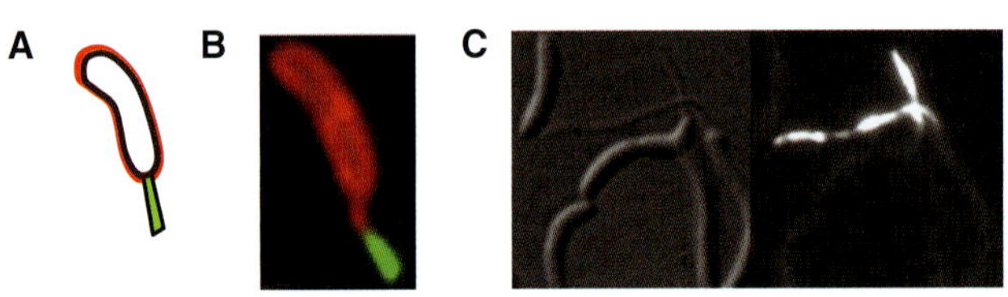

Figure 5. Discovery of stalked protein X (StpX) of *Caulobacter crescentus* from a genome-wide localization screen (Hughes et al. 2010). (*A*) A schematic of the *C. crescentus* cell in panel B with a fluorescent polar stalk. (*B*) Fluorescence micrograph from cells expressing StpX-GFP and a red-fluorescent membrane stain (picture taken by P. Viollier). (*C*) A mutant with a mispositioned stalk containing StpX-GFP. The left image is a DIC (differential interference contrast) micrograph, and the image on the right represents the corresponding (green) fluorescent image.

mystery. It is possible that PodJ localization is cued by an analogous mechanism, as that of TipN (see above). If so, then PodJ must also recognize a cytokinetic signal left behind at the newly formed cell poles in the wake of division. To be recognized by PodJ, this cytokinetic signal must remain at the newborn poles until new PodJ protein is synthesized after the daughter cells separate. At the time of division, only a proteolytic fragment of PodJ is present, while the synthesis of new (full-length) PodJ occurs during the inter-division period (Viollier et al. 2002b; Hinz et al. 2003; Chen et al. 2005; Chen et al. 2006). In support of the idea that localization of PodJ to the new pole is cued by a cytokinetic signal, PodJ can be seen at the division septum when it is constitutively synthesized (P Viollier, unpubl.). Moreover, overexpression of C-terminally truncated form of PodJ interferes with division and is lethal (Crymes et al. 1999). These results are consistent with the view that PodJ, akin to TipN, interacts with the division apparatus or an ancillary division protein. PodJ is still polar in the absence of TipN, indicating that the signal at the newborn pole is not dependent on TipN.

WHAT WE KNOW, WE DON'T KNOW

For a large number of polar organelles or protein complexes, we do not yet have the foot in the door to the secret of their positioning mechanism(s). One such case is that of the polar stalk of *C. crescentus* (Fig. 5A,B) that is elaborated by the cylindrical polar outgrowth of the cell envelope. Attempts to identify structural genes for the stalk have been futile, perhaps because of the essential nature of proteins required for stalk synthesis (Wagner et al. 2005). Alternatively, the imposed genetic selections or bioinformatic criteria might have been suboptimal. Comprehensive and genome-wide protein localization efforts may provide an alternative to such recalcitrant cases. Indeed, they hold the potential to directly identify the players involved with a cell biological rather than a genetic bias and, at the very least, can yield new screening tools that represent entry points for the dissection of the organelle positioning and other localization mechanisms (Werner et al. 2009; Hughes et al. 2010; Ingerson-Mahar et al. 2010) (Fig. 5C).

To put the underlying mechanisms addressed above into perspective, one must ask why cells have developed such extensive regulatory pathways to ensure the polar placement of organelles and protein complexes. An appealing model posits that polar localization serves simply to ensure the proper partitioning of functions into the corresponding daughter cell that inherits a particular pole. This is of course widely accepted in the case of sister chromosomes that must be segregated into both daughter cells, but the same principle should apply to polar organelles or protein complexes that are present in too low a copy number to rely on stochastic distribution into daughters (see Fig. 2B). Perhaps not coincidentally, relatives of ParA, a protein that localizes chromosomal regions and plasmids to polar regions to ensure their segregation (Gerdes et al. 2010), have been appropriated by bacteria to fulfill key roles in the polar positioning of organelles and, by inference, their partitioning to progeny at division.

ACKNOWLEDGMENTS

We thank Tâm Mignot, Yong Zhang, and Jeff Skerker for providing micrographs. Funding support is from the U.S. Department of Energy, Office of Science (Biological and Environmental Research, Grant # DE-FG02-05ER64136), the Swiss National Science Foundation (Grant # 31003A_127287), and the Human Frontiers Science Program (Program Grant # RGP0051/2010).

REFERENCES

Atmakuri K, Cascales E, Burton OT, Banta LM, Christie PJ. 2007. *Agrobacterium* ParA/MinD-like VirC1 spatially coordinates early conjugative DNA transfer reactions. *EMBO J* **26:** 2540–2551.

Balaban M, Joslin SN, Hendrixson DR. 2009. FlhF and its GTPase activity are required for distinct processes in flagellar gene regulation and biosynthesis in *Campylobacter jejuni*. *J Bacteriol* **191:** 6602–6611.

Bange G, Petzold G, Wild K, Parlitz RO, Sinning I. 2007. The crystal structure of the third signal-recognition particle

GTPase FlhF reveals a homodimer with bound GTP. *Proc Natl Acad Sci U S A* **104:** 13621–13625.

Bhattacharjee MK, Kachlany SC, Fine DH, Figurski DH. 2001. Nonspecific adherence and fibril biogenesis by *Actinobacillus actinomycetemcomitans*: TadA protein is an ATPase. *J Bacteriol* **183:** 5927–5936.

Boyd JM. 2000. Localization of the histidine kinase PilS to the poles of Pseudomonas aeruginosa and identification of a localization domain. *Mol Microbiol* **36:** 153–162.

Buckstein MH, He J, Rubin H. 2008. Characterization of nucleotide pools as a function of physiological state in *Escherichia coli*. *J Bacteriol* **190:** 718–726.

Bulyha I, Schmidt C, Lenz P, Jakovljevic V, Hone A, Maier B, Hoppert M, Sogaard-Andersen L. 2009. Regulation of the type IV pili molecular machine by dynamic localization of two motor proteins. *Mol Microbiol* **74:** 691–706.

Carballido-Lopez R, Formstone A, Li Y, Ehrlich SD, Noirot P, Errington J. 2006. Actin homolog MreBH governs cell morphogenesis by localization of the cell wall hydrolase LytE. *Dev Cell* **11:** 399–409.

Carlsson F, Joshi SA, Rangell L, Brown EJ. 2009. Polar localization of virulence-related Esx-1 secretion in mycobacteria. *PLoS Pathog* **5:** e1000285.

Chen JC, Hottes AK, McAdams HH, McGrath PT, Viollier PH, Shapiro L. 2006. Cytokinesis signals truncation of the PodJ polarity factor by a cell cycle-regulated protease. *EMBO J* **25:** 377–386.

Chen JC, Viollier PH, Shapiro L. 2005. A membrane metalloprotease participates in the sequential degradation of a *Caulobacter* polarity determinant. *Mol Microbiol* **55:** 1085–1103.

Christen B, Fero MJ, Hillson NJ, Bowman G, Hong SH, Shapiro L, McAdams HH. 2010a. High-throughput identification of protein localization dependency networks. *Proc Natl Acad Sci U S A* **107:** 4681–4686.

Christen M, Kulasekara HD, Christen B, Kulasekara BR, Hoffman LR, Miller SI. 2010b. Asymmetrical distribution of the second messenger c-di-GMP upon bacterial cell division. *Science* **328:** 1295–1297.

Cole JL, Hardy GG, Bodenmiller D, Toh E, Hinz A, Brun YV. 2003. The HfaB and HfaD adhesion proteins of *Caulobacter crescentus* are localized in the stalk. *Mol Microbiol* **49:** 1671–1683.

Correa NE, Peng F, Klose KE. 2005. Roles of the regulatory proteins FlhF and FlhG in the *Vibrio cholerae* flagellar transcription hierarchy. *J Bacteriol* **187:** 6324–6332.

Cowles KN, Gitai Z. 2010. Surface association and the MreB cytoskeleton regulate pilus production, localization and function in *Pseudomonas aeruginosa*. *Mol Microbiol* **76:** 1411–1426.

Crymes WB Jr., Zhang D, Ely B. 1999. Regulation of *podJ* expression during the *Caulobacter crescentus* cell cycle. *J Bacteriol* **181:** 3967–3973.

de Boer PA, Crossley RE, Hand AR, Rothfield LI. 1991. The MinD protein is a membrane ATPase required for the correct placement of the *Escherichia coli* division site. *Embo J* **10:** 4371–4380.

Driessen AJ, Nouwen N. 2008. Protein translocation across the bacterial cytoplasmic membrane. *Annu Rev Biochem* **77:** 643–667.

Dworkin J. 2009. Cellular polarity in prokaryotic organisms. *Cold Spring Harb Perspect Biol* **1:** a003368.

Dye NA, Pincus Z, Theriot JA, Shapiro L, Gitai Z. 2005. Two independent spiral structures control cell shape in *Caulobacter*. *Proc Natl Acad Sci U S A* **102:** 18608–18613.

Ewing CP, Andreishcheva E, Guerry P. 2009. Functional characterization of flagellin glycosylation in *Campylobacter jejuni* 81–176. *J Bacteriol* **191:** 7086–7093.

Freese E, Heinze JE, Galliers EM. 1979. Partial purine deprivation causes sporulation of *Bacillus subtilis* in the presence of excess ammonia, glucose and phosphate. *J Gen Microbiol* **115:** 193–205.

Gerdes K, Howard M, Szardenings F. 2010. Pushing and pulling in prokaryotic DNA segregation. *Cell* **141:** 927–942.

Gitai Z, Dye N, Shapiro L. 2004. An actin-like gene can determine cell polarity in bacteria. *Proc Natl Acad Sci U S A* **101:** 8643–8648.

Goldberg MB, Barzu O, Parsot C, Sansonetti PJ. 1993. Unipolar localization and ATPase activity of IcsA, a *Shigella flexneri* protein involved in intracellular movement. *Infect Agents Dis* **2:** 210–211.

Golding I, Cox EC. 2006. Physical nature of bacterial cytoplasm. *Phys Rev Lett* **96:** 098102.

Goon S, Kelly JF, Logan SM, Ewing CP, Guerry P. 2003. Pseudaminic acid, the major modification on *Campylobacter* flagellin, is synthesized via the Cj1293 gene. *Mol Microbiol* **50:** 659–671.

Green JC, Kahramanoglou C, Rahman A, Pender AM, Charbonnel N, Fraser GM. 2009. Recruitment of the earliest component of the bacterial flagellum to the old cell division pole by a membrane-associated signal recognition particle family GTP-binding protein. *J Mol Biol* **391:** 679–690.

Hardy GG, Allen RC, Toh E, Long M, Brown PJ, Cole-Tobian JL, Brun YV. 2010. A localized multimeric anchor attaches the *Caulobacter* holdfast to the cell pole. *Mol Microbiol* **76:** 409–427.

Hengge R. 2009. Principles of c-di-GMP signalling in bacteria. *Nat Rev Microbiol* **7:** 263–273.

Hinz AJ, Larson DE, Smith CS, Brun YV. 2003. The *Caulobacter crescentus* polar organelle development protein PodJ is differentially localized and is required for polar targeting of the PleC development regulator. *Mol Microbiol* **47:** 929–941.

Hughes HV, Huitema E, Pritchard S, Keiler KC, Brun YV, Viollier PH. 2010. Protein localization and dynamics within a bacterial organelle. *Proc Natl Acad Sci U S A* **107:** 5599–5604.

Huitema E, Pritchard S, Matteson D, Radhakrishnan SK, Viollier PH. 2006. Bacterial birth scar proteins mark future flagellum assembly site. *Cell* **124:** 1025–1037.

Ingerson-Mahar M, Briegel A, Werner JN, Jensen GJ, Gitai Z. 2010. The metabolic enzyme CTP synthase forms cytoskeletal filaments. *Nat Cell Biol*.

Jacobs C, Domian IJ, Maddock JR, Shapiro L. 1999. Cell cycle-dependent polar localization of an essential bacterial histidine kinase that controls DNA replication and cell division. *Cell* **97:** 111–120.

Jain S, van Ulsen P, Benz I, Schmidt MA, Fernandez R, Tommassen J, Goldberg MB. 2006. Polar localization of the autotransporter family of large bacterial virulence proteins. *J Bacteriol* **188:** 4841–4850.

Janakiraman RS, Brun YV. 1999. Cell cycle control of a holdfast attachment gene in *Caulobacter crescentus. J Bacteriol* **181:** 1118–1125.

Jaumouille V, Francetic O, Sansonetti PJ, Tran Van Nhieu G. 2008. Cytoplasmic targeting of IpaC to the bacterial pole directs polar type III secretion in Shigella. *EMBO J* **27:** 447–457.

Jenal U, Malone J. 2006. Mechanisms of Cyclic-di-GMP Signaling in Bacteria. *Annu Rev Genet* **40:** 385–407.

Judd PK, Kumar RB, Das A. 2005. Spatial location and requirements for the assembly of the *Agrobacterium tumefaciens* type IV secretion apparatus. *Proc Natl Acad Sci U S A* **102:** 11498–11503.

Kaiser D, Robinson M, Kroos L. 2010. Myxobacteria, Polarity, and Multicellular Morphogenesis. *Cold Spring Harb Perspect Biol* **2:** a000380.

Kitagawa M, Ara T, Arifuzzaman M, Ioka-Nakamichi T, Inamoto E, Toyonaga H, Mori H. 2005. Complete set of ORF clones of *Escherichia coli* ASKA library (a complete set of E. coli K-12 ORF archive): unique resources for biological research. *DNA Res* **12:** 291–299.

Komeili A, Li Z, Newman DK, Jensen GJ. 2006. Magnetosomes are cell membrane invaginations organized by the actin-like protein MamK. *Science* **311:** 242–245.

Kusumoto A, Nishioka N, Kojima S, Homma M. 2009. Mutational analysis of the GTP-binding motif of FlhF which regulates the number and placement of the polar flagellum in *Vibrio alginolyticus. J Biochem* **146:** 643–650.

Kusumoto A, Shinohara A, Terashima H, Kojima S, Yakushi T, Homma M. 2008. Collaboration of FlhF and FlhG to regulate polar-flagella number and localization in *Vibrio alginolyticus. Microbiology* **154:** 1390–1399.

Lam H, Schofield WB, Jacobs-Wagner C. 2006. A landmark protein essential for establishing and perpetuating the polarity of a bacterial cell. *Cell* **124:** 1011–1023.

Laus MC, Logman TJ, Lamers GE, Van Brussel AA, Carlson RW, Kijne JW. 2006. A novel polar surface polysaccharide from *Rhizobium leguminosarum* binds host plant lectin. *Mol Microbiol* **59:** 1704–1713.

Le Quere B, Ghigo JM. 2009. BcsQ is an essential component of the *Escherichia coli* cellulose biosynthesis apparatus that localizes at the bacterial cell pole. *Mol Microbiol* **72:** 724–740.

Leclerc G, Wang SP, Ely B. 1998. A new class of *Caulobacter crescentus* flagellar genes. *J Bacteriol* **180:** 5010–5019.

Leonard TA, Butler PJ, Lowe J. 2005. Bacterial chromosome segregation: structure and DNA binding of the Soj dimer–a conserved biological switch. *EMBO J* **24:** 270–282.

Leonardy S, Freymark G, Hebener S, Ellehauge E, Sogaard-Andersen L. 2007. Coupling of protein localization and cell movements by a dynamically localized response regulator in *Myxococcus xanthus. EMBO J* **26:** 4433–4444.

Leonardy S, Miertzschke M, Bulyha I, Sperling E, Wittinghofer A, Sogaard-Andersen L. 2010. Regulation of dynamic polarity switching in bacteria by a Ras-like G-protein and its cognate GAP. *EMBO J* **29:** 2276–2289.

Levi A, Jenal U. 2006. Holdfast formation in motile swarmer cells optimizes surface attachment during *Caulobacter crescentus* development. *J Bacteriol* **188:** 5315–5318.

Lindner AB, Madden R, Demarez A, Stewart EJ, Taddei F. 2008. Asymmetric segregation of protein aggregates is associated with cellular aging and rejuvenation. *Proc Natl Acad Sci U S A* **105:** 3076–3081.

Macnab RM. 2003. How bacteria assemble flagella. *Annu Rev Microbiol* **57:** 77–100.

Maddock JR, Shapiro L. 1993. Polar location of the chemoreceptor complex in the *Escherichia coli* cell. *Science* **259:** 1717–1723.

Margolin W. 2005. FtsZ and the division of prokaryotic cells and organelles. *Nat Rev Mol Cell Biol* **6:** 862–871.

Martin PR, Watson AA, McCaul TF, Mattick JS. 1995. Characterization of a five-gene cluster required for the biogenesis of type 4 fimbriae in *Pseudomonas aeruginosa. Mol Microbiol* **16:** 497–508.

Mattick JS. 2002. Type IV pili and twitching motility. *Annu Rev Microbiol* **56:** 289–314.

Mauriello EM, Mouhamar F, Nan B, Ducret A, Dai D, Zusman DR, Mignot T. 2010. Bacterial motility complexes require the actin-like protein, MreB and the Ras homologue, MglA. *EMBO J* **29:** 315–326.

Merker RI, Smit J. 1988. Characterization of the adhesive holdfast of marine and freshwater *caulobacters. Appl Environ Microbiol* **54:** 2078–2085.

Mignot T, Merlie JP Jr, Zusman DR. 2005. Regulated pole-to-pole oscillations of a bacterial gliding motility protein. *Science* **310:** 855–857.

Mignot T, Shaevitz JW, Hartzell PL, Zusman DR. 2007. Evidence that focal adhesion complexes power bacterial gliding motility. *Science* **315:** 853–856.

Murat D, Quinlan A, Vali H, Komeili A. 2010. Comprehensive genetic dissection of the magnetosome gene island reveals the step-wise assembly of a prokaryotic organelle. *Proc Natl Acad Sci U S A* **107:** 5593–5598.

Murray TS, Kazmierczak BI. 2006. FlhF is required for swimming and swarming in *Pseudomonas aeruginosa. J Bacteriol* **188:** 6995–7004.

Newell PD, Monds RD, O'Toole GA. 2009. LapD is a bis-(3',5')-cyclic dimeric GMP-binding protein that regulates surface attachment by *Pseudomonas fluorescens* Pf0–1. *Proc Natl Acad Sci U S A* **106:** 3461–3466.

Niehus E, Gressmann H, Ye F, Schlapbach R, Dehio M, Dehio C, Stack A, Meyer TF, Suerbaum S, Josenhans C. 2004. Genome-wide analysis of transcriptional hierarchy and feedback regulation in the flagellar system of *Helicobacter pylori. Mol Microbiol* **52:** 947–961.

Nudleman E, Wall D, Kaiser D. 2006. Polar assembly of the type IV pilus secretin in *Myxococcus xanthus. Mol Microbiol* **60:** 16–29.

Obuchowski PL, Jacobs-Wagner C. 2007. PflI, a protein involved in flagellar positioning in *Caulobacter crescentus. J Bacteriol* **190:** 1718–1729.

Pandza S, Baetens M, Park CH, Au T, Keyhan M, Matin A. 2000. The G-protein FlhF has a role in polar flagellar placement and general stress response induction in *Pseudomonas putida. Mol Microbiol* **36:** 414–423.

Patryn J, Allen K, Dziewanowska K, Otto R, Hartzell PL. 2010. Localization of MglA, an essential gliding motility protein in *Myxococcus xanthus. Cytoskeleton (Hoboken)* **67:** 322–337.

Paul R, Jaeger T, Abel S, Wiederkehr I, Folcher M, Biondi EG, Laub MT, Jenal U. 2008. Allosteric regulation of histidine kinases by their cognate response regulator determines cell fate. *Cell* **133:** 452–461.

Perez BA. 2007. The *Actinobacillus actinomycetemcomitans tad* locus: Defining the Genetic Requirements for Biofilm Formation and Characterization of the TadZ Protein. in *Department of Microbiology & Immunology.* Columbia University, New York.

Ptacin JL, Lee SF, Garner EC, Toro E, Eckart M, Comolli LR, Moerner WE, Shapiro L. 2010. A spindle-like apparatus guides bacterial chromosome segregation. *Nat Cell Biol* **12:** 791–798.

Rudner DZ, Losick R. 2010. Protein subcellular localization in bacteria. *Cold Spring Harb Perspect Biol* **2:** a000307.

Russell JH, Keiler KC. 2008. Screen for localized proteins in *Caulobacter crescentus. PLoS One* **3:** e1756.

Salvetti S, Ghelardi E, Celandroni F, Ceragioli M, Giannessi F, Senesi S. 2007. FlhF, a signal recognition particle-like GTPase, is involved in the regulation of flagellar arrangement, motility behaviour and protein secretion in *Bacillus cereus. Microbiology* **153:** 2541–2552.

Schirm M, Soo EC, Aubry AJ, Austin J, Thibault P, Logan SM. 2003. Structural, genetic and functional characterization of the flagellin glycosylation process in *Helicobacter pylori. Mol Microbiol* **48:** 1579–1592.

Schofield WE, Lim HC, Jacobs-Wagner C. 2010. Cell cycle coordination and regulation of bacterial chromosome segregation dynamics by polarly localized proteins. *Embo J* doi:10.1038/emboj.2010.207.

Scott ME, Dossani ZY, Sandkvist M. 2001. Directed polar secretion of protease from single cells of *Vibrio cholerae* via the type II secretion pathway. *Proc Natl Acad Sci U S A* **98:** 13978–13983.

Shapiro L, McAdams HH, Losick R. 2009. Why and how bacteria localize proteins. *Science* **326:** 1225–1228.

Shih YL, Kawagishi I, Rothfield L. 2005. The MreB and Min cytoskeletal-like systems play independent roles in prokaryotic polar differentiation. *Mol Microbiol* **58:** 917–928.

Shih YL, Le T, Rothfield L. 2003. Division site selection in *Escherichia coli* involves dynamic redistribution of Min proteins within coiled structures that extend between the two cell poles. *Proc Natl Acad Sci U S A* **100:** 7865–7870.

Skerker JM, Shapiro L. 2000. Identification and cell cycle control of a novel pilus system in *Caulobacter crescentus. Embo J* **19:** 3223–3234.

Smith CS, Hinz A, Bodenmiller D, Larson DE, Brun YV. 2003. Identification of Genes Required for Synthesis of the Adhesive Holdfast in *Caulobacter crescentus. J Bacteriol* **185:** 1432–1442.

Spiers AJ, Kahn SG, Bohannon J, Travisano M, Rainey PB. 2002. Adaptive divergence in experimental populations of *Pseudomonas fluorescens.* I. Genetic and phenotypic bases of wrinkly spreader fitness. *Genetics* **161:** 33–46.

Thanbichler M. 2010. Synchronization of chromosome dynamics and cell division in bacteria. *Cold Spring Harb Perspect Biol* **2:** a000331.

Thanbichler M, Shapiro L. 2006. MipZ, a spatial regulator coordinating chromosome segregation with cell division in *Caulobacter. Cell* **126:** 147–162.

Thompson SR, Wadhams GH, Armitage JP. 2006. The positioning of cytoplasmic protein clusters in bacteria. *Proc Natl Acad Sci U S A* **103:** 8209–8214.

Tomich M, Planet PJ, Figurski DH. 2007. The *tad* locus: postcards from the widespread colonization island. *Nat Rev Microbiol* **5:** 363–375.

Tomlinson AD, Fuqua C. 2009. Mechanisms and regulation of polar surface attachment in *Agrobacterium tumefaciens. Curr Opin Microbiol* **12:** 708–714.

Tschowri N, Busse S, Hengge R. 2009. The BLUF-EAL protein YcgF acts as a direct anti-repressor in a blue-light response of *Escherichia coli. Genes Dev* **23:** 522–534.

Viollier PH, Sternheim N, Shapiro L. 2002a. A dynamically localized histidine kinase controls the asymmetric distribution of polar pili proteins. *Embo J* **21:** 4420–4428.

Viollier PH, Sternheim N, Shapiro L. 2002b. Identification of a localization factor for the polar positioning of bacterial structural and regulatory proteins. *Proc Natl Acad Sci U S A* **99:** 13831–13836.

Wagner JK, Galvani CD, Brun YV. 2005. *Caulobacter crescentus* requires RodA and MreB for stalk synthesis and prevention of ectopic pole formation. *J Bacteriol* **187:** 544–553.

Werner JN, Chen EY, Guberman JM, Zippilli AR, Irgon JJ, Gitai Z. 2009. Quantitative genome-scale analysis of protein localization in an asymmetric bacterium. *Proc Natl Acad Sci U S A* **106:** 7858–7863.

Zhang Y, Franco M, Ducret A, Mignot T. 2010. A Bacterial Ras-Like Small GTP-Binding Protein and Its Cognate GAP Establish a Dynamic Spatial Polarity Axis to Control Directed Motility. *PLoS Biol* **8:** e1000430.

The Bacterial Cell Envelope

Thomas J. Silhavy[1], Daniel Kahne[2], and Suzanne Walker[2]

[1]Department of Molecular Biology, Princeton University, Princeton, New Jersey 08544

[2]Department of Chemistry and Chemical Biology, Harvard University, Cambridge, Massachusetts 02138; and Department of Biological Chemistry and Molecular Pharmacology, Harvard Medical School, Boston, Massachusetts 02115

Correspondence: tsilhavy@princeton.edu

The bacteria cell envelope is a complex multilayered structure that serves to protect these organisms from their unpredictable and often hostile environment. The cell envelopes of most bacteria fall into one of two major groups. Gram-negative bacteria are surrounded by a thin peptidoglycan cell wall, which itself is surrounded by an outer membrane containing lipopolysaccharide. Gram-positive bacteria lack an outer membrane but are surrounded by layers of peptidoglycan many times thicker than is found in the Gram-negatives. Threading through these layers of peptidoglycan are long anionic polymers, called teichoic acids. The composition and organization of these envelope layers and recent insights into the mechanisms of cell envelope assembly are discussed.

It has been well known since the late 1830s that all living organisms are composed of fundamental units called cells. The cell is a finite entity with a definite boundary, the plasma membrane. That means that the essence of the living state must be contained within the biological membrane; it is a defining feature of all living things. Everything that exists outside of the biological membrane is nonliving. Chemists tend to think of membranes as self assembling but biological membranes do not self assemble; they require energy to be established and maintained. In almost all cells this structure is a phospholipid bilayer that surrounds and contains the cytoplasm. In addition to lipid components biological membranes are composed of proteins; the proteins are what make each membrane unique. Despite its obvious importance, membranes and their associated functions remained poorly understood until the 1950s. Before this, many viewed the membrane as a semipermeable bag. This view persisted for a number of years, particularly in the case of bacteria, because it could not be envisioned how such a "simple" organism could have anything but a simple membrane (Rothfield 1971).

The bacterial cell envelope, i.e., the membrane(s) and other structures that surround and protect the cytoplasm, however, is anything but a simple membrane. Unlike cells of higher organisms, the bacterium is faced with an unpredictable, dilute and often hostile environment. To survive, bacteria have evolved a sophisticated and complex cell envelope that protects them,

 doi: 10.1101/cshperspect.a000414
Cite this article as *Cold Spring Harb Perspect Biol* 2010;2:a000414

but allows selective passage of nutrients from the outside and waste products from the inside. The following discussion concerns the organization, composition, and the functions of the various layers and compartments that make up this remarkable cellular structure. It is easily appreciated that a living system cannot do what it does without the ability to establish separate compartments in which components are segregated. Specialized functions occur within different compartments because the types of molecules within the compartment can be restricted. However, membranes do not simply serve to segregate different types of molecules. They also function as surfaces on which reactions can occur. Recent advances in microscopy, which are discussed in other articles on this subject, have revealed strikingly nonrandom localization of envelope components. Here, we will highlight recent advances in our understanding of how these extracellular organelles are assembled.

More than 100 years ago Christian Gram (1884) developed a staining procedure that allowed him to classify nearly all bacteria into two large groups, and this eponymous stain is still in widespread use. One group of bacteria retain Christian's stain, Gram-positive, and the other do not, Gram-negative. The basis for the Gram stain lies in fundamental structural differences in the cell envelope of these two groups of bacteria. For our discussion of the Gram-negative bacterial cell envelope we will use *Escherichia coli*, an extensively-studied organism that has served as a model for understanding a number of fundamental biological processes. In comparing Gram-negative and Gram-positive cell envelopes we will use *Staphylococcus aureus* as a reference point but will highlight specific differences between it and *Bacillus subtilis*. *B. subtilis* is the major Gram-positive model organism and a substantial knowledge base exists for it, but the cell envelope of *S. aureus* has been studied more extensively because of interest in how surface features mediate interactions with the environment in the course of infection. Care should be taken in generalizing from examples drawn from particular microorganisms. For example, *E. coli* inhabits the mammalian gut. Accordingly, *E. coli* and other enteric bacteria must have a cell envelope that is particularly effective at excluding detergents such as bile salts. This need not be a pressing issue for other Gram-negative bacteria, and their envelopes may differ in species- and environmentally specific ways. Nonetheless, the ability to use the Gram stain to categorize bacteria suggests that the basic organizational principles we present are conserved. In addition, many bacteria express an outermost coat, the S-layer, which is composed of a single protein that totally encases the organism. S-layers and capsules, which are coats composed of polysaccharides, are beyond the scope of this review.

THE GRAM-NEGATIVE CELL ENVELOPE

After more than a decade of controversy, techniques of electron microscopy were improved to the point in which they finally revealed a clearly layered structure of the Gram-negative cell envelope (Fig. 1) (Glauert and Thornley, 1969). There are three principal layers in the envelope; the outer membrane (OM), the peptidoglycan cell wall, and the cytoplasmic or inner membrane (IM). The two concentric membrane layers delimit an aqueous cellular compartment that Peter Mitchell (1961) first termed the periplasm. During a similar time frame biochemical methods were developed to isolate and characterize the distinct set of proteins found in the periplasm (Heppel, 1967), and to characterize the composition of both the inner and outer membranes (Miura and Mizushima, 1968; Osborn et al. 1972). Studies since then have only reinforced their basic conclusions.

The Outer Membrane

Starting from the outside and proceeding inward the first layer encountered is the OM. The OM is a distinguishing feature of Gram-negative bacteria; Gram-positive bacteria lack this organelle. Like other biological membranes, the OM is a lipid bilayer, but importantly, it is not a phospholipid bilayer. The OM does contain phospholipids; they are confined to the inner leaflet of this membrane. The outer

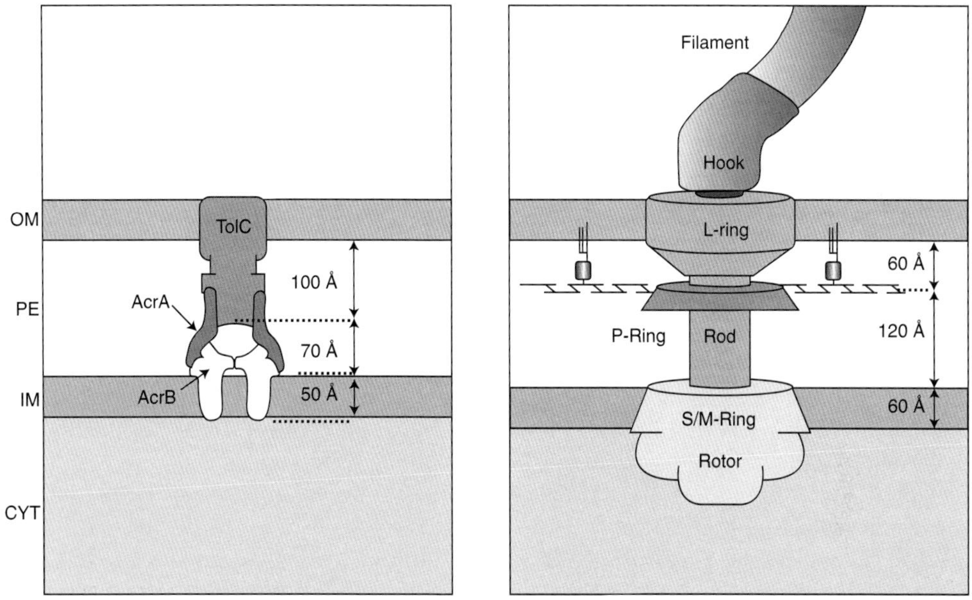

Figure 1. Transenvelope machines in the Gram-negative cell envelope. The AcrA/B proteins together with TolC form an efflux pump that expels harmful molecules such as antibiotics from the cell directly into the media (Koronakis et al. 2000; Eswaran et al. 2004; Murakami et al. 2006). The flagellar basal body hook structure connects the motor to the flagella (DePamphilis and Adler, 1971). Distances shown provide a reasonable estimate of the size of the cellular compartments shown. PE, periplasm; CYT, cytoplasm.

leaflet of the OM is composed of glycolipids, principally lipopolysaccharide (LPS) (Kamio and Nikaido 1976). LPS is an infamous molecule because it is responsible for the endotoxic shock associated with the septicemia caused by Gram-negative organisms (Raetz and Whitfield 2002). The human innate immune system is sensitized to this molecule because it is a sure indicator of infection.

With few exceptions, the proteins of the OM can be divided into two classes, lipoproteins and β-barrel proteins. Lipoproteins contain lipid moieties that are attached to an amino-terminal cysteine residue (Sankaran and Wu 1994). It is generally thought that these lipid moieties embed lipoproteins in the inner leaflet of the OM. In other words, these proteins are not thought to be transmembrane proteins. There are about 100 OM lipoproteins in *E.coli*, and the functions of most of these are not known (Miyadai et al. 2004; but see below). Nearly all of the integral, transmembrane proteins of the outer membrane assume a β-barrel conformation. These proteins are β sheets that are wrapped into cylinders, and we will refer to these outer membrane proteins as OMPs. Not surprisingly, some of these OMPs, such as the porins, OmpF, and OmpC, function to allow the passive diffusion of small molecules such as mono- and disaccharides and amino acids across the OM. These porins have 16 transmembrane β strands, they exist as trimers (Cowan et al. 1992), and they are very abundant; together they are present at approximately 250,000 copies per cell. Other OMPs, such as LamB (18 transmembrane β strands) (Schirmer et al. 1995) or PhoE (16 transmembrane β strainds) (Cowan et al. 1992), exist as trimers as well and they function in the diffusion of specific small molecules, maltose or maltodextrins and anions such as phosphate respectively, across the OM. When induced by the presence of maltose or phosphate starvation, respectively, these proteins are very abundant as well. OmpA is another abundant OMP. It is monomeric, and it is unusual in that it can exist in two different conformations (Arora et al. 2000). A minor form of the protein, with an unknown number of transmembrane strands, can function as a porin, but the major, nonporin form has only eight transmembrane strands, and the periplasmic domain of this form performs a largely structural role (see later discussion). An additional class of OMPs, which are larger β-barrels (20–24 transmembrane β strands), but are present at much lower levels, function as gated channels in the high affinity transport of large ligands such as Fe-chelates or vitamins such as vitamin B-12 (for review see Nikaido 2003).

The OM is essential for the survival of *E. coli*, but it contains only a few enzymes. For example, there is a phospholipase (PldA) (Snijder et al. 1999), a protease (OmpT) (Vandeputte-Rutten et al. 2001), and an enzyme that modifies LPS (PagP) (Hwang et al. 2002). The active site of all of these enzymes is located in the outer leaflet, or it faces the exterior of the cell (OmpT). Mutants lacking any of these enzymes exhibit no striking phenotypes. The only known function of the OM is to serve as a protective barrier, and it is not immediately obvious why this organelle is essential. But what a barrier it is. *Salmonella*, another enteric bacterium, can live at the site of bile salt production in the gall bladder (Sinnott and Teall, 1987), and it is generally true that Gram-negative bacteria are more resistant to antibiotics than are their Gram-positive cousins. Indeed, some Gram-negative bacteria, such as *Pseudomonas*, are notorious in this regard.

LPS plays a critical role in the barrier function of the OM. It is a glucosamine disaccharide with six or seven acyl chains, a polysaccharide core, and an extended polysaccharide chain that is called the O-antigen (Raetz and Whitfield 2002). Traditionally pathogenic *E. coli* are classified by the antigenic properties of their O-antigen and the major protein (flagellin, termed H) component of the flagella (see later discussion). Hence, *E. coli* O157:H7. LPS molecules bind each other avidly, especially if cations like Mg^{++} are present to neutralize the negative charge of phosphate groups present on the molecule. The acyl chains are largely saturated, and this facilitates tight packing. The nonfluid continuum formed by the LPS molecules is a very effective barrier for hydrophobic molecules. This coupled with the fact that the porins limit diffusion of hydrophilic molecules larger than about 700 Daltons, make the OM a very effective yet, selective permeability barrier (Nikaido 2003).

The Peptidoglycan Cell Wall

Bacteria do not lyse when put into distilled water because they have a rigid exoskeleton. Peptidoglycan is made up of repeating units of the disaccharide N-acetyl glucosamine-N-actyl muramic acid, which are cross-linked by pentapeptide side chains (Vollmer et al. 2008). The peptidoglycan sacculus is one very large polymer that can be isolated and viewed in a light microscope. Because of its rigidity, it determines cell shape. The enterics are rod shaped, but cell shapes can vary. For example, vibrios and caulobacters are comma shaped. Recent results suggest that the glycan chains run perpendicular to the long axis of a rod shaped cell, i.e., hoops of glycan chains around the girth of the cell (Gan et al. 2008). Agents such as enzymes or antibiotics that damage the peptidoglycan cause cell lysis owing to the turgor pressure of the cytoplasm. Lysis can be prevented in media of high osmolarity. However, without the peptidoglycan, cells lose their characteristic shape. The resulting cells are called spheroplasts. With *E. coli*, normal methods of spheroplast production produce nonviable cells, but they can continue metabolism and biosynthesis for hours. However, special methods can be used to produce L forms, which are spherical in shape and can be propagated on high osmolarity media (Joseleau-Petit et al. 2007).

The OM is basically stapled to the underlying peptidoglycan by a lipoprotein called Lpp, murein lipoprotein, or Braun's lipoprotein (Braun, 1975). The lipids attached to the amino terminus of this small protein (58 amino acids) embed it in the OM. Lpp is the most abundant protein in *E. coli*, more than 500,000 molecules per cell. The ε-amino group of the carboxy-terminal lysine residue of one third of these molecules is covalently attached to the diaminopimelate residue in the peptide cross-bridge. In addition, proteins such as OmpA bind peptidoglycan noncovalently. Nonetheless, mutants that lack Lpp shed OM vesicles (Yem and Wu 1978).

The Periplasm

The OM and IM delimit an aqueous cellular compartment called the periplasm. The periplasm is densely packed with proteins and it is more viscous than the cytoplasm (Mullineax et al. 2006). Cellular compartmentalization allows Gram-negative bacteria to sequester potentially harmful degradative enzymes such as RNAse or alkaline phosphatase. Because of this, the periplasm has been called an evolutionary precursor of the lysosomes of eukaryotic cells (De Duve and Wattiaux, 1966). Other proteins that inhabit this compartment include the periplasmic binding proteins, which function in sugar and amino acid transport and chemotaxis, and chaperone-like molecules that function in envelope biogenesis (Ehrmann, 2007; see later discussion).

The Inner Membrane

One of the hallmarks of eukaryotic cells is the presence of intracellular organelles. These organelles are defined by limiting membranes, and these organelles perform a number of essential cellular processes. The mitochondria produce energy, the smooth endoplasmic reticulum (ER) synthesize lipids, protein secretion occurs in the rough ER, and the cytoplasmic membrane contains the receptors that sense the environment and the transport systems for nutrients and waste products. Bacteria lack intracellular organelles, and consequently, all of the membrane-associated functions of all of the eukaryotic organelles are performed in the IM. Many of the membrane proteins that function in energy production, lipid biosynthesis, protein secretion, and transport are conserved in bacteria, but their cellular location is different. In bacteria, these proteins are located in the IM.

The IM is a phospholipid bilayer. In *E. coli* the principal phospholipids are phosphatidyl ethanolamine and phosphatidyl glycerol, but there are lesser amounts of phosphatidyl serine and cardiolipin. Other minor lipids include polyisoprenoid carriers (C55), which function in the translocation of activated sugar intermediates that are required for envelope biogenesis (Raetz and Dowhan 1990).

Transenvelope Machines

Certain types of surface appendages such as flagella (DePamphilis and Adler 1971; Macnab 2003), which are required for bacteria motility; Type III secretion systems (Kubori et al. 1998), which inject toxins into the cytoplasm of eukaryotic host cells during the infection process; and efflux pumps (Koronakis et al. 2000; Eswaran et al. 2004; Murakami et al. 2006; Symmons et al. 2009), which pump toxic molecules such as antibiotics from the cell clear across the cell envelope into the surrounding media and are responsible in part for much of the antibiotic resistance in pathogenic bacteria, are molecular machines that are made up of individual protein components that span the peptidoglycan and are located in all cellular compartments. The structures of some of these machines are known at sufficient resolution to provide meaningful insight into the size of the various cellular compartments in *E. coli*. As shown in Figure 1, these size predictions are in reasonable agreement. However, it should be noted that experimental measurements of the volume of the periplasm, for example, vary widely (Stock et al. 1977).

Envelope Biogenesis

All of the components of the Gram-negative cell envelope are synthesized either in the cytoplasm or at the inner surface of the IM. Accordingly, all of these components must be translocated from the cytoplasm or flipped across the IM. Periplasmic components must be released from the IM, peptidoglycan components must be released and polymerized, and OM components must be transported across the aqueous, viscous periplasm and assembled into an asymmetric lipid bilayer. All of this construction takes place outside of the cell in a potentially hostile environment that lacks an obvious energy source. It seems clear that there is no ATP out there for example. In this section we will summarize what is currently known about the assembly of the major envelope components; proteins, including lipoproteins, LPS, and phospholipids.

All proteins, of course, are synthesized in the cytoplasm. Proteins destined for the periplasm or the OM are made initially in precursor form with a signal sequence at the amino terminus. The signal sequence targets them for translocation from the cytoplasm (Driessen and Nouwen 2008). This translocation reaction is catalyzed by an essential, heterotrimeric IM protein complex called SecYEG (Van den Berg et al. 2004). The signal sequence and this heterotrimeric membrane protein complex are conserved throughout biology (Rapoport 2007). The essential ATPase SecA, together with the proton motive force, drives this translocation reaction (Zimmer et al. 2008). Periplasmic and OM proteins are generally translocated in post-translational fashion, i.e., synthesis and translocation are not coupled. Proteins must be secreted in linear fashion from the amino to the carboxy terminus like spaghetti through a hole; SecYEG cannot handle folded molecules. The cytoplasmic SecB chaperone maintains these secreted proteins in unfolded form until they can be secreted (Randall and Hardy 2002). During the secretion process the signal sequence is proteolytically removed by Signal Peptidase I (Paetzel et al. 2002). Other components of the Sec translocon, such SecD, SecF, and YajC, perform important but nonessential function(s) during translocation, perhaps facilitating release of secreted proteins into the periplasm. Once released, periplasmic proteins are home, but it seems likely that chaperones function to prevent misfolding and aggregation. For example, the periplasmic protein MalS, which contains disulfide bonds, requires the periplasmic disulfide oxidase DsbA for proper folding. In the absence of the DsbA the periplasmic protease/chaperone DegP (HtrA) can substitute (Spiess et al. 1999).

Periplasmic chaperones function to protect OMPs during their transit through the periplasm. Three such proteins have been well characterized and shown have general chaperone activity: SurA, which also functions as a peptidyl-proline isomerase (Behrens and Gross 2001; Bitto and McKay 2003), Skp (Chen and Henning 1996; Walton et al. 2009), and the aforementioned DegP (Krojer et al. 2008; Shen et al. 2009). Genetic analysis indicates that these three proteins function in parallel pathways for OMP assembly; SurA functions in one pathway; DegP/Skp function in the other. Mutants lacking either one of these pathways are viable, but cells cannot tolerate loss of both (Rizzitello et al. 2001). Mutants lacking SurA and Skp, or SurA and DegP are not viable and they show massive defects in OMP assembly. By definition then, these chaperone pathways are redundant. However, this redundancy does not reflect equal roles in OMP assembly. The major OMPs, which account for most of the protein mass of the OM, show preference for the SurA pathway (Sklar et al. 2007) as does the minor OMP LptD (Vertommen et al. 2009; see later). At present no OMP that prefers the DegP/Skp pathway has been identified. It may be that many minor OMPs show no pathway preference. The primary role of the DegP/Skp pathway may be to rescue OMPs that have fallen off the normal assembly pathway, particularly under stressful conditions. It is also possible that other periplasmic proteins have chaperone function that is important for the assembly of a subset of OMPs.

The periplasmic chaperones deliver OMPs to a recently identified assembly site in the OM termed the Bam complex (Fig. 2). This complex is composed of a large β-barrel protein, BamA (aka YaeT or Omp85), and four lipoproteins, BamBCDE (aka YfgL, NlpB, YfiO, and SmpA respectively) (Wu et al. 2005; Sklar et al. 2007). In addition to the β-barrel domain BamA has a large amino-terminal periplasmic domain composed of five POTRA (polypeptide transport associated). The structure of a large fraction of the BamA periplasmic domain has been determined (Kim et al. 2007). Each of the four visible POTRA domains has a nearly identical fold, despite the fact that the amino acid sequence identity between them is very low. In *E. coli*

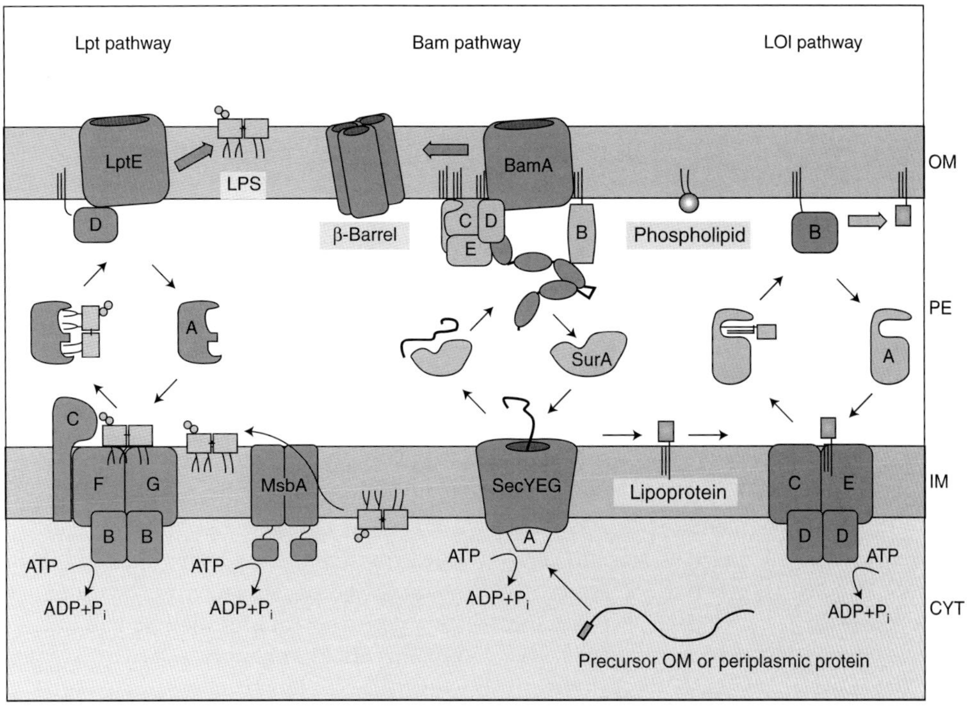

Figure 2. The cellular machineries required for OM biogenesis. The Lpt pathway, together with MsbA, transports LPS from its site of synthesis to the cell surface. β-barrel proteins and lipoproteins are made initially in the cytoplasm in precursor form with a signal sequence at the amino terminus. The signal sequence directs these precursors to the Sec machinery for translocation from the cytoplasm. Chaperones like SurA deliver beta-barrel proteins to the Bam machinery for assembly in the OM. For OM lipoproteins, after the signal sequence is removed and lipids are attached to the amino-terminal cysteine residue, the Lol machinery delivers them to the OM.

the first two POTRA domains are not essential for the life of the organism. Nevertheless, BamA is highly conserved in Gram-negative bacteria, and in these organisms there are always five POTRA domains. There are homologs of BamA in both mitochondria and chloroplasts (Moslavac et al. 2005), which are thought to be derived from Gram-negative bacteria. These homologs have one, two, or three POTRA domains, and the proteins function to assemble β-barrel proteins in the OM of these organelles. BamD is the only essential lipoprotein in the Bam complex (Malinverni et al. 2006), and it is highly conserved in Gram-negative bacteria as well. The remaining three lipoproteins are not essential, and they are conserved to varying degrees. We do not yet understand the mechanism of β-barrel folding nor do we understand the functions of the individual proteins in the Bam complex, but there is evidence suggesting that the POTRA domains of BamA may template folding by a process termed β augmentation (Kim et al. 2007).

Lipoproteins are made initially with an amino-terminal signal sequence as well, and they too are translocated by the Sec machinery. However, the signal sequence is removed by a different signal peptidase, signal peptidase II (Paetzel et al. 2002). Signal sequence processing of lipoproteins requires the formation of a thioether diglyceride at the cysteine residue, which will become the amino terminus of the mature lipoprotein. Once the signal sequence is removed, an additional fatty acyl chain is added to the cysteine amino group (Sankaran and Wu 1994). These lipid moieties tether the newly formed lipoprotein to the outer leaflet of the IM.

Some lipoproteins remain in the IM, and their biogenesis is complete after signal sequence processing and lipid addition. However, most of the lipoproteins in *E. coli* are destined for the outer membrane. The Lol system, which transports lipoproteins to the OM has been well characterized (Fig. 2) (Narita and Tokuda, 2006). There is an ABC transporter (LolCDE) in the IM that utilizes ATP hydrolysis to extract the molecule from the IM and pass it to a soluble periplasmic carrier called LolA. LolA delivers the molecule to the OM assembly site, which is the lipoprotein LolB. IM lipoproteins have a "Lol avoidance" signal so that they remain in the IM. The most common Lol avoidance signal is an aspartate residue at position two of the mature lipoprotein.

There is a second protein translocation system in the IM called Tat that translocates folded proteins (Sargent et al. 2006). *E. coli* uses the Tat system for proteins which have prosthetic groups that must be added in the cytoplasm, and this constitutes a small fraction of the secreted proteins. Other bacteria, such as thermophiles, use the Tat system extensively; presumable because it is easier to fold proteins in the cytoplasm than it is in the hostile environments they live in. In terms of components, the Tat system is remarkable simple; three components. TatB and TatC function to target proteins for translocation by TatA, but how this system recognizes that the substrate is folded, and how it accomplishes the translocation reaction are not yet understood.

Proteins destined for the IM are handled by the Sec machinery as well. However, in general, these proteins are targeted for cotranslational translocation by Signal recognition particle (SRP) and the SRP receptor FtsY (Bernstein 2000). Presumably, posttranslational translocation of these hydrophobic substrates would be inefficient and perhaps dangerous, owing to their great potential for aggregation. Prokaryotic SRP is much simpler than its eukaryotic counterpart. It contains only a single protein, Ffh (fifty four homolog) and an RNA, Ffs (four point five S RNA). Both Ffh and Ffs are GTPases, as are their more complex eukaryotic counterparts.

The transmembrane α-helices, which are characteristic of most biological membrane proteins in both prokaryotes and eukaryotes, serve as secretion signals. The first trans-membrane segment functions as a signal sequence to initiate translocation of the sequences that follow it. These transmembrane sequences tend to be longer and more hydrophobic than typical signal sequences, and this serves as the basis for SRP recognition (Hegde and Bernstein 2006). This signal sequence is not cleaved; it remains attached serving now as a typical transmembrane helix. The second transmembrane helix functions to stop the translocation reaction and this helix exits the SecYEG translocator laterally where it remains in the IM (Van den Berg et al. 2004; Driessen and Nouwen 2008). The third transmembrane helix functions again as an uncleaved signal sequence. These alternating start and stop translocation signals stitch IM proteins into the membrane in stepwise fashion.

Small IM proteins, especially those with small periplasmic domains, can be inserted into the membrane by a second IM translocase called YidC. YidC family members can be found in mitochondria and chloroplasts. Like their mitochondrial homologs, YidC plays an important role in the assembly of energy-transducing membrane proteins such as subunit c of ATPase. YidC may also play a role in the SecYEG-dependent insertion of larger IM proteins during the lateral transfer of the trans-membrane α-helices into the lipid bilayer (Xie and Dalbey 2008).

LPS, including the core polysaccharide, and the O-antigen are both synthesized on the inner leaflet of the IM. LPS is flipped to the outer leaflet of the IM by the ABC transporter MsbA. O-antigen is synthesized on a polyisoprenoid carrier, which then flips it to the outer leaflet. The O-antigen is ligated to the LPS core in the outer leaflet of the IM, a reaction catalyzed by WaaL (Raetz and Whitfield 2002). Note that the common laboratory strain *E. coli* K-12 does not make the O-antigen. Accordingly, it is termed "rough," as opposed to the wild-type "smooth" strain. In the last several years a combination of genetics, biochemistry, and bioinformatics was employed to identify seven essential proteins that are required to transport LPS from the outer leaflet of the OM to the cell surface (Fig. 2). These proteins have been termed Lpt (lipopolysaccharide transport); LptA (aka YhbN) (Sperandeo et al. 2007), LptB (aka YhbG) (Sperandeo et al. 2007), LptC (aka YrbK) (Sperandeo et al. 2008), LptD (aka Imp or OstA) (Braun and Silhavy 2002; Bos et al. 2004), LptE (RlpB) (Wu et al. 2006), LptF (aka YjgP) (Ruiz et al. 2008), and LptG (aka YjgQ) (Ruiz et al. 2008). The large β-barrel protein, LptD and the lipoprotein LptE form a complex in the OM. LptA is made with a cleavable signal sequence and resides in the periplasm. LptF and LptG are IM proteins that likely interact with the cytoplasmic protein LptB, a predicted ATPase, to form an ABC transporter that together with the bitopic IM protein LptC, extracts LPS from the IM and passes it to the periplasmic protein LptA for delivery to the OM assembly site, LptD and LptE. An alternative model proposes that all seven proteins together form a transenvelope machine that transports LPS directly from the IM to the cell surface in analogy with efflux pumps. What is clear is that if any of the seven proteins are removed, LPS accumulates in the outer leaflet of the IM (Sperandeo et al. 2008, Ruiz et al. 2008).

Like LPS, phospholipids are synthesized in the inner leaflet of the IM. MsbA can flip these molecules to the outer leaflet of the IM (Doerrler et al. 2004), but it is likely that other mechanisms to flip these molecules also exist. How phospholipids reach the OM is not known. What is known is that phospholipids added into the OM reach the IM very quickly (Jones and Osborn 1977). This is true even for lipids like cholesterol which are not naturally found in bacteria. This could suggest sites of IM-OM fusion, or hemi-fusion, that allow intermembrane phospholipid trafficking by diffusion, a hypothesis made by Manfred Bayer (Bayer 1968) long ago that has remained highly controversial.

THE GRAM-POSITIVE CELL ENVELOPE

The Gram-positive cell envelope differs in several key ways from its Gram-negative counterpart (Fig. 3). First and foremost, the outer membrane is absent. The outer membrane plays a major role in protecting Gram-negative organisms from the environment by excluding toxic molecules and providing an additional stabilizing layer around the cell. Because the outer membrane indirectly helps stabilize the inner membrane, the peptidoglycan mesh surrounding Gram-negative cells is relatively thin. Gram-positive bacteria often live in harsh environments just as *E. coli* does—in fact, some live in the gut along with *E. coli*—but they lack a

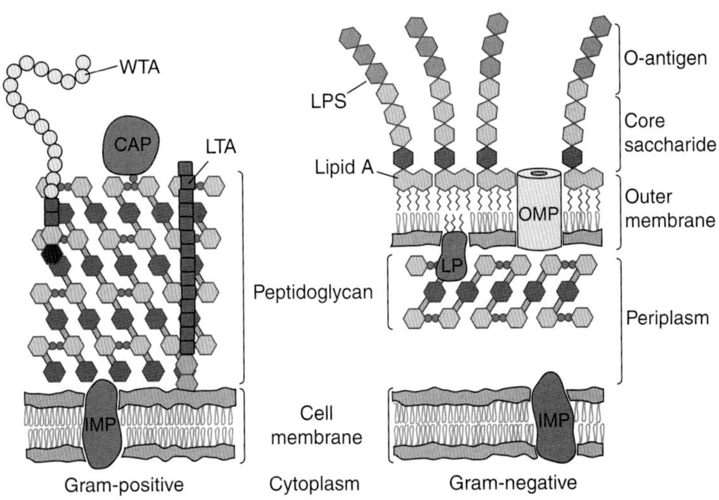

Figure 3. Depiction of Gram-positive and Gram-negative cell envelopes: CAP = covalently attached protein; IMP, integral membrane protein; LP, lipoprotein; LPS, lipopolysaccharide; LTA, lipoteichoic acid; OMP, outer membrane protein; WTA, wall teichoic acid.

protective outer membrane. To withstand the turgor pressure exerted on the plasma membrane, Gram-positive microorganisms are surrounded by layers of peptidoglycan many times thicker than is found in *E. coli*. Threading through these layers of peptidoglycan are long anionic polymers, called teichoic acids, which are composed largely of glycerol phosphate, glucosyl phosphate, or ribitol phosphate repeats. One class of these polymers, the wall teichoic acids, are covalently attached to peptidoglycan; another class, the lipoteichoic acids, are anchored to the head groups of membrane lipids (Neuhaus 2003). Collectively, these polymers can account for over 60% of the mass of the Gram-positive cell wall, making them major contributors to envelope structure and function. In addition to the TAs, the surfaces of Gram-positive microorganisms are decorated with a variety of proteins, some of which are analogous to proteins found in the periplasm of Gram-negative organisms (Dramsi et al. 2008). Because there is no outer membrane in Gram-positive organisms to contain extracellular proteins, all these proteins feature elements that retain them in or near the membrane. Some contain membrane-spanning helices and some are attached to lipid anchors inserted in the membrane. Others are covalently attached to or associated tightly with peptidoglycan (Scott and Barnett 2006). Still others bind to teichoic acids. Studies on *S. aureus* have shown that the composition of surface-expressed proteins can change dramatically depending on environmental cues or growth conditions, reflecting the important role of the cell envelope in adapting to the local environment (Pollack and Neuhaus 1994). The major structural elements of Gram-positive cell walls, excluding capsules, will be described below.

Gram-positive Peptidoglycan

The chemical structure of peptidoglycan in Gram-positive organisms is similar to that in Gram-negatives in that it is composed of a disaccharide-peptide repeat coupled through glycosidic bonds to form linear glycan strands, which are crosslinked into a meshlike framework through the peptide stems attached to the disaccharide repeat. The major difference between Gram-positive and Gram-negative peptidoglycan involves the thickness of the layers surrounding the plasma membrane. Whereas Gram-negative peptidoglycan is only a few nanometers thick, representing one to a few layers, Gram-positive peptidoglycan is 30–100 nm thick and contains many layers.

There are many differences among Gram-positive organisms with respect to the details of peptidoglycan structure, but perhaps the most notable difference relates to the peptide crosslinks between glycan strands (Vollmer 2008; Vollmer et al. 2008). *S. aureus* contains crosslinks in which the peptides are connected through a pentaglycine branch extending from the third amino acid of one of the stem peptides. This pentaglycine branch is assembled by a set of nonribosomal peptidyl transferases known as FemA, B, and X (Ton-That et al. 1998; Rohrer and Berger-Bachi 2003). Staphylococci can tolerate, albeit with difficulty, the loss of FemA or B, but not of FemX, which attaches the first glycine unit to the stem peptide (Hegde and Shrader 2001; Hubscher et al. 2007). Many Gram-positive organisms contain branched stem peptides, but *B. subtilis* does not; the stem peptides and crosslinks in this organism are identical in structure to those found in *E. coli*.

Branched stem peptides in *S. aureus* and other Gram-positive organisms play a variety of roles. Chief among these roles, they serve as attachment sites for covalently-associated proteins (discussed in more detail later). They have also been implicated in resistance to beta lactam antibiotics (Chambers 2003). Beta lactams inactivate transpeptidases that catalyze the peptide crosslinking step of peptidoglycan synthesis by reacting with the active site nucleophile of transpeptidases. Transpeptidases that couple branched stem peptides are mechanistically similar to those that couple unbranched stem peptides; however, their substrate specificity is sufficiently different that they only recognize unbranched stem peptides, and some of them are resistant to beta lactams (Rohrer and Berger-Bachi 2003; Pratt 2008; Sauvage et al. 2008). For example, methicillin-resistant *S. aureus* strains express a transpeptidase, PBP2A, that couples *only* pentaglycine-branched substrates. Many other Gram-positive organisms are also thought to harbor low affinity PBPs that preferentially recognize and couple branched stem peptides. It is thus speculated that the evolution of branched peptides in the peptidoglycan biosynthetic pathway may be an adaptation that enables escape from beta lactams. As described in the following section, however, branched stem peptides also play other important roles.

Surface Proteins

S. aureus colonizes human skin and mucosal surfaces. Breaches in the epithelium occasionally result in invasive *S. aureus* infections that are extremely serious. The ability to adhere to host tissue is a crucial first step in effective colonization by *S. aureus*, and a variety of surface factors are involved in this process. These factors include teichoic acids, which are discussed in the following section, as well as surface proteins that recognize components of host extracellular matrix such as fibronectin, fibrinogen, and elastin (Clarke and Foster 2006).

Some of these surface proteins, called adhesins, are attached via noncovalent ionic interactions to peptidoglycan or teichoic acids, but many are attached covalently to stem peptides within the peptidoglycan layers (Dramsi et al. 2008; Sjoquist et al. 1972; Fischetti et al. 1990). Proteins destined for covalent surface display contain an amino-terminal signal sequence that enables secretion through the cytoplasmic membrane and a carboxy-terminal pentapeptide cell wall sorting motif, which is commonly LPXTG (DeDent 2008). Enzymes called sortases catalyze a transpeptidation reaction between these sorting motifs and the glycine branch of the stem peptide of peptidoglycan precursors. The transpeptidation reaction is thought to occur in two steps: in the first, a nucleophile in the sortase active site attacks the amide bond between the threonine and glycine of the sorting motif to produce a covalent intermediate in which the amino-terminal portion of the protein substrate is attached to the enzyme; in the second, the glycine branch of the stem peptide enters the active site and the nucleophilic amino terminus of the glycine branch attacks the acyl- enzyme intermediate, regenerating the enzyme and forming a new TG amide bond that anchors the protein to the peptidoglycan precursor. The protein-modified peptidoglycan precursor is then incorporated into peptidoglycan (Dramsi et al. 2008; Marraffini et al. 2006).

In *S. aureus*, more than twenty protein substrates for the major sortase, sortase A, have been identified. In addition to adhesins, these protein substrates include proteins involved in immune system evasion, internalization, and phage binding. A minor sortase, sortase B, is responsible for surface display of proteins involved in iron acquisition, which is necessary for pathogenesis because iron is required for the function of many bacterial enzymes. In other Gram-positive organisms, sorting enzymes similar to SrtA and SrtB covalently couple proteins that comprise pili.

In addition to covalent attachment, Gram-positive organisms have other ways of retaining cell surface proteins. Many proteins involved in peptidoglycan biosynthesis are anchored to the cytoplasmic membrane by membrane spanning helices. Some proteins required for cell-wall degradation are associated noncovalently with peptidoglycan; others appear to be scaffolded and/or activated by teichoic acids or other types of cell surface polymers.

Teichoic Acids

Teichoic acids are anionic cell surface polymers found in a wide range of Gram-positive organisms, including *S. aureus* and *B. subtilis*. There are two major types of teichoic acids: wall teichoic acids (WTAs), which are coupled to peptidoglycan, and lipoteichoic acids (LTAs), which are anchored to the cell membrane. Wall teichoic acids are attached via a phosphodiester linkage to the C6 hydroxyl of occasional MurNAc residues in peptidoglycan. Although the structural variations are considerable, the most common WTAs are composed of a disaccharide linkage unit to which is appended a polyribitol phosphate (polyRboP) or polyglycerol phosphate (polyGroP) chain containing as many as 60 repeats. WTAs extend perpendicularly through the peptidoglycan mesh into what has been characterized as a "fluffy" layer beyond. *S. aureus* produces polyRboP WTAs; *B. subtilis* produces either polyRboP and polyGroP depending on the strain. The hydroxyls on the RboP or GroP repeats are tailored with other groups, typically D-alanyl esters or glycosyl moieties, and the nature and extent of the tailoring modifications significantly affect the properties and functions of WTAs (see later) (Neuhaus, 2003).

LTAs are similar to WTAs in that they are composed of polyGroP polymers that are often functionalized with D-alanine or a sugar moiety; however, they also differ in a number of ways. For example, they contain glycerol-phosphate repeats of opposite chirality to those found in WTAs. Furthermore, rather than being attached to peptidoglycan, they are anchored to membrane-embedded glycolipids and typically contain fewer GroP repeats. Thus, they extend from the cell surface into the peptidoglycan layers rather than through and beyond. Together, the LTAs and WTAs comprise what has been concisely described as a "continuum of anionic charge" that originates at the Gram-positive cell surface but extends well beyond the peptidoglycan barrier. The importance of this continuum of negative charge is underscored by the fact that Gram-positive organisms lacking WTAs (either because they do not contain the gene clusters or because they are grown under phosphate-limiting conditions) produce other types of polyanionic polymers in which the negative charges are supplied by carboxylate or sulfate groups. Furthermore, although neither LTAs nor WTAs are essential, deleting the pathways for the biosynthesis of either of these polymers produces organisms that have cell division and morphological defects as well as other, less serious growth defects. Moreover, it is not possible to delete both pathways because the genes involved are synthetic lethals (Oku et al. 2009; Morath et al. 2005).

Teichoic acids account for a significant fraction of the cell wall mass in producing organisms and their functions are many, varied, and species-dependent. Because they are anionic they bind cations and thus play a role in cation homeostasis (Marquis et al. 1976). Networks of metal cations between WTAs also influence the rigidity and porosity of the cell wall. The negative charge density on WTAs can be modulated by tailoring modifications that introduce positive charges along the polymer backbone, and these modifications can have a profound effect on the interactions of bacteria with other cells or molecules. For example, in *S. aureus*, a D-alanine transferase couples D-alanine moieties to free hydroxyls on the polyribitol phosphate backbone. *S. aureus* strains lacking D-alanine esters are more susceptible to antimicrobial cationic peptides and to lytic enzymes produced by host neutrophils (Collins et al. 2002; Peschel et al.1999; Peschel et al. 2000). They also display reduced autolysin activity, suggesting a role for functionalized WTAs in scaffolding or activating hydrolytic enzymes involved in cell wall synthesis and degradation.

THE CELL ENVELOPE OF *CORYNEBACTERINEAE*

The *Corynebacterineae* are a group of bacteria that includes the very important pathogens *Mycobacterium tuberculosis* and *Mycobacterium leprae*. These bacteria are generally classified as high G+C Gram positives, however their cell envelope has characteristics of both Gram-positive and Gram-negative bacteria. Indeed, a genome-based phylogeny places them in between Gram positives and Gram negatives (Fu and Fu-Liu, 2002).

The cell envelope of the *Corynebacterineae* is very complex and this complexity contributes substantially to their virulence. The peptidoglycan layer that surrounds a standard IM contains covalently attached arabinogalactan and this is covalently attached to mycolic acids (Minnikin, 1982). These mycolic acids have very long alkyl side chains (up to C_{90}) that give the bacteria a waxy appearance and account for their resistance to acid decolorization during staining procedures (acid-fast).

Unlike other Gram-positive bacteria, *Corynebacterineae* have an OM. This OM appears to be symmetrical unlike the Gram-negative OM.

Mycolic acids appear essential for this OM, but how they are organized remains unclear (Hoffmann et al. 2008; Zuber et al. 2008). Mycobacteria have porin proteins in their OM, but the structure of the main porin, MspA, from *Mycobacterium smegmatis* is quite different from the typical porin proteins of Gram-negative bacteria (Faller et al. 2004). Indeed, mycobacteria have no obvious homologs of the Bam complex members. This might suggest a novel mechanism for the assembly of proteins in the OM of *Corynebacterineae*.

CONCLUSION

The cell envelopes of bacteria are complex, dynamic structures that play a variety of protective and adaptive roles. The major conserved component of all bacterial cell envelopes is peptidoglycan, which is essential for stabilizing cell membranes against high internal osmotic pressures. But peptidoglycan alone is not enough to enable bacteria to survive in their different environments. In addition to peptidoglycan, the outer membrane of Gram-negative bacteria the, dense array of negatively charged polymers embedded in Gram-positive peptidoglycan and the complex outer layers of the *Corynebacterineae* play important roles in cell envelope integrity. One of the major challenges in the next decade will be to define the mechanisms by which these complex structures are assembled and regulated in response to changing environmental conditions.

ACKNOWLEDGMENTS

We thank Juliana Malinverni for help with figures and Susan DiRenzo for help preparing the manuscript. The authors were each supported by grants from the National Institute of General Medical Sciences.

REFERENCES

Arora A, Rinehart D, Szabo G, Tamm LK. 2000. Refolded outer membrane protein A of *Escherichia coli* formsion channels with two conductance states in planar lipid bilayers. *J Biol Chem* **275:** 1594–1600.

Bayer ME. 1968. Areas of adhesion between wall and membrane of *Escherichia coli*. *J Gen Microbiol* **53:** 395–404.

Behrens S, Maier R, de Cock H, Schmid FX, Gross CA. 2001. The SurA periplasmic PPIase lacking its parvulin domains functions in vivo and has chaperone activity. *EMBO J* **20:** 285–294.

Bernstein HD. 2000. The biogenesis and assembly of bacterial membrane proteins. *Curr Opin Microbiol* **3:** 203–209.

Bitto E, McKay DB. 2003. The periplasmic molecular chaperone protein SurA binds a peptide motif that is characteristic of integral outer membrane proteins. *J Biol Chem* **278:** 49316–49322.

Bos MP, Tefsen B, Geurtsen J, Tommassen J. 2004. Identification of an outer membrane protein required for the transport of lipopolysaccharide to the bacterial cell surface. *Proc Natl Acad Sci* **101:** 9417–9422.

Braun V. 1975. Covalent lipoprotein from the outer membrane of *Escherichia coli*. *Biochim Biophys Acta* **415:** 335–377.

Braun M, Silhavy TJ. 2002. Imp/OstA is required for cell envelope biogenesis in *Escherichia coli*. *Mol Microbiol* **45:** 1289–1302.

Chambers HF. 2003. Solving staphylococcal resistance to beta-lactams. *Trends in Microbiol* **11:** 145–148.

Chen R, Henning U. 1996. A periplasmic protein (Skp) of *Escherichia coli* selectively binds a class of outer membrane proteins. *Mol Microbiol* **19:** 1287–1294.

Clarke SR, Foster SJ. 2006. In *Advances in Microbial Physiology, Vol* ***51***, pp. 187–225.

Collins LV, Kristian S, Weidenmaier C, Faigle M, Van Kessel KP, Van Strijp JA, Götz F, Neumeister B, Peschel A. 2002. *Staphylococcus aureus* strains lacking D-alanine modifications of teichoic acids are highly susceptible to human neutrophil killing and are virulence attenuated in mice. *J Infect Dis* **186:** 214–219.

Cowan SW, Schirmer T, Rummel G, Steiert M, Ghosh R, Pauptit RA, Jansonius JN, Rosenbusch JP. 1992. Crystal structures explain functional properties of two *E. coli*porins. *Nature* **358:** 727–733.

DeDent A, Bae T, Missiakas DM, Schneewind O. 2008. Signal peptides direct surface proteins to two distinct envelope locations of Staphylococcus aureus. *EUBO J* **27:** 2656–2668.

De Duve C, Wattiaux R. 1966. Functions of lysosomes. *Annu Rev Physiol* **28:** 435–492.

DePamphilis ML, Adler J. 1971. Fine structure and isolation of the hook-basal body complex of flagella from *Escherichia coli* and *Bacillus subtilis*. *J Bacteriol* **105:** 384–395.

Doerrler WT, Gibbons HS, Raetz CR. 2004. MsbA-dependent translocation of lipids across the inner membrane of *Escherichia coli*. *J Biol Chem* **279:** 45102–45109.

Dramsi S, Magnet S, Davison S, Arthur M. 2008. Covalent attachment of proteins to peptidoglycan. *FEMS Microbiol Rev* **32:** 307–320.

Driessen AJ, Nouwen N. 2008. Protein translocation across the bacterial cytoplasmic membrane. *Annu Rev Biochem* **77:** 643–667.

Erhmann M. 2007. *The periplasm*. ASM Press, Washington, D.C.

Eswaran J, Koronakis E, Higgins MK, Hughes C, Koronakis V. 2004. Three's company: Component structures bring a closer view of tripartite drug efflux pumps. *Current Opinion in Structural Biol* **14:** 741–747.

Faller M, Niederweis M, Schulz GE. 2004. The structure of a mycobacterial outer-membrane channel. *Science* **303:** 1189–1192.

Fischetti VA, Pancholi V, Schneewind O. 1990. Conservation of a hexapeptide sequence in the anchor region of surface-proteins from Gram-postive cocci. *Mol Micro* **4:** 1603–1605.

Fu LM, Fu-Liu CS. 2002. Is mycobacterium tuberculosis a closer relative to Gram-positive or Gram-negative bacterial pathogens? *Tuberculosis* **82:** 85–90.

Gan L, Chen S, Jensen GJ. 2008. Molecular organization of Gram-negative peptidoglycan. *Proc Natl Acad Sci* **105:** 18953–18957.

Glauert AM, Thornley MJ. 1969. The topography of the bacterial cell wall. *Annu Rev Microbiol* **23:** 159–198.

Gram HCJ. 1884. Uber die isolirte Farbung der Schizomyceten in Schnitt- and Trockenpraparaten. *Fortschritte der Medizin* **2:** 185–189.

Hegde RS, Bernstein HD. 2006. The surprising complexity of signal sequences. *Trends Biochem Sci* **31:** 563–571.

Hegde SS, Shrader TE. 2001. FemABX family members are novel nonribosomal peptidyltransferases and important pathogen-specific drug targets. *J Boil Chem* **276:** 6998–7003.

Heppel LA. 1967. Selective release of enzymes from bacteria. *Science* **156:** 1451–1455.

Hoffmann C, Leis A, Niederweis M, Plitzko JM, Engelhardt H. 2008. Disclosure of the mycobacterial outer membrane: Cryo-electron tomography and vitreous sections reveal the lipid bilayer structure. *Proc Natl Acad Sci* **105:** 3963–3967.

Hubscher J, Jansen A, Kotte O, Schafer J, Majcherczyk PA, Harris LG, Bierbaum G, Heinemann M, Berger-Bachi B. 2007. Living with an imperfect cell wall: Compensation of *femAB* inactivation in *Staphylococcus aureus*. *BMC Genomics* **8:** 307–320.

Hwang PM, Choy WY, Lo EI, Chen L, Forman-Kay JD, Raetz CR, Prive GG, Bishop RE, Kay LE. 2002. Solution structure and dynamics of the outer membrane enzyme PagP by NMR. *Proc Natl Acad Sci* **99:** 13560–13565.

Jones NC, Osborn MJ. 1977. Translocation of phospholipids between the outer and inner membranes of *Salmonella typhimurium*. *J Biol Chem* **252:** 7405–7412.

Joseleau-Petit D, Liebart JC, Ayala JA, D'Ari R. 2007. Unstable Escherichia coli L forms revisited: Growth requires peptidoglycan synthesis. *J Bacteriol* **189:** 6512–6520.

Kamio Y, Nikaido H. 1976. Outer membrane of *Salmonella typhimurium*: Accessibility of phospholipid head groups to phospholipase c and cyanogen bromide activated dextran in the external medium. *Biochemistry* **15:** 2561–2570.

Kaper JB, Nataro JP, Mobley HL. 2004. Pathogenic Escherichia coli. *Nat Rev Microbiol* **2:** 123–140.

Kim S, Malinverni JC, Sliz P, Silhavy TJ, Harrison SC, Kahne D. 2007. Structure and function of an essential component of the outer membrane protein assembly machine. *Science* **317:** 961–964.

Koronakis V, Sharff A, Koronakis E, Luisi B, Hughes C. 2000. Crystal structure of the bacterial membrane protein TolC central to multidrug efflux and protein export. *Nature* **405:** 914–919.

Krojer T, Sawa J, Schafer E, Saibil HR, Ehrmann M, Clausen T. 2008. Structural basis for the regulated protease and chaperone function of DegP. *Nature* **453:** 885–890.

Kubori T, Matsushima Y, Nakamura D, Uralil J, Lara-Tejero M, Sukhan A, Galan JE, Aizawa SI. 1998. Supramolecular structure of the *Salmonella typhimurium* type III protein secretion system. *Science* **280:** 602–605.

Macnab RM. 2003. How bacteria assemble flagella. *Annu Rev Microbiol* **57:** 77–100.

Malinverni JC, Werner J, Kim S, Sklar JG, Kahne D, Misra R, Silhavy TJ. 2006. YfiO stabilizes the YaeT complex and is essential for outer membrane protein assembly in Escherichia coil. *Mol Microbiol* **61:** 151–164.

Marquis RE, Mayzel K, Carstensen EL. 1976. Cation exchange in cell walls of Gram-positive bacteria. *Can J Microbiol* **22:** 975–982.

Marraffini LA, DeDent AC, Schneewind O. 2006. Sortases and the art of anchoring proteins to the envelopes of gram-positive bacteria. *Microbiol Molec Biol Rev* **70:** 192–221.

Mitchell P. 1961. Approaches to the analysis of specific membrane transport. In *Biological structure and function* (ed. T.W. Goodwin, O. Lindberg), pp. 581–603. Academic Press, New York.

Minnikin DE. 1982. Lipids: Complex lipids, their chemistry, biosynthesis and roles, p. 95–184. *In* Ratledge C., Stanford J. (ed.). *The biology of the mycobacteria*, vol 1. Physiology, identification and classification. Academic Press, Inc., New York, NY.

Miura T, Mizushima S. 1968. Separation by density gradient centrifugation of two types of membranes from spheroplast membrane of *Escherichia coli* K12. *Biochim Biophys Acta* **150:** 159–161.

Miyadai H, Tanaka-Masuda K, Matsuyama S, Tokuda H. 2004. Effects of lipoprotein overproduction on the induction of DegP (HtrA) involved in quality control in the *Escherichia coli* periplasm. *J Biol Chem* **279:** 39807–39813.

Morath S, Von Aulock S, Hartung T. 2005. Structure/function relationships of lipoteichoic acids. *J Endotoxin Res* **11:** 348–356.

Moslavac S, Mirus O, Bredemeier R, Soll J, von Haeseler A, Schleiff E. 2005. Conserved pore-forming regions in polypeptide-transporting proteins. *FEBS J* **272:** 1367–1378.

Mullineaux CW, Nenninger A, Ray N, Robinson C. 2006. Diffusion of green fluorescent protein in three cell environments in *Escherichia coli*. *J Bacteriol* **188:** 3442–3448.

Murakami S, Nakashima R, Yamashita E, Matsumoto T, Yamaguchi A. 2006. Crystal structures of a multidrug transporter reveal a functionally rotating mechanism. *Nature* **443:** 173–179.

Narita S, Tokuda H. 2006. An ABC transporter mediating the membrane detachment of bacterial lipoproteins depending on their sorting signals. *FEBS Lett* **580:** 1164–1170.

Neuhaus F. 2003. A continuum of anionic charge: Structures and functions of D-alanyl-teichoic acids in Gram-positive bacteria. *Micro Molec Biol Rev* **67:** 686–723.

Nikaido H. 2003. Molecular basis of bacterial outer membrane permeability revisited. *Microbiol Mol Biol Rev* **67:** 593–656.

Oku Y, Kurokawa K, Matsuo M, Yamada S, Lee BL, Sekimizu K. 2009. Pleiotropic roles of polyglycerolphosphate synthase of lipoteichoic acid in growth of *Staphylococcus aureus* cells. *J Bacteriol* **191:** 141–151.

Osborn MJ, Gander JE, Parisi E, Carson J. 1972. Mechinism of assembly of the outer membrane of *Salmonella typhimurium*. Isolation and characterization of cytoplasmic and outer membrane. *J Biol Chem* **247:** 3962–3972.

Paetzel M, Karla A, Strynadka NC, Dalbey RE. 2002. Signal peptidases. *Chem Rev* **102:** 4549–4580.

Pautsch A, Schulz GE. 2000. High-resolution structure of the OmpA membrane domain. *J Mol Biol* **298:** 273–282.

Peschel A, Otto M, Jack RW, Kalbacher H, Jung G, Götz F. 1999. Inactivation of the *dlt* operon in *Staphylococcus aureus* confers sensitivity to defensins, protegrins, and other antimicrobial peptides. *J Biol Chem* **274:** 8405–8410.

Peschel A, Vuong C, Otto M, Götz F. 2000. The D-alanine residues of *Staphylococcus aureus* teichoic acids alter the susceptibility to vancomycin and the activity of autolytic enzymes. *Antimicrobial Agents and Chemotherapy* **44:** 2845–2847.

Pollack JH, Neuhaus FC. 1994. Changes in wall teichoic acid during the rod- sphere transition of *Bacillus subtilis* 168. *J Bacteriol* **176:** 7252–7259.

Pratt RF. 2008. Substrate specificity of bacterial DD-peptidases (penicillin-binding proteins). *Cellular and Molecular Life Sciences* **65:** 2138–2155.

Raetz CR, Dowhan W. 1990. Biosynthesis and function of phospholipids in *Escherichia coli*. *J Biol Chem* **265:** 1235–1238.

Raetz CR, Whitfield C. 2002. Lipopolysaccharide endotoxins. *Annu Rev Biochem* **71:** 635–700.

Randall LL, Hardy SJ. 2002. SecB, one small chaperone in the complex milieu of the cell. *Cell Mol Life Sci* **59:** 1617–1623.

Rapoport TA. 2007. Protein translocation across the eukaryotic endoplasmic reticulum and bacterial plasma membranes. *Nature* **450:** 663–669.

Rizzitello AE, Harper JR, Silhavy TJ. 2001. Genetic evidence for parallel pathways of chaperone activity in the periplasm of *Escherichia coli*. *J Bacteriol* **183:** 6794–6800.

Rohrer S, Berger-Bachi B. 2003. FemABX peptidyl transferases: a link between branched-chain cell wall peptide formation and beta-lactam resistance in Gram- positive cocci. *Antimicrobial Agents and Chemotherapy* **47:** 837–846.

Rothfield LI. 1971. Biological membranes: An overview at the molecular level. In *Structure and function of biological membranes* (ed. L.I. Rothfield), pp. 3–9. Academic Press, New York.

Ruiz N, Gronenberg LS, Kahne D, Silhavy TJ. 2008. Identification of two inner-membrane proteins required for the transport of lipopolysaccharide to the outer membrane of *Escherichia coli*. *Proc Natl Acad Sci* **105:** 5537–5542.

Sankaran K, Wu HC. 1994. Lipid modification of bacterial prolipoprotein. Transfer of diacylglyceryl moiety from phosphatidylglycerol. *J Biol Chem* **269:** 19701–19706.

Sargent F, Berks BC, Palmer T. 2006. Pathfinders and trailblazers: A prokaryotic targeting system for transport of folded proteins. *FEMS Microbiol Lett* **254:** 198–207.

Sauvage E, Kerff F, Terrak M, Ayala JA, Charlier P. 2008. The penicillin-binding proteins: structure and role in peptidoglycan biosynthesis. *FEMS Microbiol Rev* **32:** 234–258.

Schirmer T, Keller TA, Wang YF, Rosenbusch JP. 1995. Structural basis for sugar translocation through maltoporin channels at 3.1 A resolution. *Science* **267:** 512–514.

Scott JR, Barnett TC. 2006. Surface proteins of Gram-positive bacteria and how they get there. *Ann Rev Microbiol* **60:** 397–423.

Shen QT, Bai XC, Chang LF, Wu Y, Wang HW, Sui SF. 2009. Bowl-shaped oligomeric structures on membranes as DegP's new functional forms in protein quality control. *Proc Natl Acad Sci* **106:** 4858–4863.

Sinnott CR, Teall AJ. 1987. Persistent gallbladder carriage of *Salmonella typhi*. *Lancet* **1:** 976.

Sjoquist J, Hjelm H, Johansso IB, Movitz J. 1972. Localization of Protein-A in bacteria. *Eur J Biochem* **30:** 190–194.

Sklar JG, Wu T, Gronenberg LS, Malinverni JC, Kahne D, Silhavy TJ. 2007. Lipoprotein SmpA is a component of the YaeT complex that assembles outer membrane proteins in *Escherichia coli*. *Proc Natl Acad Sci* **104:** 6400–6405.

Sklar JG, Wu T, Kahne D, Silhavy TJ. 2007. Defining the roles of the periplasmic chaperones SurA, Skp, and DegP in *Escherichia coli*. *Genes Dev* **21:** 2473–2484.

Snijder HJ, Ubarretxena-Belandia I, Blaauw M, Kalk KH, Verheij HM, Egmond MR, Dekker N, Dijkstra BW. 1999. Structural evidence for dimerization-regulated activation of an integral membrane phospholipase. *Nature* **401:** 717–721.

Sperandeo P, Cescutti R, Villa R, Di Benedetto C, Candia D, Deho G, Polissi A. 2007. Characterization of lptA and lptB, two essential genes implicated in lipopolysaccharide transport to the outer membrane of *Escherichia coli*. *J Bacteriol* **189:** 244–253.

Sperandeo P, Lau FK, Carpentieri A, De Castro C, Molinaro A, Deho G, Silhavy TJ, Polissi A. 2008. Functional analysis of the protein machinery required for transport of lipopolysaccharide to the outer membrane of *Escherichia coli*. *J Bacteriol* **190:** 4460–4469.

Spiess C, Beil A, Ehrmann M. 1999. A temperature-dependent switch from chaperone to protease in a widely conserved heat shock protein. *Cell* **97:** 339–347.

Stock JB, Rauch B, Roseman S. 1977. Periplasmic space in Salmonella typhimurium and *Escherichia coli*. *J Biol Chem* **252:** 7850–7861.

Symmons MF, Bokma E, Koronakis E, Hughes C, Koronakis V. 2009. The assembled structure of a complete tripartite bacterial multidrug efflux pump. *Proc Natl Acad Sci* **106:** 7173–7178.

Ton-That H, Labischinski H, Berger-Bachi B, Scbneewind O. 1998. Anchor structure of staphylococcal surface proteins III. Role of the *femA*, *femB*, and *femX* factors in anchoring surface proteins to the bacterial cell wall. *J Biol Chem* **273:** 29143–29149.

van den Berg B, Clemons WMJr, Collinson I, Modis Y, Hartmann E, Harrison SC, Rapoport TA. 2004. X-ray structure of a protein-conducting channel. *Nature* **427:** 36–44.

Vandeputte-Rutten L, Kramer RA, Kroon J, Dekker N, Egmond MR, Gros P. 2001. Crystal structure of the outer membrane protease OmpT from *Escherichia coli* suggests a novel catalytic site. *EMBO J* **20:** 5033–5039.

Vertommen D, Ruiz N, Leverrier P, Silhavy TJ, Collet JF. 2009. Characterization of the role of the *Escherichia coli* periplasmic chaperone SurA using differential proteomics. *Proteomics* **9:** 2432–2443.

Vollmer W. 2008. Structural variation in the glycan strands of bacterial peptidoglycan. *FEMS Microbial Rev* **32:** 287–306.

Vollmer W, Blanot D, de Pedro MA. 2008. Peptidoglycan structure and architecture. *FEMS Microbiol Rev* **32:** 149–167.

Walton TA, Sandoval CM, Fowler CA, Pardi A, Sousa MC. 2009. The cavity-chaperone Skp protects its substrate from aggregation but allows independent folding of substrate domains. *Proc Natl Acad Sci* **106:** 1772–1777.

Wu T, Malinverni J, Ruiz N, Kim S, Silhavy TJ, Kahne D. 2005. Identification of a multicomponent complex required for outer membrane biogenesis in *Escherichia coli*. *Cell* **121:** 235–245.

Wu T, McCandlish AC, Gronenberg LS, Chng SS, Silhavy TJ, Kahne D. 2006. Identification of a protein complex that assembles lipopolysaccharide in the outer membrane of *Escherichia coli*. *Proc Natl Acad Sci* **103:** 11754–11759.

Xie K, Dalbey RE. 2008. Inserting proteins into the bacterial cytoplasmic membrane using the Sec and YidC translocases. *Nat Rev Microbiol* **6:** 234–244.

Yem DW, Wu HC. 1978. Physiological characterization of an Escherichia coli mutant altered in the structure of murein lipoprotein. *J Bacteriol* **133:** 1419–1426.

Zimmer J, Nam Y, Rapoport TA. 2008. Structure of a complex of the ATPase SecA and the protein-translocation channel. *Nature* **455:** 936–943.

Zuber B, Chami M, Houssin C, Dubochet J, Griffiths G, Daffe M. 2008. Direct visualization of the outer membrane of mycobacteria and corynebacteria in their native state. *J Bacteriol* **190:** 5672–5680.

Cell Biology of Prokaryotic Organelles

Dorothee Murat, Meghan Byrne, and Arash Komeili

Department of Plant and Microbial Biology, University of California, Berkeley, Berkeley, California 94720-3102

Correspondence: komeili@berkeley.edu

Mounting evidence in recent years has challenged the dogma that prokaryotes are simple and undefined cells devoid of an organized subcellular architecture. In fact, proteins once thought to be the purely eukaryotic inventions, including relatives of actin and tubulin control prokaryotic cell shape, DNA segregation, and cytokinesis. Similarly, compartmentalization, commonly noted as a distinguishing feature of eukaryotic cells, is also prevalent in the prokaryotic world in the form of protein-bounded and lipid-bounded organelles. In this article we highlight some of these prokaryotic organelles and discuss the current knowledge on their ultrastructure and the molecular mechanisms of their biogenesis and maintenance.

The emergence of eukaryotes in a world dominated by prokaryotes is one of the defining moments in the evolution of modern day organisms. Although it is clear that the central metabolic and information processing machineries of eukaryotes and prokaryotes share a common ancestry, the origins of the complex eukaryotic cell plan remain mysterious. Eukaryotic cells are typified by the presence of intracellular organelles that compartmentalize essential biochemical reactions whereas their prokaryotic counterparts generally lack such sophisticated subspecialization of the cytoplasmic space. In most cases, this textbook categorization of eukaryotes and prokaryotes holds true. However, decades of research have shown that a number of unique and diverse organelles can be found in the prokaryotic world raising the possibility that the ability to form organelles may have existed before the divergence of eukaryotes from prokaryotes (Shively 2006).

Skeptical readers might wonder if a prokaryotic structure can really be defined as an organelle. Here we categorize any compartment bounded by a biological membrane with a dedicated biochemical function as an organelle. This simple and broad definition presents cells, be they eukaryotes or prokaryotes, with a similar set of challenges that need to be addressed to successfully build an intracellular compartment. First, an organism needs to mold a cellular membrane into a desired shape and size. Next, the compartment must be populated with the proper set of proteins that carry out the activity of the organelle. Finally, the cell must ensure the proper localization, maintenance and segregation of these compartments across the cell cycle. Eukaryotic cells perform these difficult mechanistic steps using dedicated molecular pathways. Thus, if connections exist between prokaryotic and eukaryotic organelles it seems likely that relatives of these molecules may be involved in

 doi: 10.1101/cshperspect.a000422
Cite this article as *Cold Spring Harb Perspect Biol* 2010;2:a000422

the biogenesis and maintenance of prokaryotic organelles as well.

Prokaryotic organelles can be generally divided into two major groups based on the composition of the membrane layer surrounding them. First are the cellular structures bounded by a nonunit membrane such a protein shell or a lipid monolayer (Shively 2006). Well-known examples of these compartments include lipid bodies, polyhydroxy butyrate granules, carboxysomes, and gas vacuoles. The second class consists of those organelles that are surrounded by a lipid-bilayer membrane, an arrangement that is reminiscent of the canonical organelles of the eukaryotic endomembrane system. Therefore, this article is dedicated to a detailed exploration of three prokaryotic lipid-bilayer bounded organelle systems: the magnetosomes of magnetotactic bacteria, photosynthetic membranes, and the internal membrane structures of the *Planctomycetes*. In each case, we present the most recent findings on the ultrastructure of these organelles and highlight the molecular mechanisms that control their formation, dynamics, and segregation. We also highlight some protein-bounded compartments to present the reader with a more complete view of prokaryotic compartmentalization.

Magnetosomes: Bacterial Compasses

The **magnetosomes** of magnetotactic bacteria (MB) are one of the most fascinating prokaryotic compartments (Fig. 1). MB are a phylogenetically diverse group of microorganisms with the ability to use geomagnetic field lines as guides in their search for their preferred redox conditions (Bazylinski and Frankel 2004; Komeili 2007). This behavior is achieved through the use of a unique magnetic organelle termed the magnetosome. A magnetosome consists of a lipid bilayer membrane that houses an approximately 50-nanometer crystal of the magnetic mineral magnetite (Fe_3O_4) or greigite (Fe_3S_4). Individual magnetosomes are arranged into one or more chains within the cell where they act passively to orient the bacterium within a magnetic field. The unusual properties of these magnetic minerals and their potential to be exploited in a variety of applied settings have made them center of most studies on magnetosomes (Bazylinski and Frankel 2004).

From a cell biological perspective, however, it is the often-neglected magnetosome membrane that may hold the key to understanding fundamental properties of prokaryotic organelles. Detailed electron microscopic (EM) work and biochemical studies have shown that the magnetosome membrane has the cytological and chemical properties of a lipid bilayer membrane (Gorby et al. 1988; Grünberg et al. 2004). Additionally, numerous proteomic studies have shown that this compartment contains a unique mix of soluble and transmembrane domain-containing proteins, implying the existence of a dedicated protein sorting pathway (Okuda et al. 1996; Grünberg et al. 2001; Grünberg et al. 2004; Tanaka et al. 2006). The magnetosome membrane loaded with its protein cohort is present before crystal formation and serves as the site of biomineralization further confirming that it is an independent organelle (Komeili et al. 2004). The organization of magnetosomes into one or multiple chains also suggests that mechanisms must exist for the proper localization and division of this structure within the cell. This already detailed view of the magnetosome has been pushed to the next level with two recent imaging studies that describe the use of cryo-electron tomography (CET) to obtain high resolution three-dimensional images of MB (Komeili et al. 2006; Scheffel et al. 2006). In CET a series of two-dimensional images of a specimen, taken by tilting the stage of an electron microscope at various angles relative to the electron beam, is translated into a three-dimensional image using a specific algorithm. This technique provides such a detailed view of a cell that disruptive fixing and staining treatments common in other EM techniques are not needed. As a result one can prepare a sample by a simple rapid freezing method and subsequently image a cell at high resolution in a near-native state (Milne and Subramaniam 2009). This combination of rapid preservation, minimal disruption of cellular features, and nanometer scale resolution revealed features of magnetosomes that had not been visualized in

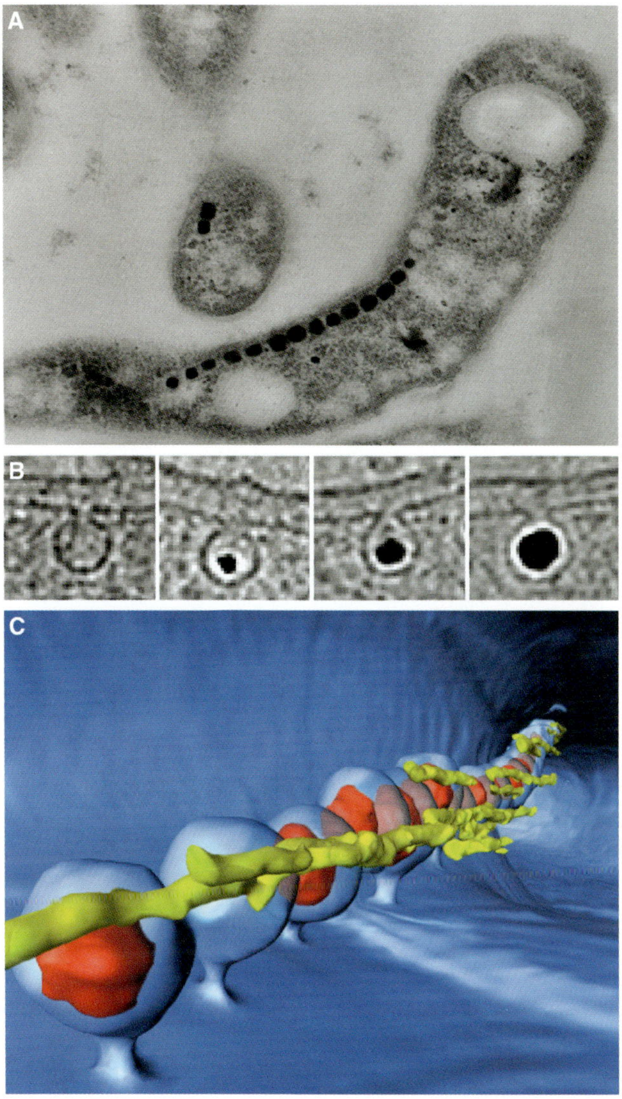

Figure 1. Magnetosomes can be easily visualized with various forms of electron microscopy. The electron-dense magnetite crystals are seen as a chain running through the cell in (*A*). Cryo-electron tomography was instrumental in demonstrating that the magnetosome membrane is an invagination of the inner cell membrane (*B*) and cytoskeletal filaments surround the magnetosome chain (*C*). (*A*, Reprinted, with permission from Komeili et al. 2004 [© National Academy of Sciences]; *B*, reprinted with permission from Komeili et al. 2006 [© AAAS]; *C*, image courtesy of Zhuo Li and Grant Jensen.)

more than 30 years of work on MB. Most striking was the finding that in *Magnetospirillum magneticum* AMB-1, individual magnetosomes are not separated into vesicles and are instead invaginations of the inner cell membrane (Fig. 1B). This state was observed in empty magnetosomes as well as those that contained fully formed crystals implying that this organelle is an invagination of the inner membrane at all times (Komeili et al. 2006). Although such an organization might seem puzzling at first it does make sense in the context of magnetosome function and magnetite biomineralization. Because the primary job of the magnetosome chain is to orient the cell in external magnetic fields the organelle must be attached to the rest of the cell and by integrating the magnetosome into the cell membrane no additional machinery is needed to achieve proper orientation in magnetic fields. It has also been hypothesized that the biomineralization of magnetite may involve the formation of precursor minerals such as ferrihydrite in the periplasmic space (Frankel et al. 1983). In such a case the small opening between the magnetosome lumen and the periplasm would provide a simple path for the transport of these precursor minerals. The CET imaging of *Magnetospirillum gryphiswaldense* MSR-1, an organism closely related to AMB-1, did not specifically explore the existence of any connections between the magnetosome membrane and the inner cell membrane (Scheffel et al. 2006). However, in this organism the magnetosomes were found juxtaposed against the cell membrane consistent with the possibility that they are also invaginations of the inner cell membrane (Scheffel et al. 2006). These tremendous imaging studies have revealed the organization and ultrastructure of the magnetosome at nanometer scales and recent studies are beginning to define the molecular basis of magnetosome formation and organization.

MB are fastidious and slow growing organisms but offer multiple advantages as model systems for the molecular study of organelle formation in prokaryotes. Multiple MB genomes have been sequenced in the past few years, magnetosomes can be readily purified from cell extracts using simple magnetic columns and most importantly, magnetosomes are not essential for cell survival under laboratory growth conditions, opening the door to the use of genetics as a tool for uncovering the steps involved in magnetosome formation. A combination of these approaches has led to the identification of a large list of genes thought to be involved in the formation and function of magnetosomes. Surprisingly, most of these genes are organized into a coherent and unstable genomic region whose core components are conserved across multiple species of magnetotactic bacteria (Ullrich et al. 2005; Fukuda et al. 2006; Richter et al. 2007; Jogler et al. 2009). This region, termed the magnetosome island or MAI, carries signature features of other genomic islands found in bacteria and encompasses a substantial portion of the genome. For instance, in AMB-1 the MAI is predicted to contain over one hundred genes accounting for approximately 2% of the organism's gene content (Fukuda et al. 2006). From an evolutionary perspective the organization of core magnetosome genes into an unstable genomic segment implies that the appearance of this organelle in diverse bacterial species was accomplished through lateral transfer of the MAI (Jogler et al. 2009). What makes the MAI intriguing to cell biologists is the possibility that it contains the unique functions required to build a magnetosome. Biochemical and genetic studies have shown that a number of MAI genes encode proteins that can influence the size and morphology of magnetite crystals (Arakaki et al. 2003; Scheffel et al. 2008; Murat et al. 2010). Other factors, such as the MamA protein, appear to function in activating or priming preformed magnetosomes for biomineralization (Komeili et al. 2004). And, as described later, one core region of the MAI is essential for the formation of the magnetosome membrane, protein sorting to this organelle and its specific localization within the cell (Komeili et al. 2006; Scheffel et al. 2006; Murat et al. 2010).

At the heart of the MAI is the *mamABE* operon, a gene cluster conserved in multiple species of MB. A comprehensive genetic analysis of the MAI showed that in the absence of the *mamABE* operon, AMB-1 is nonmagnetic and fails to even form empty magnetosome membranes (Murat et al. 2010). An analysis of individual deletions of each of the 18 genes of this cluster revealed a range of mutant phenotypes with defects at every step of magnetosome formation. Interestingly, four genes, *mamI*, *mamL*, *mamQ* and *mamB*, seem to be essential for the formation of the magnetosome membrane (Murat et al. 2010). None of these genes encodes for proteins with homology to known membrane deformation factors found in eukaryotes. However, they contain intriguing features that may hint at a potential mechanism for magnetosome formation. MamB and MamQ share homology with large families of membrane proteins whereas MamI and MamL are unique to MB. These two latter proteins are small (~70 amino acid) polypeptides with two predicted transmembrane domains. MamI does not possess any distinguishing structural features but MamL contains a cytoplasmic tail that is rich in positively charged residues. One potential model for membrane deformation is that this tail interacts with one leaflet of the inner cell membrane creating an asymmetry that favors the bending of the membrane. Another finding of this work is that membrane formation can be decoupled from the sorting of at least a subset of magnetosome proteins. When the putative protease, MamE, is absent, empty magnetosome membranes are still formed although a number of magnetosome proteins are mislocalized arguing for a step-wise assembly of this organelle (Murat et al. 2010).

One of the central genes of the *mamABE* operon, *mamK*, is homologous to the large and diverse family of bacterial actin-like proteins discovered in the last decade (Carballido-López 2006). When *mamK* is deleted in AMB-1, the resulting mutants are not defective in magnetosome membrane formation or biomineralization of magnetite. Instead, magnetosomes are no longer organized into chains and are spread out across the cell membrane (Komeili et al. 2006). The CET imaging studies of AMB-1 and MSR-1 had also discovered that the magnetosome chain is surrounded by a network of cytoskeletal filaments with dimensions similar to bacterial actin-like filaments (Komeili et al. 2006; Scheffel et al. 2006) (Fig. 1C). Interestingly, these filaments are no longer present when *mamK* is deleted (Komeili et al. 2006). Together with recent observations that MamK can form filaments in heterologous systems and in vitro, these results suggest that this actin-like protein constitutes the structural component of the magnetosome-specific cytoskeleton (Pradel et al. 2006; Taoka et al. 2007). MamJ, a highly acidic protein encoded by a gene directly upstream of *mamK*, seems to play a crucial role in the organization of magnetosome chain as well. When *mamJ* is deleted in MSR-1, the magnetosome chain collapses into a ball within the cell (Scheffel et al. 2006). In this mutant, structures similar to the magnetosome-specific cytoskeleton can still be seen by CET but they are no longer associated with magnetosomes. Given that MamJ can associate with MamK in a bacterial two-hybrid system a simple and attractive model has been proposed whereby MamJ can anchor MamK to the magnetosome membrane allowing it to organize individual organelles into a chain (Scheffel et al. 2006; Scheffel and Schüler 2007). Taken together these results suggest that similar to eukaryotic cells, prokaryotes can take advantage of cytoskeletal elements to position and organize subcellular compartments.

As can be seen, progress in the study of magnetosomes has been rapid in the last few years and the incredible gains made from the ultrastructural characterization of this organelle are beginning to be matched with molecular studies. Yet, much work remains to be performed. Although the discovery and genetic analysis of the MAI provides a potential "parts list" for the magnetosome formation machinery, the specific mechanisms that control membrane biogenesis and protein sorting have yet to be defined. Moreover, even though the discovery of the MamK/MamJ system for chain formation is a breakthrough advance in this field, its mechanism of action remains elusive. A resolution of these key issues is necessary before evolutionary comparisons can be drawn between eukaryotic organelles and magnetosomes. The elucidation of these key cellular mechanisms will also provide new modes for exploitation of magnetosomes in a variety of applied settings.

Photosynthetic Membranes: Variations on a Theme

Photosynthetic membranes are perhaps the most thoroughly studied of all prokaryotic organelles (Fig. 2). They fall into three general categories and each has unique and intriguing characteristics that make it ideal for the study of membrane dynamics and intracellular organization in prokaryotes. The first category, historically referred to as **chromatophores**, contains the various **intracytoplasmic membrane** (ICM)

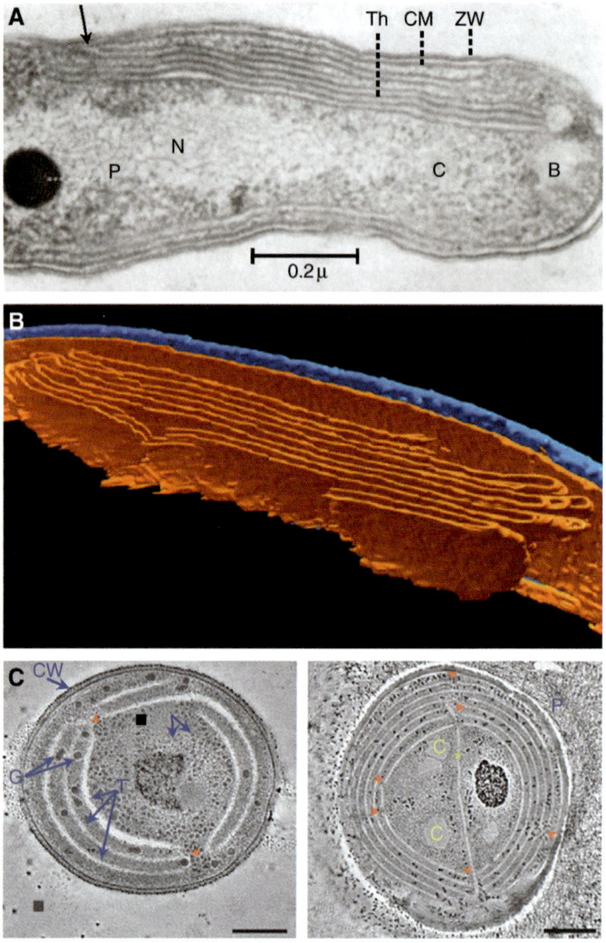

Figure 2. Photosynthetic membranes were the first of bacterial organelles to be imaged with electron microscopy. (*A*) is an image from a 1967 imaging study of *Rhodopseudomonas palustris.* The photosynthetic membranes (Th) are arranged as ribbon-like structures that are clearly continuous with the inner cell membrane (CM) at the point indicated by the arrow. These features are revealed in three dimensions in a surface rendered reconstruction of *Rhodopseudomonas viridis* in (*B*). Thylakoid membranes of cyanobacteria (*C*) are arranged in several circular layers and display species-specific morphologies. In contrast with photosynthetic membranes of purple bacteria, thylakoids appear to be fully separated from the inner cell membrane. (*A*, Reprinted, with permission, from Tauschel and Drews 1967 [© Springer]; *B*, reprinted, with permission, from Konorty et al. 2008 [© Elsevier]; *C*, reprinted, with permission, from Nevo et al. 2007 [© Nature Publishing Group].)

structures that house the photosynthetic protein complexes of the purple photosynthetic bacteria. The second category consists of the numerous examples of **thylakoid** membrane compartments found in cyanobacteria. The **chlorosome** compartments of green photosynthetic bacteria constitute the third major category of bacterial photosynthetic compartment. All of these organelle systems act to maximize the efficiency of photosynthesis by increasing the number of available photosynthetic protein complexes, maximizing the size of the light-exposed membrane surface and by providing an idealized subcellular environment for this vital reaction. However, despite their functional relatedness these organelles differ in fundamental ways that impact the mechanisms by which they are formed and maintained.

Chromatophores were first studied biochemically where it was shown that a defined fraction of cellular extract was capable of carrying out certain light-dependent reactions in vitro. Elegant EM imaging of different species of purple phototrophic bacteria showed the presence of extensive intracellular membranes with distinct and species-specific morphological characteristics (Oelze and Drews 1972). For instance, the photosynthetic membranes of *Rhodopseudomonas palustris* appear as neatly folded membrane stacks that are continuous with the cell membrane (Tauschel and Drews 1967) (Fig. 2A). In contrast, in the *Rhodobactericiae* species these structures are spherical invaginations of the inner membrane where multiple bubbles are connected to one another. These early ultrastructural studies have recently been augmented with the CET imaging of *Rhodopseudomonas viridis*, an organism that forms membranes that are similar in morphology to those observed in *R. palustris* (Konorty et al. 2008) (Fig. 2B). The near-native state preserved by cryofixation reveals much of the same features observed in traditional electron microscopic imaging of the same organism. Membranes are folded in an accordion-like structure and they are invaginations of the inner membrane with a distinct 128 nm wide opening to the periplasmic space (Konorty et al. 2008). In many organisms, including *R. viridis*, chromatophores are produced only under photosynthetic growth conditions. The CET imaging of *R. viridis* at early time points after switch to photosynthetic growth reveals the presence of small vesicular structures adjacent to the cell membrane (Konorty et al. 2008). Presumably these early compartments eventually mature into the membrane stacks seen in cells grown under continuous photosynthetic conditions.

How are these exquisite and species-specific membrane morphologies generated? The answer appears to be a simple and elegant mechanism in which the inherent properties of photosynthetic protein complexes determine the resulting shape of the membrane. The chromatophores of purple bacteria such as *R. sphaeroides* house the major components of the photosynthetic machinery (Tavano and Donohue 2006). These protein complexes include the light harvesting 2 (LH2) protein complex and the "core" complex consisting of the multimeric light harvesting 1 (LH1) and reaction center (RC) polypeptides. In some species, dimers of core complexes are formed through a linkage with the PufX protein. These protein complexes are first assembled in the inner cell membrane at sites that will invaginate to form chromatophores (Tavano and Donohue 2006). Genetic studies with *R. sphaeroides* have shown that in the absence of LH2 the chromatophore membranes lose their characteristic stacked spherical shape and instead turn into long tubules within the cell (Tavano and Donohue 2006). Interestingly, when *pufX* is deleted in these LH2-deficient strains the membrane tubules disappear and instead large spherical internal membranes are observed (Tavano and Donohue 2006). Thus, the major protein components of chromatophores play a decisive role in determining the morphology of the membrane. Recent structural and biophysical modeling studies of photosynthetic membrane proteins have built on these functional studies to provide a mechanistic basis for the remodeling of the cell membrane into chromatophores. When chromatophores are imaged by atomic force microscopy (AFM) ordered arrays of photosynthetic protein complexes are seen to densely pack the membrane surface (Sturgis et al. 2009). Based on the known structures of these complexes distinct domains containing dimers of the RC-LH1-PufX complex as well as rings of LH2 can be placed in the AFM images (Sturgis et al. 2009). Interestingly, three-dimensional electron microscopic reconstruction of negatively stained single particles reveals that the dimers of RC-LH1-PufX form a complex that is bent toward the lumen of the chromatophore (Qian et al. 2008). An in silico model of chromatophores based on this bent structure of the core complex predicts the formation of long membrane tubules with dimensions similar to that observed in mutants lacking LH2 (Qian et al. 2008). Using these functional and structural results as a guide, other modeling studies have also supported the hypothesis that the biophysical properties of photosynthetic proteins and their long-range interactions with each other would be sufficient to produce curved membrane structures in vivo (Chandler et al. 2008). Despite these convincing arguments in favor of a self-assembly model for chromatophore formation, it is possible that other factors may be involved in the process. For instance, a recent proteomic analysis of *R. sphaeroides* has shown that a number of proteins outside the major photosynthetic complexes may also be present within chromatophores raising the prospects that novel factors may have roles in the development of this organelle (Zeng et al. 2007).

The thylakoid membranes of cyanobacteria are the evolutionary precursors of chloroplasts. As with chromatophores these organelles are responsible for some of the central light-dependent reactions of photosynthesis. However, the morphology and subcellular arrangement of thylakoids is markedly different than that of chromatophores (Fig. 2C). Several recent CET studies have corroborated earlier EM studies of thylakoids and provided additional insights into the organization and species-specific diversity of this fascinating organelle (van de Meene et al. 2006; Nevo et al. 2007; Ting et al. 2007). In most cases thylakoids appear as several flattened and stacked layers of lipid-bilayer membrane that encircle the cell. The number of layers and the spacing between them follows a species-specific arrangement (Nevo et al. 2007). Although these layers cover much of the cytoplasmic space there is still substantial flow of cellular components in between the thylakoid stacks. This is because of the presence of numerous perforations within the thylakoid membrane and in CET images a number of macromolecules such as ribosomes and storage granules are seen within these openings (Nevo et al. 2007). The three-dimensional images provided by CET also reveal numerous bridges and fusions formed by membranes that traverse the different stacks of thylakoids (Nevo et al. 2007). Finally, large cytoplasmic vesicles are seen near and at times fused to the thylakoids. The highly networked nature of this membrane system suggests that long-range communication and transport may occur throughout the whole organelle (Nevo et al. 2007). CET studies as well as other attempts to reconstruct the cellular arrangement of thylakoids have also revealed that, in contrast with the chromatophore membrane, the inner cell membrane and the thylakoid membrane are not continuous with each other (Liberton et al. 2006; van de Meene et al. 2006; Nevo et al. 2007; Ting et al. 2007). The lack of connections between thylakoids and the cell membranes has also been shown through the use of various fluorescent membrane dyes (Schneider et al. 2007). FM1-43 is a hydrophobic dye that fluoresces once incorporated into membranes and is thought to be incapable of diffusing past the inner cell membrane. When it is used to stain the cyanobacterium *Synechocystis* sp. PCC 6803 the inner membrane and outer membrane are labeled but the thylakoids do not incorporate the dye indicating that these membrane systems are separate entities or that a physical barrier prevents the migration of the dye to the thylakoids. In contrast when Mitotracker, a membrane dye that can diffuse past cellular membranes, is used as a marker all membranes including the thylakoids are stained in this organism. Long incubations with FM1-43 initially stain intracellular structures resembling vesicles and eventually highlight the thylakoid membranes indicating a mode for transfer of lipids and proteins from the cell membrane to this organelle (Schneider et al. 2007). Thus, similar to eukaryotes, cyanobacteria form membrane structures that are discontinuous from the cell membrane implying the presence of mechanisms for bending and fission of cellular membranes.

Given their evolutionary connections, one clue to the mechanisms of thylakoid membrane formation has come from examining the pathways of chloroplast biogenesis in plants. The vesicular inducing protein in plastid 1, Vipp1, is a protein implicated in membrane remodeling and vesicular trafficking in chloroplasts in *Arabidopsis* (Kroll et al. 2001). Cyanobacteria contain homologs of Vipp1 and its absence in *Synechocystis* results in the loss of stacks of thylakoid membranes (Westphal et al. 2001). These findings had suggested a possible role for Vipp1 in the biogenesis of thylakoid membranes but a recent study suggests that this defect may have less to do with membrane biogenesis than it does with the assembly of photosynthetic complexes (Gao and Xu 2009). Using a repressible promoter, Vipp1 was depleted to levels in which cells could no longer perform photosynthesis. Under these conditions the thylakoid membranes had a wild-type appearance suggesting that Vipp1 may function at a step downstream of membrane biogenesis (Gao and Xu 2009). Another fascinating possibility has come from the observation that homologs of eukaryotic dynamin can be found in several species of cyanobacteria (Low and Lowe 2006). In eukaryotes dynamin and dynamin-like proteins are important for membrane fission and tubulation in processes ranging from endocytosis to cytokinesis (Praefcke and McMahon 2004). As with eukaryotic dynamins, the putative dynamin homolog found in cyanobacteria is also a GTPase, can bind liposomes in vitro, and localizes to cellular membranes in vivo (Low and Lowe 2006). More strikingly, the three-dimensional structure of prokaryotic dynamin is remarkably similar to that of eukaryotic dynamin (Low and Lowe 2006). Given these similarities in structure and biochemical activity it has been postulated that cyanobacterial dynamins may play a role in establishing the complex assemblies of thylakoid membranes (Low and Lowe 2006). However, dynamin-like proteins are not found in all cyanobacteria and in the strains where they do exist, no functional data exists to suggest that they have a dedicated role in thylakoid membrane biogenesis.

Chlorosomes are the largest light-harvesting systems found thus far in photosynthetic organisms, and they have been shown to allow cells to harvest light energy at extremely low light intensities (Frigaard and Bryant 2006). A striking example of this is a chlorosome-containing obligate phototroph that was found 2391 meters below the surface of the Pacific Ocean and thought to extract the energy necessary for growth from the infrared radiation of a geothermal vent (Beatty et al. 2005). Chlorosomes are found in all *Chlorobi* or green sulfur bacteria and some *Chloroflexi* or green filamentous anoxygenic phototrophs. Recently, chlorosomes were discovered in an acidobacterium isolated from a microbial mat community in Yellowstone National Park making it the first photosynthetic bacterium that has been identified in the phylum *Acidobacteria* (Bryant et al. 2007).

Chlorosomes are flattened, ellipsoidal structures that are connected to the cytoplasmic membranes by a relatively thick baseplate (Fig. 4A). The chlorosome envelope is 3–5 nm thick and electron opaque, as seen by thin-layer transmission electron microscopy (Cohen-Bazire et al. 1964; Staehelin et al. 1980). This layer is thinner than the cytoplasmic membrane (8 nm), indicating it is not a lipid bilayer. However, lipids have been identified in purified chlorosomes, and the chlorosome envelope fractures in freeze-fracture electron microscopy in a manner characteristic of lipids, suggesting that the envelope is a lipid monolayer (Staehelin et al. 1980; Frigaard and Bryant 2006).

Chlorosomes primarily contain bacteriochorophyll (BChl) c, d, or e, which can number 150,000–300,000 molecules in a single organelle. Ten proteins have been purified from *Chlorobium tepidum* chlorosomes, and all of them have been shown to be susceptible to cleavage by proteases, suggesting they are surface exposed. Antisera to these proteins can precipitate chlorosomes, further supporting the model that these proteins are in the chlorosome envelope (Chung and Bryant 1996; Vassilieva et al. 2002). A number of these envelope proteins show similarity with each other leading to the hypothesis that they perform redundant functions. This idea is supported by genetic studies in which individual deletions of 9 of the 10 chlorosome genes had virtually no effect on chlorosome structure or function (Frigaard et al. 2004). However, when double, triple and quadruple mutants were created in which combinations of genes predicted to be in the same family were deleted, dramatic phenotypes in the size and morphology of chlorosomes were uncovered suggesting that the protein content of the organelle determines its ultrastructural properties (Li and Bryant 2009). The 10th gene, *csmA*, has been proposed to act in the flow of energy from the antenna to the reaction center. Interestingly, in the aforementioned study *csmA* could not be deleted, suggesting that it is essential to the cells (Frigaard et al. 2004).

The discovery of chlorosome proteins and the directed functional studies detailed earlier are important steps in understanding the mechanism of chlorosome formation. The unique arrangement of lipids and envelope proteins suggests that this mechanism will be different than the one used to form other lipid-bounded organelles. To account for their architecture and composition a recent hypothesis suggests that a self-assembly process is responsible for the formation of chlorosomes (Hohmann-Marriott and Blankenship 2007). According to this model, bacteriochlorophylls and other pigment molecules accumulate in between the two leaflets of the inner membrane creating a growing bubble surrounded by a single lipid layer. In fact, when the gene encoding for bacteriochlorophyll synthase *c* was deleted in *Chlorobium tepidum* normal chlorosomes were not formed and instead smaller deflated structures containing other pigments were seen within the cell (Frigaard et al. 2002). Within this monolayer, glycosyl diacylglycerides are enriched because of their preferred interactions with the accumulated pigments. Finally, chlorosome proteins are recruited because of their preference for these chlorosome components. A combination of genetic and biochemical studies are now needed to directly test this simple self-assembly model for chlorosome biogenesis.

Planctomycete Membrane Compartments: True Ancestors of Eukaryotic Organelles?

The examples discussed thus far represent the broad spectrum of intracellular compartmentalization that can be found in the prokaryotic world. These structures, however, do not resemble the characteristic organelles that define the endomembrane system of eukaryotes making it difficult to draw any evolutionary parallels. The members of the *Planctomycetes*, a deep branching phylum of the Bacteria, however, may contain the bacterial ancestors of eukaryotic organelles. Most species of this phylum are characterized by extensive and truly unique compartmentalization of their cytoplasmic space (Fuerst 2005). The simplest configuration is found in organisms such as those of the genus *Pirellula* in which a large lipid-bilayer bounded compartment contains and separates the chromosome and ribosomes from other cellular components. This organelle, termed the **pirellulosome**, is surrounded by a small area of cytoplasmic space known as the **paryphoplasm** (Fig. 3B). Unlike the periplasmic space of Gram-negative bacteria macromolecules such as RNA can be found in the paryphoplasm (Lindsay et al. 1997).

In some *Planctomycetes*, more complicated forms of compartmentalization have been observed in which the pirellulosome is further subdivided into smaller and more specialized compartments. The most dramatic example is found in species such as *Gemmata obscuriglobus* in which a compacted chromosome is surrounded by a double lipid-bilayer membrane to form a nuclear body (Lindsay et al. 2001) (Fig. 3A). Ribosomes are found both within the nuclear body and throughout the rest of the pirellulosome indicating that some translational activity may be separated from transcription. The unusual membrane architecture and the partial separation of transcription from translation are reminiscent of the eukaryotic nucleus thus raising the possibility that the *Planctomycetes* may represent the early forms of compartmentalization that has come to define the eukaryotes. This arrangement also implies that communication and transport of macromolecules must occur between the various compartments of *G. obscuriglobus*. Although molecular pathways and evidence for such transport have not been found, microscopic examination has revealed that the folding of the lipid bilayer membrane surrounding the nuclear body creates a small opening that may be a portal for transport of macromolecules (Lindsay et al. 2001). Time-lapse microscopy experiments have also helped to elucidate the steps involved in the segregation of nuclei and biogenesis of organelles during cell division. Many of the *Planctomycetes*, including *G. obscuriglobus*, divide by budding rather than the binary fission mechanism often seen in bacteria (Lee et al. 2009). During early

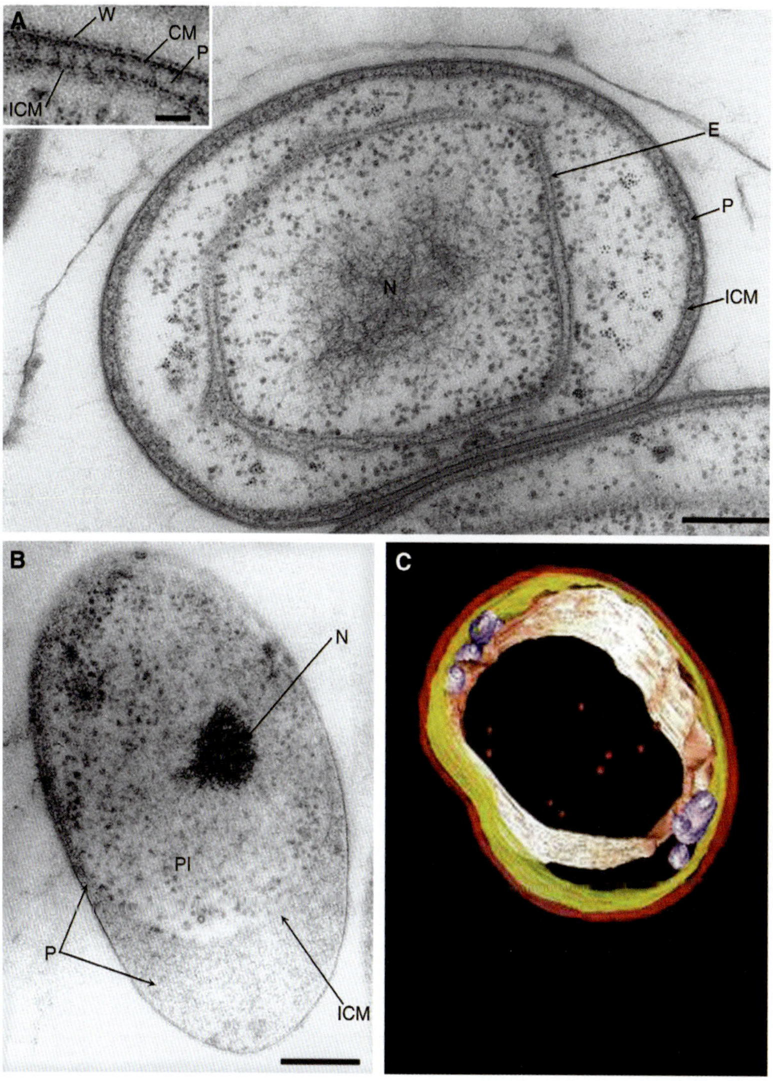

Figure 3. The nucleus-like organelle of *Gemmata obscuriglobus* is shown in (*A*). The nuclear envelope (E) is a double lipid-bilayer membrane containing the chromosome (N). The inset highlights the intracytoplsmic membrane (ICM) that separates the riboplasm from the paryphoplasm (P) compartment. A simpler organization is seen in organisms such as *Pirellula marina* in which the intracytoplsmic membrane (ICM) differentiates the pirellulosome (PI) from the paryphoplasm (P) (*B*). Many of the *Planctomycetes* contain another unique organelle called the anammoxosome (*C*). Here a CET reconstruction of *Brocadia fulgida* is shown. The anammoxosome is the central compartment of this cell and iron particles (red) are found within it. (*A*, *B*, Reprinted, with permission, from Lindsay et al. 2001 [© Springer]; *C*, reprinted, with permission, from Niftrik et al. 2008a [© ASM].)

stages of the budding process the newly divided nucleoid unbound by any membranes can be seen in a relatively young bud. As the bud grows a complex migration of the mother cell inner membrane and the daughter cell inner membrane are followed by membrane fusion events to create the new nuclear envelope (Lee et al. 2009). At present little is known about the molecular mechanisms of organelle formation in these organisms and the studies of this fascinating topic are hampered by a lack of robust genetic tools. However, recent sequencing of several *Planctomycete* genomes may help in identification of novel gene products with a unique role in organelle assembly and dynamics (Studholme et al. 2004; Staley et al. 2005). One such clue has emerged from the genome of *Gemmata* Wa-1 in which a homolog of the eukaryotic Gle2 protein, a component of the nuclear pore complex, has been discovered (Staley et al. 2005). A recent study conducted a more directed search for bacterial proteins that contain signatures of eukaryotic membrane coat proteins, which play key roles in vesicle trafficking and organelle maintenance in eukaryotes (Santarella-Mellwig et al. 2010). These proteins are typified by an unusual combination of structural domains where a specialized arrangement of β-sheets, called a β-propeller, is followed by an α-helical structure termed an α solenoid. These proteins are ubiquitous among the eukaryotes but when the genomes of all sequenced bacteria where queried only species within the *Planctomycete-Verrucomicrobia-Chlamydiae* phyla contained genes encoding for eukaryotic coatlike proteins. Interestingly one of these candidates found in *G. obscuriglobus* was seen to localize to the organism's internal membrane structures (Santarella-Mellwig et al. 2010). These results provide molecular evidence for the possible ancestral link between Planctomycete compartments and eukaryotic organelles.

Other species of the *Planctomycetes* have an additional membrane-bound compartment called the **anammoxosome** capable of anaerobic ammonium oxidation (Strous et al. 1999) (Fig. 3C). For decades, this anammox reaction had been hypothesized to exist based on thermodynamic calculations but had never been associated with a living organism (Broda 1977). The anammoxosome is located within the pirellulosome and it is the only *Planctomycete* organelle that can be purified, which has facilitated its study (Lindsay et al. 2001). Among the proteins found in the anammoxosome membrane is hydroxylamine oxidoreductase, a unique enzyme that catalyzes ammonium oxidation (Schalk et al. 2000). Analysis of the anammoxosome composition has also revealed that its membrane is enriched in an unusual type of concatenated lipids, never before found in nature (Sinninghe Damsté et al. 2002). These molecules, termed ladderane lipids form a denser and more impermeable barrier than regular biological membranes that may prevent the diffusion of the toxic intermediates produced during the anammox reaction. The diffusion barrier provided by this organelle is also thought to help in retaining the intermediates of the slow anammox reaction within the cell (Sinninghe Damsté et al. 2002). A recent study of this organelle by CET has revealed that its membrane is highly curved leading to the proposal that the curvature could optimize the membrane surface and thus the membrane-associated metabolic processes that happen in the anammoxosome (van Niftrik et al. 2008a; van Niftrik et al. 2008b). Some anammox bacteria are also distinguished by their unique mode of cell and organelle division. In *Kuenenia stuttgartiensis* cell division follows the typical binary fission mode observed in other bacteria (van Niftrik et al. 2009). As a result, the anammoxosome is divided in half during each division cycle and segregated equally among the two daughter cells. EM and CET imaging reveal the presence of a distinct cytokinetic ring apparatus in the outermost compartment of this organism. Most bacteria use the tubulin-like protein FtsZ to form a division ring but the genome of *K. stuttgartiensis* is devoid of any homolog to *ftsZ*. Instead, another GTPase, named *kustd1438*, was found to specifically localize to the cytokinetic ring of this organism (van Niftrik et al. 2009). The observation that *kust1438* homologs are not found outside of the anammox bacteria also hints at its unique and important function in this process. However, further functional studies are required to determine a direct role for this protein in cell division and organelle partitioning. Ultimately, development of robust genetic systems will help to further define the molecular mechanisms of organelle formation in the *Planctomycetes.*

PROTEIN-BOUNDED COMPARTMENTS

Carboxysomes are one of the best-known examples of protein-bounded organelles in bacteria (Yeates et al. 2008). They occur in all cyanobacteria as well as chemoautolithotrophs where they serve as the site for the first step of the Calvin cycle. The major catalytic components of carboxysomes are the enzymes Ribulose-1,5-bisphosphate carboxylase oxygenase (RuBisCO) and carbonic anhydrase. RuBisCO catalyzes the reaction of CO_2 with ribulose bisphosphate to two molecules of 3-phosphoglyceric acid (3PGA) and carbonic anhydrase catalyzes the conversion of bicarbonate to CO_2. By increasing the local concentration of RuBisCO and the CO_2 substrate, carboxysomes are likely increasing the efficiency of the productive carbon fixation reaction (Yeates et al. 2008). This idea is supported by recent electron cryotomography studies, which show that each carboxysome (measuring 80 to 150 nm) contains over 200 RuBisCO enzyme complexes arranged in concentric layers (Schmid et al. 2006; Iancu et al. 2007) (Fig. 4B).

Only a few genes, found in one or more operons, are involved in the formation of carboxysomes. In *Halothiobacillus neopolitans*, the carboxysome genes encode for the large and small RuBisCO subunits, three small shell proteins that share high homology, a large shell protein, carbonic anhydrase, and two unknown proteins that seem to have a regulatory function. Other bacteria that form carboxysomes have slightly different genes in their operons, but all contain homologs of the small shell protein genes and genes that encode for the RuBisCO subunits. Recently, small shell proteins from both a cyanobacterium and a chemolithoautotrophic bacterium have been crystallized, which has provided valuable insights into how the protein shell of the carboxysomes may assemble (Kerfeld et al. 2005; Tsai et al. 2007). These crystal structures reveal that the proteins, purified individually, self-assemble into hexamers that bind edge-to-edge to form monolayer sheets. These sheets of protein have been proposed to make up the walls of the carboxysome. The crystal structures also revealed a positively charged pore at the center of the hexamers. This pore could allow for the passage of negatively charged molecules such as bicarbonate while blocking the entrance of O_2 creating another way in which the carboxysome could increase the efficiency of RuBisCO and the fixation of carbon.

The defined set of proteins found in these operons is likely to be the minimal components required to build a carboxysome. However, recent results show that the proper organization and segregation of carboxysomes across the cell cycle require it to interface with other cellular components (Savage et al. 2010). Fluorescent protein fusions to either a shell protein or to a RuBisCO component revealed that carboxysomes are linearly spaced throughout the cell. The most relevant consequence of this arrangement is that during cell division approximately equal numbers of carboxysomes will be partitioned to each daughter cell (Savage et al. 2010). This arrangement relies on cytoskeletal systems as disruptions of either *mreB* (a bacterial actin-like protein) or *parA* lead to a disorganization of carboxysomes within the cell. In the *parA* mutants, some daughter cells do not receive any carboxysomes meaning that they have to build their carbon fixation machinery de novo, which in turn causes a significant lengthening of their doubling times (Savage et al. 2010). This fascinating study establishes a clear link between the cytoskeleton and carboxysome organization. However, the specific connections between this organelle and ParA, as well as the mechanisms by which the proper spacing of carboxysomes is achieved remain to be elucidated.

Carboxysomes are actually part of a larger family of protein-bounded compartments, which are all related through homology between their shell proteins. One such organelle is the **1,2-propanediol use (Pdu) compartment** found in *Salmonella enterica*. Similar to carboxysomes, Pdu compartments house specific enzymes that are important for their cellular function. Interestingly, a recent report has shown that these enzymes all share a 20 amino acid amino-terminal sequence that is necessary for their packaging within the Pdu compartment (Fan et al. 2010). Furthermore, these amino-terminal sequences are also sufficient to target heterologous proteins such as GFP to Pdu compartments. Such amino-terminal sequence extensions were also detected in enzymes thought to be associated with other

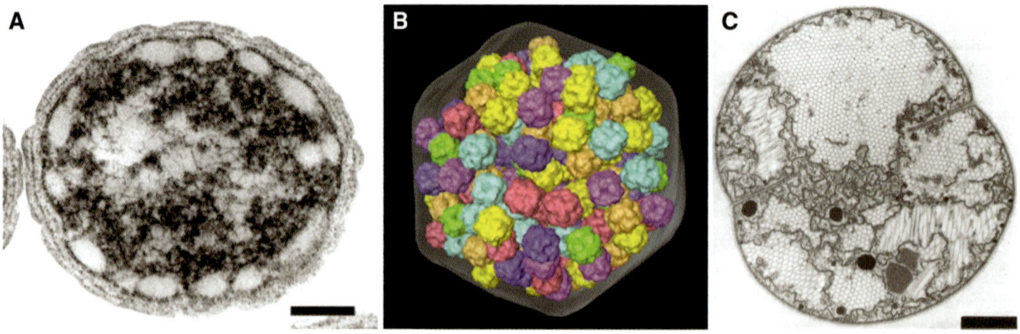

Figure 4. Chlorosomes of *Chlorobium tepidum* appear as flattened ovals arranged around the cell periphery (*A*). A representation of a single carboxysome based on CET imaging. The interior of the carboxysome appears to be packed with RuBisCO based on similarities between the known crystal structure of the enzyme and electron-dense entities seen in CET reconstructions (*B*). A TEM image of ta cyanobacterial cell reveals that the cytoplasmic space is filled with gas vesicles sectioned in two different orientations (*C*). (*A*, Reprinted, with permission, from Frigaard et al. 2002 [© ASM]; *B*, reprinted, with permission, from Iancu et al. 2007 [© Elsevier]; *C*, reprinted, with permission, from Walsby 1994 [©ASM].)

microcompartments making it likely that this mode of protein localization is universal among protein-bounded organelles (Fan et al. 2010). Beyond their relevance to understanding the cell biology of organelles, this finding also provides a method for engineering protein compartments in bacteria through the specific targeting of heterologous enzymes.

Another unique protein-bounded organelle in bacteria is the **gas vesicle** (Fig. 4C). Gas vesicles are gas-filled, protein-bound organelles that function to modulate the buoyancy of cells (Walsby 1994). They are found in a number of bacteria and archaea including halophilic and methanogenic archaea and phototrophic and heterotrophic bacteria. Most bacteria and archaea that have been shown to form gas vesicles are found in aqueous environments and are nonmotile. The proteinacious walls of gas vesicles are freely permeable to gas molecules. Water is also able to enter the gas vesicles but cannot form droplets on the inner surface because of its highly hydrophobic nature. Thus, any water that enters the gas vesicles evaporates (Walsby 1994), and the gas-filled vesicles decrease the overall density of the cells, allowing them to float upward. By controlling the formation of the gas vesicles, these organisms can specify their position in the water column to regulate their exposure to light, salt, nutrients and other environmental stimuli. Gas vesicles are cylindrical or spindle-shaped and the size of gas vesicles varies between species. Cells that grow at greater depths have gas vesicles that are narrower in width and are able to withstand greater hydrostatic pressure.

Ten to fourteen gas vesicle protein (*gvp*) genes, depending on the species, have been identified as being involved in gas vesicle formation. In *Halobacterium halobium*, at least ten *gvp* genes were found to be required for gas vesicle formation (DasSarma et al. 1994), and eight *gvp* genes in the halophilic archaeon *Halobacterium salinarum* are necessary and sufficient for gas vesicle formation (Offner et al. 2000). One of the essential genes encodes GvpA, the main vesicle wall component and one of the most hydrophobic proteins known. The crystal structure of GvpA has not been solved, mainly because GvpA aggregates and cannot be dissolved without denaturation. Nonetheless, the structure of gas vesicles has been investigated by X-ray analysis and atomic force microscopy (Blaurock and Walsby 1976; Blaurock and Wober 1976; McMaster et al. 1996), which revealed that the proteins form very ordered ribs and that the protein subunits align at an angle of 54° to the rib axis. Interestingly, 54° is close to the angle at which transverse and longitudinal stresses are equal in the wall of a cylindrical structure (Walsby 1994).

Much work has been performed to understand the physical properties of gas vesicles including their structure, their ability to withstand hydrostatic pressure, their ability to exclude water, and their permeability to gas. However, how the gas vesicle proteins interact to form the gas vesicles, and how gas vesicle formation is regulated in response to environmental cues remains largely unknown. Finally, it is possible that gas vesicle-like structures are found in other bacteria that exist in nonaqueous environments. Homologs of gas vesicle genes have also been found in actinomycetes that live in the soil yet no gas vesicle-like structures and no buoyancy phenotype has been seen (van Keulen et al. 2005).

CONCLUDING REMARKS

In conclusion, compartmentalization is not a feature limited to the eukaryotic world and numerous examples of highly complex and dynamic organelle systems can be found among the prokaryotes. The limited knowledge of the molecular mechanisms that control the biogenesis of these prokaryotic organelles does not allow for a direct mechanistic and evolutionary comparison to their well-studied eukaryotic counterparts. In many cases attempts to study prokaryotic organelles are hampered by their small size and a lack of molecular and genetic tools. With the advent of high-resolution imaging systems such as CET and the availability of numerous genome sequences, some of the barriers to the study of prokaryotic organelle biology are beginning to fade. Although the molecular understanding of organelle formation in prokaryotes is still at a relatively immature stage some general rules can be seen in the recent findings detailed in this article. First, it is clear that the proteins that can influence organelle formation are unique to each of the organelle systems discussed here. MamI and MamL are found only within the magnetotactic bacteria, the putative eukaryotic-like proteins found in the *Planctomycetes* are also unique to a limited group of bacteria and photosynthetic membranes are formed through a self-assembly mechanism using the photosynthetic proteins. These findings may imply that among bacteria membrane-bounded organelles evolved multiple times independently. Second, self-assembly may be a common mode of organelle biogenesis in both lipid-bounded organelles such as photosynthetic membranes and chlorosomes and protein-bounded compartments such as carboxysomes. At the moment, the restrictions placed on the cell by this mode of organelle formation remain unknown. For instance, can newly synthesized enzymes still be targeted to carboxysomes after the shell has closed? There are clear exceptions to this rule as well. In the case of magnetosomes, a large number of magnetosome proteins can be eliminated and yet the initial stages of membrane formation can still occur. Finally, cytoskeletal elements are used in multiple divergent systems as a means to organize and divide organelles. The magnetosomes of magnetotactic bacteria and the protein-bounded carboxysomes both require cytoskeletal proteins for accurate placement in the cell, which aids in proper function and segregation of these organelles during division.

Beyond the establishment of model systems and robust tools, a change in perspective may also be needed to move the understanding of these organelles to the next level. For decades the major focus of research in the study of prokaryotic organelles has been to uncover the enzymatic basis of their function and to take advantage of the biochemical products of these reactions for applied purposes. We would like to suggest that a dedicated focus on the cell biology of these organelles is needed to move forward. The approach should be similar to that taken by cell biologists studying organelle formation in eukaryotes in which experiments are focused on understanding the mechanisms that allow for membrane bending, protein sorting, and organelle division. By defining the cellular basis for organelle formation in prokaryotes we may then be able to directly tackle the evolutionary basis of compartmentalization across the various domains of life. Furthermore, this avenue of research will shed light on the general mechanisms used by prokaryotes to build large macromolecular assemblies and organize their cytoplasmic space. Finally, understanding the cell biology of prokaryotic organelles will allow for a more rational approach to their re-engineering in biotechnological and biomedical applications.

ACKNOWLEDGMENTS

A. Komeili is supported by grants from the David and Lucille Packard Foundation, The Hellman Family Foundation and the National Institutes of Health.

REFERENCES

Arakaki A, Webb J, Matsunaga T. 2003. A novel protein tightly bound to bacterial magnetic particles in *Magnetospirillum magneticum* strain AMB-1. *J Biol Chem* **278:** 8745–8750.

Bazylinski DA, Frankel RB. 2004. Magnetosome formation in prokaryotes. *Nat Rev Micro* **2:** 217–230.

Beatty JT, Overmann J, Lince MT, Manske AK, Lang AS, Blankenship RE, Van Dover CL, Martinson TA, Plumley FG. 2005. An obligately photosynthetic bacterial anaerobe from a deep-sea hydrothermal vent. *Proc Natl Acad Sci* **102:** 9306–9310.

Blaurock AE, Walsby AE. 1976. Crystalline structure of the gas vesicle wall from Anabaena flos-aquae. *J Mol Biol* **105:** 183–199.

Blaurock AE, Wober W. 1976. Structure of the wall of *Halobacterium halobium* gas vesicles. *J Mol Biol* **106:** 871–878.

Broda E. 1977. Two kinds of lithotrophs missing in nature. *Zeitschrift für allgemeine Mikrobiologie* **17:** 491–493.

Bryant DA, Costas AM, Maresca JA, Chew AG, Klatt CG, Bateson MM, Tallon LJ, Hostetler J, Nelson WC, Heidelberg JF, et al. 2007. *Candidatus Chloracidobacterium thermophilum*: An aerobic phototrophic *Acidobacterium*. *Science* **317:** 523–526.

Carballido-López R. 2006. The bacterial actin-like cytoskeleton. *Microbiol Mol Biol Rev* **70:** 888–909.

Chandler DE, Hsin J, Harrison CB, Gumbart J, Schulten K. 2008. Intrinsic curvature properties of photosynthetic proteins in chromatophores. *Biophys J* **95:** 2822–2836.

Chung S, Bryant DA. 1996. Characterization of csmB genes, encoding a 7.5-kDa protein of the chlorosome envelope, from the green sulfur bacteria *Chlorobium vibrioforme* 8327D and Chlorobium tepidum. *Arch Microbiol* **166:** 234–244.

Cohen-Bazire G, Pfennig N, Kunisawa R. 1964. The fine structure of green bacteria. *J Cell Biol* **22:** 207–225.

DasSarma S, Arora P, Lin F, Molinari E, Yin LR. 1994. Wild-type gas vesicle formation requires at least ten genes in the gvp gene cluster of *Halobacterium halobium* plasmid pNRC100. *J Bacteriol* **176:** 7646–7652.

Fan C, Cheng S, Liu Y, Escobar CM, Crowley CS, Jefferson RE, Yeates TO, Bobik TA. 2010. Short N-terminal sequences package proteins into bacterial microcompartments. *Proc Natl Acad Sci* **107:** 7509–7514.

Frankel RB, Papaefthymiou GC, Blakemore RP, Obrien W. 1983. Fe3o4 precipitation in magnetotactic bacteria. *Biochim Biophys Acta* **763:** 147–159.

Frigaard NU, Bryant DA. 2006. Chlorosomes: antenna organelles in green photosynthetic bacteria. In *Complex intracellular structures in prokaryotes (Microbiology Monographs)* (ed. JM Shively), pp. 79–114. Springer.

Frigaard NU, Voigt GD, Bryant DA. 2002. Chlorobium tepidum mutant lacking bacteriochlorophyll c made by inactivation of the bchK gene, encoding bacteriochlorophyll c synthase. *J Bacteriol* **184:** 3368–3376.

Frigaard NU, Li H, Milks KJ, Bryant DA. 2004. Nine mutants of *Chlorobium tepidum* each unable to synthesize a different chlorosome protein still assemble functional chlorosomes. *J Bacteriol* **186:** 646–653.

Fuerst JA. 2005. Intracellular compartmentation in planctomycetes. *Annu Rev Microbiol* **59:** 299–328.

Fukuda Y, Okamura Y, Takeyama H, Matsunaga T. 2006. Dynamic analysis of a genomic island in *Magnetospirillum* sp. strain AMB-1 reveals how magnetosome synthesis developed. *FEBS Lett* **580:** 801–812.

Gao H, Xu X. 2009. Depletion of Vipp1 in *Synechocystis* sp. PCC 6803 affects photosynthetic activity before the loss of thylakoid membranes. *FEMS Microbiol Lett* **292:** 63–70.

Gorby YA, Beveridge TJ, Blakemore RP. 1988. Characterization of the bacterial magnetosome membrane. *J Bacteriol* **170:** 834–841.

Grünberg K, Müller EC, Otto A, Reszka R, Linder D, Kube M, Reinhardt R, Schüler D. 2004. Biochemical and proteomic analysis of the magnetosome membrane in *Magnetospirillum gryphiswaldense*. *Appl Environ Microbiol* **70:** 1040–1050.

Grünberg K, Wawer C, Tebo BM, Schüler D. 2001. A large gene cluster encoding several magnetosome proteins is conserved in different species of magnetotactic bacteria. *Appl Environ Microbiol* **67:** 4573–4582.

Hohmann-Marriott MF, Blankenship RE. 2007. Hypothesis on chlorosome biogenesis in green photosynthetic bacteria. *Febs Letters* **581:** 800–803.

Iancu CV, Ding HJ, Morris DM, Dias DP, Gonzales AD, Martino A, Jensen GJ. 2007. The structure of isolated *Synechococcus* strain WH8102 carboxysomes as revealed by electron cryotomography. *J Mol Biol* **372:** 764–773.

Jogler C, Kube M, Schübbe S, Ullrich S, Teeling H, Bazylinski DA, Reinhardt R, Schüler D. 2009. Comparative analysis of magnetosome gene clusters in magnetotactic bacteria provides further evidence for horizontal gene transfer. *Environ Microbiol* **11:** 1267–1277.

Kerfeld CA, Sawaya MR, Tanaka S, Nguyen CV, Phillips M, Beeby M, Yeates TO. 2005. Protein structures forming the shell of primitive bacterial organelles. *Science* **309:** 936–938.

Komeili A. 2007. Molecular mechanisms of magnetosome formation. *Annu Rev Biochem* **76:** 351–366.

Komeili A, Li Z, Newman DK, Jensen GJ. 2006. Magnetosomes are cell membrane invaginations organized by the actin-like protein MamK. *Science* **311:** 242–245.

Komeili A, Vali H, Beveridge TJ, Newman DK. 2004. Magnetosome vesicles are present before magnetite formation, and MamA is required for their activation. *Proc Natl Acad Sci* **101:** 3839–3844.

Konorty M, Kahana N, Linaroudis A, Minsky A, Medalia O. 2008. Structural analysis of photosynthetic membranes by cryo-electron tomography of intact *Rhodopseudomonas viridis* cells. *J Struct Biol* **161:** 393–400.

Kroll D, Meierhoff K, Bechtold N, Kinoshita M, Westphal S, Vothknecht UC, Soll J, Westhoff P. 2001. VIPP1, a nuclear gene of *Arabidopsis thaliana* essential for thylakoid membrane formation. *Proc Natl Acad Sci* **98:** 4238–4242.

Lee KC, Webb RI, Fuerst JA. 2009. The cell cycle of the planctomycete *Gemmata obscuriglobus* with respect to cell compartmentalization. *Bmc Cell Biol* **10:** 4.

Li H, Bryant DA. 2009. Envelope proteins of the CsmB/CsmF and CsmC/CsmD motif families influence the size, shape, and composition of chlorosomes in *Chlorobaculum tepidum*. *J Bacteriol* **191:** 7109–7120.

Liberton M, Howard Berg R, Heuser J, Roth R, Pakrasi H. 2006. Ultrastructure of the membrane systems in the unicellular cyanobacterium *Synechocystis* sp. strain PCC 6803. *Protoplasma* **227:** 129–138.

Lindsay MR, Webb RI, Fuerst JA. 1997. Pirellulosomes: A new type of membrane-bounded cell compartment in planctomycete bacteria of the genus *Pirellula*. *Microbiology-Uk* **143:** 739–748.

Lindsay M, Webb R, Strous M, Jetten M, Butler M, Forde R, Fuerst J. 2001. Cell compartmentalisation in planctomycetes: Novel types of structural organisation for the bacterial cell. *Arch Microbiol* **175:** 413–429.

Low HH, Lowe J. 2006. A bacterial dynamin-like protein. *Nature* **444:** 766–769.

McMaster TJ, Miles MJ, Walsby AE. 1996. Direct observation of protein secondary structure in gas vesicles by atomic force microscopy. *Biophys J* **70:** 2432–2436.

Milne JLS, Subramaniam S. 2009. Cryo-electron tomography of bacteria: Progress, challenges and future prospects. *Nat Rev Microbiol* **7:** 666–675.

Murat D, Quinlan A, Vali H, Komeili A. 2010. Comprehensive genetic dissection of the magnetosome gene island reveals the step-wise assembly of a prokaryotic organelle. *Proc Natl Acad Sci* **107:** 5593–5598.

Nevo R, Charuvi D, Shimoni E, Schwarz R, Kaplan A, Ohad I, Reich Z. 2007. Thylakoid membrane perforations and connectivity enable intracellular traffic in cyanobacteria. *EMBO J* **26:** 1467–1473.

Oelze J, Drews G. 1972. Membranes of photosynthetic bacteria. *Biochim Biophys Acta* **265:** 209–239.

Offner S, Hofacker A, Wanner G, Pfeifer F. 2000. Eight of fourteen gvp genes are sufficient for formation of gas vesicles in halophilic archaea. *J Bacteriol* **182:** 4328–4336.

Okuda Y, Denda K, Fukumori Y. 1996. Cloning and sequencing of a gene encoding a new member of the tetratricopeptide protein family from magnetosomes of *Magnetospirillum magnetotacticum*. *Gene* **171:** 99–102.

Pradel N, Santini CL, Bernadac A, Fukumori Y, Wu LF. 2006. Biogenesis of actin-like bacterial cytoskeletal filaments destined for positioning prokaryotic magnetic organelles. *Proc Natl Acad Sci* **103:** 17485–17489.

Praefcke GJK, McMahon HT. 2004. The dynamin superfamily: Universal membrane tubulation and fission molecules? *Nat Rev Mol Cell Biol* **5:** 133–147.

Qian P, Bullough PA, Hunter CN. 2008. Three-dimensional reconstruction of a membrane-bending complex: the RC-LH1-PufX core dimer of *Rhodobacter sphaeroides*. *J Biol Chem* **283:** 14002–14011.

Richter M, Kube M, Bazylinski DA, Lombardot T, Glöckner FO, Reinhardt R, Schüler D. 2007. Comparative genome analysis of four magnetotactic bacteria reveals a complex set of group-specific genes implicated in magnetosome biomineralization and function. *J Bacteriol* **189:** 4899–4910.

Santarella-Mellwig R, Franke J, Jaedicke A, Gorjanacz M, Bauer U, Budd A, Mattaj IW, Devos DP. 2010. The compartmentalized bacteria of the planctomycetes-verrucomicrobia-chlamydiae superphylum have membrane coat-like proteins. *Plos Biol* **8:** e1000281.

Savage DF, Afonso B, Chen AH, Silver PA. 2010. Spatially ordered dynamics of the bacterial carbon fixation machinery. *Science* **327:** 1258–1261.

Schalk J, de Vries S, Kuenen J, Jetten M. 2000. Involvement of a novel hydroxylamine oxidoreductase in anaerobic ammonium oxidation. *Biochemistry* **9:** 635–642.

Scheffel A, Schüler D. 2007. The acidic repetitive domain of the *Magnetospirillum* gryphiswaldense MamJ protein displays hypervariability but is not required for *magnetosome* chain assembly. *J Bacteriol* **189:** 6437–6446.

Scheffel A, Gruska M, Faivre D, Linaroudis A, Plitzko JM, Schüler D. 2006. An acidic protein aligns magnetosomes along a filamentous structure in magnetotactic bacteria. *Nature* **440:** 110–114.

Scheffel A, Gärdes A, Grünberg K, Wanner G, Schüler D. 2008. The major magnetosome proteins MamGFDC are not essential for magnetite biomineralization in *Magnetospirillum gryphiswaldense* but regulate the size of magnetosome crystals. *J Bacteriol* **190:** 377–386.

Schmid MF, Paredes AM, Khant HA, Soyer F, Aldrich HC, Chiu W, Shively JM. 2006. Structure of *Halothiobacillus neapolitanus* carboxysomes by cryo-electron tomography. *J Mol Biol* **364:** 526–535.

Schneider D, Fuhrmann E, Scholz I, Hess WR, Graumann PL. 2007. Fluorescence staining of live cyanobacterial cells suggest non-stringent chromosome segregation and absence of a connection between cytoplasmic and thylakoid membranes. *BMC Cell Biol* **8:** 39.

Shively J. 2006. Complex intracellular structures in prokaryotes. Springer-Verlag, Berlin Heidelberg, Germany.

Sinninghe Damsté JS, Strous M, Rijpstra WIC, Hopmans EC, Geenevasen JAJ, van Duin ACT, van Niftrik LA, Jetten MSM. 2002. Linearly concatenated cyclobutane lipids form a dense bacterial membrane. *Nature* **419:** 708–712.

Staehelin LA, Golecki JR, Drews G. 1980. Supramolecular organization of chlorosomes (chlorobium vesicles) and of their membrane attachment sites in *Chlorobium limicola*. *Biochim Biophys Acta* **589:** 30–45.

Staley JT, Bouzek H, Jenkins C. 2005. Eukaryotic signature proteins of *Prosthecobacter dejongeii* and *Gemmata* sp

Wa-1 as revealed by in silico analysis. *Fems Microbiology Letters* **243:** 9–14.

Strous M, Fuerst JA, Kramer EH, Logemann S, Muyzer G, van de Pas-Schoonen KT, Webb R, Kuenen JG, Jetten MS. 1999. Missing lithotroph identified as new planctomycete. *Nature* **400:** 446–449.

Studholme DJ, Fuerst JA, Bateman A. 2004. Novel protein domains and motifs in the marine planctomycete *Rhodopirellula baltica*. *Fems Microbiology Letters* **236:** 333–340.

Sturgis JN, Tucker JD, Olsen JD, Hunter CN, Niederman RA. 2009. Atomic force microscopy studies of native photosynthetic membranes. *Biochemistry* **48:** 3679–3698.

Tanaka M, Okamura Y, Arakaki A, Tanaka T, Takeyama H, Matsunaga T. 2006. Origin of magnetosome membrane: Proteomic analysis of magnetosome membrane and comparison with cytoplasmic membrane. *Proteomics* **6:** 5234–5247.

Taoka A, Asada R, Wu L, Fukumori Y. 2007. Polymerization of the actin-like protein mamk, which is associated with magnetosomes. *J Bacteriol* **189:** 8737–8740.

Tauschel HD, Drews G. 1967. [Morphogenesis of thylakoids in Rhodopseudomonas palustris]. *Archiv für Mikrobiologie* **59:** 381–404.

Tavano CL, Donohue TJ. 2006. Development of the bacterial photosynthetic apparatus. *Curr Opin Microbiol* **9:** 625–631.

Ting CS, Hsieh C, Sundararaman S, Mannella C, Marko M. 2007. Cryo-electron tomography reveals the comparative three-dimensional architecture of *Prochlorococcus*, a globally important marine cyanobacterium. *J Bacteriol* **189:** 4485–4493.

Tsai Y, Sawaya MR, Cannon GC, Cai F, Williams EB, Heinhorst S, Kerfeld CA, Yeates TO. 2007. Structural analysis of CsoS1A and the protein shell of the *Halothiobacillus neapolitanus* carboxysome. *Plos Biol* **5:** e144.

Ullrich S, Kube M, Schübbe S, Reinhardt R, Schüler D. 2005. A hypervariable 130-kilobase genomic region of *Magnetospirillum gryphiswaldense* comprises a magnetosome island which undergoes frequent rearrangements during stationary growth. *Journal of Bacteriology* **187:** 7176–7184.

van de Meene AML, Hohmann-Marriott MF, Vermaas WFJ, Roberson RW. 2006. The three-dimensional structure of the cyanobacterium *Synechocystis* sp. PCC 6803. *Arch Microbiol* **184:** 259–270.

van Keulen G, Hopwood DA, Dijkhuizen L, Sawers RG. 2005. Gas vesicles in actinomycetes: old buoys in novel habitats? *Trends Microbiol* **13:** 350–354.

van Niftrik L, Geerts WJC, van Donselaar EG, Humbel BM, Webb RI, Fuerst JA, Verkleij AJ, Jetten MSM, Strous M. 2008a. Linking ultrastructure and function in four genera of anaerobic ammonium-oxidizing bacteria: Cell plan, glycogen storage, and localization of cytochrome c proteins. *Journal of Bacteriology* **190:** 708–717.

van Niftrik L, Geerts WJC, van Donselaar EG, Humbel BM, Webb RI, Harhangi HR, Camp HJMOd, Fuerst JA, Verkleij AJ, Jetten MSM, et al. 2009. Cell division ring, a new cell division protein and vertical inheritance of a bacterial organelle in *Anammox planctomycetes*. *Mol Microbiol* **73:** 1009–1019.

van Niftrik L, Geerts WJC, van Donselaar EG, Humbel BM, Yakushevska A, Verkleij AJ, Jetten MSM, Strous M. 2008b. Combined structural and chemical analysis of the anammoxosome: A membrane-bounded intracytoplasmic compartment in anammox bacteria. *J Structural Biol* **161:** 401–410.

Vassilieva EV, Stirewalt VL, Jakobs CU, Frigaard NU, Inoue-Sakamoto K, Baker MA, Sotak A, Bryant DA. 2002. Subcellular localization of chlorosome proteins in *Chlorobium tepidum* and characterization of three new chlorosome proteins: CsmF, CsmH, and CsmX. *Biochemistry* **41:** 4358–4370.

Walsby AE. 1994. Gas Vesicles. *Microbiol Rev* **58:** 94–144.

Westphal S, Heins L, Soll J, Vothknecht UC. 2001. Vipp1 deletion mutant of Synechocystis: A connection between bacterial phage shock and thylakoid biogenesis? *Proc Natl Acad Sci* **98:** 4243–4248.

Yeates TO, Kerfeld CA, Heinhorst S, Cannon GC, Shively JM. 2008. Protein-based organelles in bacteria: Carboxysomes and related microcompartments. *Nature Reviews Microbiology* **6:** 681–691.

Zeng X, Zeng X, Roh JH, Roh JH, Callister SJ, Callister SJ, Tavano CL, Tavano CL, Donohue TJ, Donohue TJ, et al. 2007. Proteomic characterization of the *Rhodobacter sphaeroides* 2.4.1 photosynthetic membrane: Identification of new proteins. *J Bacteriol* **189:** 7464.

Bacterial Nanomachines: The Flagellum and Type III Injectisome

Marc Erhardt[1], Keiichi Namba[2], and Kelly T. Hughes[1]

[1]Department of Biology, University of Fribourg, Fribourg 1700, Switzerland

[2]Graduate School of Frontier Biosciences, Osaka University, 1-3 Yamadaoka, Suita, Osaka 565-0871 Japan

Correspondence: hughes@biology.utah.edu

The bacterial flagellum and the virulence-associated injectisome are complex, structurally related nanomachines that bacteria use for locomotion or the translocation of virulence factors into eukaryotic host cells. The assembly of both structures and the transfer of extracellular proteins is mediated by a unique, multicomponent transport apparatus, the type III secretion system. Here, we discuss the significant progress that has been made in recent years in the visualization and functional characterization of many components of the type III secretion system, the structure of the bacterial flagellum, and the injectisome complex.

Bacteria swim in their environment by rotating a rigid, helical filament, the flagellum (Berg and Anderson 1973). The flagellum is a sophisticated molecular nanomachine made of about 25 different proteins (Berg and Anderson 1973; Namba and Vonderviszt 1997; Macnab 2003; Kojima and Blair 2004; Chevance and Hughes 2008) (Fig. 1A). Closely related to the flagellum is the type III injectisome, another complex nanomachine that allows Gram-negative bacteria to deliver effector proteins into eukaryotic host cells (Galán and Collmer 1999; Cornelis and Van Gijsegem 2000; Cornelis 2006; Galán and Wolf-Watz 2006) (Fig. 1B).

Most extracellular components of the flagellum and the injectisome are exported via a specific type III secretion system (T3SS). Importantly, the export apparatus of both the flagellum and the injectisome are closely related on structural and functional levels (Fig. 2 and Table 1).

STRUCTURAL AND FUNCTIONAL SIMILARITIES OF THE FLAGELLUM AND INJECTISOME

Electron micrographs (EM) of the injectisome and the flagellum revealed clear structural similarities (Fig. 1). However, in place of the flagellar hook and helical filament, injectisomes have usually a short, straight needle. In addition to the structural resemblance, there is strong conservation in the constituent proteins. Of the roughly 20 proteins needed to form the flagellar basal body, about half have clear counterparts in the injectisome (Table 1).

Importantly, most components of T3SS are conserved in both flagellar and virulence-associated T3SS and many proteins are homologous in sequence or function (Kubori et al. 1998; Blocker et al. 2003; Macnab 2004; Cornelis 2006) (Table 1).

 doi: 10.1101/cshperspect.a000299
Cite this article as *Cold Spring Harb Perspect Biol* 2010;2:a000299

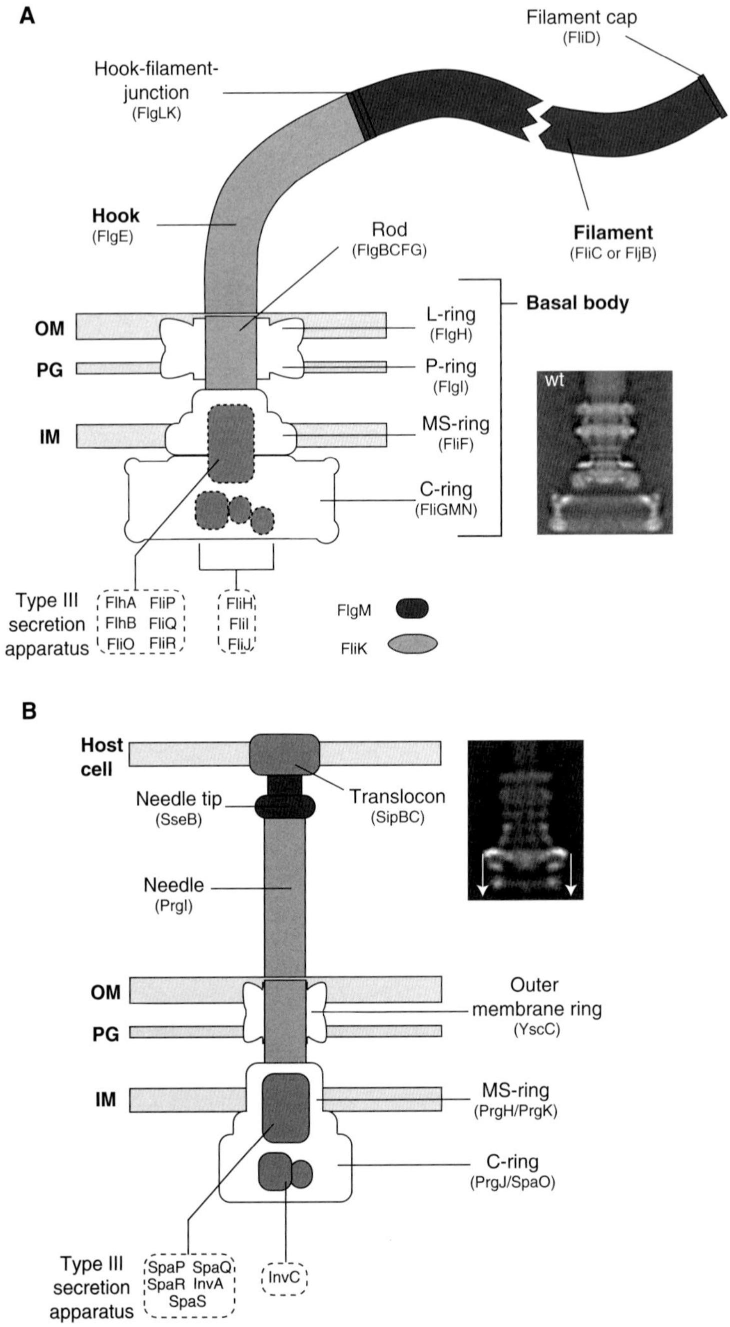

Figure 1. Schematic comparison of the flagellum and type III injectisome of *Salmonella*. (*A*) Schematic overview of the flagellum of *Salmonella*. The structure of the flagellum consists of three parts: 1) a basal body with a flagellar-specific type III secretion system within the inner membrane ring; 2) a flexible hook acting as a universal joint to 3) the rigid filament. Dashed boxes illustrate proteins with functions in flagellar type III secretion. Also indicated are FlgM, the negative regulator of late substrate gene expression that is secreted after hook-basal-body secretion, and the hook-length regulator FliK that measures rod-hook length and ultimately determines the secretion substrate specificity switch. (*See facing page for legend.*)

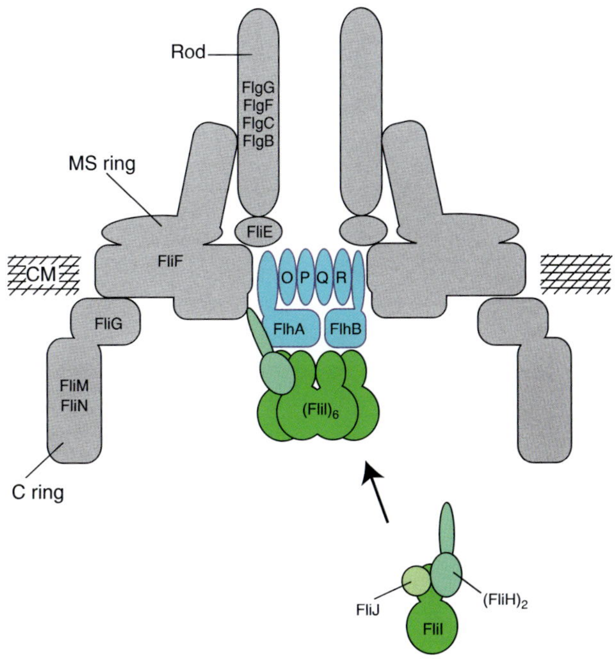

Figure 2. Schematic of the flagellar type III secretion apparatus. FlhA, FlhB, FliO, FliP, FliQ, and FliR are integral membrane components of the flagellar type III secretion apparatus. The membrane components of the T3S apparatus are believed to assemble within a central pore of the MS-ring. FliI, the flagellar-specific ATPase, FliH, the regulator of FliI and FliJ, a general chaperone, are soluble proteins. FliI forms a heterotrimer together with the homodimer FliH in the cytoplasm. After docking to the membrane components of the T3S apparatus, FliI forms a functional hexamer. The protein translocation is dependent on the proton motive force and the substrates have to be secreted in an unfolded state through a narrow channel of about 2.0 nm in diameter. It has been proposed that FliI facilitates substrate unfolding in an ATP-dependent manner. CM = cytoplasmic membrane. (Reprinted, with permission, from Minamino et al. 2008a [Royal Society of Chemistry].)

Common features of all T3SS are the very fast operation (in case of the flagellar T3SS, almost 20 molecules of 55-kDa flagellin are transported per second in early stages of filament assembly [Iino 1974]) and stringent substrate discrimination (only a few proteins are exported, from among the many hundreds present in the cell). Additionally, substrates of T3SS feature an amino (N)-terminal peptide secretion signal that is usually structurally disordered (Namba and Vonderviszt 1997; Namba 2001) and distinct secretion chaperones stabilize or

Figure 1. (*Continued*). OM = outer membrane; PG = peptidoglycan; IM = inner membrane. The inlay EM (electron micrograph) picture shows an isolated hook-basal-body complex of *Salmonella* (Thomas et al. 2001). (*B*) Schematic overview of the SPI-1 injectisome of *Salmonella*. Many components of the flagellum and injectisome are structurally and/or functionally related. The injectisome structure can also be divided into three main parts: 1) the basal body with the type III secretion apparatus within the inner membrane ring; 2) a straight needle connecting the bacterial secretion system to 3) the translocon complex that forms a pore in the membrane of eukaryotic host cells. The inlay EM picture shows an injectisome of *Salmonella enterica* (Marlovits et al. 2004). (The arrows on this image served in the original paper to explain aspects of the reconstruction procedure.) (*A* [inlay], Reprinted, with permission, from Thomas et al. 2001 [© ASM]; *B* [inlay] reprinted, with permission, from Marlovits et al. 2004 [© AAAS].)

Table 1. Comparison of structurally and/or functionally related components of the flagellum and injectisome systems of *Salmonella enterica, Yersinia ssp., Shigella ssp.*, and enteropathogenic *Escherichia coli*.

Flagellum (*S. enterica*)	Injectisome (*S. enterica* SPI-1)	Injectisome (*Yersinia* ssp. Ysc)	Injectisome (*Shigella* ssp.)	Injectisome (enteropathogenic *Escherichia coli* EPEC)	Structure/Function
FliF	PrgH/PrgK	YscJ	MxiJ	MxiJ	MS-ring inner-membrane ring
FliI	InvC	YscN	Spa47	EscN	ATPase
FliJ		YscO			T3S chaperone
FliGMN	PrgJ/SpaO	YscQ	Spa33		C-ring cytoplasmic ring (HrcQ in Pseudomonas)
FliP	SpaP	YscR	Spa24		T3S apparatus inner-membrane protein
FliQ	SpaQ	YscS	Spa33		T3S apparatus inner-membrane protein
FliR	SpaR	YscT	Spa29		T3S apparatus inner-membrane protein
FlhA	InvA	YscV	MxiA		T3S apparatus inner-membrane protein
FlhB	SpaS	YscU	Spa40		T3S apparatus inner-membrane protein
N/A	InvG	YscC	MxiD		outer-membrane ring
FlgE?	PrgI	YscF	MxiH	EscF	extracellular needle
N/A	SipBC	YopBD	IpaBC	EsoBD	translocation pore
FliC?	SseB	LcrV	IpaD	EspA	needle extension
FliH?		YscL	MxiN		ATPase regulator
FliK	InvJ	YscP	Spa32		hook/needle length regulator

target their cognate secretion substrates to the secretion apparatus (Parsot et al. 2003). Recently, it has been unraveled that translocation of the substrates across the inner membrane is dependent on the proton motive force (PMF) (Minamino and Namba 2008; Paul et al. 2008) and is presumably coupled to ATP-dependent substrate release and unfolding (Akeda and Galán 2005).

Another interesting fact is that the flagellar T3SS is believed to be the ancestor of all T3SS. Flagellar T3SS are present in both Gram-positive and Gram-negative bacteria, and it has been proposed that type III secretion required for pathogenesis evolved from flagellar-specific T3SS (Hueck 1998; Macnab 2004). However, recent phylogenetic studies indicate that both the flagellum and the injectisome share a common ancestor (Gophna et al. 2003).

THE FLAGELLAR TYPE III SECRETION APPARATUS

The flagellar-specific T3S apparatus is believed to assemble within the MS-ring at the basis of the flagellar basal body and consists of six integral membrane proteins (FlhA, FlhB, FliO, FliP, FliQ, FliR) and three cytoplasmic proteins (FliH, FliI, FliJ) (Minamino and Macnab 1999), which are essential for the export of rod-type, hook-type, and filament-type secretion substrates (Minamino 1999; Minamino 2008a). A schematic of the flagellar-specific T3S apparatus is displayed in Figure 2.

It is important to note that secretion substrates are translocated by the flagellar T3S apparatus across the cytoplasmic membrane through a narrow channel of about 2.0 nm in diameter (Yonekura et al. 2003); therefore, the proteins are most likely exported in an unfolded or partially folded state.

The soluble components FliH, FliI, and FliJ of the flagellar T3S system are thought to facilitate delivery and unfolding of secretion substrates to the export apparatus (Macnab 2003). FliI is an ATPase (Fan and Macnab 1996) that can form a hexameric ring-shaped structure (Claret et al. 2003) and that is presumed to couple ATP hydrolysis to some energy-utilizing step(s). FliI was thought to energize the transmembrane transport process, but other studies indicate that ATP hydrolysis energizes steps in substrate delivery instead, probably facilitating the unfolding of the substrate and its release from the chaperone (Akeda and Galán 2005).

Recent findings have shown that the transport process itself is energized by the proton motive force and does not require FliI (Minamino and Namba 2008; Paul et al. 2008). The two other soluble components of the T3S apparatus are FliH, which regulates the ATP-hydrolyzing activity of FliI, and FliJ, a general chaperone for flagellar secretion substrates.

The integral membrane components FliOPQR of the flagellar-specific T3SS are relatively small (FliO = 13.1 kDa; FliP = 26.8 kDa; FliQ = 9.6 kDa; FliR = 28.9 kDa) and are predicted to have one to eight membrane spanning helices. Little is known about possible functions of FliO, FliP, FliQ, and FliR respectively, but the proteins are essential for flagellar type III secretion. FlhA and FlhB are much larger integral membrane proteins (75 kDa and 42 kDa, respectively) and also have large cytoplasmic domains, where the proteins interact with the soluble components of the T3S apparatus, FliH, FliI, and FliJ (Minamino and Macnab 2000). FlhB is furthermore controlling the switch of substrate specificity to export late structural subunits upon completion of early flagellar structures (Ferris and Minamino 2006). Although the major players in export have probably all been identified, the molecular mechanism of flagellar T3SS is still poorly understood. A reasonable, general proposal for the mechanism is that FliH/FliI/FliJ complexes bind to substrate proteins and target them to the membrane-bound export apparatus. Subsequently, the membrane components of the apparatus then translocate the substrate across the membrane and into the channel, using energy from the electrochemical potential gradient of protons, which is the sum of the membrane potential $\Delta\psi$ and the proton concentration gradient ΔpH (Minamino and Namba 2008; Paul et al. 2008).

STRUCTURE AND ASSEMBLY OF THE BACTERIAL FLAGELLUM

The structure of the flagellum can be divided into three parts: I) the basal body as the rotary motor (Berg and Anderson 1973; Silverman and Simon 1974) and the type III export apparatus; II) the flexible hook that couples the rotary motor to III) the rigid filament (DePamphilis and Adler 1971a,b) (Fig. 1A). The filament is about 10 to 15 µm long, yet only 120–240 nm in diameter, and the rotary machine turns at hundreds of revolutions per second, utilizing both $\Delta\psi$ and ΔpH (Manson et al. 1977; Matsuura et al. 1977).

Flagellum assembly initiates with the formation of the MS-ring (consisting of FliF) in the inner membrane, followed by attachment of the rotor/switch complex (C-ring proteins FliG, FliM, and FliN) at the cytoplasmic face of the MS-ring. An overview of the assembly process is given in Figure 3. Upon completion of the C-ring, the motor proteins MotA and MotB are assembled in the inner membrane. MotA and MotB form the stator complex, which also forms the pathway for proton influx, and attach noncovalently to the peptidoglycan layer via the C-terminal periplasmic domain of MotB, whereas the rotor (FliG) is noncovalently attached to the MS-ring. Together, the stator and the rotor form the flagellar motor, whose rotation is energized by the proton motive force (Macnab 2003; Chevance and Hughes 2008). The periplasmic domain of MotB is thought to go through a relatively large conformational change to bind to the peptidoglycan layer and open the proton pathway upon assembly of the MotAB complex to the motor (Kojima et al. 2008).

After assembly of the C-ring, the flagellar-specific T3SS (consisting of FlhA, FlhB, FliH,

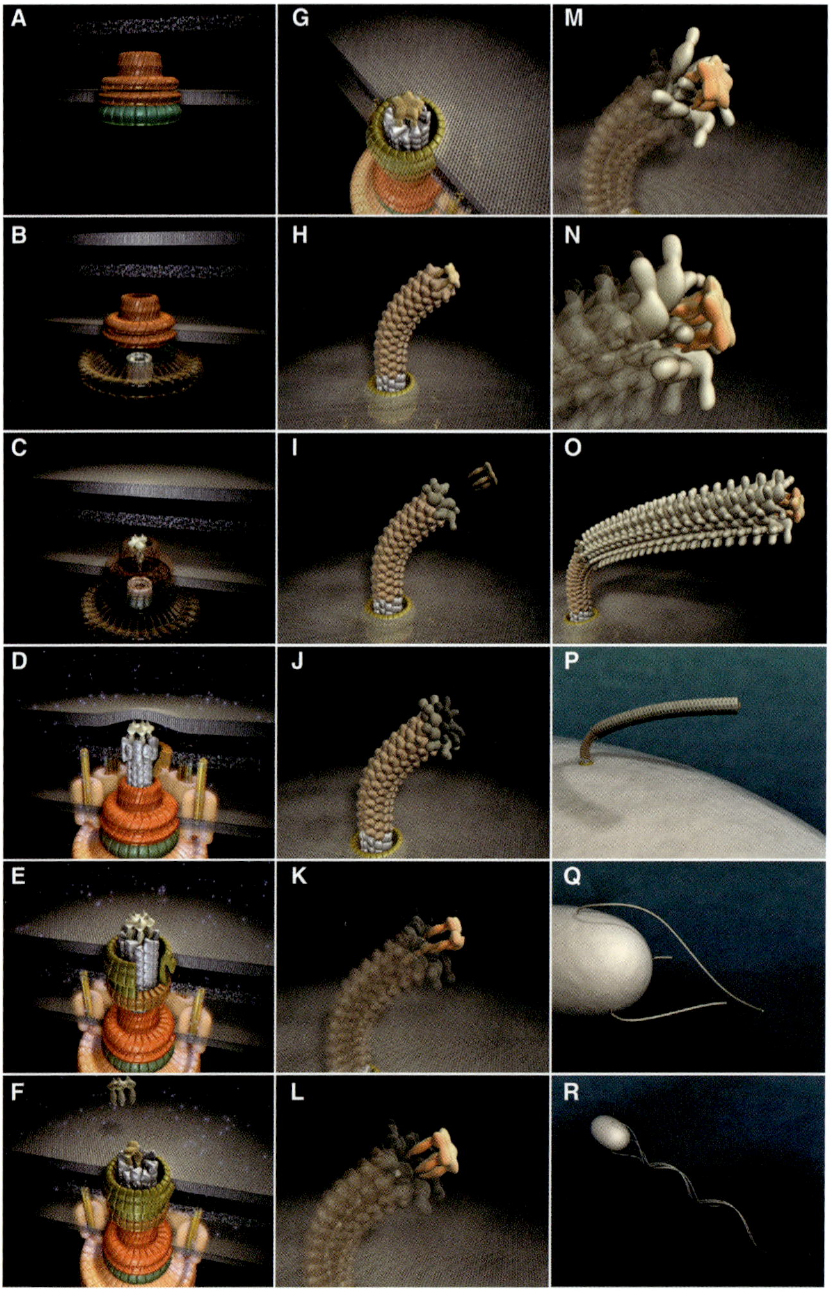

Figure 3. Steps in assembly of the bacterial flagellum. The self-assembly process of the bacterial flagellum starts from the top left (*A*) and proceeds to the bottom right (*R*). Assembly starts with the formation of the MS-ring in the cytoplasmic membrane (*A*). Afterward, the cytoplasmic C-ring is attached to the MS-ring and the flagellar-specific type III secretion apparatus assembles within a central pore of the MS-ring (*B*). Flagellar secretion substrates are now secreted specifically via the T3S apparatus. Flagellar proteins that reach the distal end of the growing structure, self-assemble onto the existing structure with the help of distal cap proteins (shown as pentamers at the tips of the axial structures) (*C*–*O*). First, the rod acting as a driveshaft is assembled beneath the rod-cap, which is also a muramidase to allow penetration through the peptidoglycan layer (*D*). (*See facing page for legend.*)

FliI, FliJ, FliO, FliP, FliQ, and FliR) assembles within the putative central pore of the MS-ring at the basis of the basal body (Aizawa 1996), facilitating the PMF-dependent export of most extra cytoplasmic components of the flagellum (Paul et al. 2008). Subsequently, the rod components FliE, FlgB, FlgC, FlgF (proximal rod), and FlgG (distal rod) are secreted by the T3S apparatus and progressively assembled (Minamino et al. 2000). The P-ring protein FlgI and the L-ring lipoprotein FlgH are probably exported by the Sec-dependent pathway (Homma et al. 1987), but the assembly of the periplasmic P-ring and the L-ring in the outer membrane is dependent upon the construction of the preceding flagellar rod structure. Upon completion of the rod and the rings, the hook (consisting of FlgE) is assembled to an approximate length of 55 nm, which is controlled by the hook-length regulator FliK (Patterson-Delafield et al. 1973; Hirano et al. 1994). The last steps of flagellum assembly include secretion of the anti-σ^{28} factor FlgM (Hughes et al. 1993), which results in σ^{28}-dependent expression of flagellar genes that are under Class III promotor regulation, and the assembly of the hook-associated proteins FlgK, FlgL, and FliD (Homma et al. 1990). Subsequently, the filament is constructed with the subunit FliC or FljB (flagellin) as the last structural component of the flagellum, where the assembly of flagellin occurs at the distal end of the growing filament under the cap composed of FliD, as described later. A single filament can be constructed of as many as 20,000–30,000 flagellin subunits (Macnab 2003; Chevance and Hughes 2008), which represents a significant amount of the total protein mass of the cell.

STRUCTURE AND ASSEMBLY OF THE INJECTISOME

The injectisome is used by many Gram-negative pathogens to pump virulence-effector proteins into host cells (Kubori et al. 1998; Blocker et al. 2003; Journet et al. 2005; Cornelis 2006; Galán and Wolf-Watz 2006). Members of the injectisome family are found in many Gram-negative plant and animal pathogens, including *Salmonella sp.*, *Yersinia sp.*, enteropathogenic *Escherichia coli* (EPEC), *Shigella sp.*, *Chlamydia sp.*, and *Erwinia carotovora* (Blocker et al. 2003; Galán and Wolf-Watz 2006). Injected effector proteins alter host-cell functions and membrane cytoskeletons to promote invasion, survival, and growth of the bacterium, or in some cases are used to facilitate symbiosis.

The overall structure of the injectisome is very similar to the flagellum: I) a basal, cylindrical structure spanning the inner and outer membrane and containing the T3SS and II) an extracellular, hollow tubular structure (Tampakaki et al. 2004; Cornelis 2006) (Fig. 1B). The extracellular appendages have different structures and length, depending on the family of injectisomes; for example, a stiff needle with an approximate length of 58 nm in *Yersinia enterocolitica*, a filament with a length up to 600 nm in enteropathogenic *Escherichia coli*, or a Hrp pilus with a length of several μm in *Pseudomonas syringae* (Cornelis 2006).

Similar to the assembly of the flagellum, it has been assumed that the injectisome also assembles in a sequential manner. The basal structure of the injectisome includes the T3SS, a pair of rings in the inner and outer membranes, and the rod. In *Salmonella ssp.*, the inner ring consists of the proteins PrgH and

Figure 3. (*Continued*). In addition, motor force generators that couple proton flow to torque generation assemble in the cytoplasmic membrane and interact with C-ring components (*D*). Rod assembly ends with the formation of the PL-ring bushing, outer membrane penetration, and the replacement of the rod scaffold with the hook scaffold (*E–G*). Afterward, hook subunits are secreted and the hook grows to a final length of 55 nm (Hirano et al. 1994), triggering a secretion specificity switch from rod-hook-type substrates to late-secretion substrates (*H*). Upon completion of the hook-basal-body complex, hook-filament junction proteins (*I–J*), the filament cap (*K–L*), and filament subunits are secreted, and the filament grows to a maximal length of about 10–15 μM (*M–R*). (Reprinted, with permission, from Minamino et al. 2008a [Royal Society of Chemistry].)

PrgK that are homologous to the MS-ring protein FliF (Kimbrough and Miller 2000; Kubori et al. 2000). In the flagellar system, the C-ring (FliGMN) is attached to the MS-ring on the cytoplasmic face. Although no EM pictures of injectisome show any attached C-ring-like structure, the proteins of the YscQ family share significant similarity to FliN and FliM (Cornelis 2006). The membrane-embedded components YscR, YscS, YscT, YscU, and YscV are located within the inner membrane ring of the injectisome T3SS. Although these inner-membrane components of the injectisome T3SS share high sequence homology to the inner-membrane proteins FliPQR and FlhAB of the flagellum, no further structural data is available. As it has been shown for the flagellar T3SS (Minamino and Namba 2008; Paul et al. 2008), the T3SS of *Yersinia enterolitica* is also dependent on the proton-motive force (Wilharm et al. 2004), indicating also a functional similarity. At the base of the inner membrane ring, the hexameric ATPase of the YscN family shares striking similarity to the flagellar ATPase FliI (Imada et al. 2007; Zarivach et al. 2007). It has been shown for the *Salmonella* homolog InvC that the ATPase energizes the secretion substrate release from its cognate cytoplasmic chaperone prior to PMF-dependent secretion (Akeda and Galán 2005).

A protruding rod structure assembles on top of the MS-ring in the flagellar basal body, and it has been suggested that the inner rod of the *Salmonella* injectisome is composed of PrgJ (Kubori et al. 2000; Marlovits et al. 2004). The last components of the basal structure are a pair of rings associated with the peptidoglycan layer and the outer membrane made of the YscC family of Secretin proteins (Koster et al. 1997; Kubori et al. 2000).

The extracellular structures of the *Yersinia* injectisome are exported via the membrane-embedded T3SS and can be divided into three parts: I) a needle, II) a needle-extension forming the tip complex, and III) the translocation pore that presumably forms a channel in the host cell membrane. The injectisome needle is made of approximately 100–150 molecules of the YscF family proteins that polymerize as a helical assembly into a filamentous structure (Kubori et al. 2000; Cornelis 2006). The molecular ruler YscP controls needle length in *Yersinia* to about 58 ± 10 nm (Journet et al. 2003). At the top of the needle, the tip complex formed by LcrV is located (Mueller et al. 2005; Deane et al. 2006). It has been suggested that the tip complex forms a scaffold for the assembly of the translocator pore consisting of YopB and YopD, which will be inserted into the membrane of host cells (Cornelis 2006).

ELECTRON MICROGRAPH RECONSTRUCTIONS OF THE FLAGELLAR HOOK-BASAL BODY COMPARED TO THE NEEDLE COMPLEX STRUCTURE

The core basal complexes of both the flagellar hook-basal-body and injectisome are structurally very similar. Figure 4 compares 3D-EM image reconstructions of both the hook-basal-body (Derosier 2006) and the needle complex (Marlovits et al. 2004) of *Salmonella enterica* serovar *Typhimurium*. The structure of the needle complex from *Shigella* has also been studied by 3D-EM image reconstruction (Hodgkinson et al. 2009). The 3D images of those purified needle complexes revealed a cylindrical basal body and a hollow, needle-like structure that spans both membranes with a central channel of 20–30 Å diameter (Blocker et al. 2003). The hook-basal-body structure of the flagellum also displays a cylindrical basal body with rings in the inner and outer membrane, as well as in the peptidoglycan layer. The main differences are the diameter of the cytoplasmic ring of the flagellar basal body, which is significantly wider (approx. 40 nm) (Macnab 2003) compared to the ring of the injectisomes (approx. 20 nm) (Marlovits et al. 2004; Hodgkinson et al. 2009), and the lack of a C-ring-like structure in the injectisome.

Although these EM studies provide an excellent overview of the architecture of both the flagellar and injectisome basal bodies, the resolution of these EM maps is too low to allow detailed studies of the localization of individual structural elements. Additionally, because of harsh sample preparation methods, the resulting structures lack essential components, like the soluble components of the T3SS and the C-ring homolog of the injectisome.

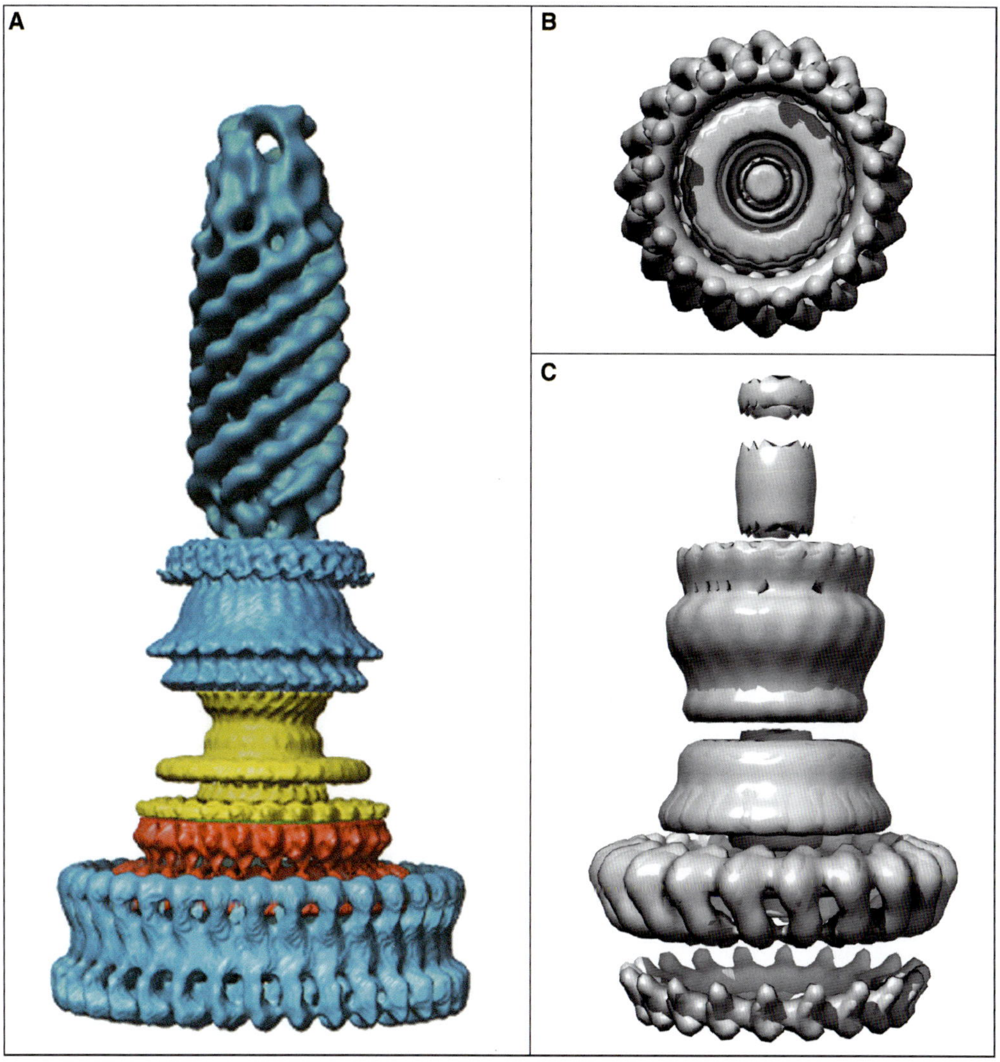

Figure 4. EM reconstructions of basal bodies of the flagellum and type III injectisome of *Salmonella*. (*A*) Model of the bacterial hook-basal-body complex based on EM reconstructions (Derosier 2006). (*B*, *C*) Surface renderings of the injectisome basal body and needle structure based on EM reconstruction data EmDep Database accession number EMD1100 (Marlovits et al. 2004). (*B*) View of the injectisome basal body complex from the cytoplasm and (*C*) side-view of the injectisome basal body with attached needle. (*A*, Reprinted, with permission, from Derosier 2006 [© Elsevier]; *B*, *C*, based on Marlovits et al. 2004 [© AAAS].)

A POSSIBLE ATOMIC MODEL OF THE INNER MEMBRANE RING ESCJ OF THE INJECTISOME

The formation of a ring complex in the inner membrane seems to be one of the first steps in the assembly of both the flagellum and injectisome basal structure. The inner membrane ring is thought to act as a scaffold for the assembly of other sub-structures and houses the membrane-embedded components of the T3SS.

The inner membrane ring of the flagellum is made of a single protein, FliF, forming the MS-ring. In case of the injectisome, the inner membrane ring is composed of the highly conserved YscJ family (PrgK in *Salmonella*, EscJ in EPEC).

Recently, the atomic structure of EscJ has been determined by X-ray crystallography, and a symmetrical 24-subunit model has been constructed (Yip et al. 2005) (Fig. 5). The overall dimensions of the modeled EscJ ring (diameter 180 Å, height 52 Å) are similar to previous estimations of the diameter of the inner ring based on EM reconstructions of the injectisome (Marlovits et al. 2004). Intriguingly, the EscJ ring model can be fitted into existing 3D-EM maps of the injectisome basal body of *Salmonella enterica* (Moraes et al. 2008).

CRYSTAL STRUCTURES OF THE C-RING COMPONENTS FliN, FliG, AND HrcQ

At the cytoplasmic face of the flagellar MS-ring, a cytoplasmic ring (C-ring) composed of FliG, FliM, and FliN is assembled. The C-ring components of the flagellum are assumed to be part of the rotor and have functions in torque generation, switching of the rotational direction, and flagellar export (Vogler et al. 1991; Sockett et al. 1992; Lloyd et al. 1996). It has been estimated that about 26 copies of FliF and FliG, about 34 copies of FliM, and about 130 copies of FliN are located in the flagellar basal body structure (Thomas et al. 1999, 2001, 2006; Paul and Blair 2006).

The crystal structure of the middle and C-terminal domains of FliG is available (Brown et al. 2002) and importantly can be fitted into a 3D-EM structure of the C-ring (Brown et al. 2007), where both the middle and C-terminal domains of FliG can interact with FliM (Fig. 6A).

In the case of the injectisome, no EM images are available that prove the presence of a C-ring-like structure. However, biochemical evidence

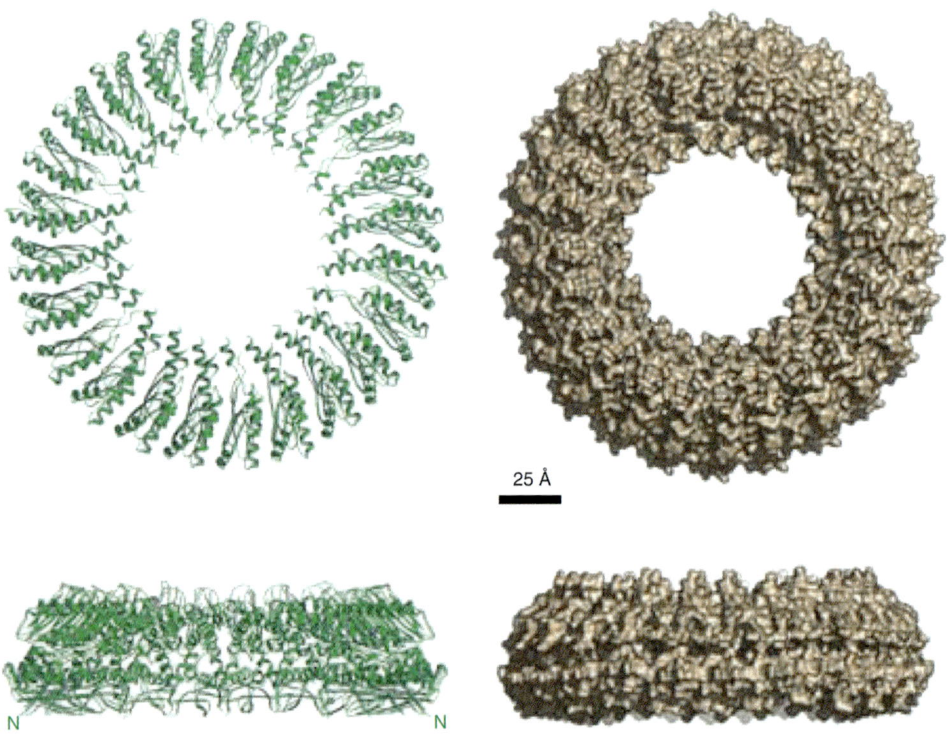

Figure 5. Model of the injectisome inner membrane ring, EscJ. Shown are ribbon and surface representations of the 24-subunit EscJ ring model. The sideview illustrates the two-layered exterior structure that closely resembles the flagellar MS-ring structure based on cryo-EM studies. The N-termini of the EscJ subunits are located at the wide face of the ring (also labeled N in the ribbon diagrams sideview). Scale bar 25 Å. (Reprinted, with permission, from Yip et al. 2005 [Nature Publishing Group].)

argues strongly toward the existence of a C-ring in injectisomes made of proteins of the YscQ family (Cornelis 2006). Additionally, the YscQ family shows significant sequence similarity to the flagellar C-ring components FliM and FliN, and high-resolution structures of both FliN with the N-terminal 22 residues truncated (Brown et al. 2005), and the C-terminal part of HrcQ of the YscQ family of *Pseudomonas syringae* (Fadouloglou et al. 2004) are available. The overall structures of HrcQ and FliN are remarkably similar and both proteins form dimers composed mostly of β-sheets (Fig. 6B–D). Two HrcQ dimers form a homotetramer that has been proposed to be the repeating unit forming a C-ring-like structure in the injectisome basal body (Fadouloglou et al. 2004), and a FliN tetramer can also be modeled to fit into the flagellar basal body C-ring (Paul and Blair 2006) (Fig. 6A).

COMPARISON OF THE STRUCTURES OF ATPase SUBUNITS OF THE T3SS, FliI, AND EscN

At the basis of the basal body of both the flagellum and the injectisome, the T3SS is localized. The T3SS is composed of several integral-membrane and soluble components. Soluble components are a T3SS-specific ATPase, regulators of the ATPase, and T3SS chaperones. As noted above, the ATPase subunit FliI of the flagellar T3SS is thought to unfold secretion substrates prior to the PMF-dependent export (Akeda and Galán 2005; Minamino et al. 2008a; Paul et al. 2008).

In the case of the injectisome, the YscN family of proteins represents the ATPase subunit. The ATPase subunit forms hexameric oligomers that are associated with the inner membrane as it has been shown for both FliI and HrcN, the ATPase of *Pseudomonas syringae* (Claret et al. 2003; Pozidis et al. 2003).

Recently, the atomic structures of EscN of EPEC (Zarivach et al. 2007) and FliI of *Salmonella* (Imada et al. 2007) have been determined (Fig. 7C,D). Although the probable function of the T3SS ATPase is to disassemble itself and secretion chaperones from their cognate secretion substrates, the high-resolution structures of both EscN and FliI reveal a structural homology to the α/β-subunit of F_1-ATPase rather than to the AAA ATPase as previously suggested (Akeda and Galán 2005). Based on the atomic structures of EscN and FliI, respectively, hexameric models have been constructed that reveal the close structural homology to the α_3/β_3-heterohexamer of F_1-ATPase (Fig. 7A,B).

As displayed in Figure 7, the overall structures of EscN and FliI are highly similar to each other. The EscN structure is composed of two domains corresponding to the ATPase and C-terminal domains of F_1-ATPase, respectively, while the FliI structure contains all three domains of F_1-ATPase. The N-terminal domain containing a six-stranded β-barrel might function in oligomerization (Okabe et al. 2009). The ATPase domain has Walker A/B motifs and the Rossman fold for nucleotide binding. The C-terminal domain shares less homology to that of F_1-ATPase and might function in interaction with T3SS chaperones.

STRUCTURES OF EXTRACELLULAR APPENDAGES: THE FLAGELLAR HOOK, FILAMENT, AND INJECTISOME NEEDLE

The injectisome needle is a straight, hollow tube of approximately 60 nm in length and 7 nm in diameter and is made of 100–150 subunits of proteins of the YscF family (Cornelis 2006). YscF proteins are relatively small with about 9 kD, compared to the flagellar hook subunit FlgE (42 kDa) and filament subunit FliC (51 kDa).

Although YscF proteins share no sequence homology to the flagellar filament or hook, the atomic resolution structure of MxiH, the needle subunit of *Shigella ssp.* (Deane et al. 2006), revealed structural features similar to the D0 domain of flagellin and EspA, the needle extension protein (Fig. 8A). MxiH consists of two antiparallel α-helices that are connected via a short turn, and the same helix bundles can be found in both EspA and flagellin. Furthermore, the atomic modeling of an injectisome needle was possible by docking the MxiH structure onto a 3D-EM reconstruction

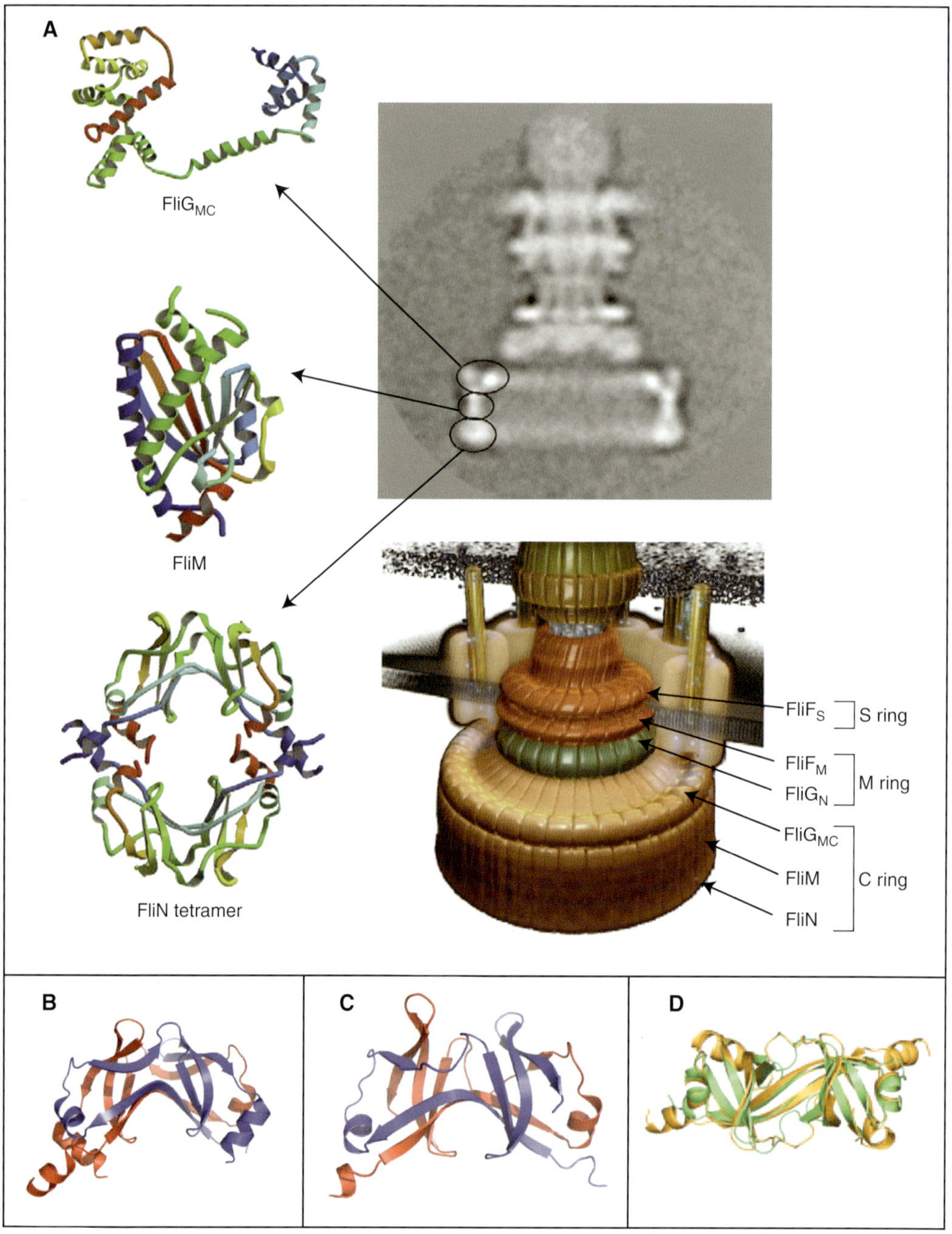

Figure 6. Crystal structures of the components of the C-ring, FliG$_{MC}$, FliM, and FliN. (*A*) Left panel: Ribbon representation of the crystal structures of FliG$_{MC}$ (middle and C-terminal domains of FliG), FliM, and the doughnut-like tetramer of FliN. Upper right panel: cyro-EM image of a hook-basal-body complex. The arrows illustrate the positions of the FliG, FliM, and FliN proteins in the C-ring structure. Lower right panel: Model of the flagellar basal body complex with proposed locations of FliF, FliG, FliM, and FliN forming the MS-ring and C-ring, respectively (Minamino et al. 2008b). (*B,C*) Comparison of the homologous structures of FliN and HrcQB$_C$ (Fadouloglou et al. 2004). (*B*) Crystal structure of the C-terminal fragment of FliN of *Thermotoga maritima* (PDB 1O6A). (*See facing page for legend.*)

of the *Shigella* needle (Cordes et al. 2003; Deane et al. 2006) (Fig. 8B).

Importantly, the overall helical properties of the needles isolated from *Shigella* are remarkably similar to the helical properties of the flagellar hook and filament (5.6 subunits per turn of the 1-start helix and a helical pitch of 2.4 nm), highlighting the structural similarities of both the flagellum and the injectisome (Cordes et al. 2003; Cornelis 2006).

The flagellar hook acts as a universal joint connecting the motor with the helical filament and thereby allows for transmission of torque even if the motor and the helical filament are not coaxially orientated (Berg and Anderson 1973). The hook is composed of about 120 subunits of FlgE and assembles, like the filament, into 11 circularly arranged protofilaments. Importantly, overall hook-length is relatively tightly controlled to a length of 55 nm $\pm$ 6 nm (Hirano et al. 1994; Macnab 2003; Chevance and Hughes 2008). The structure of the FlgE31 fragment, which is missing both terminal regions of the hook protein FlgE, has been solved (Samatey et al. 2004). The FlgE31 structure consists of two domains, D1 and D2, each having an oval shape, and is predominantly made of β-sheets. Although the structure of the hook subunit FlgE is very different compared to that of the filament subunit flagellin, their helical symmetries are strikingly similar to each other as stated above (Morgan et al. 1993; Mimori et al. 1995).

A partial atomic hook model was obtained by docking the crystal structure of the FlgE31 fragment into a 3D-EM density map of a straight hook (Samatey et al. 2004; Shaikh et al. 2005) (Fig. 9A). By deforming the helical lattice of the straight hook it was possible to model a curved hook, as shown in Figure 9B,C. Using this curved hook model, the molecular mechanism of the universal joint function of the flagellar hook has been studied in detail by molecular dynamics simulations (Samatey et al. 2004). This curved hook model revealed that, while the hook subunits are tightly packed in the inner core domains, the outer two domains are loosely packed in the axial intersubunit interactions in the protofilament structure.

The bacterial flagellar filament is much longer than the injectisome needle and grows up to 10–15 μm long, consisting of about 20,000–30,000 subunits of flagellin per filament (Namba and Vonderviszt 1997; Berg 2003). The filament structure is composed of 11 protofilaments, each of which can be in either a left-handed (L-type) or a right-handed (R-type) conformation (Asakura 1970). Only when all protofilaments are in the same conformation, the filament morphology is straight (Kamiya et al. 1979), and accordingly the filament can adopt a variety of helical morphologies by the supercoiling of the filament structure due to the mixture of the two protofilament conformations in different number ratios (Asakura 1970; Calladine 1975, 1978). The left-handed supercoiled filaments allow the formation of a bundle that propels the cells through the media. During a process called chemotaxis, the transition to a few different types of right-handed supercoiled forms upon reversal of the motor rotation results in disruption of the bundle and tumbling of the cell (Macnab and Ornston 1977).

The complete atomic structure of the filament and the individual flagellin subunit has been solved by cryo-EM image analysis and X-ray crystallography (Samatey et al. 2001; Yonekura et al. 2003) (Fig. 10). The structure of the filament has a diameter of about 23 nm with a central channel of about 2 nm (Fig. 10B,C). The filament structure is stabilized by interaction of the α-helical coiled-coils of the

Figure 6. (*Continued*). (*C*) Crystal structure of the C-terminal domain of $HrcQB_C$ of *Pseudomonas syringae* (PDB 1O9Y) (Fadouloglou et al. 2004). (*D*) Ribbon representation of the FliN dimer (gold) superimposed on half of the $HrcQB_C$ tetramer (green) illustrating the highly homologous structure (Brown et al. 2005). (*A*, reprinted, with permission, from Minamino et al. 2008b [© Elsevier]; *B*, Protein data base accession number 1O6A, based on Fadouloglou et al. 2004; *C*, Protein data base accession number 1O9Y, based on Fadouloglou et al. 2004; *D*, reprinted, with permission, from Brown et al. 2005 [© ASM].)

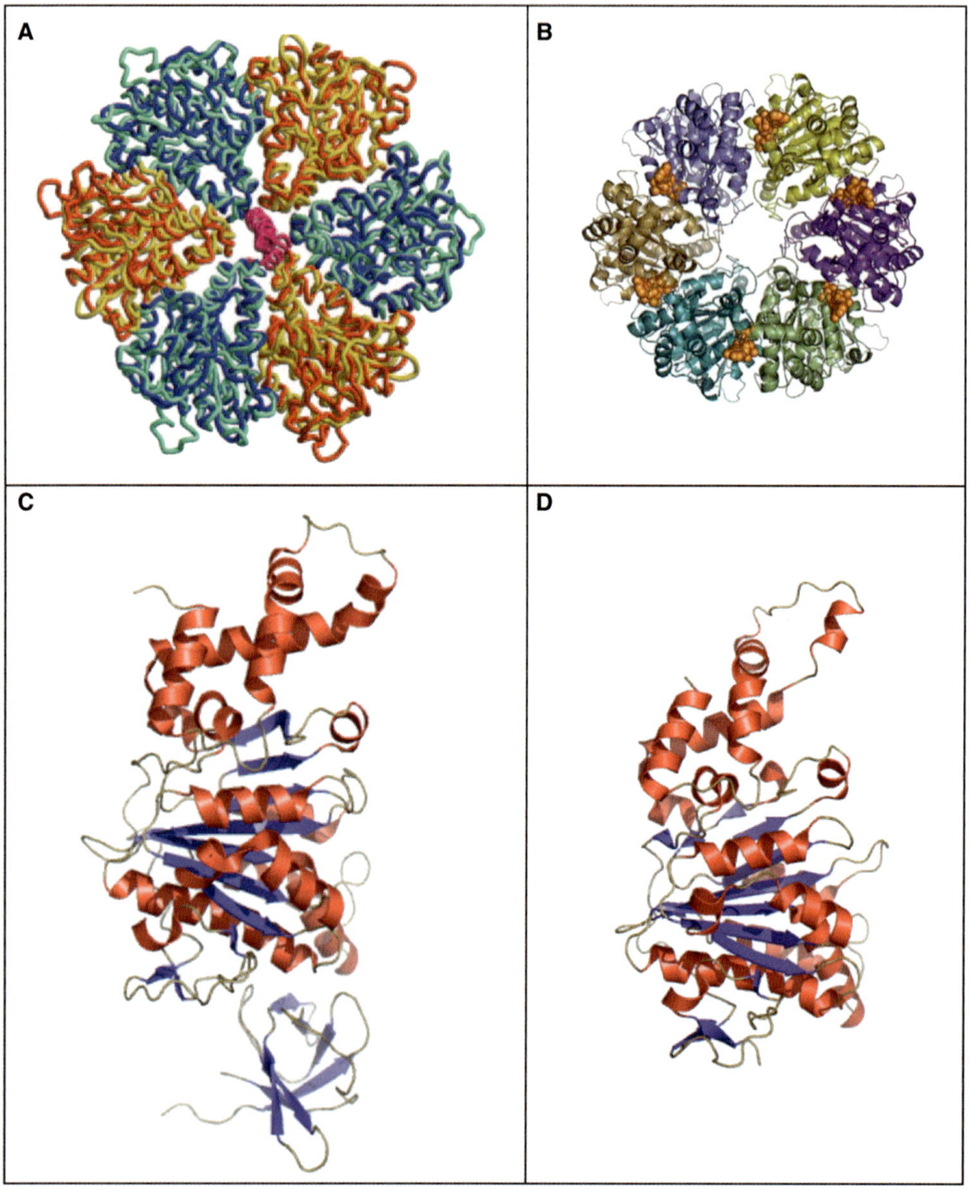

Figure 7. Crystal structures of the type III secretion ATPases, FliI, and EscN. (*A*) Model of the ATPase domain of the FliI hexamer (blue and yellow) superimposed onto the ATPase domain (blue-green) and (orange) of F1-ATPase (Imada et al. 2007). (*B*) Top-view of the EscN hexamer model; position of the ATP shown in a van-der-Waals representation in gold (Zarivach et al. 2007). (*C*) Crystal structure of the flagellar type III ATPase FliI (Δ1–18) missing the first 18 residues (PDB 2DPY). (*D*) Crystal structure of the C-terminal domain, residues 103–446 of the injectisome ATPase EscN (PDB 2OBL). (*A*, reprinted, with permission, from Imada et al. 2007 [© National Academy of Sciences]; *B*, reprinted, with permission, from Zarivach et al. 2007 [Nature Publishing Group]; *C*, Protein database accession number 2DPY, based on Imada et al. 2007; *D*, Protein database accession number 2OBL, based on Zarivach et al. 2007.)

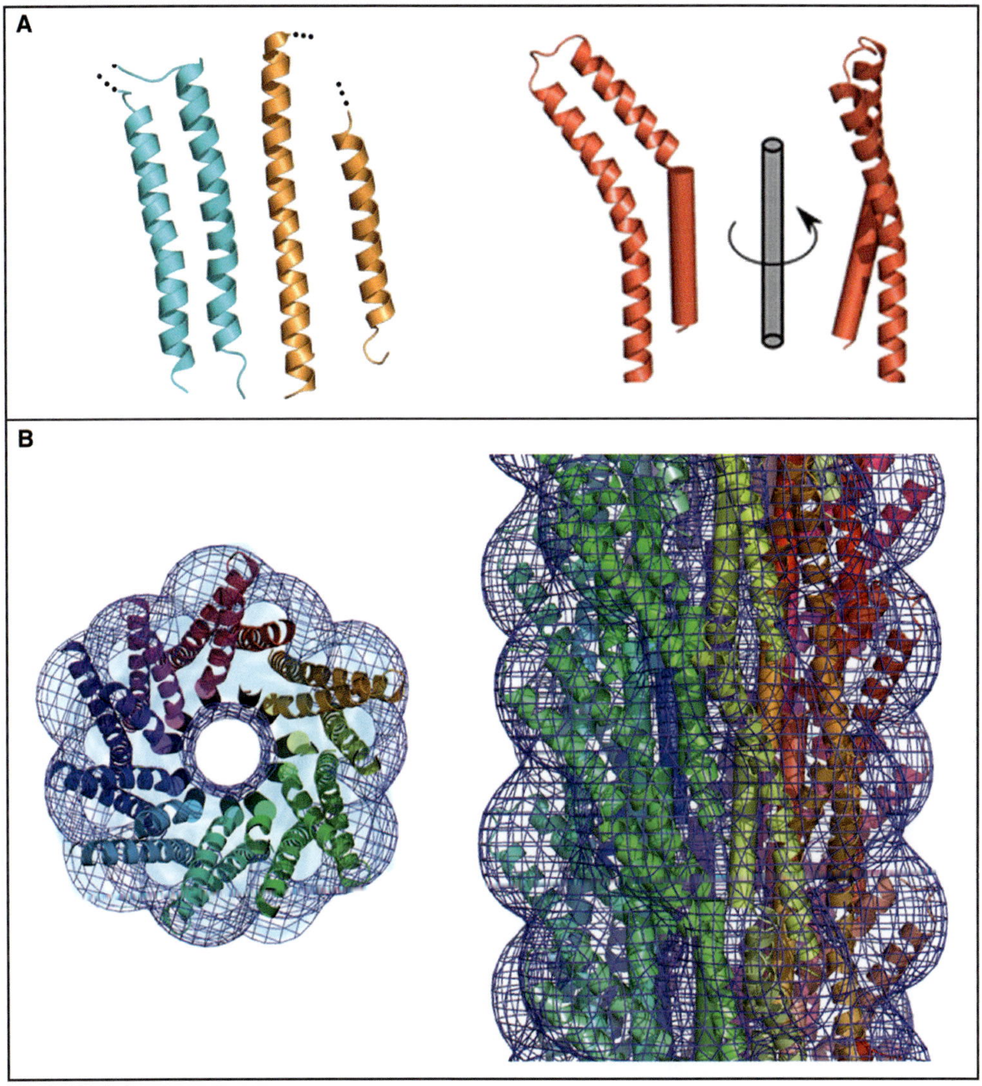

Figure 8. Crystal structure and molecular model of a type III secretion needle. (*A*) Ribbon representation of the D0 domain of flagellin (cyan), the ordered (chaperone-bound) region of EspA (orange) and molecule A of MxiH (red). The modeled N-terminal helix of MxiH is also shown rotated by 90° about the long axis of the molecule. (*B*) Model of the type III secretion needle obtained by docking of the atomic model of MxiH into the EM density of the *Shigella* T3SS needle. Left panel: end-on view of a 40-Å-thick section of the assembled needle. Each MxiH monomer is colored differently. The EM density is shown as a blue mesh. Right panel: side-view of the assembled needle, with colors as in the left panel. Note that the needle models in the left and right panel are not shown at the same scale. (Reprinted, with permission, from Deane et al. 2006 [© National Academy of Sciences].)

D0 and D1 domains of flagellin in the inner core of the filament structure, in which the structure of the D0 domain is strikingly similar to the anti-parallel α-helices of the needle protein MxiH (Deane et al. 2006) (Fig. 8A).

CAP STRUCTURES, THE FILAMENT TIP FliD, AND THE NEEDLE TIP LrcV

The flagellar filament grows at its distal end by self-assembly of flagellin subunits that have to be transported in an unfolded conformation

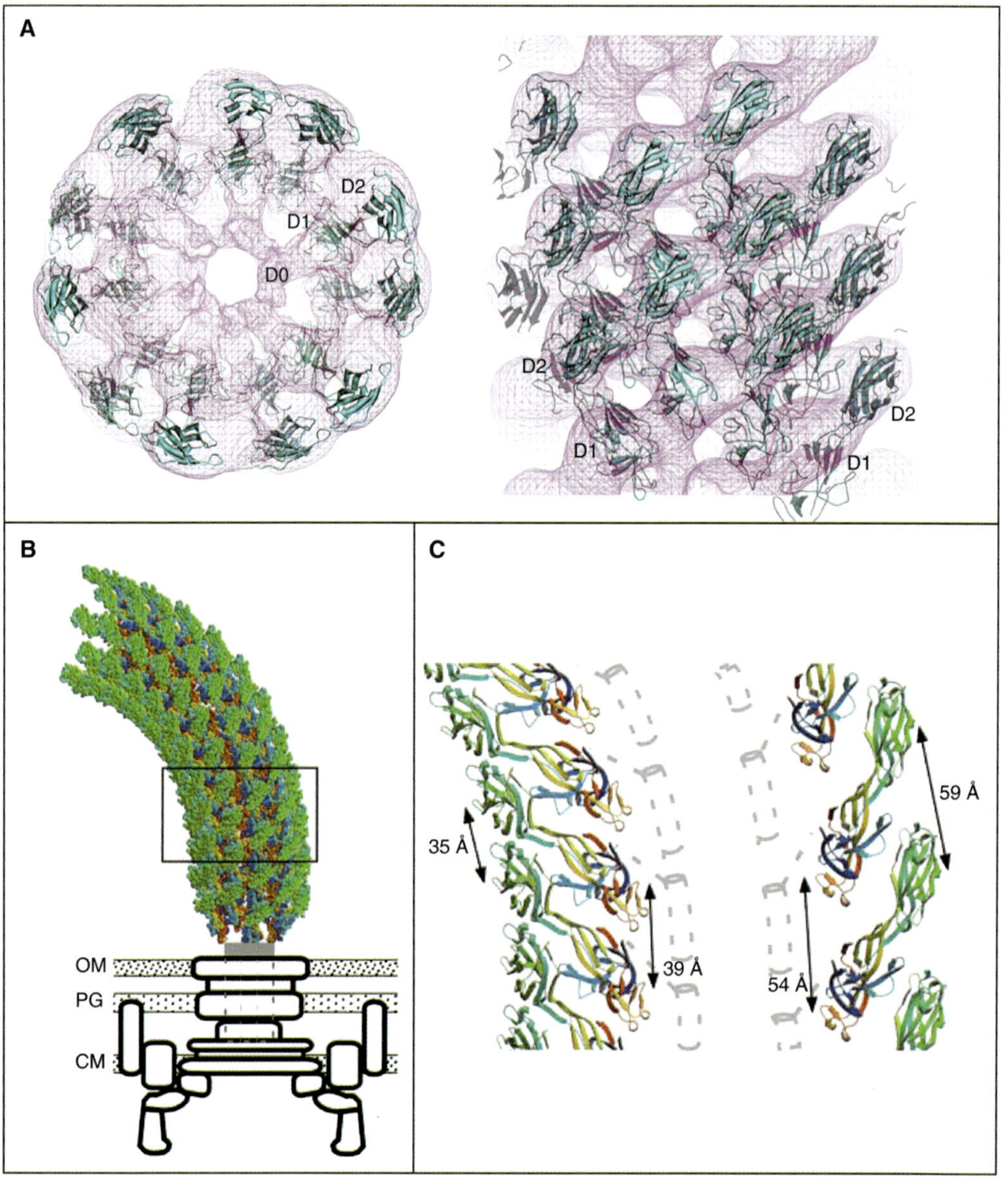

Figure 9. Model of the supercoiled flagellar hook and crystal structure of the hook subunit fragment, FlgE31. (*A*) Docking of the atomic model of the crystal structure of the hook subunit fragment FlgE31 into the outer two domains of the hook obtained by cyro-EM. Left panel: end-on-view; right panel: side-view. The hook EM density is shown as a purple mesh. (*B*) Atomic model of the supercoiled hook and a schematic diagram of the basal body complex. OM, outer membrane; PG, peptidoglycan layer; CM, cytoplasmic membrane. (*C*) Magnified, atomic model of the supercoiled hook shown in (*B*). The innermost and the outermost protofilaments are displayed on the left and right, respectively. Dotted gray lines represent the central channel. (Reprinted, with permission, from Samatey et al. 2004 [Nature Publishing Group].)

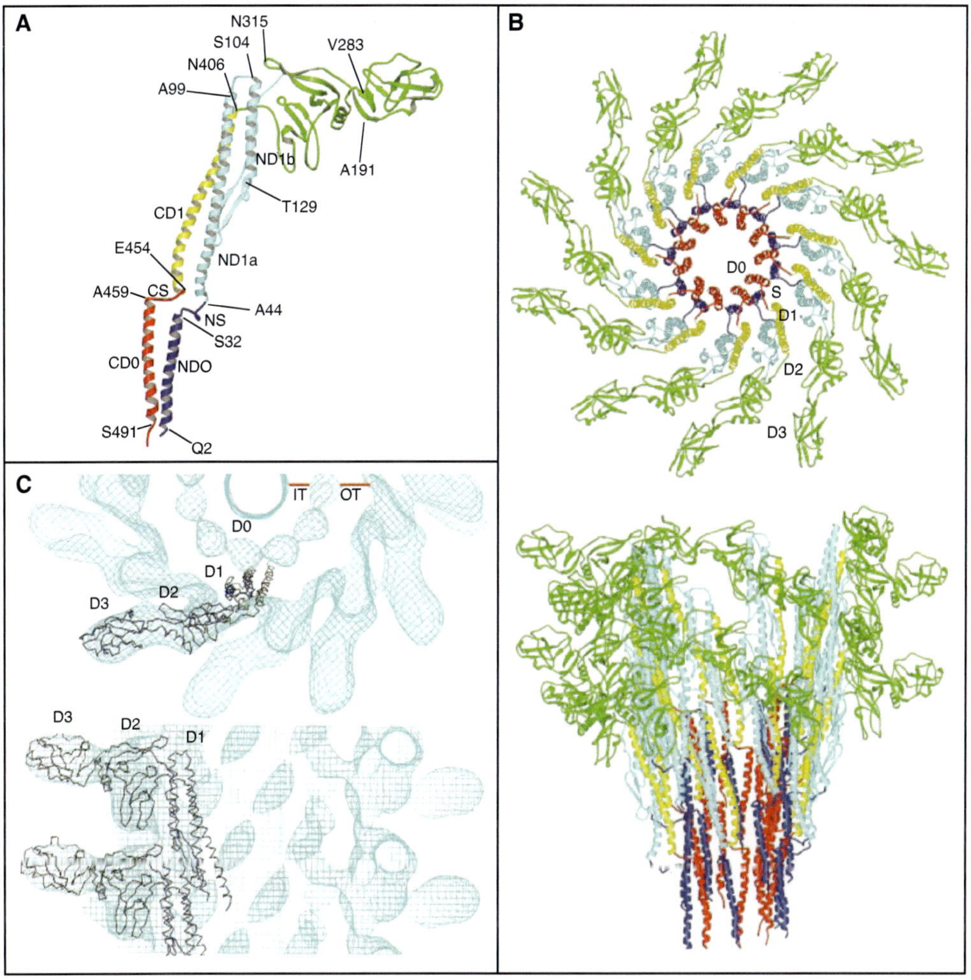

Figure 10. Crystal structure and model of the flagellar filament. (*A*) Atomic model of the flagellar filament obtained by cyro-EM. Ribbon representation of the filament subunit (Yonekura et al. 2003). (*B*) Ribbon representation of the model of the flagellar filament. Upper panel: end-on-view from the distal end of the filament displaying 11 subunits. Lower panel: side-view from outside of the filament (Yonekura et al. 2003). (*C*) Docking of a protofilament into the electron density map of the filament. Upper panel: end-on-view from the top. Bottom panel: Side-view. D0, D1, D2, and D3 indicate domains of the flagellin protein (Samatey et al. 2001). (*A, B,* Reprinted, with permission, from Yonekura et al. 2003; *C*, reprinted, with permission, from Samatey et al. 2001 [all Nature Publishing Group].)

through the narrow channel of the filament. At the very end of the filament, an essential cap structure made of the FliD protein promotes flagellin self-assembly and polymerization by a rotation mechanism. EM studies revealed that the pentameric cap is attached at the distal end of the filament via its five leg-shaped anchor domains (Yonekura et al. 2000) (Fig. 11A). The five domains are separated from each other, thereby forming five differently sized gaps. One significantly larger gap is thought to represent the exit site as well as the binding site for individual flagellin subunits during the assembly process. Insertion of a flagellin subunit presumably forces the legs of the cap subunits to transition into the next stable conformation, thereby providing another gap for the subsequent assembly of the following

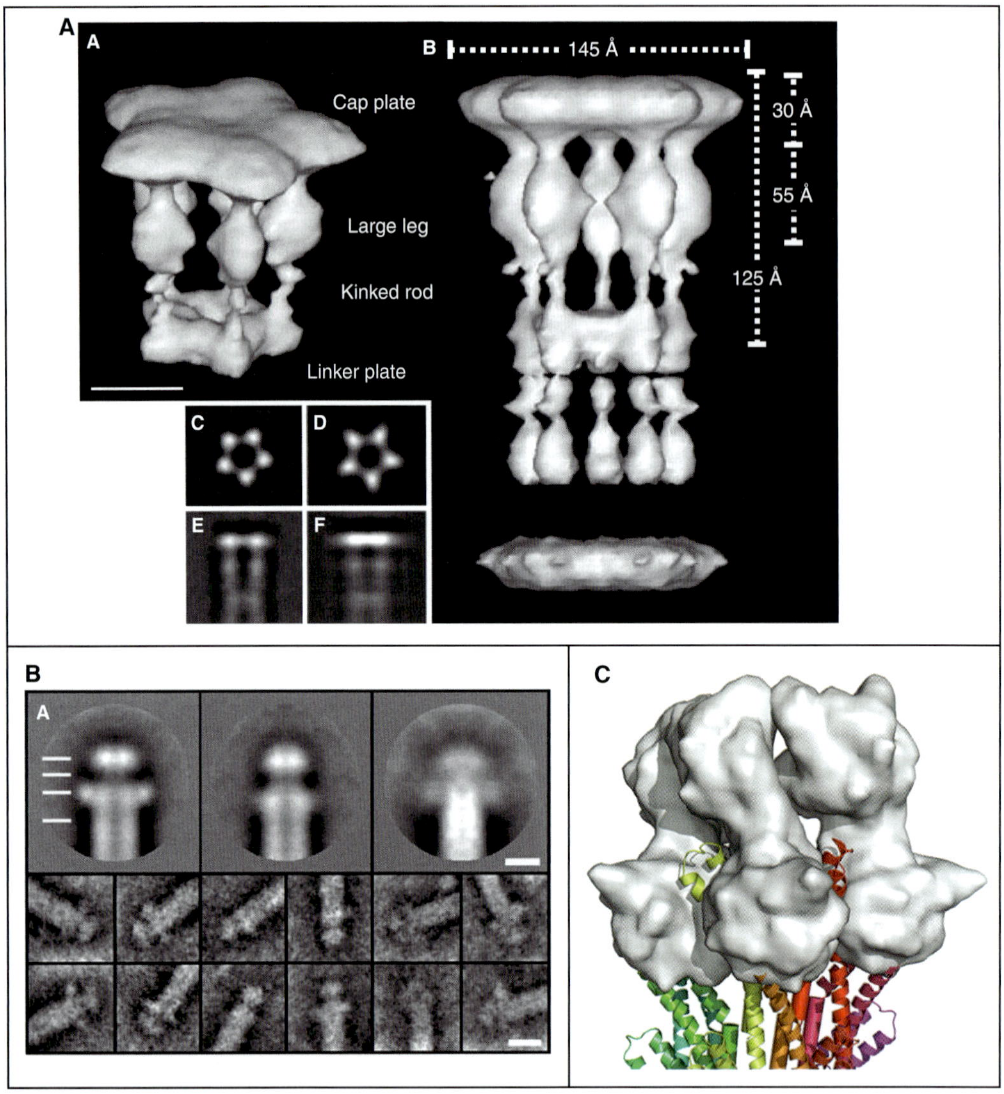

Figure 11. EM images of the flagellar filament cap, needle tip and model of the needle-tip interaction. (*A*) 3D-EM image reconstruction of the flagellar cap pentamer (labeled A in the image) and decamer (labeled B in the image). On the lower left side of the image are shown: averaged end-on-view (labeled C); projection of the 3D reconstruction along the 5-fold axis (labeled D); averaged side-view (labeled E), and projection of the 3D reconstruction perpendicular to the 5-fold axis (labeled F) (Maki-Yonekura et al. 2000). (*B*) Upper three panels: averaged EM images of the needle tip complexes formed by LcrV (left; resolution 1.5 nm), PcrV (center; resolution 1.5 nm), and AcrV (right; resolution 2.5 nm). Also visible is the central channel of both the needle and tip complex. Lower images on the bottom display typical single images (Mueller et al. 2005). (*C*) The LcrV tip complex modeled onto the distal end of a injectisome needle. The LcrV tip complex is displayed in surface representation (gray) (Deane et al. 2006). (*A*, Reprinted, with permission, from Maki-Yonekura et al. 2003 [© National Academy of Sciences]; *B*, reprinted, with permission, from Mueller et al. 2005 [© AAAS]; *C*, reprinted, with permission, from Deane et al. 2006 [© National Academy of Sciences].)

flagellin subunit. This conformational rearrangement sequence has been described as a walk of the pentameric cap along the helical steps of the filament end. Accordingly, the assembly of approximately 55 flagellin subunits results in a complete rotation of the cap (Yonekura et al. 2000; Maki-Yonekura et al. 2003).

In the case of the injectisome needle, the tip proteins are thought to be responsible for the correct insertion of the translocon complex that forms a pore in eukaryotic cell membranes. In *Yersinia*, LrcV forms the tip complex at the distal end of the needle (Mueller et al. 2005). It has been suggested that LrcV functions as a scaffold or platform for the proteins of the translocation pore (Cornelis 2006).

However, it cannot be ruled out that the LrcV tip complex also promotes the assembly of individual needle subunits during needle growth, similarly to the function of the filament cap described above. The overall structure of the needle tip complex and the filament cap are indeed comparable. Similar to the filament cap, the LrcV tip complex seems to form a bell-shaped structure at the distal end of the needle, as visualized by EM studies of *Yersinia* needles (Mueller et al. 2005) (Fig. 11B). A pentameric complex has been suggested by superimposing the LrcV structure with the C-terminal helices of the needle subunit MxiH (Deane et al. 2006). A maximum of five LrcV molecules could be placed at the tip of the needle, and the resulting bell-shaped structures are strikingly reminiscent of the flagellar cap complex (Yonekura et al. 2003; Deane et al. 2006) (Fig. 11C).

CONCLUSIONS AND FUTURE PERSPECTIVES

Significant progress has been made in the visualization of various components of both the flagellum and the type III injectisome. Importantly, high-resolution structures of several key components of the highly homologous systems, like the cytoplasmic ATPase, the inner membrane ring, the flagellar C-ring proteins, the injectisome needle, the flagellar hook, and the flagellar filament are now available. Together with the 3D-EM image reconstructions of both the flagellar hook-basal-body and the injectisome basal body complex, we now have a relatively comprehensive picture of the overall structure of these fascinating nanomachines. However, high-resolution 3D images that allow the visualization of secondary structures are still essential to build more reliable atomic models to obtain deeper insights into their mechanisms.

The other main challenges in the field are the structure determination of several missing components like the flagellar rod and C-ring of the injectisome. Importantly, no structural information about the membrane components of the T3SS is available. Recent studies emphasized the importance of the membrane components in the PMF-dependent protein transport process, and obtaining structural information about how proton influx is coupled to the protein export process remains the most challenging goal.

ACKNOWLEDGMENTS

This work was supported by PHS grant GM056141 from the National Institutes of Health. We thank the Hughes lab for useful comments and discussions of the manuscript. M.E. gratefully acknowledges scholarship support of the Boehringer Ingelheim Fonds.

REFERENCES

Aizawa SI. 1996. Flagellar assembly in *Salmonella typhimurium*. *Mol Microbiol* **19:** 1–5.

Akeda Y, Galán JE. 2005. Chaperone release and unfolding of substrates in type III secretion. *Nature* **437:** 911–915.

Asakura S. 1970. Polymerization of flagellin and polymorphism of flagella. *Adv Biophys* **1:** 99–155.

Berg HC, Anderson RA. 1973. Bacteria swim by rotating their flagellar filaments. *Nature* **245:** 380–382.

Berg H. 2003. The rotary motor of bacterial flagella. *Annu Rev Biochem* **72:** 19–54.

Blocker A, Komoriya K, Aizawa S. 2003. Type III secretion systems and bacterial flagella: insights into their function from structural similarities. *Proc Natl Acad Sci U S A* **100:** 3027–3030.

Brown PN, Hill CP, Blair DF. 2002. Crystal structure of the middle and C-terminal domains of the flagellar rotor protein FliG. *EMBO J* **21:** 3225–3234.

Brown P, Mathews M, Joss L, Hill C, Blair D. 2005. Crystal structure of the flagellar rotor protein FliN from *Thermotoga maritima*. *J Bacteriol* **187:** 2890–2902.

Brown PN, Terrazas M, Paul K, Blair DF. 2007. Mutational analysis of the flagellar protein FliG: sites of interaction with FliM and implications for organization of the switch complex. *J Bacteriol* **189:** 305–312.

Calladine CR. 1975. Construction of bacterial flagella. *Nature* **255:** 121–124.

Calladine CR. 1978. Change of waveform in bacterial flagella: the role of mechanics at the molecular level. *J Mol Biol* 457–479.

Chevance FFV, Hughes KT. 2008. Coordinating assembly of a bacterial macromolecular machine. *Nat Rev Micro* **6:** 455–465.

Claret L, Calder SR, Higgins M, Hughes C. 2003. Oligomerization and activation of the FliI ATPase central to bacterial flagellum assembly. *Mol Microbiol* **48:** 1349–1355.

Cordes FS, Komoriya K, Larquet E, Yang S, Egelman EH, Blocker A, Lea SM. 2003. Helical structure of the needle of the type III secretion system of *Shigella flexneri*. *J Biol Chem* **278:** 17103–17107.

Cornelis GR. 2006. The type III secretion injectisome. *Nat Rev Microbiol* **4:** 811–825.

Cornelis GR, Van Gijsegem F. 2000. Assembly and function of type III secretory systems. *Annu Rev Microbiol* **54:** 735–774.

Deane JE, Roversi P, Cordes FS, Johnson S, Kenjale R, Daniell S, Booy F, Picking WL, Blocker AJ, Lea SM. 2006. Molecular model of a type III secretion system needle: Implications for host-cell sensing. *Proc Natl Acad Sci USA* **103:** 12529–12533.

DePamphilis ML, Adler J. 1971a. Purification of intact flagella from *Escherichia coli* and *Bacillus subtilis*. *J Bacteriol* **105:** 376–383.

DePamphilis ML, Adler J. 1971b. Fine structure and isolation of the hook-basal body complex of flagella from *Escherichia coli* and *Bacillus subtilis*. *J Bacteriol* **105:** 384–395.

Derosier D. 2006. Bacterial flagellum: visualizing the complete machine in situ. *Curr Biol* **16:** R928–930.

Fadouloglou VE, Tampakaki AP, Glykos NM, Bastaki MN, Hadden JM, Phillips SE, Panopoulos NJ, Kokkinidis M. 2004. Structure of HrcQB-C, a conserved component of the bacterial type III secretion systems. *Proc Natl Acad Sci U S A* **101:** 70–75.

Fan F, Macnab RM. 1996. Enzymatic characterization of FliI. An ATPase involved in flagellar assembly in *Salmonella typhimurium*. *J Biol Chem* **271:** 31981–31988.

Ferris HU, Minamino T. 2006. Flipping the switch: bringing order to flagellar assembly. *Trends Microbiol* **14:** 519–526.

Galán JE, Collmer A. 1999. Type III secretion machines: bacterial devices for protein delivery into host cells. *Science* **284:** 1322–1328.

Galán JE, Wolf-Watz H. 2006. Protein delivery into eukaryotic cells by type III secretion machines. *Nature* **444:** 567–573.

Gophna U, Ron EZ, Graur D. 2003. Bacterial type III secretion systems are ancient and evolved by multiple horizontal-transfer events. *Gene* **312:** 151–163.

Hirano T, Yamaguchi S, Oosawa K, Aizawa S. 1994. Roles of FliK and FlhB in determination of flagellar hook length in *Salmonella typhimurium*. *J Bacteriol* **176:** 5439–5449.

Hodgkinson JL, Horsley A, Stabat D, Simon M, Johnson S, da Fonseca PC, Morris EP, Wall JS, Lea SM, Blocker AJ. 2009. Three-dimensional reconstruction of the *Shigella* T3SS transmembrane regions reveals 12-fold symmetry and novel features throughout. *Nat Struct Mol Biol* **16:** 477–485.

Homma M, Komeda Y, Iino T, Macnab RM. 1987. The *flaFIX* gene product of *Salmonella typhimurium* is a flagellar basal body component with a signal peptide for export. *J Bacteriol* **169:** 1493–1498.

Homma M, DeRosier DJ, Macnab RM. 1990. Flagellar hook and hook-associated proteins of *Salmonella typhimurium* and their relationship to other axial components of the flagellum. *J Mol Biol* **213:** 819–832.

Hueck C. 1998. Type III protein secretion systems in bacterial pathogens of animals and plants. *Microbiol Mol Biol Rev* **62:** 379–433.

Hughes KT, Gillen KL, Semon MJ, Karlinsey JE. 1993. Sensing structural intermediates in bacterial flagellar assembly by export of a negative regulator. *Science* **262:** 1277–1280.

Iino T. 1974. Assembly of *Salmonella* flagellin in vitro and in vivo. *J Supramol Struct* **2:** 372–384.

Imada K, Minamino T, Tahara A, Namba K. 2007. Structural similarity between the flagellar type III ATPase FliI and F1-ATPase subunits. *Proc Natl Acad Sci USA* **104:** 485–490.

Journet L, Agrain C, Broz P, Cornelis GR. 2003. The needle length of bacterial injectisomes is determined by a molecular ruler. *Science* **302:** 1757–1760.

Journet L, Hughes KT, Cornelis GR. 2005. Type III secretion: a secretory pathway serving both motility and virulence (review). *Mol Membr Biol* **22:** 41–50.

Kamiya R, Asakura S, Wakabayashi K, Namba K. 1979. Transition of bacterial flagella from helical to straight forms with different subunit arrangements. *J Mol Biol* **131:** 725–742.

Kimbrough TG, Miller SI. 2000. Contribution of *Salmonella typhimurium* type III secretion components to needle complex formation. *Proc Natl Acad Sci U S A* **97:** 11008–11013.

Kojima S, Blair DF. 2004. The bacterial flagellar motor: structure and function of a complex molecular machine. *Int Rev Cytol* **233:** 93–134.

Kojima S, Furukawa Y, Matsunami H, Minamino T, Namba K. 2008. Characterization of the periplasmic domain of MotB and implications for its role in the stator assembly of the bacterial flagellar motor. *J Bacteriol* **190:** 3314–3322.

Koster M, Bitter W, de Cock H, Allaoui A, Cornelis GR, Tommassen J. 1997. The outer membrane component, YscC, of the Yop secretion machinery of *Yersinia enterocolitica* forms a ring-shaped multimeric complex. *Mol Microbiol* **26:** 789–797.

Kubori T, Matsushima Y, Nakamura D, Uralil J, Lara-Tejero M, Sukhan A, Galán JE, Aizawa SI. 1998. Supramolecular structure of the *Salmonella typhimurium* type III protein secretion system. *Science* **280:** 602–605.

Kubori T, Sukhan A, Aizawa SI, Galán JE. 2000. Molecular characterization and assembly of the needle complex of the *Salmonella typhimurium* type III protein secretion system. *Proc Natl Acad Sci U S A* **97:** 10225–10230.

Lloyd SA, Tang H, Wang X, Billings S, Blair DF. 1996. Torque generation in the flagellar motor of *Escherichia coli*: evidence of a direct role for FliG but not for FliM or FliN. *J Bacteriol* **178:** 223–231.

Macnab R. 2003. How bacteria assemble flagella. *Annu Rev Microbiol* **57:** 77–100.

Macnab RM. 2004. Type III flagellar protein export and flagellar assembly. *Biochim Biophys Acta* **1694:** 207–217.

Macnab RM, Ornston MK. 1977. Normal-to-curly flagellar transitions and their role in bacterial tumbling. Stabilization of an alternative quaternary structure by mechanical force. *J Mol Biol* **112:** 1–30.

Maki-Yonekura S, Yonekura K, Namba K. 2003. Domain movements of HAP2 in the cap-filament complex formation and growth process of the bacterial flagellum. *Proc Natl Acad Sci U S A* **100:** 15528–15533.

Manson M, Tedesco P, Berg H, Harold F, Van der Drift C. 1977. A protonmotive force drives bacterial flagella. *Proc Natl Acad Sci* **74:** 3060–3064.

Marlovits TC, Kubori T, Sukhan A, Thomas DR, Galán JE, Unger VM. 2004. Structural insights into the assembly of the type III secretion needle complex. *Science* **306:** 1040–1042.

Matsuura A, Shioi JI, Imae Y. 1977. Motility in *Bacillus subtilis* driven by an artificial protonmotive force. *FEBS Lett* **82:** 187–190.

Mimori Y, Yamashita I, Murata K, Fujiyoshi Y, Yonekura K, Toyoshima C, Namba K. 1995. The structure of the R-type straight flagellar filament of *Salmonella* at 9 A resolution by electron cryomicroscopy. *J Mol Biol* **249:** 69–87.

Minamino T, Macnab RM. 1999. Components of the *Salmonella* flagellar export apparatus and classification of export substrates. *J Bacteriol* **181:** 1388–1394.

Minamino T, Macnab RM. 2000. FliH, a soluble component of the type III flagellar export apparatus of *Salmonella*, forms a complex with FliI and inhibits its ATPase activity. *Mol Microbiol* **37:** 1494–1503.

Minamino T, Namba K. 2008. Distinct roles of the FliI ATPase and proton motive force in bacterial flagellar protein export. *Nature* **451:** 485.

Minamino T, Yamaguchi S, Macnab RM. 2000. Interaction between FliE and FlgB, a proximal rod component of the flagellar basal body of *Salmonella*. *J Bacteriol* **182:** 3029–3036.

Minamino T, Imada K, Namba K. 2008a. Mechanisms of type III protein export for bacterial flagellar assembly. *Mol Biosyst* **4:** 1105–1115.

Minamino T, Imada K, Namba K. 2008b. Molecular motors of the bacterial flagella. *Curr Opin Struct Biol* **18:** 693–701.

Moraes TF, Spreter T, Strynadka NC. 2008. Piecing together the type III injectisome of bacterial pathogens. *Curr Opin Struct Biol* **18:** 258–266.

Morgan DG, Macnab RM, Francis NR, DeRosier DJ. 1993. Domain organization of the subunit of the *Salmonella typhimurium* flagellar hook. *J Mol Biol* **229:** 79–84.

Mueller CA, Broz P, Muller SA, Ringler P, Erne-Brand F, Sorg I, Kuhn M, Engel A, Cornelis GR. 2005. The V-antigen of *Yersinia* forms a distinct structure at the tip of injectisome needles. *Science* **310:** 674–676.

Namba K. 2001. Roles of partly unfolded conformations in macromolecular self-assembly. *Genes Cells* **6:** 1–12.

Namba, Vonderviszt. 1997. Molecular architecture of bacterial flagellum. *Q Rev Biophys* **30:** 1–65.

Okabe M, Minamino T, Imada K, Namba K, Kihara M. 2009. Role of the N-terminal domain of FliI ATPase in bacterial flagellar protein export. *FEBS Lett* **583:** 743–748.

Parsot C, Hamiaux C, Page AL. 2003. The various and varying roles of specific chaperones in type III secretion systems. *Curr Opin Microbiol* **6:** 7–14.

Patterson-Delafield J, Martinez RJ, Stocker BA, Yamaguchi S. 1973. A new *fla* gene in *Salmonella typhimurium* – *flaR* – and its mutant phenotype-superhooks. *Archiv für Mikrobiologie* **90:** 107–120.

Paul K, Blair DF. 2006. Organization of FliN subunits in the flagellar motor of *Escherichia coli*. *J Bacteriol* **188:** 2502–2511.

Paul K, Erhardt M, Hirano T, Blair DF, Hughes KT. 2008. Energy source of flagellar type III secretion. *Nature* **451:** 489–492.

Pozidis C, Chalkiadaki A, Gomez-Serrano A, Stahlberg H, Brown I, Tampakaki AP, Lustig A, Sianidis G, Politou AS, Engel A, et al. 2003. Type III protein translocase: HrcN is a peripheral ATPase that is activated by oligomerization. *J Biol Chem* **278:** 25816–25824.

Samatey, Imada, Nagashima, Vonderviszt, Kumasaka, Yamamoto M, Namba K. 2001. Structure of the bacterial flagellar protofilament and implications for a switch for supercoiling. *Nature* **410:** 331–337.

Samatey FA, Matsunami H, Imada K, Nagashima S, Shaikh TR, Thomas DR, Chen JZ, Derosier DJ, Kitao A, Namba K. 2004. Structure of the bacterial flagellar hook and implication for the molecular universal joint mechanism. *Nature* **431:** 1062–1068.

Shaikh TR, Thomas DR, Chen JZ, Samatey FA, Matsunami H, Imada K, Namba K, Derosier DJ. 2005. A partial atomic structure for the flagellar hook of *Salmonella typhimurium*. *Proc Natl Acad Sci USA* **102:** 1023–1028.

Silverman M, Simon M. 1974. Flagellar rotation and the mechanism of bacterial motility. *Nature* **249:** 73–74.

Sockett H, Yamaguchi S, Kihara M, Irikura VM, Macnab RM. 1992. Molecular analysis of the flagellar switch protein FliM of *Salmonella typhimurium*. *J Bacteriol* **174:** 793–806.

Tampakaki A, Fadouloglou V, Gazi A, Panopoulos N, Kokkinidis M. 2004. Conserved features of type III secretion. *Cell Microbiol* **6:** 805–816.

Thomas DR, Morgan DG, DeRosier DJ. 1999. Rotational symmetry of the C ring and a mechanism for the flagellar rotary motor. *Proc Natl Acad Sci U S A* **96:** 10134–10139.

Thomas D, Morgan DG, DeRosier DJ. 2001. Structures of bacterial flagellar motors from two FliF-FliG gene fusion mutants. *J Bacteriol* **183:** 6404–6412.

Thomas DR, Francis NR, Xu C, DeRosier DJ. 2006. The three-dimensional structure of the flagellar rotor from a clockwise-locked mutant of *Salmonella enterica* serovar Typhimurium. *J Bacteriol* **188:** 7039–7048.

Vogler AP, Homma M, Irikura VM, Macnab RM. 1991. *Salmonella typhimurium* mutants defective in flagellar filament regrowth and sequence similarity of FliI to F0F1, vacuolar, and archaebacterial ATPase subunits. *J Bacteriol* **173:** 3564–3572.

Wilharm G, Lehmann V, Krauss K, Lehnert B, Richter S, Ruckdeschel K, Heesemann J, Trülzsch K. 2004. *Yersinia enterocolitica* type III secretion depends on the proton motive force but not on the flagellar motor components MotA and MotB. *Infect Immun* **72:** 4004–4009.

Yip CK, Kimbrough TG, Felise HB, Vuckovic M, Thomas NA, Pfuetzner RA, Frey EA, Finlay BB, Miller SI, Strynadka NC. 2005. Structural characterization of the molecular platform for type III secretion system assembly. *Nature* **435:** 702–707.

Yonekura K, Maki S, Morgan D, DeRosier D, Vonderviszt F, Imada K, Namba K. 2000. The bacterial flagellar cap as the rotary promoter of flagellin self-assembly. *Science* **290:** 2148–2152.

Yonekura K, Maki-Yonekura S, Namba K. 2003. Complete atomic model of the bacterial flagellar filament by electron cryomicroscopy. *Nature* **424:** 643–650.

Zarivach R, Vuckovic M, Deng W, Finlay BB, Strynadka NC. 2007. Structural analysis of a prototypical ATPase from the type III secretion system. *Nature Struct & Mol Biol* **14:** 131–137.

Membrane-associated DNA Transport Machines

Briana Burton[1] and David Dubnau[2]

[1]Department of Molecular and Cellular Biology, Harvard University, Cambridge, Massachusetts

[2]Public Health Research Center, New Jersey Medical School, University of Medicine and Dentistry of New Jersey, Newark, New Jersey

Correspondence: dubnauda@umdnj.edu

DNA pumps play important roles in bacteria during cell division and during the transfer of genetic material by conjugation and transformation. The FtsK/SpoIIIE proteins carry out the translocation of double-stranded DNA to ensure complete chromosome segregation during cell division. In contrast, the complex molecular machines that mediate conjugation and genetic transformation drive the transport of single stranded DNA. The transformation machine also processes this internalized DNA and mediates its recombination with the resident chromosome during and after uptake, whereas the conjugation apparatus processes DNA before transfer. This article reviews these three types of DNA pumps, with attention to what is understood of their molecular mechanisms, their energetics and their cellular localizations.

The transport of DNA across membranes by bacteria occurs during sporulation, during cytokinesis, directly from other cells and from the environment. This review addresses the question "how is the DNA polyanion transferred processively across the hydrophobic membrane barrier"?

DNA transport must occur through water-filled channels, at least conceptually addressing the problem posed by the hydrophobic membrane. DNA transporters presumably use metabolic energy directly or a coupled-flow (symporter or antiporter) mechanism to drive DNA processively through the channel. It is possible that a Brownian ratchet mechanism, in which directionality is imposed on a diffusive process, also contributes to transport.

In this article, we will consider several DNA transport systems. We will begin with the simplest one, namely the FtsK/SpoIIIE system that is involved in cell division and sporulation. We will then turn to the more complex, multiprotein DNA uptake systems that accomplish genetic transformation (the uptake of environmental DNA from the environment) and the conjugation systems of Gram-negative bacteria that mediate the unidirectional transfer of DNA between cells. In each case we will discuss the proteins involved, their actions and the sources of energy that drive transport. Space limitations prevent discussion of other relevant topics, such as DNA transport during bacteriophage infection and more than a brief reference to conjugation in Gram-positive bacteria.

 doi: 10.1101/cshperspect.a000406
Cite this article as *Cold Spring Harb Perspect Biol* 2010;2:a000406

CELL DIVISION AND SPORULATION

General Features of DNA Transport in Cell Division

Proper chromosome segregation is essential for successful cell division, and yet under rapid growth conditions bacterial cells can form a division septum before completion of DNA segregation. To accomplish this, the cells have dedicated proteins of the FtsK/SpoIIIE family that move the remainder of the chromosome into the appropriate daughter-cell compartment. Given that a single gene product codes for all the transport functionality, this is the simplest DNA transport system discussed here.

The simplicity of the single polypeptide chain belies the fact that these transporters function in the division septum, one of the most topologically challenging environments of the bacterial cell. When chromosomes have not fully segregated into daughter compartments, the constricting septal membranes constitute an obstacle for the final resolution of the cellular genetic material. In many of the bacteria that contain FtsK/SpoIIIE proteins, the chromosomes are circular molecules. This means that any portion of the chromosome that is on the wrong side of the division plane will have a loop of dsDNA that must first be resolved properly from its sister chromosome, and second be moved in the correct direction across the division plane.

Information about the FtsK/SpoIIIE family of transporters is derived largely from the work on two founding members of the protein family. FtsK from *Escherichia coli* has yielded much functional information about the role of these proteins in terminus resolution during chromosome segregation (a more detailed account of its recombination function may be found in Thanbichler 2009). SpoIIIE, which has a specialized function during the asymmetric division of sporulating *Bacillus subtilis* cells, has produced complementary information about these transporters. The focus of this section is on how the ATPases of the FtsK/SpoIIIE protein family transport chromosomal DNA in the context of cellular division membranes.

The FtsK/SpoIIIE Proteins

The FtsK/SpoIIIE family of DNA transporters is conserved among the bacteria (Iyer et al. 2004). Proteins of the FtsK/SpoIIIE family have diverse functions including the export of virulence factors, conjugative transfer of plasmid DNA and chromosome partitioning. Those members involved in chromosome partitioning share a functional domain called γ (discussed later) (Wang et al. 2006; Ausmees et al. 2007; Le Bourgeois et al. 2007; Wang et al. 2007; Val et al. 2008; Dedrick et al. 2009) and have the distinction of functioning at a unique interface within cells—the constricting division plane.

Until recently, it was thought that all chromosomal transporters in this family harbor amino-terminal transmembrane domains. New studies on a SpoIIIE paralog in *B. subtilis* have prompted a revision of the group to include soluble FtsK proteins. Interestingly, all of these FtsK/SpoIIIE proteins localize to cell division membranes, and they do so only at the time of division (Wu and Errington 1997; Biller and Burkholder 2009; Kaimer et al. 2009). The newly characterized soluble protein, SftA, localizes to predivisional sites, presumably through direct protein–protein interactions with FtsZ, rather than through membrane tethering (Biller and Burkholder 2009). FtsK localizes to vegetative cell division septa, as does SpoIIIE, although the latter does so rarely (Wu and Errington 1997; Wang and Lutkenhaus 1998; Yu et al. 1998). In the case of sporulation, SpoIIIE localization requires the presence of DNA in the septal plane (Ben-Yehuda et al. 2003). The presence of FtsK at division planes is enhanced under conditions that produce an increased number of bisected nucleoids (Yu et al. 1998). The amino-terminal transmembrane segments of both proteins are required for proper septal localization (Begg et al. 1995; Draper et al. 1998; Yu et al. 1998; Sharp and Pogliano 1999) and although there is little sequence conservation in this region, the transmembrane segments appear to mediate interaction with and recruitment of other division machinery proteins (Di Lallo et al. 2003). Furthermore, the transmembrane domains are required for proper function of SpoIIIE (Sharp and Pogliano 2003), although the precise role of the FtsK transmembrane domains for DNA transport activity is still unknown.

Little is understood about the precise role of the variable length linker domain that follows the transmembrane segments. However studies in *E. coli* using domain deletions and chimeras with the *Haemophilus influenzae* FtsK show that the linker portion of the protein does play a role in proper cell division (Bigot et al. 2004). The absence of the linker domain correlates with increased filamentation in these strains, and this phenotype could be functionally separated from a previously characterized role for the linker in chromosome dimer resolution (Aussel et al. 2002).

The carboxy-terminal regions of these proteins contain the conserved portions of the proteins including the ATPase motifs and the directionality determining gamma domain. TrwB, a conjugation protein from the plasmid R388, provided the first structural insights into the ATPase domain of this protein family (Gomis-Ruth et al. 2001). The structure of the soluble domain of TrwB (lacking the amino-terminal transmembrane region) revealed two domains. First, an NBD similar to RecA and ring helicases, and second a domain entirely composed of α-helices. A solution structure of the soluble ATPase domain of FtsK confirmed the presence of a RecA-like fold, but also revealed a unique jawlike domain connected to the RecA fold (Massey et al. 2006). Comparison of two crystal forms suggests a hinge-like movement between the two domains that could lead to a rotary-inchworm translocation mechanism. The structural studies also showed that FtsK has a large central pore, about 25 Å in diameter that could accommodate a dsDNA molecule (Massey et al. 2006). In contrast, TrwB which may transport ssDNA (see later), has a smaller, 8 Å diameter, central pore (Gomis-Ruth et al. 2001).

The soluble ATPase domains of both SpoIIIE and FtsK have been shown to translocate along double stranded DNA in an ATP dependent manner (Bath et al. 2000). In fact, in both cases, the ATPase activity is dependent on the presence of double stranded DNA. In single-molecule experiments, FtsK produced an average translocation rate of ~5 Kb/s (Saleh et al. 2004; Pease et al. 2005). EM of the FtsK motor domain in the presence of double-stranded DNA suggest that a hexamer of the protein forms rings around the DNA (Aussel et al. 2002; Massey et al. 2006). Further evidence that these rings are hexamers comes from size exclusion chromatography where the soluble domain of FtsK migrates as a hexamer (Aussel et al. 2002).

Recent studies have focused on SftA, the soluble FtsK/SpoIIIE-like DNA translocase from *B. subtilis* mentioned earlier (Biller and Burkholder 2009; Kaimer et al. 2009). In contrast to SpoIIIE, which assembles at late stages of cell division, SftA assembles at nascent division septa with FtsZ and before membrane scission (Biller and Burkholder 2009) (Fig. 1a). SftA is required for chromosome partitioning early in cell division, in part through increased chromosome dimer resolution, thus making it more similar to FtsK in terms of its role in chromosome segregation. SpoIIIE, on the other hand, serves as a back-up mechanism for completing chromosome segregation postseptationally (Biller and Burkholder 2009; Kaimer et al. 2009). Interestingly, some bacterial species such as clostridia carry only a single γ domain containing paralog. Those with two FtsK/SpoIIIE ATPases containing γ domains, cluster in the low GC-Gram-positive bacteria, α-proteobacteria, and β-proteobacteria (Biller and Burkholder 2009). Many of these organisms are naturally closely linked to soil and plants but are not endospore formers, suggesting that the presence of the SpoIIIE-like back-up protein is not tightly correlated with spore formation.

A remarkable feature of these chromosome transporters is their ability to move the DNA in the appropriate direction with astounding accuracy. This ability depends on sequence motifs in the chromosomal DNA (Bigot et al. 2005; Levy et al. 2005; Pease et al. 2005; Ptacin et al. 2008; Biller and Burkholder 2009). Short eight-nucleotide sequences, known as KOPS and SRS sequences for FtsK- and SpoIIIE-recognition respectively, are overrepresented on the plus strand of the chromosome and are recognized by the γ domain on the very

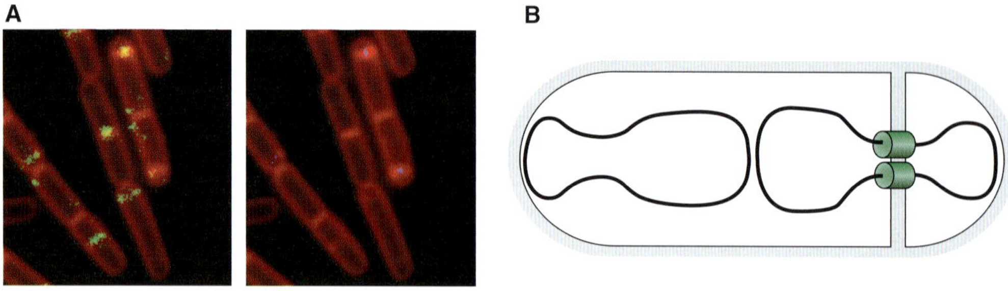

Figure 1. Localization of chromosome segregation proteins, SftA and SpoIIIE, in *Bacillus subtilis*. (*A*) SftA localizes to early division septa in *B. subtilis*, whereas SpoIIIE localizes late in division and largely sporulation septa. This image shows fluorescence overlayed from FM4-64 membrane-staining (red), a SftA-YFP fusion (green) and a SpoIIIE-CFP fusion (blue), of cells from a sporulating culture. SftA fluorescence is apparent at predivisional and vegetative divsion sites, whereas SpoIIIE fluorescence colocalizes with fully formed sporulation division membranes. (Images courtesy of J. Liu.) (*B*) Schematic of the chromosome crossing the division septum through two independent DNA-conducting channels formed by SpoIIIE in a sporulating *B. subtilis* cell. At the end of the transport process, the final loop of DNA in the large compartment must be resolved across the division plane, by an as yet undetermined mechanism.

carboxyl terminus of the FtsK/SpoIIIE proteins. This domain recognizes these sequences and uses them as a guide to accurately move the chromosome into the appropriate daughter cell (Ptacin et al. 2006; Sivanathan et al. 2006). The co-crystal structure of three γ domains bound to KOPS sequence DNA provides structural support for biochemical experiments, suggesting that KOPS DNA establishes directional DNA translocation through oriented protein loading (Bigot et al. 2006; Lowe et al. 2008). Further cytological and functional evidence suggests that SpoIIIE follows this same oriented loading paradigm (Sharp and Pogliano 2002a; Sharp and Pogliano 2002b; Becker and Pogliano 2007; Ptacin et al. 2008).

Although these insights have helped to answer the major conundrum in the field regarding how the FtsK/SpoIIIE proteins faithfully segregate chromosomal material into the correct daughter cell, a lingering question remains: How do the proteins resolve loops of DNA at the division plane? In bacteria with circular chromosomes, it is predicted based on the pore size of the ATPase domain, that each independent hexameric transporter will encompass one double stranded "arm" of the chromosome (Fig. 1B). For the proteins that act predivisionally, such as SftA and presumably FtsK, the scenario is relatively simple to imagine. A small loop of DNA remains on the "wrong" side of the division plane, and when the transporters reach the loop, they might disengage the DNA. This release would allow the loop of DNA to move into the appropriate daughter cell through simple Brownian motion facilitated by the act of chromosome organization and compaction on the daughter cell side. In contrast, the resolution of a loop in a postseptational system, such as that of SpoIIIE, must involve one of two possible mechanisms (Becker and Pogliano 2007; Burton et al. 2007). One possibility is that the chromosomal DNA is linearized and rejoined after transport. Another is that a rearrangement of the proteins to form a single, larger channel could allow the loop to move across the septum uncleaved. Lateral opening of the channels could be driven by the mechanical force that the loop of DNA exerts on the two hexameric complexes. This sort of mechanism is appealing because it would allow for resolution of any DNA loop caught in the septal space, not just the last loop near the chromosome terminus. Further studies on the link between directional transport and the ultimate resolution of local DNA structures at the division plane will be needed to address this issue.

B. subtilis Sporulation as Model System

B. subtilis SpoIIIE, is required for active transport of chromosomal DNA during *B. subtilis* sporulation (Wu and Errington 1998). During sporulation, *B. subtilis* cells divide asymmetrically, creating a larger mother cell and a smaller forespore. Two-thirds of the chromosome destined for the spore compartment remains on the mother cell side of the newly formed septum, and must be actively transported into the spore (Wu and Errington 1998). Cells lacking SpoIIIE are unable to move the chromosome into the spore compartment, and are therefore completely defective for sporulation (Wu and Errington 1994). Similarly cells specifically defective for SpoIIIE ATPase activity are unable to complete the sporulation program (Sharp and Pogliano 2002b).

SpoIIIE is expressed in vegetative cells and is distributed uniformly throughout the membrane (Foulger and Errington 1989). During sporulation, SpoIIIE protein levels do not change. However, at the onset of sporulation, SpoIIIE moves to the septum, localizing to the midpoint of the converging lipid bilayers (Wu and Errington 1997). The generation of this SpoIIIE focus requires the presence of DNA in the septal annulus (Ben-Yehuda et al. 2003). The SpoIIIE focus remains at the septum until the entire chromosome has moved into the forespore (Sharp and Pogliano 2003). Interestingly, SpoIIIE has been implicated in a second function during sporulation, apparently unrelated to chromosome transport. After transport is complete, the protein relocalizes to the pole of the sporulating cell, where it plays an additional role in fusing the membranes of the mother cell engulfing the prespore (Sharp and Pogliano 1999).

What about the proteins that normally decorate a chromosome? Fluorescence microscopy experiments provided evidence that not only are the FtsK/SpoIIIE proteins efficient movers of DNA, they also have the ability to remove virtually every protein roadblock in their way, leaving the transported DNA to enter the daughter cell naked (Marquis et al. 2008). The case of sporulation, where over three megabases of DNA are transported to the daughter cell, highlights this best as transcription factors, chromosome compaction proteins, and other site-specifically bound proteins, are all removed during DNA transport (Marquis et al. 2008). How exactly these translocases accomplish such a daunting task remains to be determined.

Protein–Protein Interactions

The most striking feature of the FtsK/SpoIIIE proteins is that despite their complex functions, the actual DNA transport activity is seemingly accomplished entirely by these proteins. Whereas the next nucleic acids transporter systems to be discussed, competence and transformation, are composed of numerous gene products, FtsK and SpoIIIE appear to function entirely on their own. Nevertheless, it remains to be seen whether there are still other accessory proteins to be discovered that play a role in the DNA transport activity, or whether these proteins really function as lone powerhouses for their chromosome transport function.

CONJUGATION

General Features of Gram-Negative Conjugation Systems

Conjugation refers to the transfer of DNA from one cell to another, by a process requiring cell–cell contact. As with transformation, the transferred molecule is ssDNA. Conjugation and Type 4 protein secretion are mediated by similar complex machines, which in at least some cases mediate the transfer of both DNA and protein substrates. Conjugation involves the participation of pilins, but these are different from the Type 4 pilins needed for transformation. Our discussion will emphasize the *Agrobacterium tumefaciens* system that transfers DNA into plant cells and the *E. coli* R388 plasmid conjugation system. For more detailed descriptions of conjugation, the reader is referred to recent reviews (Chen et al. 2005; Christie et al. 2005; Chandran et al. 2009).

The Conjugation Proteins

The core proteins required for DNA transfer during conjugation are usually divided for convenience into two groups; the DNA processing and transfer (DTR) and the mating pair formation (MPF) proteins. DTR proteins process the DNA to be transferred and resemble the proteins needed for rolling circle replication by bacteriophage ΦX174 and by certain plasmids. It has been proposed that the MPF proteins are needed to form a pilus that connects the donor and recipient cells, retracting to bring the cell surfaces into more intimate contact (Panicker and Minkley 1985). Recent evidence suggests that at least in the case of the F-pilus, single stranded DNA may travel through the pilus, into the recipient cell (Babic et al. 2008; Shu et al. 2008). However, the role of pili in conjugation is not clear and some Type 4 secretion systems appear to lack them. MPF proteins are also needed to form structures that traverse the inner membrane, the periplasm and the outer membrane, forming a channel for the passage of DNA. In the conjugation systems, a Type 4 coupling (T4C) protein mediates interaction between ssDNA produced by the DTR proteins and the MPF apparatus.

DTR and Coupling Proteins

Among the R388 DTR proteins are the relaxase/helicase TrwC, the host IHF protein and TrwA, which also functions as a transcriptional repressor of the *dtr* genes. These proteins form the relaxosome, which binds to the *oriT* region of the super-coiled plasmid DNA (Christie 1997; Gomis-Ruth and Coll 2006). TrwC nicks one strand and covalently attaches by its catalytic tyrosine residue to the 5′ end of the strand destined for transfer. The helicase activity of TrwC then unwinds the DNA and DNA polymerase III replaces the transferred strand by rolling circle replication. The IHF and TrwA components of the relaxosome probably play architectural roles, shaping the DNA into a conformation that favors TrwC binding and activity. The covalently attached TrwC then acts as a "pilot protein" accompanying the T-DNA-protein complex into the recipient cell.

Although recognition by the DNA transfer machinery requires motifs present on the relaxase, rather than a specific DNA sequence, this is not a mere protein secretion system, with DNA coming along for the ride. The length of the transferred strand demands that active DNA transport must take place. The Type 4 coupling (T4CP) protein (TrwB from R388 and VirD4 from *A. tumefaciens*) assembles into a hexamer and has an amino-terminal membrane-embedded domain on each protomer, as well as a large cytosolic moiety. The structure of this moiety has been solved and consists of two domains, a nucleotide-binding domain distal to the membrane surface and a membrane-proximal α-helical domain (Gomis-Ruth et al. 2001). Running through the cytosolic moiety is a central channel that is ~20 Å wide in TrwB but decreases to ~8 Å at the cytosolic extremity. It is not known if this channel continues through the membrane domain and accommodates the ssDNA during transfer, because the structure was derived from a truncated derivative lacking the membrane-embedded moiety. The narrow entrance at the cytoplasmic end may carry out a gating function as transfer is initiated. Aside from its suggestive structure and cellular location, several other attributes support the role of TrwB as a DNA pump. For instance, TrwB is a DNA-dependent ATPase, and shows conformational changes in the surface residues of the central channel depending on its ATP-binding state (Tato et al. 2005). In addition, the closest structural analog to TrwB is stated to be FtsK, although the latter protein has a wider central channel, presumably to allow the passage of dsDNA. It is worth noting that members of the HerA/FtsK/SpoIIIE protein family are also required for conjugation in Gram-positive bacteria (Iyer et al. 2004; Berkman et al. 2010) The nucleotide-binding domain of TrwB also resembles that of the ring helicases, which are known to use the energy of ATP hydrolysis to translocate along DNA (Gomis-Ruth et al. 2001).

MPF Proteins

The MPF proteins are numerous, 11 in the case of *A. tumefaciens* where they are best characterized (Christie et al. 2005; Fronzes et al. 2009a). In addition to the proteins that assemble the pilus (VirB2 and VirB5) additional proteins (VirB4, VirB6, VirB7, VirB8, VirB9, and VirB10) assemble across the bacterial envelope. Biochemical investigation established that VirB8, VirB10, and the NTPase VirB4, are anchored in the inner membrane, VirB9 is located in either the periplasm or in association with the outer membrane, and VirB7 is probably associated with the outer membrane. These proteins presumably establish a channel extending from the T4CP protein (VirD4) to the exterior surface. The VirB11 ATPase is recruited to the inner face of the inner membrane where it is needed for the assembly of the secretion channel. This proposed role for VirB11, in mediating assembly of a trans-envelope structure perhaps containing a pilus subunit, is analogous to that proposed for the secretion ATPases of the T4 pilus-related systems (see later), and in fact VirB11 has been classified as a secretion (or traffic) ATPase. In addition to VirD4 and VirB11, VirB4 is also an ATPase and together these three proteins, which contact one another, presumably power the assembly of the conjugation nano-machine as well as secretion of DNA.

The role of the pilin proteins is unclear. A filamentous structure composed of pilin subunits, perhaps a mini- or pseudopilus, may be an obligatory part of the conjugation machinery, even providing a channel for DNA transport. Another possibility is that pilin subunits may assemble with MPF proteins to form a transport structure whereas pili per se may be an alternative structure, requiring the MPF proteins for its formation.

Recently, two major advances have contributed to an enhanced understanding of these complex systems. The first is the so-called transfer DNA immunoprecipitation (TrIP) assay. This technique, based on cross-linking of DNA to proteins using immunoprecipitation and PCR, has identified proteins that are proximal to the DNA during transport (Cascales and Christie 2004). This has established a provisional transport pathway and has suggested the identity of channel subunits. As expected, VirD4 is the first protein contacted by the DNA following interaction with the relaxosome. Next the DNA contacts the VirB11 ATPase and this interaction depends on the presence of other core components, but not on the Walker A and B residues of VirB11. Why VirB11 interacts with DNA is unknown, but it may have a direct role during transport, aside from its morphogenetic role in building the core. In two subsequent steps, close contacts are established between the DNA and VirB6 and VirB8, located in the inner membrane, and then with the outer membrane/periplasmic VirB2 pilin and with VirB9. These valuable findings constrain model building, based on the second major advance, the acquisition of high-resolution structural data.

Although several individual component structures had been solved (Yeo et al. 2000; Yeo et al. 2003; Terradot et al. 2005; Bayliss et al. 2007), a big step forward derives from the 15-Å-resolution cryoelectron microscopy structure of a complex between the VirB7-VirB9-VirB10 orthologs from the plasmid pK101 (TraN, TraO, and TraF respectively) (Fronzes et al. 2009b). This 1.05 MDa "core complex" is composed of 14 copies of each protein, forming a double-walled two-chambered channel, proposed to span both membranes. The inner and outer chambers contain a channel that extends 105 Å, sufficient to bridge the two membranes. The channel is open on the cytoplasmic side and constricted near the other end. The outer surface of the structure is composed mostly of TraF/VirB10, whereas TraN/VirB7 and TraO/VirB9 form the inner wall of the channel. It is suggested that the orthologs of the VirB4 and VirB11 ATPases may regulate the opening at the distal end, communicating across the inner membrane and periplasm by some unknown mechanism, permitting passage of the DNA. This beautiful structure provides no explicit role for these ATPases, the T4C protein, or for the orthologs of VirB2, VirB6, and VirB8. Presumably one or more of these proteins decorates the outside of the structure, interacts with it within the membrane or may even be located in one of the large chambers.

More recently, the remarkable crystal structure of a 0.6 MDa fragment of this core structure, derived from chymotrypsin treatment, has been reported (Chandran et al. 2009). This structure corresponds to the outer chamber of the cryo-EM structure. Unexpectedly, the VirB10 protein, which crosses the inner membrane, also comprises the outer membrane channel. It may thus represent the first bacterial protein that traverses both membranes. Figure 2 shows the crystal structure of the outer membrane complex superimposed on the cryo-EM structure. Detailed comparison of the two structures reveals suggestive differences, which may reflect conformational changes underlying DNA transport (Chandran et al. 2009). It has been suggested that the large chambers may be sites of assembly of the MPF proteins not represented in the core structure, and that these may form the actual channel for substrate transport (Chandran et al. 2009; Christie 2009). Naturally many questions remain concerning the detailed mechanisms of secretion and the energetics of the process. One wonders if the showed requirement for the VirB2 pilin for transport hints at a dynamic role for a pilus, or for a pseudopilus, analogous to that proposed below for the competence pseudopilus? The mechanism by which the DNA is transported into the recipient cell is also obscure, although a role for a pilus has been suggested (Backert et al. 2008). Although these and other major questions are unresolved, the VirB7-9-10 structure represents a breakthrough that should lead to rapid advances in understanding.

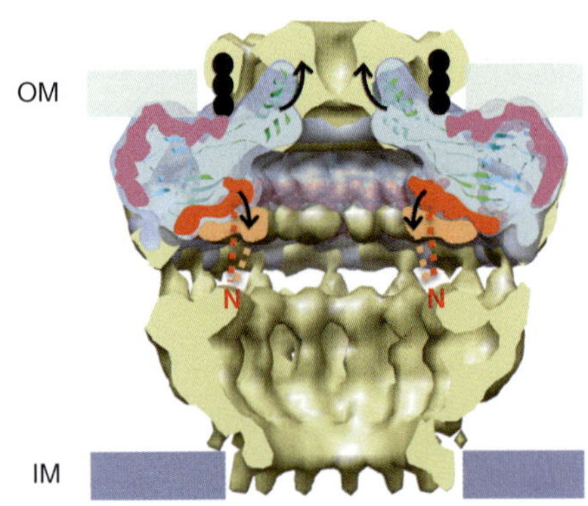

Figure 2. Proposed structure of the core complex for T4S system (pKM101)-mediated conjugation. The cryo-EM structure (yellow) is superimposed on the crystal structure of the outer-membrane complex. The location of lipid associated with TraN/VirB7 is shown with black dots. The arrows illustrate proposed conformational changes (Chandran et al. 2009). OM and IM indicate outer and inner membrane, respectively. (Adapted, with permission, from Chandran et al. 2009 [© Nature Publishing Group].)

TRANSFORMATION

General Features of Bacterial Transformation

Transformation refers to the transport of environmental DNA to the cytosolic compartment, where it can meet several fates. If the incoming DNA is complementary to the resident chromosome, it can recombine. If it contains the information needed for independent replication, it can do so as a plasmid or a bacteriophage. As with conjugation, the transported DNA is single-stranded. The nontransforming strand is degraded, most likely by a mechanism that is closely connected to transport across the cell membrane (Mejean and Claverys 1993).

These shared features reflect the existence of a common set of transformation proteins (Dubnau 1999). Gram-positive and Gram-negative bacteria share the core proteins, except that the latter group has additional components enabling DNA to cross the outer membrane. Among these conserved transformation proteins are those that are needed for Type 4 pili (T4P) and Type 2 secretion (T2S). *Helicobacter pylori* is unusual because it uses a conjugation-like system for DNA transport, rather than the usual shared transformation proteins (Hofreuter et al. 2000).

In most bacteria, complex signal transduction pathways regulate the proteins that mediate DNA uptake (Macfadyen 2000; Hamoen et al. 2003; Claverys et al. 2006; Leisner et al. 2008). Cells that synthesize these transformation proteins are referred to as "competent." Although the core transformation proteins are mostly conserved, the regulatory mechanisms are not. This is a common theme; nature often uses a common set of tools, but uses species-specific instructions for their use. We will not further describe the regulation of competence gene expression. This article reflects the bias of the authors by emphasizing the *B. subtilis* system, although information from other bacteria will be included when needed. The *B. subtilis* nomenclature for proteins and genes will be used throughout. Several reviews are available that emphasize transformation in other bacteria or that offer more detail (Berge et al. 2002; Friedrich et al. 2002; Hamilton and Dillard 2006; Claverys et al. 2009).

The Fate of Transforming DNA

It is useful to follow a DNA molecule from its existence in solution outside the cell, until it becomes integrated in the resident genome of the transformant (reviewed in Berge et al. 2002; Chen and Dubnau 2004; Chen et al. 2005). Double stranded DNA (dsDNA) first binds to the *B. subtilis* cell surface without sequence specificity and is then fragmented by the action of a membrane-localized nuclease (NucA) that presumably acts at the points of attachment (Provvedi et al. 2001). After binding and fragmentation, one strand is degraded, releasing phosphorylated products into the medium and the transforming strand is converted into a form that is no longer accessible to added DNAse. DNAse resistance is taken as an indication of transport across the cell membrane. Indeed no difference in the kinetics of appearance of transformants, DNAase resistance and degradation products has been detected. In *S. pneumoniae*, the transforming strand is taken up with 3′ to 5′ polarity and the nontransforming strand is degraded with 5′ to 3′ polarity, consistent with close coordination of the degradation and uptake steps (Méjean and Claverys 1988; Mejean and Claverys 1993). (For a possible difference in *B. subtilis*, see Vagner et al. 1990.) 3′ ssDNA termini initiate recombination in all systems and in the case of transformation, the uptake mechanism itself generates this required substrate. Internalized ssDNA thus interacts with the resident chromosome and RecA-mediated recombination results in formation of a heteroduplex in which one strand is from the donor and the other from the recipient chromosome. This heteroduplex molecule is at first stabilized by non-covalent bonds, after which ligation occurs. The heteroduplex is resolved by replication and in *S. pneumoniae* by mismatch repair as well (Claverys and Lacks 1986; Majewski and Cohan 1998).

In Gram-negative bacteria, e.g., *H. influenzae* and *Neisseria gonorrhoeae*, dsDNA presumably passes across the outer membrane through ring-shaped assemblies of proteins known as secretins (Tomb et al. 1991; Collins et al. 2004). It has been reported that the secretin complex of *N. gonorrhoeae* contains the initial binding site for DNA during transformation (Assalkhou et al. 2007). On entry to the periplasm, DNA is thought to become resistant to exogenously added DNAse. Transport into the cytosol is more difficult to detect in Gram-negative bacteria and "uptake" is (operationally) defined differently than in the Gram-positives. In *Haemophilus* and *Neisseria*, specific sequences are required for efficient transformation (Danner et al. 1980; Elkins et al. 1991), but the receptor for this sequence is unknown. More typically, DNA uptake shows no discernible sequence specificity.

Single molecule experiments (Maier et al. 2004) have shown that DNA uptake in *B. subtilis* is processive, with no detectable pausing at a temporal resolution down to 1 s. Uptake was observed to occur at a rate of about 80 bp/s. (A similar rate of (~90 bp/s) has been estimated for *S. pneumococus* (Mejean and Claverys 1993)). Uptake remained processive and its velocity was invariant against large external forces, up to ~40 pN. These measurements establish both the DNA transformation motor and the FtsK/SpoIIIE motors as uniquely powerful. They also set the parameters that must be explained by any proposed mechanism.

The Transformation Proteins

In *B. subtilis* and *S. pneumoniae*, several cytosolic proteins under competence-specific control are required for optimal transformation; RecA, SsbB (YwpH), DprA (Smf) and YjbF (CoiA) (reviewed in Claverys et al. 2009). RecA is required for integration of transforming DNA and in its absence, little or no transformation for chromosomal markers is detected, although transformation by plasmid DNA proceeds normally. When the competence-specific ssDNA binding proteins SsbB or DprA are absent, incoming DNA is degraded (Berge et al. 2003). DprA appears to play a role in loading RecA onto ssDNA to form RecA filaments, preparatory to recombination (Mortier-Barriere et al. 2007). The molecular role of YjbF is obscure. Inactivation of any one of these cytosolic proteins decreases transformation, but does not prevent the uptake of transforming DNA, measured by the DNAse resistance of radiolabeled DNA. Nevertheless, it is possible that they do play a role in transport, perhaps by binding to incoming ssDNA, thereby preventing backward slippage. Consequently, their absence may affect the rate or processivity of uptake.

Two membrane proteins are needed for the appearance of DNAse resistant radiolabeled DNA but are not required for binding of DNA to the cell. One of these, ComEC, is a large polytopic transmembrane protein, which is likely to form at least part of the pore for DNA uptake (Draskovic and Dubnau 2005). In *B. subtilis*, inactivation of *comEC* prevents not only uptake of DNA, but also the degradation of the nontransforming strand. In contrast, degradation proceeds normally in the equivalent *S. pneumoniae* mutant (Berge et al. 2002). In *S. pneumoniae*, a membrane endonuclease (EndA) has been shown to be required for nontransforming strand degradation (Lacks et al. 1974). The *endA* gene is not under competence control and no ortholog of *endA* exists in *B. subtilis*. These observations suggest an important difference in the coupling of DNA uptake and degradation in the two bacteria. Another membrane-associated protein, ComFA, possesses the conserved motifs typical of Super family 2 helicase/translocases and is also needed for DNA uptake but not for DNA binding. Mutation of the ComFA Walker A motif completely inactivates the protein for uptake (Londono-Vallejo and Dubnau 1994) and ComFA may therefore contribute energetically to DNA uptake by translocation or may mediate strand separation before degradation of the nontransforming strand. These are attractive ideas, because a membrane-anchored translocase could drive DNA into the cell and particularly because Super family 2 proteins can track along a single strand and do so mostly with 3′ to 5′ polarity (Singleton et al. 2007). A null mutant of a third membrane protein, ComEA, does not internalize DNA and was reported to reduce binding to the cell surface, measured by incubation with radiolabeled DNA followed by centrifugal washing (Inamine and Dubnau 1995). ComEA possesses two helix-loop-helix motifs near its carboxyl terminus, and is inserted in the membrane with these motifs located externally. Such motifs are known to bind DNA nonspecifically, and indeed the corresponding part of ComEA binds dsDNA (Provvedi and Dubnau 1999). An in-frame deletion of a central portion of ComEA, has a minor effect on DNA binding, but eliminates uptake. Apparently, although the dsDNA binding activity of ComEA is essential, ComEA is also needed for transport.

An important group of proteins essential for transformation is related to Type 4 pilus (Craig et al. 2004) and Type 2 secretion (T2S) proteins (Johnson et al. 2006) of Gram-negative bacteria (Albano et al. 1989; Hobbs and Mattick 1993). Type 4 pili (T4P) are filamentous helical assemblies of pilin proteins that emerge through a secretin pore in the outer membrane. The assembly and disassembly of the T4P drives twitching motility (Merz et al. 2000; Skerker and Berg 2001; Burrows 2005). By analogy, it has been proposed that cycles of assembly and disassembly of a "pseudopilus" provide the mechanical force for T2S through a secretin pore and for the uptake of DNA during transformation (Mattick 2002). A T2S pseudopilus has been characterized (Skerker and Berg 2001; Kohler et al. 2004), but no direct evidence for its cyclical assembly and disassembly has been presented.

Typically, the T4P, T2S and competence systems encode several pilin or pseudopilin proteins, which contain conserved, hydrophobic, α-helical domains that traverse the membrane (Sauvonnet et al. 2000; Mattick 2002; Burrows 2005; Chen et al. 2005; Proft and Baker 2009). These systems also require dedicated, membrane-localized proteases, a conserved polytopic integral membrane protein, and at least one hexameric, P-loop "traffic ATPase" that is peripherally associated with the inner face of the cell membrane. This class of ATPases is commonly associated with processes that involve the transport of macromolecules and is a subclass of AAA+ proteins. Traffic ATPases are required for formation of the pili and pseudopili. After removal of several amino-terminal amino acid residues from the pilin proteins by the dedicated protease, the mature pilins are translocated from the membrane to the periplasm/wall and the T4P or Type 2 pseudopili are assembled so that the hydrophobic α-helices of the (pseudo)pilins are buried in the interfaces between the subunits of these structures (Forest and Tainer 1997; Craig et al. 2004).

This general picture applies in the case of *B. subtilis* and almost certainly the other transformation systems. The *B. subtilis* orthologs of T4P and T2S genes are encoded by *comC* (the protease) and by the *comG* operon (reviewed in Chen and Dubnau 2004). This operon contains seven genes, encoding the secretion ATPase (ComGA), the conserved polytopic membrane protein (ComGB) and four typical pseudopilins. Thus all the core components of these systems are present and a competence pseudopilus, composed of the major pseudopilin, ComGC, has been described (Chen et al. 2006). Immediately after synthesis, ComGC is located as an integral membrane protein with its carboxyl terminus extending outward from the membrane (Chung et al. 1998). The competence-specific thiol oxido-reductase proteins, BdbD and BdbC then oxidize the two cysteine residues of ComGC to form an intramolecular disulfide bond, and cleavage by ComC takes place (Fig. 3A). The processed ComGC is then translocated to a location outside the membrane, and the competence pseudopilus is assembled (Chen et al. 2006). In addition to *comC* and *bdbDC*, each of the *comG* ORFs are required for formation of the pseudopilus. This structure is asymmetric and highly polydisperse. Remarkably, the ComGC subunits are held together not only by noncovalent interactions, but also by intermolecular disulfide bonds, which presumably form by exchange with the pre-existing intramolecular S-S bonds. The average length of the competence pseudopilus is apparently sufficient to cross the cell wall of *B. subtilis*. (It is interesting that in other, unrelated pili of Gram-positive bacteria, the subunits are covalently linked by isopeptide bonds (Mandlik et al. 2008)).

FRET measurements and microscopic colocalization studies on fluorescent fusion proteins have shown that many of the competence proteins are in contact or in close proximity (Hahn et al. 2005; Kidane and Graumann 2005; Kramer et al. 2007; Tadesse and Graumann 2007). Even the cytosolic ("soluble") proteins are in intimate association with the membrane components of the transformation apparatus. Specifically, ComFA, ComEA, ComGA, SsbB, YjbF, DprA, and RecA have been shown to colocalize at the poles of *B. subtilis* (Fig. 4). This colocalization is dynamic, at least for DprA and RecA (Kidane and Graumann 2005; Tadesse and Graumann 2007). DNA binding takes place preferentially at the poles, close to foci of competence protein concentrations, and less often near laterally located foci, which are helically distributed. Single molecule experiments indicate that DNA uptake, like binding, also occurs preferentially at or near the cell poles (Hahn et al. 2005). These studies show that the transformation proteins function in concert and that a multicomponent protein machine carries out transformation, from the initial binding step all the way to integration.

Given the striking similarities of the proteins involved in transformation, T4P and T2S, the filamentous structures they form and the requirement in each system for movement (of DNA, a cell, and protein respectively), it is attractive to consider that a mechanism like the one that powers twitching motility also drives DNA uptake for transformation. In fact, T4P-like proteins are required for transformation in every system that has been studied, both Gram-negative and -positive, with the exception of *Helicobacter pylori* (see later). The single molecule DNA uptake studies show that the force-velocity characteristics of DNA uptake are similar to those of T4P retraction; both are processive and operate with constant velocity

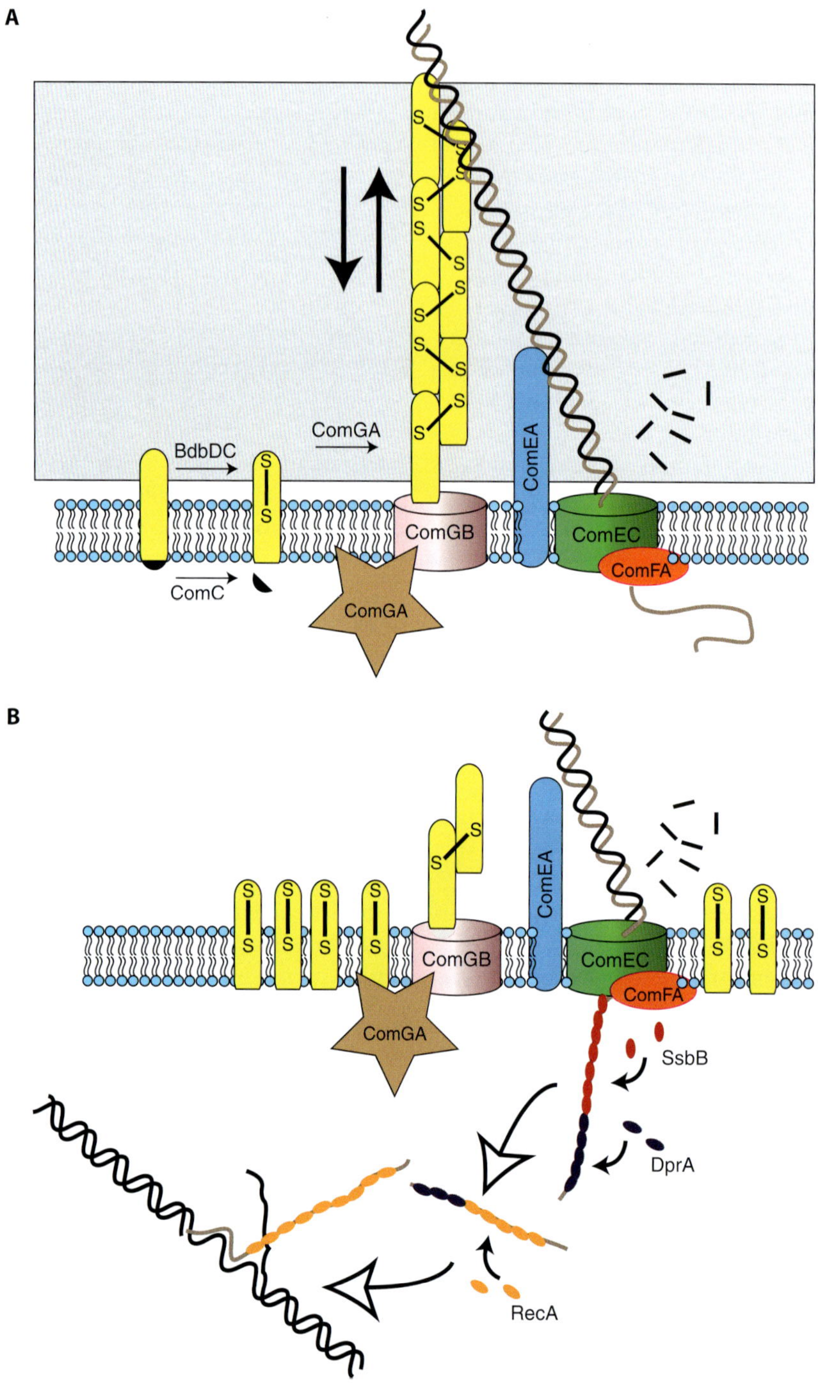

Figure 3. A model for transformation in Gram positive bacteria. ComGC is processed by the protease ComC and oxidized to form an intramolecular disulfide bond by BdbDC. (*A*) It is then translocated from the membrane in a step that requires the traffic ATPase ComGA and assembled into a pseudopilus in which the intramolecular disulfide bonds are replaced by links between the subunits. ComGB is represented at the base of the pseudopilus without direct evidence. It is proposed that assembly and disassembly of the pseudopilus brings transforming DNA to the ComEC channel and that interaction of the DNA with the C-terminal part of

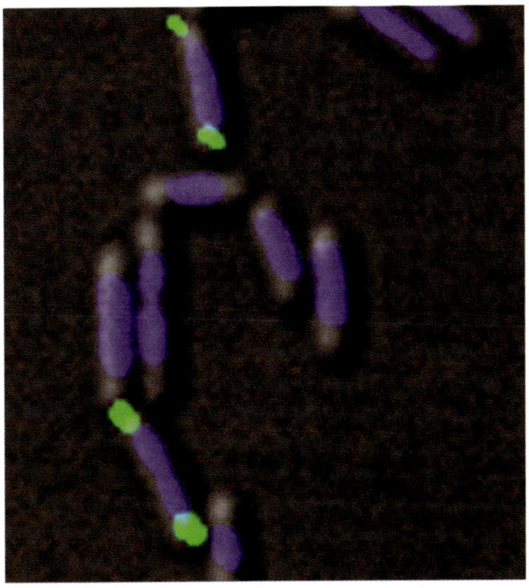

Figure 4. ComGA-GFP localizes to the poles of competent *B. subtilis*. This image shows fluorescence from DAPI-staining (blue), and a ComGA-GFP fusion (green), overlayed on a DIC image of ten cells from a competent culture. Two competent cells are present, in both of which the ComGA fluorescence is localized near the cell poles. (Photo courtesy of J. Hahn.)

against relatively high loads (Maier et al. 2002; Maier et al. 2004). No direct evidence yet exists demonstrating a role for pseudopilus assembly and disassembly in DNA transport, but evidence from analysis of *comGA* point mutations has shown that the ATPase ComGA, and probably the pseudopilus, are needed for transport (K. Briley and D. Dubnau, unpublished). ComGA, but not the pseudopilus, is also needed for DNA binding to the cell. Because ComGA is located on the inner surface of the membrane, the ComGA requirement for competence-specific DNA binding suggests that this protein communicates across the membrane with an unidentified DNA receptor. This receptor is not ComEA, and in fact when a filtration technique rather than the more disruptive centrifugal washing measures binding of radiolabeled DNA, inactivation of *comEA* has very little effect on total DNA bound. It is likely that DNA binding takes place first to the unknown receptor and to ComEA at a later step.

The Transformation Model

The following model for the transformation process is consistent with the available evidence (Fig. 3). dsDNA first binds to an unknown receptor that communicates with ComGA. The DNA is cleaved by the membrane nuclease NucA, providing an end for the initiation of transport through a pore formed at least in part by ComEC. The pseudopilus then drives uptake by cycles of assembly and disassembly. The highly processive nature of uptake requires either the participation of several pseudopili acting in concert, or of another DNA binding protein that prevents the DNA from slipping backwards through the pore. ComEA is an attractive candidate for this role. An unknown nuclease in *B. subtilis* (EndA in the case of *S. pneumoniae*) degrades one strand during transport, and ComFA may unwind the DNA before this degradation step and/or may provide part of the driving force for uptake after degradation. Following transport, ssDNA is protected by binding to SsbB and DprA and also interacts with RecA. Because the core competence proteins are conserved in (nearly) all transformation systems, the essential features of this model are likely to be universal. An alternative view, compatible with the evidence, is

Figure 3. (*Continued*) ComEA is required for uptake, which is assisted by the ATPase ComFA. The transforming strand (gray) enters the cytosol and the nontransforming strand is degraded with the products exiting the cell. The nomenclature is that of *B. subtilis*. In *S. pneumoniae*, EndA (not shown) carries out degradation of the non-transforming strand. Also not shown is NucA, which introduces double strand breaks in the bound DNA. The large gray square represents cell wall material. (B) The pseudopilus has disassembled and the ComGC subunits have returned to the membrane where they may form a pool available for further cycles of assembly/disassembly. As the ssDNA enters the cytosol, it interacts with SsbB and DprA. These proteins protect the DNA and DprA also assists in loading RecA to form a filament, which is proposed to carry out the homology search leading to recombination with the resident chromosome.

that the pseudopilus functions to bring the DNA across the cell wall and periplasm, into contact with the membrane-localized proteins that then mediate transport to the cytosol. In this view, transformation would be a two-step process.

The cellular locations of DprA and RecA vary during transformation (Kidane and Graumann 2005; Tadesse and Graumann 2007). DprA, which shows only occasional polar localization before transformation, becomes dramatically localized to the poles when DNA is added. In contrast, RecA departs from the poles when DNA is added, forming filaments that then diffuse rapidly. A reasonable hypothesis is that ssDNA engages SsbB at the ComEC pores, which serves to protect the DNA and aid assembly of ssDNA-DprA complexes. ssDNA thus acts to localize DprA near the cell poles, although protein–protein contacts probably also participate. DprA assists in the loading of RecA on the ssDNA (Mortier-Barriere et al. 2007), forming filaments, which then show dynamic behavior, conducting a diffusive search for sites of homology on the chromosome. Thus, a picture is emerging of a transformation machine, with moving parts that mediate not only the binding and uptake of DNA, but also the interaction of ssDNA with the resident chromosome. This coordinated action of DNA uptake and recombination proteins constitutes strong evidence that transformation has evolved for the acquisition of genetic information.

What are the sources of energy for uptake? Older experiments with inhibitors indicate a role for the pH component of proton motive force (PMF) (van Nieuwenhoven et al. 1982; Bremer et al. 1984). Consistent with this is the demonstration in single DNA molecule experiments that uncouplers (DNP and CCCP) arrest uptake well before the ATP pool is depleted (Maier et al. 2004), suggesting that a proton symport mechanism may drive transport. However, it is fully possible that ATP also plays a direct role consistent with the essentiality of the Walker A motif of the helicase-like ComFA protein for uptake (Londono-Vallejo and Dubnau 1994).

In the T4P systems, there are two traffic ATPases, one of which appears to drive assembly and the other disassembly of the pilus (Herdendorf et al. 2002; Burrows 2005; Chiang et al. 2008; Jakovljevic et al. 2008). However, the competence and T2S systems each possess single known ATPases, which more closely resemble the assembly ATPases of T4P. Perhaps PMF directly disassembles the pseudopili or the single ATPase can operate in reverse. Recent laser tweezer analyses of T4P retraction in *N. gonorrhoeae* and *Myxococcus xanthus* have revealed the presence of two retraction motors; a high-force, low-velocity motor dependent on the retraction ATPase, PilT and a high-velocity, low-force motor that is PilT-independent (at least in *M. xanthus*) (Clausen et al. 2009a; Clausen et al. 2009b). The stalling force of the high velocity motor (70 pN) is suggestive of the *B. subtilis* DNA uptake system, in which measurements hinted at a stalling force somewhat in excess of 40 pN (Maier et al. 2004). These findings are consistent with a mode of competence pseudopilus retraction that is not dependent on a traffic ATPase, perhaps driven instead by PMF.

Other Dynamic Aspects of the Transformation Machine

During the development of competence in *B. subtilis*, many of the relevant proteins accumulate at the poles of the cells and in foci that appear to be helically distributed in association with the cell membrane. More or less concomitantly with a loss of transformability, these proteins leave the poles (Hahn et al. 2005). Time-lapse microscopy and the use of inhibitors suggest that movement to the poles depends on diffusion, with capture at the poles by an anchoring site (Hahn et al. 2009). In contrast, departure from the poles is energy-driven, and depends on the action of the proteins McsA and B. McsA and B are known to act together as adaptors for the ClpCP and ClpEP proteases (Kirstein et al. 2007) and inactivation of these degradative proteins (McsA, McsB, ClpP, or of ClpC+ClpE) prevents delocalization of competence proteins. It is postulated that the degradative complexes degrade an anchor protein, releasing the competence proteins and in agreement with this, McsB, ClpC, and ClpP move to the poles of competent cells (Hahn et al. 2009). The localization and delocalization of competence proteins is clearly regulated, but there is no information as to the signaling pathways involved. It is also not known why assembly of the transformation machinery and DNA uptake occur at the poles. Perhaps localization favors low affinity assembly interactions, and the poles simply provide a convenient address. Alternatively, unique properties of the wall/membrane at the poles facilitate assembly and insertion of the competence machine.

Transformation in *Helicobacter pylori*

The *Helicobacter* transformation system is a unique case in which core inner membrane and cytosolic constituents (notably the channel protein ComEC and *dprA*) are conserved (Ando et al. 1999; Smeets et al. 2000; Yeh et al. 2003) whereas the T4P-like proteins are replaced by conjugation proteins, encoded by the *comB2-B4* and *comB6-B10* operons (Karnholz et al. 2006). *Helicobacter* is therefore a hybrid between the more common transformation systems and the conjugation machines.

An instructive recent paper (Stingl et al. 2010) shows that the *comB* operons of *Helicobacter* mediates high velocity, low-force uptake of double stranded DNA into the periplasm, whereas *comEC* is needed for transport of single stranded DNA into the cytosol, presumably by a high-force mechanism akin to that used by *B. subtilis*. This study has at least two important implications. First, it is the clearest demonstration that transformation in Gram-negative bacteria is a two-step process, with uptake into the periplasm temporally uncoupled from transport into the cytosol. Second, it implies that the T4P-like and the conjugation-like transformation proteins play analogous roles, conveying double stranded DNA into the periplasm and across the cell wall to provide accessibility of this DNA to the conserved inner membrane transport apparatus. This view of competence is similar to that proposed above for the roles of the pseudopilus and membrane proteins in *B. subtilis*.

COMPARISON OF THE CONJUGATION, TRANSFORMATION AND FTSK/SPOIIIE SYSTEMS: UNANSWERED QUESTIONS

The conjugation and transformation systems use ssDNA as substrate and require the assembly of elaborate machines composed of several proteins, including those that span the cell envelope. In both systems cytosolic proteins process the DNA substrate either before transfer or following import and in both, these proteins interact with and coordinate their activities with the membrane transporters. Another striking similarity between these systems is the involvement of pilus-like structures. However, given our present state of knowledge, the resemblance between these systems ends here, because there is little obvious similarity between the conjugation and transformation proteins, other than the prominent involvement of P-loop NTPases. It is likely that this reflects our ignorance and that analogous roles for individual proteins will become apparent.

Because of this ignorance, many questions remain unanswered. Among the most important unresolved issues are the following. What are the immediate sources of energy driving the movement of DNA across the membrane(s)? What proteins does the DNA contact, particularly during transformation, and at what stages during transport? Is assembly and disassembly of a pilus-like structures involved in transport? In the Gram-negative transformation and conjugation systems, is there a role for pili per se in the transport process, or do the pilus proteins assemble into alternative DNA-transport specific structures? Finally, what is the meaning of the positioning at the cell poles of not only transformation, but also of the conjugation proteins of some bacteria (Judd et al. 2005; Teng et al. 2008; Berkman et al. 2010). Clearly structural information on the component proteins will contribute in a major way to our understanding. In the case of transformation, the interactions of transport and recombination proteins, analogous to those of the DTR proteins and membrane components of the conjugation apparatus, continue to be of major interest.

The FtsK system, whereas functionally quite distinct from the conjugation and transformation machinery, does share the central and defining role of the P-loop ATPase activity, and in fact as mentioned earlier, several conjugation proteins are members of the HerA/FtsK family. In contrast to conjugation and transformation, the FtsK/SpoIIIE proteins use dsDNA as a substrate, and the barrier across which FtsK/SpoIIIE translocates the DNA is variable. Thus, whereas ostensibly a more simple system than the multicomponent machineries of conjugation and transformation, the current state of knowledge about FtsK and SpoIIIE raise several challenging mechanistic questions. For example, how are loops of DNA resolved? How do these proteins clear other DNA binding proteins away from the transported DNA so rapidly and effectively? And ultimately, what is the architecture of these complexes in the context of the cell division machinery and septal membranes? These questions continue to fuel inquiry from many angles.

ACKNOWLEDGMENTS

Work cited from the BB lab was supported by a Massachusetts Life Science Center Grant, and from the DD lab by National Institutes of Health grants GM057720 and GM043756.

REFERENCES

Albano M, Breitling R, Dubnau DA. 1989. Nucleotide sequence and genetic organization of the *Bacillus subtilis comG* operon. *J Bacteriol* **171:** 5386–5404.

Ando T, Israel DA, Kusugami K, Blaser MJ. 1999. HP0333, a member of the dprA family, is involved in natural transformation in *Helicobacter pylori*. *J Bacteriol* **181:** 5572–5580.

Assalkhou R, Balasingham S, Collins RF, Frye SA, Davidsen T, Benam AV, Bjoras M, Derrick JP, Tonjum T. 2007. The outer membrane secretin PilQ from *Neisseria meningitidis* binds DNA. *Microbiology* **153:** 1593–1603.

Ausmees N, Wahlstedt H, Bagchi S, Elliot MA, Buttner MJ, Flardh K. 2007. SmeA, a small membrane protein with multiple functions in *Streptomyces* sporulation including targeting of a SpoIIIE/FtsK-like protein to cell division septa. *Mol Microbiol* **65:** 1458–1473.

Aussel L, Barre FX, Aroyo M, Stasiak A, Stasiak AZ, Sherratt D. 2002. FtsK Is a DNA motor protein that activates chromosome dimer resolution by switching the catalytic state of the XerC and XerD recombinases. *Cell* **108:** 195–205.

Babic A, Lindner AB, Vulic M, Stewart EJ, Radman M. 2008. Direct visualization of horizontal gene transfer. *Science* **319:** 1533–1536.

Backert S, Fronzes R, Waksman G. 2008. VirB2 and VirB5 proteins: Specialized adhesins in bacterial type-IV secretion systems? *Trends Microbiol* **16:** 409–413.

Bath J, Wu LJ, Errington J, Wang JC. 2000. Role of *Bacillus subtilis* SpoIIIE in DNA transport across the mother cell-prespore division septum. *Science* **290:** 995–997.

Bayliss R, Harris R, Coutte L, Monier A, Fronzes R, Christie PJ, Driscoll PC, Waksman G. 2007. NMR structure of a complex between the VirB9/VirB7 interaction domains of the pKM101 type IV secretion system. *Proc Natl Acad Sci* **104:** 1673–1678.

Becker EC, Pogliano K. 2007. Cell-specific SpoIIIE assembly and DNA translocation polarity are dictated by chromosome orientation. *Mol Microbiol* **66:** 1066–1079.

Begg KJ, Dewar SJ, Donachie WD. 1995. A new *Escherichia coli* cell division gene, ftsK. *J Bacteriol* **177:** 6211–6222.

Ben-Yehuda S, Rudner DZ, Losick R. 2003. Assembly of the SpoIIIE DNA translocase depends on chromosome trapping in *Bacillus subtilis*. *Curr Biol* **13:** 2196–2200.

Berge M, Mortier-Barriere I, Martin B, Claverys JP. 2003. Transformation of *Streptococcus pneumoniae* relies on DprA- and RecA-dependent protection of incoming DNA single strands. *Mol Microbiol* **50:** 527–536.

Berge M, Moscoso M, Prudhomme M, Martin B, Claverys JP. 2002. Uptake of transforming DNA in Gram-positive bacteria: A view from *Streptococcus pneumoniae*. *Mol Microbiol* **45:** 411–421.

Berkman MB, Lee CA, Loveday E-K, Grossman AD. 2010. Polar positioning of a conjugation protein from the integrative and conjugative element ICEBs1 of *Bacillus subtilis*. *J Bacteriol* **192:** 38–45.

Bigot S, Corre J, Louarn JM, Cornet F, Barre FX. 2004. FtsK activities in Xer recombination DNA mobilization and cell division involve overlapping and separate domains of the protein. *Mol Microbiol* **54:** 876–886.

Bigot S, Saleh OA, Cornet F, Allemand JF, Barre FX. 2006. Oriented loading of FtsK on KOPS. *Nat Struct Mol Biol* **13:** 1026–1028.

Bigot S, Saleh OA, Lesterlin C, Pages C, El Karoui M, Dennis C, Grigoriev M, Allemand JF, Barre FX, Cornet F. 2005. KOPS: DNA motifs that control *E. coli* chromosome segregation by orienting the FtsK translocase. *Embo J* **24:** 3770–3780.

Biller SJ, Burkholder WF. 2009. The *Bacillus subtilis* SftA (YtpS) and SpoIIIE DNA translocases play distinct roles in growing cells to ensure faithful chromosome partitioning. *Mol Microbiol* **74:** 790–809.

Bremer W, Kooistra J, Hellingwerf KJ, Konings WN. 1984. Role of the electrochemical proton gradient in genetic transformation of *Haemophilus influenzae*. *J Bacteriol* **157:** 868–873.

Burrows LL. 2005. Weapons of mass retraction. *Mol Microbiol* **57:** 878–888.

Burton BM, Marquis KA, Sullivan NL, Rapoport TA, Rudner DZ. 2007. The ATPase SpoIIIE transports DNA across fused septal membranes during sporulation in *Bacillus subtilis*. *Cell* **131:** 1301–1312.

Cascales E, Christie PJ. 2004. Definition of a bacterial type IV secretion pathway for a DNA substrate. *Science* **304:** 1170–1173.

Chandran V, Fronzes R, Duquerroy S, Cronin N, Navaza J, Waksman G. 2009. Structure of the outer membrane complex of a type IV secretion system. *Nature* **462:** 1011–1015.

Chen I, Dubnau D. 2004. DNA uptake during bacterial transformation. *Nat Rev Microbiol* **2:** 241–249.

Chen I, Christie PJ, Dubnau D. 2005. The ins and outs of DNA transfer in bacteria. *Science* **310:** 1456–1460.

Chen I, Provvedi R, Dubnau D. 2006. A macromolecular complex formed by a pilin-like protein in competent *Bacillus subtilis*. *J Biol Chem* **281:** 21720–21727.

Chiang P, Sampaleanu LM, Ayers M, Pahuta M, Howell PL, Burrows LL. 2008. Functional role of conserved residues in the characteristic secretion NTPase motifs of the *Pseudomonas aeruginosa* type IV pilus motor proteins PilB, PilT and PilU. *Microbiology* **154:** 114–126.

Christie PJ. 1997. *Agrobacterium tumefaciens* T-complex transport apparatus: A paradigm for a new family of multifunctional transporters in eubacteria. *J Bacteriol* **179:** 3085–3094.

Christie PJ. 2009. Structural biology: Translocation chamber's secrets. *Nature* **462:** 992–994.

Christie PJ, Atmakuri K, Krishnamoorthy V, Jakubowski S, Cascales E. 2005. Biogenesis, architecture, and function of bacterial type IV secretion systems. *Annu Rev Microbiol* **59:** 451–485.

Chung YS, Breidt F, Dubnau D. 1998. Cell surface localization and processing of the ComG proteins, required for DNA binding during transformation of *Bacillus subtilis*. *Mol Microbiol* **29:** 905–913.

Clausen M, Koomey M, Maier B. 2009b. Dynamics of type IV pili is controlled by switching between multiple states. *Biophys J* **96:** 1169–1177.

Clausen M, Jakovljevic V, Sogaard-Andersen L, Maier B. 2009a. High-force generation is a conserved property of type IV pilus systems. *J Bacteriol* **191:** 4633–4638.

Claverys JP, Lacks SA. 1986. Heteroduplex deoxyribonucleic acid base mismatch repair in bacteria. *Microbiol Rev* **50:** 133–165.

Claverys JP, Martin B, Polard P. 2009. The genetic transformation machinery: Composition, localization, and mechanism. *FEMS Microbiol Rev* **33:** 643–656.

Claverys JP, Prudhomme M, Martin B. 2006. Induction of competence regulons as a general response to stress in gram-positive bacteria. *Annu Rev Microbiol* **60:** 451–475.

Collins RF, Frye SA, Kitmitto A, Ford RC, Tonjum T, Derrick JP. 2004. Structure of the *Neisseria meningitidis* outer membrane PilQ secretin complex at 12 A resolution. *J Biol Chem* **279:** 39750–39756.

Craig L, Pique ME, Tainer JA. 2004. Type IV pilus structure and bacterial pathogenicity. *Nat Rev Microbiol* **2:** 363–378.

Danner DB, Deich RA, Sisco KL, Smith HO. 1980. An eleven-base-pair sequence determines the specificity of DNA uptake in *Haemophilus transformation*. *Gene* **11:** 311–318.

Dedrick RM, Wildschutte H, McCormick JR. 2009. Genetic interactions of smc, ftsK, and parB genes in *Streptomyces coelicolor* and their developmental genome segregation phenotypes. *J Bacteriol* **191:** 320–332.

Di Lallo G, Fagioli M, Barionovi D, Ghelardini P, Paolozzi L. 2003. Use of a two-hybrid assay to study the assembly of a complex multicomponent protein machinery: Bacterial septosome differentiation. *Microbiology* **149:** 3353–3359.

Draper GC, McLennan N, Begg K, Masters M, Donachie WD. 1998. Only the N-terminal domain of FtsK functions in cell division. *J Bacteriol* **180:** 4621–4627.

Draskovic I, Dubnau D. 2005. Biogenesis of a putative channel protein, ComEC, required for DNA uptake: Membrane topology, oligomerization and formation of disulphide bonds. *Mol Microbiol* **55:** 881–896.

Dubnau D. 1999. DNA uptake in bacteria. *Annu Rev Microbiol* **53:** 217–244.

Elkins C, Thomas CE, Seifert HS, Sparling PF. 1991. Species-specific uptake of DNA by gonococci is mediated by a 10-base- pair sequence. *J Bacteriol* **173:** 3911–3913.

Forest KT, Tainer JA. 1997. Type-4 pilus-structure: Outside to inside and top to bottom–a minireview. *Gene* **192:** 165–169.

Foulger D, Errington J. 1989. The role of the sporulation gene spoIIIE in the regulation of prespore-specific gene expression in *Bacillus subtilis*. *Mol Microbiol* **3:** 1247–1255.

Friedrich A, Prust C, Hartsch T, Henne A, Averhoff B. 2002. Molecular analyses of the natural transformation machinery and identification of pilus structures in the extremely thermophilic bacterium *Thermus thermophilus* strain HB27. *Appl Environ Microbiol* **68:** 745–755.

Fronzes R, Christie PJ, Waksman G. 2009a. The structural biology of type IV secretion systems. *Nat Rev Microbiol* **7:** 703–714.

Fronzes R, Schafer E, Wang L, Saibil HR, Orlova EV, Waksman G. 2009b. Structure of a type IV secretion system core complex. *Science* **323:** 266–268.

Gomis-Ruth FX, Coll M. 2006. Cut and move: Protein machinery for DNA processing in bacterial conjugation. *Curr Opin Struct Biol* **16:** 744–752.

Gomis-Ruth FX, Moncalian G, Perez-Luque R, Gonzalez A, Cabezon E, de la Cruz F, Coll M. 2001. The bacterial conjugation protein TrwB resembles ring helicases and F1-ATPase. *Nature* **409:** 637–641.

Hahn J, Kramer N, Briley KJr, Dubnau D. 2009. McsA and B mediate the delocalization of competence proteins from the cell poles of *Bacillus subtilis*. *Mol Microbiol* **72:** 202–215.

Hahn J, Maier B, Haijema BJ, Sheetz M, Dubnau D. 2005. Transformation proteins and DNA uptake localize to the cell poles in *Bacillus subtilis*. *Cell* **122:** 59–71.

Hamilton HL, Dillard JP. 2006. Natural transformation of *Neisseria gonorrhoeae*: from DNA donation to homologous recombination. *Mol Microbiol* **59:** 376–385.

Hamoen LW, Venema G, Kuipers OP. 2003. Controlling competence in *Bacillus subtilis*: Shared use of regulators. *Microbiology* **149:** 9–17.

Herdendorf TJ, McCaslin DR, Forest KT. 2002. *Aquifex aeolicus* PilT, homologue of a surface motility protein, is a thermostable oligomeric NTPase. *J Bacteriol* **184:** 6465–6471.

Hobbs M, Mattick JS. 1993. Common components in the assembly of type-4 fimbriae DNA transfer systems, filamentous phage and protein secretion apparatus; a general system for the formation of surface-associated protein complexes. *Mol Microbiol* **10:** 233–243.

Hofreuter D, Odenbreit S, Puls J, Schwan D, Haas R. 2000. Genetic competence in *Helicobacter pylori*: Mechanisms and biological implications. *Res Microbiol* **151:** 487–491.

Inamine GS, Dubnau D. 1995. ComEA, a *Bacillus subtilis* integral membrane protein required for genetic transformation, is needed for both DNA binding and transport. *J Bacteriol* **177:** 3045–3051.

Iyer LM, Makarova KS, Koonin EV, Aravind L. 2004. Comparative genomics of the FtsK-HerA superfamily of pumping ATPases: Implications for the origins of chromosome segregation, cell division and viral capsid packaging. *Nucleic Acids Res* **32:** 5260–5279.

Jakovljevic V, Leonardy S, Hoppert M, Sogaard-Andersen L. 2008. PilB and PilT are ATPases acting antagonistically in type IV pilus function in *Myxococcus xanthus. J Bacteriol* **190:** 2411–2421.

Johnson TL, Abendroth J, Hol WG, Sandkvist M. 2006. Type II secretion: From structure to function. *FEMS Microbiol Lett* **255:** 175–186.

Judd PK, Kumar RB, Das A. 2005. Spatial location and requirements for the assembly of the *Agrobacterium tumefaciens* type IV secretion apparatus. *Proc Natl Acad Sci* **102:** 11498–11503.

Kaimer C, Gonzalez-Pastor JE, Graumann PL. 2009. SpoIIIE and a novel type of DNA translocase, SftA, couple chromosome segregation with cell division in *Bacillus subtilis. Mol Microbiol* **74:** 810–825.

Karnholz A, Hoefler C, Odenbreit S, Fischer W, Hofreuter D, Haas R. 2006. Functional and topological characterization of novel components of the *comB* DNA transformation competence system in *Helicobacter pylori. J Bacteriol* **188:** 882–893.

Kidane D, Graumann PL. 2005. Intracellular protein and DNA dynamics in competent *Bacillus subtilis* cells. *Cell* **122:** 73–84.

Kirstein J, Dougan DA, Gerth U, Hecker M, Turgay K. 2007. The tyrosine kinase McsB is a regulated adaptor protein for ClpCP. *EMBO J* **26:** 2061–2070.

Kohler R, Schafer K, Muller S, Vignon G, Diederichs K, Philippsen A, Ringler P, Pugsley AP, Engel A, Welte W. 2004. Structure and assembly of the pseudopilin PulG. *Mol Microbiol* **54:** 647–664.

Kramer N, Hahn J, Dubnau D. 2007. Multiple interactions among the competence proteins of Bacillus subtilis. *Mol Microbiol* **65:** 454–464.

Lacks S, Greenberg B, Neuberger M. 1974. Role of a deoxyribonuclease in the genetic transformation of *Diplococcus pneumoniae. Proc Natl Acad Sci* **71:** 2305–2309.

Le Bourgeois P, Bugarel M, Campo N, Daveran-Mingot ML, Labonte J, Lanfranchi D, Lautier T, Pages C, Ritzenthaler P. 2007. The unconventional Xer recombination machinery of Streptococci/Lactococci. *PLoS Genet* **3:** e117.

Leisner M, Stingl K, Frey E, Maier B. 2008. Stochastic switching to competence. *Curr Opin Microbiol* **11:** 553–559.

Levy O, Ptacin JL, Pease PJ, Gore J, Eisen MB, Bustamante C, Cozzarelli NR. 2005. Identification of oligonucleotide sequences that direct the movement of the *Escherichia coli* FtsK translocase. *Proc Natl Acad Sci* **102:** 17618–17623.

Londono-Vallejo JA, Dubnau D. 1994. Mutation of the putative nucleotide binding site of the *Bacillus subtilis* membrane protein ComFA abolishes the uptake of DNA during transformation. *J Bacteriol* **176:** 4642–4645.

Lowe J, Ellonen A, Allen MD, Atkinson C, Sherratt DJ, Grainge I. 2008. Molecular mechanism of sequence-directed DNA loading and translocation by FtsK. *Mol Cell* **31:** 498–509.

Macfadyen LP. 2000. Regulation of competence development in *Haemophilus influenzae. J Theor Biol* **207:** 349–359.

Maier B, Chen I, Dubnau D, Sheetz MP. 2004. DNA transport into *Bacillus subtilis* requires proton motive force to generate large molecular forces. *Nat Struct Mol Biol* **11:** 643–649.

Maier B, Potter L, So M, Long CD, Seifert HS, Sheetz MP. 2002. Single pilus motor forces exceed 100 pN. *Proc Natl Acad Sci* **99:** 16012–16017.

Majewski J, Cohan FM. 1998. The effect of mismatch repair and heteroduplex formation on sexual isolation in *Bacillus. Genetics* **148:** 13–18.

Mandlik A, Swierczynski A, Das A, Ton-That H. 2008. Pili in Gram-positive bacteria: Assembly, involvement in colonization and biofilm development. *Trends Microbiol* **16:** 33–40.

Marquis KA, Burton BM, Nollmann M, Ptacin JL, Bustamante C, Ben-Yehuda S, Rudner DZ. 2008. SpoIIIE strips proteins off the DNA during chromosome translocation. *Genes Dev* **22:** 1786–1795.

Massey TH, Mercogliano CP, Yates J, Sherratt DJ, Lowe J. 2006. Double-stranded DNA translocation: Structure and mechanism of hexameric FtsK. *Mol Cell* **23:** 457–469.

Mattick JS. 2002. Type iv pili and twitching motility. *Annu Rev Microbiol* **56:** 289–314.

Méjean V, Claverys J. 1988. Polarity of DNA entry in transformation of *Streptococcus pneumoniae. Mol Gen Genet* **213:** 444–448.

Mejean V, Claverys JP. 1993. DNA processing during entry in transformation of *Streptococcus pneumoniae. J Biol Chem* **268:** 5594–5599.

Merz AJ, So M, Sheetz MP. 2000. Pilus retraction powers bacterial twitching motility. *Nature* **407:** 98–102.

Mortier-Barriere I, Velten M, Dupaigne P, Mirouze N, Pietrement O, McGovern S, Fichant G, Martin B, Noirot

P, Le Cam, et al. 2007. A key presynaptic role in transformation for a widespread bacterial protein: DprA conveys incoming ssDNA to RecA. *Cell* **130:** 824–836.

Panicker MM, Minkley EGJr. 1985. DNA transfer occurs during a cell surface contact stage of F sex factor-mediated bacterial conjugation. *J Bacteriol* **162:** 584–590.

Pease PJ, Levy O, Cost GJ, Gore J, Ptacin JL, Sherratt D, Bustamante C, Cozzarelli NR. 2005. Sequence-directed DNA translocation by purified FtsK. *Science* **307:** 586–590.

Proft T, Baker EN. 2009. Pili in Gram-negative and Gram-positive bacteria—structure, assembly and their role in disease. *Cell Mol Life Sci* **66:** 613–635.

Provvedi R, Dubnau D. 1999. ComEA is a DNA receptor for transformation of competent Bacillus subtilis. *Mol Microbiol* **31:** 271–280.

Provvedi R, Chen I, Dubnau D. 2001. NucA is required for DNA cleavage during transformation of *Bacillus subtilis. Mol Microbiol* **40:** 634–644.

Ptacin JL, Nollmann M, Becker EC, Cozzarelli NR, Pogliano K, Bustamante C. 2008. Sequence-directed DNA export guides chromosome translocation during sporulation in *Bacillus subtilis. Nat Struct Mol Biol* **15:** 485–493.

Ptacin JL, Nollmann M, Bustamante C, Cozzarelli NR. 2006. Identification of the FtsK sequence-recognition domain. *Nat Struct Mol Biol* **13:** 1023–1025.

Saleh OA, Perals C, Barre FX, Allemand JF. 2004. Fast DNA-sequence independent translocation by FtsK in a single-molecule experiment. *Embo J* **23:** 2430–2439.

Sauvonnet N, Vignon G, Pugsley AP, Gounon P. 2000. Pilus formation and protein secretion by the same machinery in *Escherichia coli. Embo J* **19:** 2221–2228.

Sharp MD, Pogliano K. 1999. An in vivo membrane fusion assay implicates SpoIIIE in the final stages of engulfment during *Bacillus subtilis* sporulation. *Proc Natl Acad Sci* **96:** 14553–14558.

Sharp MD, Pogliano K. 2002a. MinCD-dependent regulation of the polarity of SpoIIIE assembly and DNA transfer. *Embo J* **21:** 6267–6274.

Sharp MD, Pogliano K. 2002b. Role of cell-specific SpoIIIE assembly in polarity of DNA transfer. *Science* **295:** 137–139.

Sharp MD, Pogliano K. 2003. The membrane domain of SpoIIIE is required for membrane fusion during *Bacillus subtilis* sporulation. *J Bacteriol* **185:** 2005–2008.

Shu AC, Wu CC, Chen YY, Peng HL, Chang HY, Yew TR. 2008. Evidence of DNA transfer through F-pilus channels during Escherichia coli conjugation. *Langmuir* **24:** 6796–6802.

Singleton MR, Dillingham MS, Wigley DB. 2007. Structure and mechanism of helicases and nucleic acid translocases. *Annu Rev Biochem* **76:** 23–50.

Sivanathan V, Allen MD, de Bekker C, Baker R, Arciszewska LK, Freund SM, Bycroft M, Lowe J, Sherratt DJ. 2006. The FtsK γ domain directs oriented DNA translocation by interacting with KOPS. *Nat Struct Mol Biol* **13:** 965–972.

Skerker JM, Berg HC. 2001. Direct observation of extension and retraction of type IV pili. *Proc Natl Acad Sci* **98:** 6901–6904.

Smeets LC, Bijlsma JJ, Kuipers EJ, Vandenbroucke-Grauls CM, Kusters JG. 2000. The dprA gene is required for natural transformation of *Helicobacter pylori. FEMS Immunol Med Microbiol* **27:** 99–102.

Stingl K, Muller S, Scheidgen-Kleyboldt G, Clausen M, Maier B. 2010. Composite system mediates two-step DNA uptake into Helicobacter pylori. *Proc Natl Acad Sci* **107:** 1184–1189.

Tadesse S, Graumann PL. 2007. DprA/Smf protein localizes at the DNA uptake machinery in competent *Bacillus subtilis* cells. *BMC Microbiol* **7:** 105.

Tato I, Zunzunegui S, de la Cruz F, Cabezon E. 2005. TrwB, the coupling protein involved in DNA transport during bacterial conjugation, is a DNA-dependent ATPase. *Proc Natl Acad Sci* **102:** 8156–8161.

Teng WL, Bannam TL, Parsons JA, Rood JI. 2008. Functional characterization and localization of the TcpH conjugation protein from *Clostridium perfringens. J Bacteriol* **190:** 5075–5086.

Terradot L, Bayliss R, Oomen C, Leonard GA, Baron C, Waksman G. 2005. Structures of two core subunits of the bacterial type IV secretion system, VirB8 from *Brucella suis* and ComB10 from *Helicobacter pylori. Proc Natl Acad Sci* **102:** 4596–4601.

Thanbichler M. 2009. Synchronization of chromosome dynamics and cell division in bacteria. *Cold Spring Harb Perspect Biol* **1:** 000331.

Tomb J-F, El-Hajj H, Smith HO. 1991. Nucleotide sequence of a cluster of genes involved in the transformation of *Haemophilus influenzae* RD. *Gene* **104:** 1–10.

Vagner V, Claverys J-P, Ehrlich SD, Méjean V. 1990. Direction of DNA entry in competent cells of *Bacillus subtilis. Mol Microbiol* **4:** 1785–1788.

Val ME, Kennedy SP, El Karoui M, Bonne L, Chevalier F, Barre FX. 2008. FtsK-dependent dimer resolution on multiple chromosomes in the pathogen *Vibrio cholerae. PLoS Genet* **4:** e1000201.

van Nieuwenhoven MH, Hellingwerf KJ, Venema G, Konings WN. 1982. Role of proton motive force in genetic transformation of *Bacillus subtilis. J Bacteriol* **151:** 771–776.

Wang L, Lutkenhaus J. 1998. FtsK is an essential cell division protein that is localized to the septum and induced as part of the SOS response. *Mol Microbiol* **29:** 731–740.

Wang SC, West L, Shapiro L. 2006. The bifunctional FtsK protein mediates chromosome partitioning and cell division in *Caulobacter. J Bacteriol* **188:** 1497–1508.

Wang L, Yu Y, He X, Zhou X, Deng Z, Chater KF, Tao M. 2007. Role of an FtsK-like protein in genetic stability in *Streptomyces coelicolor* A3(2). *J Bacteriol* **189:** 2310–2318.

Wu LJ, Errington J. 1994. Bacillus subtilis spoIIIE protein required for DNA segregation during asymmetric cell division. *Science* **264:** 572–575.

Wu LJ, Errington J. 1997. Septal localization of the SpoIIIE chromosome partitioning protein in *Bacillus subtilis. Embo J* **16:** 2161–2169.

Wu LJ, Errington J. 1998. Use of asymmetric cell division and spoIIIE mutants to probe chromosome orientation and organization in *Bacillus subtilis*. *Mol Microbiol* **27:** 777–786.

Yeh YC, Lin TL, Chang KC, Wang JT. 2003. Characterization of a ComE3 homologue essential for DNA transformation in *Helicobacter pylori*. *Infect Immun* **71:** 5427–5431.

Yeo HJ, Savvides SN, Herr AB, Lanka E, Waksman G. 2000. Crystal structure of the hexameric traffic ATPase of the *Helicobacter pylori* type IV secretion system. *Mol Cell* **6:** 1461–1472.

Yeo HJ, Yuan Q, Beck MR, Baron C, Waksman G. 2003. Structural and functional characterization of the VirB5 protein from the type IV secretion system encoded by the conjugative plasmid pKM101. *Proc Natl Acad Sci* **100:** 15947–15952.

Yu XC, Tran AH, Sun Q, Margolin W. 1998. Localization of cell division protein FtsK to the *Escherichia coli* septum and identification of a potential N-terminal targeting domain. *J Bacteriol* **180:** 1296–1304.

Myxobacteria, Polarity, and Multicellular Morphogenesis

Dale Kaiser[1], Mark Robinson[2], and Lee Kroos[2]

[1]Departments of Biochemistry and of Developmental Biology, Stanford University School of Medicine, Stanford, California 94305

[2]Department of Biochemistry and Molecular Biology, Michigan State University, East Lansing, Michigan 48824

Correspondence: adkaiser@stanford.edu

Myxobacteria are renowned for the ability to sporulate within fruiting bodies whose shapes are species-specific. The capacity to build those multicellular structures arises from the ability of *M. xanthus* to organize high cell-density swarms, in which the cells tend to be aligned with each other while constantly in motion. The intrinsic polarity of rod-shaped cells lays the foundation, and each cell uses two polar engines for gliding on surfaces. It sprouts retractile type IV pili from the leading cell pole and secretes capsular polysaccharide through nozzles from the trailing pole. Regularly periodic reversal of the gliding direction was found to be required for swarming. Those reversals are generated by a G-protein switch which is driven by a sharply tuned oscillator. Starvation induces fruiting body development, and systematic reductions in the reversal frequency are necessary for the cells to aggregate rather than continue to swarm. Developmental gene expression is regulated by a network that is connected to the suppression of reversals.

THE ORIGIN OF MYXOBACTERIA

Roughly 50 species of myxobacteria are recognized in the soil and on the bark of trees, around the earth (Reichenbach 1984). As chemo-organotrophs that feed on particulate organic matter, many myxobacteria are proteolytic—a habit associated with predation. *Myxococcus xanthus* hydrolyzes proteins and nucleic acids, to obtain essential amino acids and phosphate. *M. xanthus* growth is not enhanced by mono- or disaccharides (Dworkin 1962). Initially surprising, Dworkin's observation reflects their use of the Embden-Myerhof pathway for gluconeogenesis starting from amino acids during growth, rather than using the pathway for glycolysis. Amino acids are also converted to acetyl-CoA, which enters the TCA cycle to provide energy. This metabolic sketch of *M. xanthus* finds support in a recent genomic analysis that also shows their capacity to scavenge purines and pyrimidines (Curtis and Shimkets 2008).

Sorangium cellulosum exemplifies a myxobacterial cellulolytic feeding habit. Digestion of cellulose fibers (filter paper) yields glucose, which enters glycolysis and the TCA cycle to supply its carbon and energy (Schneiker et al. 2007). All myxobacteria are copious secretors of polysaccharide slime (the reason for the

 doi: 10.1101/cshperspect.a000380
Cite this article as *Cold Spring Harb Perspect Biol* 2010;2:a000380

prefix "myxo"). Glucose from cellulose would, via the pentose phosphate pathway, give rise to many other hexoses and pentoses that are used to synthesize polysaccharides, including their capsule of slime and their lipopolysaccharide. Nitrate, ammonia, or amino acids can serve as nitrogen sources for *S. cellulosum*.

Representing a third habit, *Stigmatella aurantiaca* can efficiently digest chitin (Reichenbach 1993). Unlike the proteolytic *M. xanthus*, growth of *S. aurantiaca* is enhanced by the addition of glucose (Reichenbach and Dworkin 1969), indicating that glycolysis feeds pyruvate to its TCA cycle. Other particulate polysaccharides, such as agar, starch, pectin, and keratin can serve as carbon and energy sources for other myxobacteria, whose metabolic sketches have yet to be made. Overall, the ability to glycolyze sugars is correlated with a habit of feeding on polysaccharides, whereas gluconeogenesis is perfectly correlated with feeding on proteins.

Roland Thaxter, a cryptogamic botanist, was the first to describe myxobacterial cells. He reported (Thaxter 1892) that *Chondromyces crocatus* was a bacterium, which was shocking at the time because *C. crocatus* had for more than 20 years been considered an imperfect fungus: Its slime and morphologically complex fruiting bodies recall the slime molds (Fig. 1). Thaxter had discovered the unicellular vegetative stage of *C. crocatus*. The cells he found divided by binary fission; clearly the growing cells were not a fungal mycelium. Thaxter also described their multicellular behavior: locomotion, swarming, aggregation, and the process of fruiting body formation of *C. crocatus* and its relatives. Thaxter clearly saw the process of fructification as a "course of development" because it was "a definitely recurring aggregation of individuals capable of concerted action toward a definite end." The remarkable morphogenetic capacity of *M. xanthus* is outlined in Figure 2.

Hans Reichenbach recorded myxobacterial swarming and fruiting body development in a most informative series of time-lapse movies, documenting the complex yet orderly behavior of cells in several different species of myxobacteria (Reichenbach 1966; Reichenbach 1968; Reichenbach 1974). The movies show that swarming, which enables their rapid spreading over surfaces such as agar, is systematically modified by starvation to enable their sporulation within species-specific fruiting bodies.

The evolutionary origin of the myxobacteria, which are monophyletic, is found within the delta subgroup of proteobacteria, according to the sequence of their 16S ribosomal RNA (Sproer et al. 1999). Among the proteobacteria, only the myxobacteria have the capacity to develop multicellular fruiting bodies whose form is species specific. *M. xanthus* shares the ability to lyse and digest other bacteria with another δ–proteobacterium, *Bdellovibrio bacteriovorus* (Evans et al. 2008). Indeed, predatory feeding may have evolved within the δ-proteobacteria. Myxobacteria are abundant in cultivated top-soils around the earth because a wide variety of Gram- and Gram+ bacteria

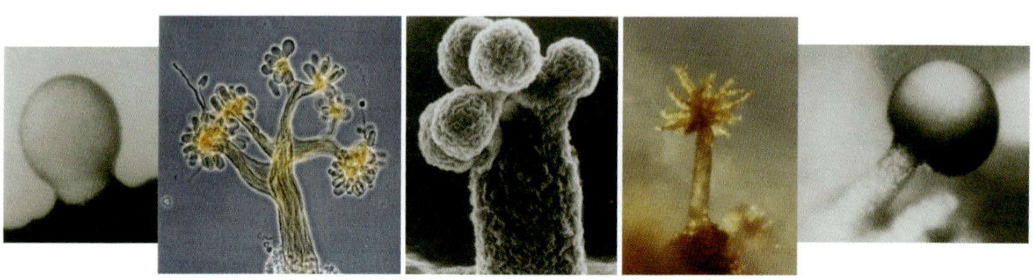

Figure 1. Fruiting bodies of five different species of myxobacteria are shown, illustrating differences in morphology. From left to right: *Myxococcus xanthus*, *Chondromyces crocatus*, *Stigmatella aurantiaca*, *Chondromyces apiculatus*, *Myxococcus stipitatus*. (Reprinted, with permission, from Kaiser 2008b [Annual Reviews].)

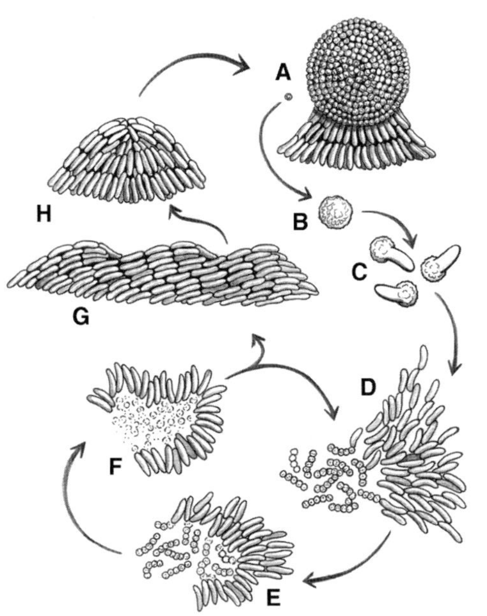

Figure 2. The lifecycle of *Myxococcus xanthus*. A group of swarming and interacting cells can have either of two fates depending on the swarm's environment. A *Myxococcus* fruiting body (*A*) is a spherical structure of approximately 1×10^5 cells that contains stress-resistant spores. The fruiting body is small (1/10 mm), sticky, and its spores are tightly packed. When a fruiting body receives nutrients, the individual spores germinate and thousands of *M. xanthus* cells emerge together as an "instant" swarm (*B–D*). When prey or other nutrient is available, the swarm becomes a predatory collective that moves and feeds cooperatively, pooling extracellular enzymes to lyse and consume prey bacteria (*E–F*). Multicellular fruiting body development is advantageous given the collective hunting behavior of swarms. Nutrient-poor conditions elicit a unified starvation stress-response that leads to the formation of fruiting bodies. Other multicellular behaviors include wave formation (*G*), streaming into aggregates (*H*), and mound building (*I*). (Adapted from Goldman et al. 2006.)

are found in association with plants. For predation, *M. xanthus*, is seen to surround a prey colony, as diagrammed in steps D–F of Figure 2, to make direct contact with the prey cell (Berleman et al. 2006), and to lyse the prey envelope. Then, it is thought to extract the cytoplasmic proteins of the prey, digest them to peptides and amino acids before consuming them for carbon, nitrogen and energy.

Apart from the Myxobacteria, the δ-proteobacterial genomes range from 3.66 to 5.01 Mb. Because the *M. xanthus* genome is 9.14 Mb, there has been a genome enlargement of 4–5 Mb from its inferred δ proteobacterial ancestor. Genome expansion specific to the myxobacterial lineage is strongly suggested by the almost identical genome sizes of *S. aurantica* (9.5 Mb), *Stigmatella erecta* (9.8 Mb), and *M. xanthus* (9.14 Mb). More than 1500 gene duplications, representing 15% of the genome, have been identified in *M. xanthus* that are lineage-specific (Goldman et al. 2006). Genes were not duplicated at random. Instead, genes for cell–cell signaling, small molecule sensing, and integrative control of transcription were amplified selectively (Goldman et al. 2006). Evidently additional expansion within the *Myxobacterial* lineage led to *Sorangium cellulosum*, whose 13 Mbp genome (Schneiker 2007) encodes the largest number of serine/threonine protein kinases of any organism (Pérez et al. 2008).

Electron microscopy reveals that *M. xanthus* has a typical Gram-negative cell structure, which averages 7 microns in length and 0.5 microns in width, and that the peptidoglycan (murein) forms a rod-shaped sacculus. Nevertheless, their muropeptide profile is unusual, and the glycan strands are 1/3 to 1/2 the length of those in *Escherichia coli* (Bui et al. 2009). These compositional differences are correlated with the great flexibility of *M. xanthus* cells (Wolgemuth 2005). Flexibility is important for their gliding motility on surfaces that may be curved (Kaiser 2007). In addition the cells are polarized: each has a head and a tail. The ecological success of the myxobacterial lifestyle is witnessed by the millions of myxobacterial cells found per gram of cultivated soil (Reichenbach 1984).

LOCOMOTION

S Motility

Several pili extend forward from the leading end of each *M. xanthus* cell. Among the type IV pilus proteins common to *M. xanthus*, *P. aeruginosa*, *N. gonorrhoeae*, *N. meningitigis*, and *Synecchocystis*

PCC6803, the most highly conserved include PilA, PilB, PilC, PilD, PilM, PilN, PilO, PilP, PilQ, and PilT (Table 1). The conserved *pil* genes are clustered together with their regulators in *M. xanthus*, facilitating genetic studies. Protein structure, function and intracellular location tend to be conserved, which has made it possible to unravel the molecular function of many of these proteins; see reviews (Mattick 2002; Nudleman and Kaiser 2004; Pelicic 2008).

Type IV pilins, which are encoded by the *pilA* gene, conserve only their amino terminal region of about 60 amino acids. Synthesized as a prepilin, PilA is processed by the PilD peptidase. Conservation of the PilA amino terminal sequence is thought to reflect the sequence requirements of PilD around the peptidase cleavage site (Strom et al. 1993), and the need of the amino terminal region to form an α-helix that is capable of regularly staggered coiled-coiling with copies of itself. A structure was inferred from a 2.6 Å resolution X-ray crystal structure of *N. gonorrhoeae* pilin dimers, a mapping of surface-exposed peptides, and cryo-electron microscopy (Craig et al. 2006).

Because individual pili are 3–10 μm long, whereas the rise per monomer is of the order of 1 nm, a single pilus fiber includes many thousands of copies of the processed pilin (Hansen and Forest 2006). The β-strands of one pilin monomer in the helical filament interact with the β-strands of the next monomer, lending tensile strength to the fiber. The flexibility of the α-helices in the fiber core that are held together by hydrophobic forces allows the fiber to bend, twist, and to bundle with other pilus fibers. Despite a diameter of only 6 nm, the fiber can withstand tension stresses in excess of 100 pN (Maier et al. 2002). Absence of a channel in the center of the pilus fiber implies that it is polymerized from its base in the inner membrane, where PilB and PilT are located to

Table 1. S motility genes

Protein[a]	Function	Cellular localization	Clustered[b]	Conserved in multiple species[c]
PilA	Pilin, monomer unit of the pilus filament	Assembled into the pilus fiber Stored in the Inner Membrane	+	+
PilB	Pilus extension	Inner membrane	+	+
PilT	Pilus retraction	Inner membrane	+	+
PilC	Unknown	Inner membrane	+	+
PilD	PilA leader peptidase	Inner membrane	+	+
PilG	ABC Transporter	Periplasm	+	
PilH	ABC Transporter	Periplasm	+	
PilI	ABC Transporter	Periplasm	+	
PilM	ATPase	Inner membrane	+	+
PilN	Unknown	Periplasm	+	+
PilO	Unknown	Periplasm	+	+
PilP	Unknown	Anchored in outer membrane	+	+
PilQ	Secretin	Outer membrane	+	+
PilR	Regulates transcription of pilA		+	–
PilS	Two-component sensor for pilR		+	–
PilR1	Transcriptional regulator		+	–
PilS1	Two-component sensor for pilR1		+	–
Tgl	Secretin assembly factor	Outer membrane	–	–

[a]Pil genes named as in *P. aeruginosa*.

[b]A set of 16 contiguous genes in *M. xanthus*.

[c]*M. xanthus*, *P. aeruginosa*, *N. gonorrhoeae*, *N. meningitigis* and *Synechocystis* PCC 6803 (Nudleman and Kaiser 2004).

catalyze extension and retraction, respectively (Table 1).

M. xanthus cells, separated by more than a pilus length (several micrometer) are rarely, if ever observed to move (Kaiser and Crosby 1983). It is as if pili retract with great force only if their tips have firmly attached to fibrils on a group of cells ahead of them (Kaiser 1979; Merz et al. 2000; Semmler et al. 1999). Li et al. argue that binding is less strict (Li et al. 2003). Fibrils, consisting of almost equal amounts of protein and polysaccharide that contains galactose, glucosamine, glucose, rhamnose, and xylose (Behmlander and Dworkin 1994; Dworkin 1999), make up a linked meshwork of amorphous strands, often 30 nm in diameter that bundle neighboring cells close together. Bundling is visible microscopically (Kearns and Shimkets 2001). Fibril-deficient mutant cells can accept fibrils from an extracellular polysaccharide fraction of normal cells (Lu et al. 2005).

Certain *M. xanthus* mutants, which lack S motility (called S$^-$ mutants), retain pilus fibers. Examples include the *pilT* mutants, the *dif* or *dsp* (dispersed growth) mutants, and certain lipopolysaccharide-defective mutants. Although *pilT* mutants grow in clumps, because their pili attach to fibrils, the *dsp* and the *dif* mutants grow dispersed in liquid culture. These genes are likely to encode regulators of fibril production because the mutants are fibril-deficient (Lu et al. 2005). It appears that both fibrils (and perhaps O-antigen) influence the number of pili that cells assemble. Pili are thought to work as follows: The relatively stiff pilus extends ahead of an *M. xanthus* cell, it adheres to fibrils on cells located near the pilus tip, retracts, and pulls the piliated cell forward toward the cells ahead. Once an attachment has been made the connection with fibrils on the target cell can withstand 100 pN of pulling force. The cell group to which the pilus has attached moves but slightly because the group, having more cells, adheres with greater strength to the substrate than the piliated cell which has contributed ATP for retraction. Finally, the group of cells bundled by fibrils enlarges by one cell.

Assembly of Type IV Pili

The type IV pilus engine includes more than 15 different proteins (Table 1). Some protein constituents are found in the cytoplasmic membrane, others reach across the periplasm, and may be anchored to the peptidoglycan. Still others pass through the outer membrane into the extracellular space, as diagrammed in Figure 3. Many thousands of *pilA* monomers are joined to form the helical pilus fiber. That fiber passes through the doughnut-shaped assembled PilQ secretin embedded in the outer membrane. The assembled complex of PilQ is a 12–14mer with a 60 Å diameter hole in the center, through which the pilus fiber can pass (Bitter et al. 1998; Collins et al. 2001; Collins et al. 2004; Frye et al. 2006). Absent a fiber, the hole in the assembled PilQ is closed at both ends. When a pilus fiber, growing up from its base in the inner membrane, pushes the channel gate open, the pilus is allowed to pass through the 60 Å hole. Then, PilQ acts like a sealed bushing that surrounds the pilus as it elongates or retracts.

Type IV pili are polarized; they are found only at one pole of *M. xanthus* rod-shaped cells (Kaiser 1979; Wu et al. 1997). When the PilQ

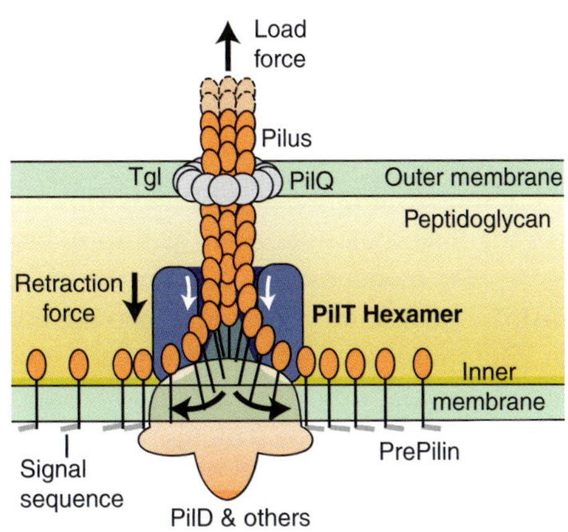

Figure 3. Type IV pilus engine. Tgl is an outer membrane lipoprotein. PilQ is the secretin protein. Once the pilus tip has attached to polysaccharide fibrils on cells ahead, PilT, a AAA ATPase causes the pilus fiber to retract. (Reprinted, with permission, from Kaiser 2008b [Annual Reviews].)

protein is extracted from whole cells by exposing them to a hot detergent solution, roughly half of the PilQ protein is extracted as 98 kDa monomers; the other half as high molecular weight, detergent-resistant multimers (Nudleman et al. 2006). Nudleman et al. found that multimer assembly requires Tgl, a 17 kDa outer membrane lipoprotein in *M. xanthus* (Table 1). Tgl is necessary for the production of pili in *M. xanthus*, and it is the only required pilus-associated protein whose gene resides outside the *pil* gene cluster (Rodriguez-Soto and Kaiser 1997a; Rodriguez-Soto and Kaiser 1997b). The primary translation product of the *tgl* gene has a signal peptidase II recognition sequence for outer membrane lipoproteins, and processed Tgl protein is found in the outer membrane (Simunovic et al. 2003). Beyond the signal peptidase recognition site, the Tgl amino acid sequence comprises six tetratrico peptide repeats (TPRs). TPR sequences are found in many proteins that assist the assembly of multiprotein complexes, and each TPR has the capacity to form an α-helix (D'Andrea and Regan 2003). Most likely the 6 TPR helices bring adjacent PilQ monomers into such close apposition that the resulting assembly resists dissociation in heated detergent. The integrity of the PilQ oligomers depends on the molecular integrity of Tgl. This role of Tgl finds support in the fact that Tgl can be exchanged by contact between cells, allowing one cell to stimulate the motility of another (Nudleman et al. 2005).

Immunolocalization experiments show that all of the PilQ molecules occur within a single large polar condensate (Nudleman et al. 2006). Condensation to a single focus implies that PilQ monomers associate with each other and perhaps with other Pil proteins in the outer membrane. However, the structures of the PilQ condensations at opposite ends of a cell differ from each other. One condensation is derived from the end that lacks pili: Those PilQ monomers are not assembled, but they are clustered. The other condensation includes several open secretin channels, each of which surrounds a single pilus fiber. Each pole has approximately half of the PilQ from whole cells (Nudleman et al. 2006). From the piliated cell ends, detergent-resistant PilQ multimers and Tgl are found. From the condensation at a nonpiliated end, PilQ would be extracted by detergent as unassembled 98 kDa monomers.

PilT, an AAA ATPase, is the motor for pilus retraction. Its structure has recently been solved (Satyshur et al. 2007), and three different monomer structures were evident in the crystalline hexamers of PilT. Those structures imply that PilT changes its conformation during retraction. *M. xanthus* PilT has been shown to develop more than 110 pN of tension during retraction (Maier et al. 2002). Retracted pilin monomers are stored in the inner membrane for reuse by PilB, another AAA ATPase found in the inner membrane that catalyzes fiber extension (Table 1).

Refering to Figure 3, what is it that keeps unassembled PilQ monomers together in a cluster at the nonpiliated pole in such a way that it can later assemble oligomers? The answer, we suggest, is a pilus-like organelle that lacks a pilus fiber. That organelle would include PilB, PilM, PilN, PilO, and PilT proteins that localize to the inner membrane and to the periplasm. Its existence is supported by the observation that PilB and PilT are bipolarly localized in *P. aeruginosa* whereas the PilU protein and the assembled pili are strictly unipolar (Chiang et al. 2005). Such an organelle would have to pass through the peptidoglycan meshwork, because PilQ is found in the outer membrane whereas PilB and PilT are in the inner membrane. The organelle would thus pierce, with an opportunity to join to, the rigid sacculus (Bui et al. 2009), which the PilB and PilT motors could use as a fulcrum. Also because the structure pierces the rigid peptidoglycan meshwork, its position would be fixed within the cell, and the associated PilQ monomers would not be expected to diffuse laterally in the membrane. Neither Tgl nor PilA would be part of the structures at the nonpiliated pole; both would be added at the time and place of secretin and pilus assembly. The precursor organelle, capped with a patch of unassembled PilQ, could reside at the nonpiliated pole, waiting for Tgl (and other proteins such as FrzS) to induce assembly of the PilQ monomers and for a pilus filament to grow through the PilQ channel.

A Motility

Two different mechanisms have been proposed for A motility. First, polysaccharide slime secretion through the hundreds of nozzles that are found clustered at the poles of each cell was observed. Many electron micrographs and light micrographs show slime extrusion from one cell pole in wild-type cells (Wolgemuth et al. 2002; Yu and Kaiser 2007). A time-lapse movie shows the moment-by-moment correlation between the deposition of a slime trail onto agar and the forward movement of the lagging cell pole, where the nozzles extruding slime are located (Kaiser 2008b). A trail of extruded polysaccharide gel sticks to an agar gel (binding to agarose in the gel), and secretion elongates the bound polysaccharide (Kaiser 2009). The higher concentration of saccharides found in the trail than in a 1% agar gel renders the trail visible in phase-contrast microscopy. Biochemical-genetic studies of the nozzles reveal their structural and functional similarity to the pores from which group 1 and 4 capsular slime is secreted from *E. coli* (Kaiser 2009; Whitfield and Roberts 1999; Yu and Kaiser 2007). Slime secretion has been specifically associated with A motility, and not with S motility (Fontes and Kaiser 1999).

A second mechanism was recently proposed—one associated with focal adhesions that contain the AglZ protein (Mignot et al. 2007). The focal adhesions are periodically distributed along the line of cell contact with the substrate, usually agar as mentioned. The focal adhesions are reported to be spaced at roughly 2-μm intervals; they first appear in the front half of the cell; then migrate toward the lagging pole as the cell moves forward (Mignot et al. 2007). Others reported that AglZ is a cytoplasmic protein with a regulatory function rather than a cell surface protein with motor function as had first been thought (Mauriello et al. 2009a; Mauriello et al. 2009b). Nevertheless, AglZ is an A motility protein (Yang et al. 2004), though nonessential (Mauriello et al. 2009b).

Instead of collaborating with the laterally arrayed engines just proposed, focal adhesions might facilitate propulsion by polar slime secretion. In particular, adhesions could relieve the elastic stress that is expected in a gel, consisting of capsular slime and LPS, interfacing a gliding cell with its agar substratum, by the mechanism proposed in Figure 3 of (Kaiser 2009). Because the interfacial gel adheres to the cell surface on one side and to the agar on the other, the gel would be deformed as the gliding cell moves relative to the agar. *Myxococcus* gliding on agar is relatively slow, about 4 μ/min (Kaiser and Crosby 1983), and the interfacial gel is viscous because of the high density of polysaccharide chains. Consequently the Reynolds number for movement is about 10^{-5}. As outlined by (Purcell 1977), the mechanical behavior of cells at very low Reynolds number is dominated by the viscous forces: Momentum can play no role. One cannot move forward by walking in thick tar, for example. Accordingly, the motion of a cell that is pushed forward by secretion from its lagging end would create viscoelastic stress in the interfacial gel layer located between the cell and the agar, as described (Kaiser 2009). Because one surface of the interfacial gel adheres to the cell whereas the other surface adheres to the agar, there is no slip on either side. The resulting viscoelastic stress would consist of stretching the formerly relaxed and compressed polysaccharide chains in the direction of cell movement, and such stress would oppose the cell's forward motion, as pointed out by Purcell (Purcell 1977). However, each focal adhesion, by rigidly connecting the cell to the agar, and remaining for a limited time, would permit the stress that had accumulated next to the adhesion to dissipate. As the stretched polysaccharide chains diffuse, and spontaneously re-compress, stress energy would be dissipated as heat. Once a cell has been pushed forward by slime secretion from its lagging end, a new, more forward adhesion can form within the gel interface that walls-off a new section of the gel for stress relief. Such a role would explain why there are several focal adhesions (to wall off the stressed gel, section by section), and why the adhesions are transient and rapidly lost as the cell moves forward (once the stress has dissipated, the adhesion will have served its purpose). In this view, focal adhesions would serve as escapements that permit the cell to move forward smoothly throughout a full reversal period. In conclusion, the two types of engines differ in their cellular location as described. The polar engines are permanent parts of the cell whereas the focal adhesions are transient parts. Finally the two mechanisms use different molecular engines: The polysaccharide gel expands as water binds to its sugar residues, and expansion provides force; protein parts of the focal adhesions that generate force are still being sought.

Figure 4. A swarm of *Myxococcus xanthus*, strain DK1622, on agar. Photo taken by Dr. Lotte Jelsbak after incubation at 32 °C. The center of the swarm has several layers of cells. At the edge of the swarm, a single layer of single cells and lateral clusters of cells are spreading outward. Slightly higher cell densities over their surroundings are indicated by the rings and spokes. (Reprinted, with permission, from Jelsbak and Kaiser 2005 [© ASM].)

SWARMING

Myxobacteria grow while they actively spread over a solid surface from which they are receiving nutrient. Called swarming, the process directs moving cells to the very edge of the swarm, where there are few cells and hence less competition for nutrients. In the swarm center, cells compete for nutrient from below and for oxygen from above; competition limits the growth rate (Kaiser 2007). The roughly 50 species of Myxobacteria form flat spreading swarms on agar. Swarm cells are in constant motion and are interacting with each other (Reichenbach 1984; Stanier 1942). Swarms of *M. xanthus* are large compared with colonies, relatively thin, and resemble an egg fried over-easy. A swarm, pictured in Figure 4, lacks the sharp edge, the heaped center, and the dome shape of a nonswarming *E. coli* colony on agar or a nonswarming *M. xanthus mgl* mutant (Kroos et al. 1988). Figure 5 shows the very edge of a swarm of *M. xanthus* at a higher magnification, revealing individual cells, slime trails, and various kinds and sizes of cell clusters, all of which are dynamic. Because of their low density, the cells at the swarm edge should have unfettered access to oxygen and nutrients, and to be able to eliminate their wastes readily. Consequently, they should be able to grow there at maximum rate without entering stationary phase, as long as fresh surface is available. Swarming depends on a strategy of random movement of long, flexible cells constrained to avoid collision with each other (Kaiser 2007). Net movement within a swarm is in the radial direction and that leads to swarm expansion. A technique used by *M. xanthus* to avoid collisions is similar to one used by pedestrians moving in opposite directions that spontaneously organize traffic lanes to slip past one another. Another behavior used by Myxococcus to promote swarming is periodically to reverse their gliding direction to recover from collisions that are unavoidable in high-density populations.

As described earlier, *M. xanthus* uses both polar type IV pili (S motility) and polar slime secretion (A motility) for their swarming movements, and the distantly related myxobacterium, *Sorangium cellulosum*, has genes for both these gliding engines (Schneiker 2007).

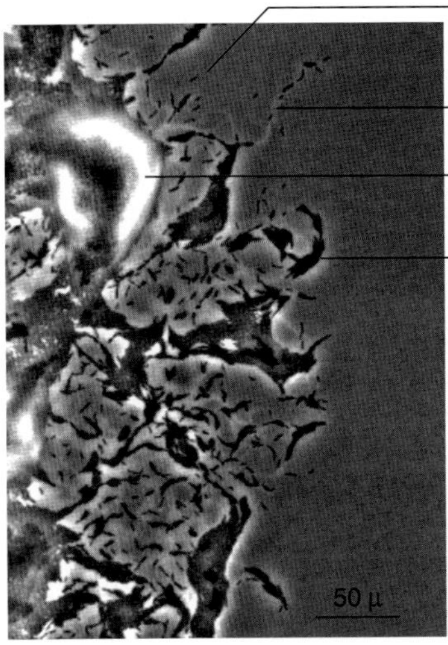

Figure 5. Distribution of cells and multicellular structures at the swarm edge. In addition to some individual cells and slime trails, multicellular rafts and multicellular mounds are labeled. The swarm is expanding in the radial direction, which is to the right in this image. (Scale bar, 50 μ). (Reprinted, with permission, from Wu et al. 2009 [National Academy of Sciences].)

Genetic studies of *M. xanthus* motility began with the isolation of mutants that produced abnormal swarms (Burchard 1970). Jonathan Hodgkin showed that the A and S engines are encoded by different gene sets (Hodgkin and Kaiser 1979a; Hodgkin and Kaiser 1979b). More than 50 double mutants, having one mutation in an A engine component and one mutation in an S engine component, were constructed, and they all failed to swarm (Hodgkin and Kaiser 1979b). By contrast, many single mutations in *mglA*, which had no effect on growth measured in suspension culture, also blocked swarming. Wild type and other single motility mutants do swarm, showing that either engine alone can support swarming. MglA mutants were common among Hodgkin's early isolates, and because they cannot swarm, their colonies have a sharp edge and a heaped center much like the colonies of nonswarming *E. coli* (Hodgkin and Kaiser 1979b).

Thirty years and several hundred new motility mutations later, MglA protein remains outside the set of protein components belonging to the A or the S engines. The A engine in an *mglA* mutant was shown to be capable of polar slime secretion (Yu and Kaiser 2007). The microscopic and genetic studies of Rosa Yu show that MglA is needed to switch both engines from one cell end to the other. Whereas all of her motile strains produced slime only at one end of each cell at any instant, *mglA* mutants secreted slime from both cell ends (Yu and Kaiser 2007). All A motile strains can produce slime from either end, and they switch the producing end when they reverse gliding direction; evidently *mglA* mutants fail to switch ends. Studies of single mutant S motile strains (A^-S^+) show that they are able to produce pili from either pole and that their pili do switch poles when cells reverse their gliding direction (Hartzell et al. 2008). Bipolar slime secretion in *mglA* mutants (which fail to switch) suggest that by trying to move in both directions simultaneously, the mutant cells are unable to make progress in either. This view of bipolar secretion explains Spormann's observation that mutant *mglA* cells are, in fact, oscillating very rapidly back and forth while moving less than 1/5 of a cell length in either direction, which would account for their sharp colony edge (Spormann and Kaiser 1999). Yu, Kaiser, and Wolgemuth suggested that these rapid oscillations are not because of signals from the reversal generator but arise from the normal statistical fluctuations in the number of active slime secretion nozzles at the two poles. Fluctuations are significant because relatively few slime filaments are observed (of order 10) at each slime-secreting end of wild-type cells (Wolgemuth et al. 2002). The speed of Δ*mglB* cells, which have partial A motility (Yu and Kaiser 2007), is observed to fluctuate from 1 to >5.5 μm/minute (Spormann and Kaiser 1999), and these fluctuations can be explained by variations in the number of active nozzles. According to this statistical hypothesis, the speed distribution of Δ*mglAB* double mutant cells should be the difference between the speeds of two independent Δ*mglB* ends. Agreement between the observed and the speeds predicted (calculation by C. Wolgemuth) by the statistical hypothesis, as reported (Kaiser 2008a), supports the hypothesis that the rapid reversals are a consequence of active slime secretion from both ends that are unable to switch.

Maintaining Engine Polarity

Both the A and the S engines are inherently polar in that they are able mechanically to produce force in one direction only: Pili only when they retract and slime only when it pushes. Comparative measurements of the rate of swarm expansion show that the A and the S engines on each cell have the same polarity (Kaiser and Crosby 1983). Under controlled conditions, a swarm, such as that shown in Figure 5, will spread outward on agar at a constant rate for many days (Burchard 1974; Kaiser and Crosby 1983). The increasing diameter of the swarm largely reflects active cell movement, because the colony of an A^-S^- mutant, which has defects in both engines, expands, because of cell growth and division, at less than 1/8 the rate of A^+S^+ swarms (Kaiser and Crosby 1983). Together, the two engines in A^+S^+ cells promote swarming at 1.6 μm/min, almost 3 times the rate of cells that have one or the other engine alone (0.6 or 0.4 μm/min) (Kaiser and Crosby 1983). Indeed, the maximum swarming rate of A^+S^+ is 60% greater than the sum of the two single engine rates (1.6 versus 0.6 + 0.4 μm/min). This synergism implies that the two engines are always arranged so as to move a cell in the same direction, and are positioned at opposite ends of the cell: The pulling pili at the leading pole of the cell and the pushing A engines at the trailing pole. This has been confirmed experimentally (Leonardy et al. 2008). The increase in swarm rate with initial cell density shows that individual cells cooperate with each other to swarm; an increase is seen both with A motility and with S motility (Wu et al. 2007; Wu et al. 2009).

The maintenance of polarity when cells divide reveals a template in each cell that is needed when the division septum creates two new cell ends. Observing the several hundred isolated cells at the edge of the A^+S^+ swarm in the published movie (Wu et al. 2009), it became apparent that both daughter cells were able to move as soon as they had separated from each other. Each daughter inherits one of her mother's engines and it must assemble new engines for its new pole. A structure that will be called the polarity template ensures that each daughter assembles the correct engine at its new pole so that a daughter always emerges with pili at its front and slime secretion at its back. The molecules that constitute the template remain to be identified, but they appear to include the peptidoglycan sacculus (Bui et al. 2009), the outer membrane at both poles, the inner membrane at both poles, and the engine rudiments at both poles. Whatever its chemical composition, the template cannot demand, "Construct a pilus engine at this pole." Or "Construct a slime engine at that pole." Template instructions must allow for the fact that all cells, including newborn cells, will reverse their gliding direction within a few minutes after birth (Blackhart and Zusman 1985a). The reversal period (of order 10 minutes) is much shorter than the generation time (roughly 3 1/2 hours in rich medium at 32°C.); for that reason reversal is not likely to be linked to cell division or to the cell cycle (Wu et al. 2009). Many reversals can also be seen in Reichenbach's movies of the swarming of several different myxobacterial species (Kuhlwein and Reichenbach 1968). Individual cells alternate between gliding in one or the other direction along their pole-to-pole axis. Finally, observation of isolated single cells reversing (Kaiser 2007) as well as the tracks of many cells at higher density (Jelsbak and Søgaard-Andersen 2002), it was concluded that cells stop momentarily before moving off in the opposite direction. When cells do reverse, head and tail switch within roughly a minute; both engines and both ends appear to be switched simultaneously.

Reversing the Engines

Although each engine includes more than 15 different proteins (Table 1 and Table 2), relatively few of those proteins are directly involved in the coordination of pole switching. At each pole, multiple copies of only one type of engine (either the A engine or the S engine) are

Table 2. A motility genes

Gene location[a]	Gene name	Phenotype of knockout mutant	Protein domain function(s)	COG#, cd#, or Ref.
MXAN1925	*mglA*	Nonmotile (A^-S^-)	Raslike small GTPase	COG1100
MXAN2541	*aglAR*	No A-motility	unknown	None
MXAN3008	*aglU*	No A-motility	WD-repeat lipoprotein, acylaminoacyl-peptidase	COG1506
MXAN3536	*aglZ*	Cond[c] A-motile	Receiver domain, coiled-coil, regulatory protein	Mauriello, 2009
MXAN3060	*cglB*	Compl[d] A-motility	Outer membrane lipoprotein, stimulatable	cd01456
MXAN2538	*cglC*	Compl[d] A-motility	Outer membrane lipoprotein, stimulatable	
MXAN0962	*cglD*	Compl[d] A-motility	Outer membrane lipoprotein, stimulatable	
MXAN4866	*cglE*	Compl[d] A-motility	Outer membrane lipoprotein, stimulatable	D. Wall, unpubl.
MXAN4868	*cglF*	Compl[d] A-motility	Outer membrane lipoprotein, stimulatable	D. Wall, unpubl.
MXAN4862	*agmX*	No A-motility	Outer membrane lipoprotein	COG2885
MXAN4863	*agmK*	No A-motility	TPR repeat protein	COG0457
	agnA		ABC transporter component	
MXAN6403	*agnB*	No A-motility	ABC transporter permease protein	COG4591
	agnC		ABC transporter component	
MXAN7296	*aglCR*	No A-motility	unknown	None
MXAN1926	*mglB*	Partial A-motile	Roadblock/LC7 exchange factor for MglA	NOG46155[b]
MXAN2050	*pglH*	Partial A-motile	TPR repeat, CheY-like receiver domain, and a DNA-binding domain	COG0745, Yu, 2007
MXAN2919	*pglJ*	Partial A-motile	Polysaccharide polymerase	COG3307, Yu, 2007
MXAN2921	*pglB*	Partial A-motile	Glycosyl transferase, RfaG	COG0438, Yu, 2007
MXAN4148	*pglK*	Partial A-motile	Transmembrane protein of unknown function	COG5662, Yu, 2007
MXAN4616	*pglF*	Partial A-motile	Glycosyl transferase	COG0463, Yu, 2007
MXAN4710	*pglN*	Partial A-motile	ADP-heptose synthase, bifunctional sugar kinase/ adenylyltransferase	COG2870, Yu, 2007
MXAN4867	*pglI*	Partial A-motile	unknown	Yu and Kaiser 2007
MXAN5319	*pglC*	Partial A-motile	TPR repeat	COG0457, Yu, 2007
MXAN5382	*aspT*	Partial A-motile	tRNA-Asp	Yu and Kaiser 2007
MXAN5585	*pglE*	Partial A-motile	4-amino-4-deoxy-L-arabinose transferase	COG1807, Yu, 2007
MXAN6501	*pglD*	Partial A-motile	GDP-mannose pyrophosphorylase	COG0836, Yu, 2007
MXAN6607	*agmT*	Partial A-motile	unknown	COG1559
MXAN7160	*pglM*	Partial A-motile	Alanine racemase	COG0787, Yu, 2007
MXAN7252	*pglA*	Partial A-motile	Exopolysaccharide synthesis, ExoD-like	COG3932, Yu, 2007

[a]MXANxxxx (Goldman et al. 2006); COG, conserved orthologous group in NCBI database; cd, conserved domain in NCBI database. Ref., publication

[b]NOG = non supervised orthologous group in STRING database (http://string.embl.de/;

[c]Cond., Conditionally A motile.

[d]Compl. = Complementable by cell end-to-end contact (Hodgkin and Kaiser 1977).

observed: Switching resembles a cooperative change from one physicochemical phase to another, each composed of similar sets of molecules and having only slightly different structures. As mentioned, wild-type cells reverse at roughly 10-min intervals (Blackhart and Zusman 1985a; Mignot et al. 2005). Yu showed that wild type and *pgl* mutants have the same reversal frequency but different cell speeds (Yu and Kaiser 2007). Each frizzy mutant of *M. xanthus* inherits a well-defined average frequency of gliding reversal (Astling et al. 2006). Among frzCD mutants, the frequency can be higher or lower than wild type (McBride et al. 1992). The *frz* system encodes a cytoplasmic circuit that controls the frequency of reversal of gliding direction (Blackhart and Zusman 1985a). Frizzy proteins are related by amino acid sequence to chemosensory proteins (Bustamante et al. 2004). Although the methyl-accepting-chemotaxis-proteins (MCPs) of *E. coli* are membrane receptors, FrzCD, the MCP homologue in the *frz* system resides in the cytoplasm (Mauriello et al. 2009b; McCleary et al. 1990). FrzCD lacks transmembrane and extracellular domains, and its amino-terminal domain has no similarity to the N-termini of chemotaxis MCPs (Bustamante et al. 2004). These differences indicate that the role of the frizzy proteins is probably not chemotaxis. FrzE is signaled by FrzCD and it has one domain homologous to a *cheA* histidine protein kinase and a second response regulator domain (Acuna et al. 1995). FrzE catalyzes its own phosphorylation and transfers that phosphate to its response regulator domain (Acuna et al. 1995; McCleary and Zusman 1990a; McCleary and Zusman 1990b). Although wild-type cells reverse at 8.8-min intervals, a *frz*E null mutant has a long reversal period. A mutant that lacks the carboxy-terminal end of FrzCD, reverses at 2-min intervals, as if methyl-FrzCD induces the phosphorylation of FrzE (Blackhart and Zusman 1985b). These observations suggest that a critical level of FrzE~P could be the signal to reverse gliding direction.

The role played by the frizzy system in swarming, where it is known to regulate the reversal frequency (Blackhart and Zusman 1985a), was discovered by Oleg Igoshin. Although Igoshin was investigating production of the traveling waves that are prominent early in aggregation for fruiting body development, he found it necessary to introduce negative feedback from FrzE~P back onto the methylation of FrzCD (Igoshin 2004). He proposed that FrzE~P inhibited the methylation of FrzCD, catalyzed by FrzF, rather than enhancing demethylation, catalyzed by FrzG (Igoshin et al. 2001). The resulting negative feedback loop would cause the levels of FrzE~P and of Me-FrzCD to oscillate out of phase with each other. Dubbed the "Frizilator", the circuit has the necessary refractory period (Igoshin et al. 2001) and the necessary time delay (Börner et al. 2002) to sustain regular oscillations. The frizilator circuit has been corroborated by quantitative studies of the reversal periods and the swarm expansion rates of frizzy gene deletion mutants (Wu et al. 2009; Kaiser D, unpubl. data). Although the Frizilator was discovered in cultures that were developing fruiting bodies, similar oscillations occur in cultures that are growing, swarming, and not yet capable of developmental signaling, which must be induced by starvation. The Frizilator circuit makes up the right-hand side of Figure 6.

MglA is Switched by the Frizilator

As mentioned, Rosa Yu observed that MglA mutants, known to be defective in both A motility and S motility, were secreting slime from both ends of its cells (Yu and Kaiser 2007). Unipolar slime secretion is essential for significant cell movement, and bipolar secretion stops A motility (Yu and Kaiser 2007). MglA is a G-protein that belongs to the extended Ras-like family of small GTPases (Hartzell and Kaiser 1991a); the crystal structure of HRAS has been solved (Holbrook and Kim 1989; Pai et al. 1989). In *M. xanthus* and many other prokaryotes, *mglA* is part of a two gene operon with *mglB*, a regulator that belongs to the Roadblock/LC7 group (Koonin and Aravind 2001). MglA protein is unstable in a Δ*mglB* mutant (Hartzell and Kaiser 1991b) as if MglA is prone to degradation, unless MglB is associated with it.

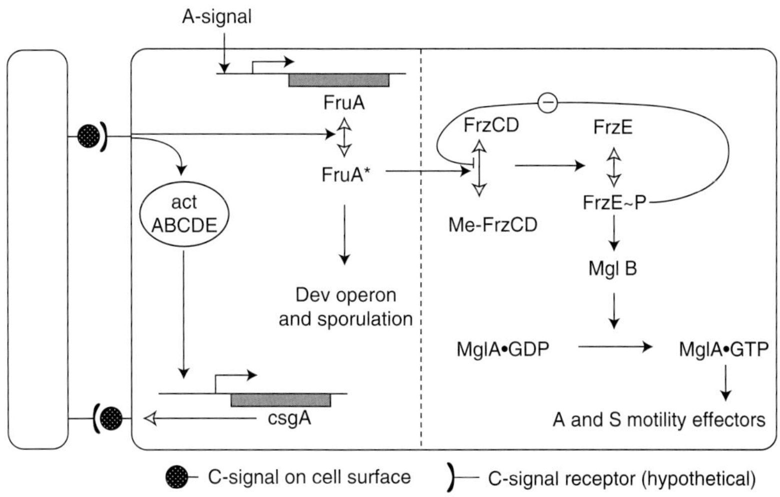

Figure 6. The Frizilator and the reversal generator are shown to the right of the dashed vertical line. That circuit is always active and it regulates swarming. The C-signal, which appears shortly after starvation induces fruiting body development, is a 17 kDa cell surface protein. Cells must make end-to-end contact with each other to transmit the signal. The C-signal transduction pathway is shown to the left of the dashed vertical line. FruA, a response regulator, that has been activated by C-signaling is symbolized FruA*. It may include FruA~P.

It is proposed that the response regulator domain of FrzE~P activates the MglAB complex to generate MglA•GTP, as shown in Figure 6. This proposal is analogous to the activation of the flagellar switch complex in enteric bacteria by CheY~P (Stock et al. 1995). Both FrzE (McCleary and Zusman 1990b) and MglA proteins (Hartzell and Kaiser 1991a) are found in the cytoplasm where they are able to interact with each other, and where MglA•GTP can be formed. Recent studies of reversal during swarming suggest that MglAB by itself is an oscillator whose natural period is about 34 min (Wu et al. 2009). We suggest further that the Frizilator evolved to entrain the MglAB oscillator to obtain an optimal reversal period in the range of 7–9 min, which relates to the observed cell speed of 4 μm/min (Wu et al. 2009).

Motility Effectors of MglAB

Reversal involves effectors of MglAB specific for A motility and other effectors specific for S motility, to switch both engines following instructions from the template. As mentioned, switching engines from one pole to the other is all or none: At either pole of the cell all the active engines are either for A motility or for S motility (type IV pili), and mixtures of different active engines do not seem to occur. As indicated in Figure 6, mutual exclusion between engines is enforced by the effector proteins that are directly regulated by the MglAB switch, usually as a result of protein-protein contact. Effector activity is graded and regulates the timing of engine assembly. Effector RomR, which is essential for A motility, is a phosphorylated response regulator. RomR protein forms clusters at both cell poles, but the cluster at the lagging pole of the cell is larger than the cluster at the leading pole (Leonardy et al. 2007). Moreover, the larger cluster switches from one pole to the other in parallel with the cell's reversal of gliding direction (Leonardy et al. 2007). Thus, at every instant, the larger cluster is found at the cell pole that is believed to be actively secreting slime, and correlating RomR~P with slime secretion. The dynamic re-localization of RomR was shown to depend on the Frz system as well as on the Mgl switch (Leonardy et al. 2007). Perhaps RomR is phosphorylated by the

M. xanthus Wzb and Wzc homologs (Kaiser 2008b), which are known to encode an autokinase/phosphatase that regulates capsule secretion in *E. coli* (Paiment et al. 2002; Wugeditsch et al. 2001).

The AglZ protein is another effector of A motility. MglA has been found to interact physically with AglZ protein in a yeast two-hybrid experiment with MglA as the bait (Yang et al. 2004). The amino-terminal 120 amino acids of AglZ is a pseudoreceiver, CheY-like domain that has been found to interact with purified MglA protein in vitro (Yang et al. 2004). The same domain of AglZ interacts with FrzCD protein in cytoplasmic clusters that lie on a helix and span the cell length (Mauriello et al. 2009b). Where the helix intersects a line drawn along the bottom of the cell, the AglZ-FrzCD clusters generate the focal adhesion sites described by Mignot et al. 2007.

CglB protein is also involved in switching A engine polarity, most likely as an effector. CglB was shown by Hodgkin to be essential for A motility and to play no role in S motility (Hodgkin and Kaiser 1979b), as is the case for RomR and AglZ. CglB is a lipoprotein with a type II signal sequence (Rodriguez and Spormann 1999) that resides in the outer membrane (Simunovic et al. 2003). Yu observed that mutants lacking CglB simultaneously secrete slime from both its ends, as if CglB normally prevents slime secretion from the nozzles found at the leading pole (Yu and Kaiser 2007). Mutant *cglC*, *cglD*, *cglE*, and *cglF* (Table 2) have motility phenotypes similar to that of *cglB* (Hodgkin and Kaiser 1977). Moreover, several other A motility genes whose mutants have a bipolar secretion phenotype (Yu and Kaiser 2007) may also be effectors, although a direct interaction with Mgl has not been shown. Also preventing bipolar secretion (and included in Table 2), AglU and AgmX are both lipoproteins, whereas AgmK is a TPR protein (White and Hartzell 2000; Youderian et al. 2003); it thus resembles Tgl, a TPR protein known to be involved in protein assembly. The lipoprotein-encoding genes *cglE*, and *cglF* belong to an operon that includes agmX, agmK, and agmU, which encode other lipoproteins essential for A motility (Table 2). Knockout mutants of AgnB, an ABC transporter permease, or of AgnA, or of AgnC also show bipolar secretion (Yu and Kaiser 2007). The need for a large array of such proteins that prevent inappropriate slime secretion may be related to the tight coupling observed between polysaccharide polymerization and its secretion (Nesper et al. 2003). This array of regulatory proteins appears to inhibit nozzles at the leading pole from secreting until such time as the next reversal is ordered and those nozzles start to secrete. Despite their substantial numbers, there are fewer effectors than nozzle structural proteins; only those engine proteins that contribute to switching from end to end need have effectors. For instance, hundreds of nozzle orifices are always present at both cell ends (Wolgemuth et al. 2002); the orifices don't switch and have no need of effectors.

S Motility Effectors of MglAB

FrzS, which regulates S motility and not A motility (Ward et al. 2000), is an effector because it responds to the Mgl switch, although direct interaction of FrzS with MglA has yet to be shown. FrzS relocates between the cell poles in apparent synchrony with RomR, a well established effector (Leonardy et al. 2007). The function of FrzS can be inferred from the fact that a deletion mutant of FrzS has a lower rate of swarm expansion and a lower reversal frequency than the wild type, although the amount of lowering is less than a deletion of the FrzE gene (Ward et al. 2000). Neither the target nor the regulatory activity of FrzS have been found.

Assembled Tgl lipoprotein might be the regulatory target of FrzS, but if it isn't, then Tgl should be the target of some other S motility effector. The *pilA* gene, which encodes the pilin monomer, is one of the most highly transcribed genes in *M. xanthus* (Jelsbak and Kaiser 2005; Wu and Kaiser 1997). When the cell reverses direction, pili disappear from one pole and extend from the other. All the while, expression of the *pilA* gene has remained constant. As a general rule, effectors regulate protein localization both post-transcriptionally and post-translationally. Regarding its location in the cell, fluorescent Tgl (Nudleman et al. 2006), like FrzS (Ward et al. 2000), is only found at the leading cell pole. Pole specificity differentiates them from the majority of Pil proteins (Table 1) which are found constantly at both poles. Tgl and assembled PilQ differentiate the leading pole of the cell from the lagging pole that lacks pili, lacks Tgl, and lacks FrzS. Unassembled PilQ in the outer membrane appears to be held together at the pole in a single patch by attachment to the top of a cylindrical complex of proteins that includes PilG, PilM, PilN, PilO, and PilP (Nudleman et al. 2006). These proteins were identified by their requirement, together with PilQ, to transport new Tgl to the outer membrane where Tgl functions (Nudleman et al. 2006). These proteins are found in the periplasm and the inner membrane with PilB (Nunn et al. 1990). This multiprotein complex, which would extend from the outer membrane, penetrate the peptidoglycan mesh to enter the inner membrane; the complex is found fixed in position at both cell poles. Newly synthesized Tgl is able to link PilQ monomers together into a detergent-resistant complex as revealed by the stimulation phenomenon, although stimulation is not required for assembly (Nudleman et al. 2005). PilQ protein monomers are conserved, and they don't switch. Because assembled PilQ resists dissociation by 1% SDS at 100°C (Nudleman et al. 2005), it seems likely that Tgl is destroyed to allow the assembled PilQ to dissociate into a cluster of monomers. The process may resemble that found recently in *Mycobacterium tuberculosis* (Pearce et al. 2008), wherein a particular protein (Tgl in this case) would be marked and targeted for destruction by a protease. PopC (Rolbetski et al. 2008) is an activatable protease located in the outer membrane of *M. xanthus*. The protease-activating system, if triggered by MglA-GTP would be considered the effector for Tgl localization. Meanwhile newly synthesized Tgl that is not yet part of assembled PilQ would escape proteolysis by a membrane protease, as it is transported to the outer membrane where it assembles PilQ at the new leading pole (Kaiser 2008b). Thus it is suggested that a protein assembly, without a pilus fiber, with no Tgl protein, and probably with no FrzS is the product of S engine inactivation. That assembly would await the addition of missing proteins to become an active pilus engine at the next polarity reversal.

RESPONDING TO STARVATION BY BUILDING A FRUITING BODY

M. xanthus has a single origin of bidirectional replication near the *dnaA* locus on its circular genome (Goldman et al. 2006). Using flow cytometry, Tzeng and Singer showed that a population of exponentially growing *M. xanthus* cells contain 1 – 2 copies of the chromosome, indicating a single DNA replication initiation event per cell cycle and that cell division follows chromosome replication (Tzeng and Singer 2005). Tzeng and Singer showed further that, as cells approach the stationary phase, the initiation of DNA replication stops when all cells have one complete chromosome.

Before it reaches stationary phase, the culture faces a choice as its growth slows with nutrient limitation: It can stop growth entirely and initiate fruiting body development, or it can continue to grow, at what may be an ever-decreasing rate dictated by the residual level of nutrient until the cells die. Because no more than 1% of the cells initiating fruiting body development become spores, and some cells prey on others, the majority of cells die whichever response is chosen. Because more than 30 new proteins must be made for aggregation and sporulation (Inouye et al. 1979), some capacity to synthesize proteins must be retained, and the population must start fruiting body construction before any essential nutrient has been totally depleted. Choosing between growing (probably with the hope of finding more nutrient soon) and fruiting body development is helped by a diffusible cell-to-cell signal, the A-signal (Figure 6). A-signal molecules, purified from medium conditioned by developing cells, proved to be a set of amino acids and peptides containing those amino acids (Kuspa et al. 1986; Kuspa et al. 1992a). Each *Myxococcus* cell that senses nutrient limitation and opts for development releases a small quantity of A-signal about two hours after it senses that starvation is immanent. Consequently the extracellular concentration of A-signal is directly proportional to the density of *M. xanthus* cells that have chosen development (Kaplan and Plamann 1996; Kuspa et al. 1992b). Cells in the population respond to A-signal only if its concentration is above a threshold. Cells respond by expressing a set of A-signal-dependent genes that include *spi*, *csgA* (the gene encoding C-signal), and *fruA*, an important developmental response regulator. A response to A-signal indicates that there are enough cells to complete one fruiting body: Those cells constitute a quorum that favors fruiting body development. Limitation for any amino acid, lack of carbon source, or too little phosphate induces *M. xanthus* to initiate fruiting body development (Manoil and Kaiser 1980a; Manoil and Kaiser 1980b). Neither the lack of purines nor of pyrimidines, which *M. xanthus* scavenges from prey nucleic acids, will induce development (Kimsey and Kaiser 1991). This suggests that a deficiency of any amino-acylated tRNA leads the population of cells to stop swarming and to initiate development. In *M. xanthus* as in many other bacteria, the absence or shortage of any one of the charged tRNAs leads a ribosome, sensing with a codon that lacks its cognate amino acylated tRNA, to synthesize guanosine tetra (and penta) phosphate, (p)ppGpp, in a reaction catalyzed by the *relA* synthase. The rise in (p)ppGpp sets off a stringent response that stops the synthesis of new ribosomes and of peptidoglycan (Cashel et al. 1996). In *M. xanthus* the stringent response also initiates expression of its developmentally regulated genes. Singer et al. showed that the accumulation of (p)ppGpp was both necessary and sufficient to trigger fruiting body development (Singer and Kaiser 1995). In *B. subtilis* and in *E. coli* DNA synthesis is arrested by their stringent responses. By contrast, *M. xanthus* chromosome replication continues during development induced by the stringent response, and fruiting body myxospores contain two complete copies of the chromosome (Tzeng et al. 2006). A complete description of these processes can be found in (Diodati et al. 2008).

Aggregation for Fruiting Body Development

As discussed earlier, swarming cells reverse periodically triggered by the Frizilator. Reversals are essential for the swarm to spread outward as a thin sheet of cells, which provides them with greater access to nutrient. Reversals also help swarm cells build different kinds and sizes of multicellular structures: Rafts of adjacent cells that have their long axes parallel, and multilayered mounds that can be seen in Figure 5 (Wu et al. 2009). When (p)ppGpp arrests growth, outward spreading stops. Then cells migrate inward causing many traffic jams some of which become foci for aggregation (Kaiser and Welch 2004). Morphogenesis of *M. xanthus* fruiting bodies at the centers is directed by C-signaling between cells. C-signal deficient mutants (*csgA*) were found to grow normally but not to aggregate or sporulate (Hagen et al. 1978; Kim and Kaiser 1990b; Shimkets et al. 1983). The C-signal is a 17-kDa cell-surface-bound protein that communicates when pairs of cells make an end-to-end contact with each other (Kim and Kaiser 1990a; Sager and Kaiser 1994). The *csgA* gene encodes a 25 kD protein that is secreted to the cell surface where it is cleaved to the active 17-kDa signal (Lobedanz and Sögaard-Andersen 2003; Rolbetski et al. 2008). The complete C-signaling circuit shown in Figure 6 was worked out from the properties of gene knockout mutants that lack circuit elements, and more detail can be found in a recent review (Søgaard-Andersen 2008). The circuit includes the reversal generator that drives swarming and was discussed above. In addition to the frizilator's negative feedback, the circuit of Figure 6 includes two positive feedback loops. For this reason the circuit can have emergent properties, and development is the most important of these. Emergent properties can be extracted by computer simulation, and a continuous three-dimensional simulation (Sozinova et al. 2005; Sozinova et al. 2006) reproduces the experimentally observed stages of fruiting body formation: asymmetric initial aggregates (traffic jams), linear streams, formation of hemispherical mounds, and finally sporulation within the mounds. Although the circuit of Figure 6 suffices to explain the sequence of stages, elements are missing, the C-signal receptor, for example.

C-Signaling and Development

As indicated in the figure, when traffic jams enlarge into fruiting bodies, C-signaling activates FruA, a developmentally important response regulator (Ellehauge et al. 1998). Simultaneously, expression of the C-signal is increased by a positive feedback loop involving the *act* operon (Gronewold and Kaiser 2001). Because of the positive feedback, there is a progressive increase in the cell surface level of C-signal. The five proteins of the *act* operon increase expression of the *csgA* gene (Gronewold and Kaiser 2007). At the start of development, there are few C-signal molecules per cell. Cells making end-to-end contact respond to signal exchange by reversing their direction of gliding which creates the traveling waves that are seen early in development (Welch and Kaiser 2001). One consequence of the feedback is that the waves are transient (Gronewold and Kaiser 2001). Each time C-signal is exchanged between cells in the crests of two colliding waves, the positive feedback loop increases expression of *csgA* and elevates the number of signal molecules on both signaling cells. The low initial level of C-signal found at the start of development produces a low level of activated FruA (FruA*) that drives the frizilator (Fig. 6) and produces a very regular 9-min period. Waves start as broad and diffuse ridges. Then, because the Frizilators in a pair of cells that are signaling to each other are brought into synchrony, they sharpen. The precise period of the FruA* driven frizilators and the synchronization of oscillation between colliding cells causes the waves to sharpen. Because of the positive feedback, more signal is produced. Higher levels of C-signal induce higher levels of FruA* (Kim and Kaiser 1991; Kruse et al. 2001; Li et al. 1992).

Aggregation

As the number of C-signal molecules per cell rises, their signaling elevates the cytoplasmic level of FruA* (Fig. 6) to a threshold. The existence of a threshold that can stop the oscillation is implicit in the negative feedback within the frizilator (Igoshin et al. 2004). Arresting the oscillation leads a developing population of cells to make the transition from traveling waves to aggregates that is evident as the waves fade away and the aggregation centers simultaneously enlarge (Kaiser and Welch 2004). The threshold level of FruA* stops the oscillation by placing all the FrzE in its nonphosphorylated state (Igoshin et al. 2004). There being no more signal to reverse, the cells continue to move in the direction they were moving before their last C-signaling event. This transition in cell reversal behavior is also observed in the tracks of individual cells (Jelsbak and Søgaard-Andersen 2002). The transition leads the cells to form streams; cells trapped in a stream are able to enlarge an aggregate. When a stream of nonreversing cells encounters a dense traffic jam that it cannot penetrate, it is deflected. It turns to glide over or around the traffic jam. Because gliding cells adhere to the surface on which they are moving, they must bend as they pass over or around it. Bending will have distorted the elastic peptidoglycan wall (Wolgemuth 2005) that has some unusual cross-links and unusually short glycan chain length (Bui et al. 2009). Persistence of the bent shape directs the cell into a circular orbit. Cells which are streaming in orbit can continue to C-signal their neighbors, and more positive feedback raises the C-signal level until this population of cells reaches the still higher threshold for the differentiation of spores, which also has been observed experimentally (Gronewold and Kaiser 2001; Kim and Kaiser 1991; Li et al. 1992). In sum, the cell movements of fruiting body morphogenesis can be explained by C-signaling. Details can be found in (Kaiser and Welch 2004).

GENE REGULATORY NETWORK

Microarray expression profiling of cells at 12 h into development indicates that nearly 30% (2134/7388) of the genes are up- or down-regulated at least 1.5-fold compared with growing cells (Caberoy et al. 2010). Several key developmental genes like *fruA*, *csgA*, and the *act* operon have been noted above in the context of the roles played by their products in fruiting body morphogenesis. How are these and the many other genes that respond to starvation and coordinate fruiting body development regulated?

Although our understanding of the gene regulatory network is far from complete, it is already clear that building a multicellular fruiting body with a defined size and shape requires a network that is highly responsive to environmental, intercellular, and intracellular signals. Analysis of the *M. xanthus* genome sequence revealed an expansion of multicomponent regulatory pathways because of lineage-specific gene duplications, whereas the number of one-component regulators was less than half that expected based on the genome size and comparison with other soil bacteria (Goldman et al. 2006). Multicomponent pathways include two or more proteins with sensory input domains (e.g., sensor domains of protein kinases and receiver domains of response regulators in the familiar two-component systems). These pathways, and connections between them, allow a complexity of signal integration similar to that in multicellular eukaryotes.

Eukaryotic-Like Components of the Network

Some components of the regulatory pathways are similar to proteins first characterized in eukaryotes. *M. xanthus* encodes 97 serine/threonine protein kinases (STPKs) and many play a role in fruiting body development (based on the phenotypes of mutants) (Inouye et al. 2008); however, the signal inputs are unknown and in most cases it is also unknown how the protein kinase exerts its effects.

Two types of eukaryotic-like transcription factors are important for the *M. xanthus* response to starvation. A family of 53 enhancer-binding proteins (EBPs) is predicted to activate transcription of particular gene sets by sigma^{54}RNA polymerase (Goldman et al. 2006). Bacterial EBPs typically bind to enhancers located 70–150 bp upstream of the transcription start site and contact sigma^{54}RNA polymerase via DNA looping to facilitate open complex formation for transcription initiation (Buck et al. 2000). At least 17 *M. xanthus* EBPs affect development, although in five of these cases motility on nutrient-rich agar is impaired (perhaps explaining the defects in fruiting body formation) (Kroos and Inouye 2008). Seven of the development-specific EBPs have amino-terminal CheY-like receiver domains that are presumably phosphorylated by histidine protein kinases (HPKs), forming two-component signal transduction systems, which are common in bacteria, but three EBPs involved in development have amino-terminal fork-head-associated (FHA) domains (Jelsbak et al. 2005). The FHA domain was first recognized in eukaryotic transcription factors and it mediates interactions between proteins by recognizing phosphothreonine (Durocher and Jackson 2002). FHA domain-containing EBPs presumably link STPK signaling pathways to activation of transcription by sigma^{54}RNA polymerase.

The second type of eukaryotic-like transcription factor involved in *M. xanthus* development is CarD, a protein also important for light-induced carotenogenesis (Elias-Arnanz et al. 2008). Bacterial orthologs of CarD have been found only in myxobacteria, and their DNA-binding domains are similar to those of eukaryotic high-mobility group A proteins or histone H1. These proteins normally modulate chromatin structure but their DNA-binding domains can substitute for that of *M. xanthus* CarD to permit fruiting body development and carotenogenesis (Garcia-Heras et al. 2009).

Modular Design

How are eukaryotic-like regulatory proteins and typical bacterial regulators organized into a signal-responsive network to govern gene expression, cell movement, and differentiation during *M. xanthus* multicellular development? Figure 7 depicts the effects of starvation, including an increase in the intracellular signal (p)ppGpp, which leads to production of the intercellular A- and C-signals. These signals provide input into 3 regulatory modules designated the EBP cascade, Mrp, and FruA in Figure 7. As described above, FruA* is proposed to drive the frizilator that controls cell reversals

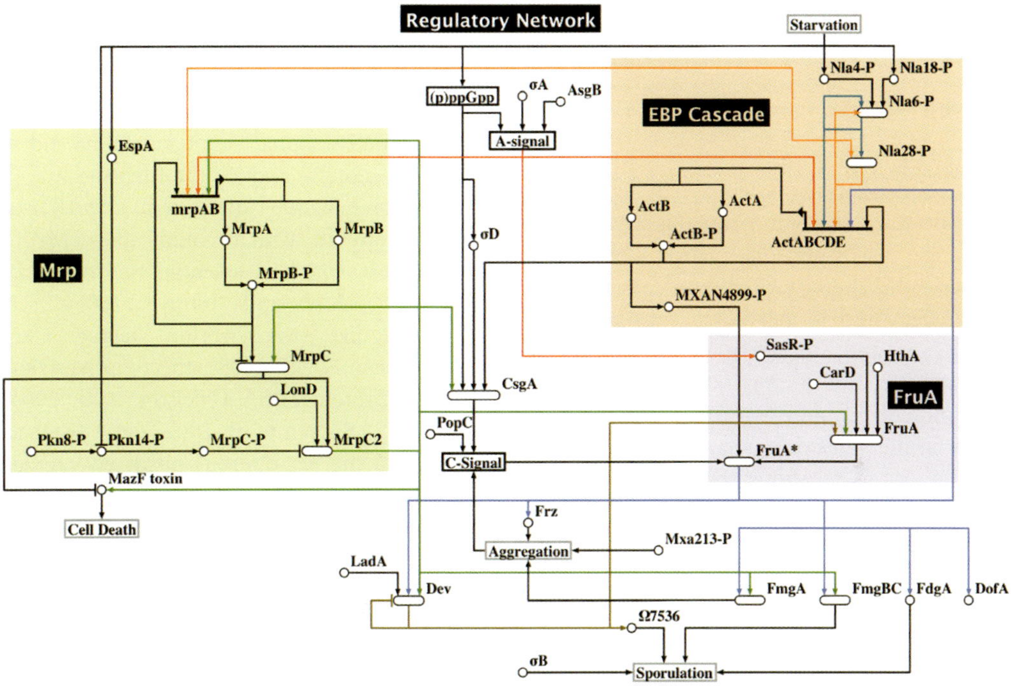

Figure 7. Network diagram of regulation during *M. xanthus* development. Arrows indicate positive regulation and lines with barred ends indicate negative regulation. Several lines are colored for emphasis. Background colors distinguish three regulatory modules.

(Fig. 6), producing waves and then aggregates. FruA* also is proposed to activate transcription of target genes, often in cooperation with MrpC2 from the Mrp module, and some of the target genes are important for aggregation or sporulation (Fig. 7). The Mrp module also governs programmed cell death. The inputs and outputs of the 3 regulatory modules are described in more detail below.

The EBP Cascade Module

Six EBPs form a four-step cascade (Caberoy et al. 2010) that responds to starvation and A-signal (Fig. 7). Nla4 and Nla18 have amino-terminal CheY-like and FHA domains that might be phosphorylated by an HPK and an STPK, respectively, in response to starvation. Both Nla4 and Nla18 bind to the promoter region of the gene that codes for Nla6, presumably activating its transcription. The CheY-like domain of Nla6 is presumably phosphorylated by an HPK (in response to an unknown signal) and activates the gene for Nla28 (whose CheY-like domain is also presumably phosphorylated in response to an unknown signal). Nla6 and Nla28 cross-activate, auto-activate, and activate the *actABCDE* operon, which codes for the HPK ActA and its presumed substrate EBP, ActB. A-signaling is required for transcription of the *act* operon (Gronewold and Kaiser 2007) but whether this is because A-signal is sensed at an earlier step in the cascade remains to be tested. ActB auto-activates and it activates the last EBP in the cascade, MXAN4899 (Caberoy et al. 2010), whose FHA domain might be phosphorylated by an STPK (Jelsbak et al. 2005). Output from this module includes (1) a connection to the Mrp module as Nla28 appears to activate *mrpAB* transcription (A.G. Garza, personal comm.), (2) ActB directly or indirectly increasing expression of *csgA* to produce C-signal (Gronewold and Kaiser 2001), and (3) MXAN4899 is proposed to facilitate activation of FruA (Jelsbak et al. 2005), providing a connection to the FruA module.

The Mrp Module

A-signaling might provide a separate input into *mrpAB* transcription (as depicted in Fig. 7) or it might do so via Nla28 if A-signal is sensed at or before the Nla28 phosphorylation step in the EBP cascade. The *mrpAB* operon codes for the HPK MrpA and its presumed substrate EBP, MrpB (Sun and Shi 2001). MrpB appears to auto-activate and it appears to activate the gene immediately downstream, *mrpC*. MrpC is a CRP-family transcription factor that is cleaved near its amino terminus to produce MrpC2 (Ueki and Inouye 2003). Starvation not only impacts *mrpAB* transcription, it affects EspA, an HPK that somehow inhibits *mrpC* expression posttranscriptionally (Higgs et al. 2008). Also, starvation inhibits accumulation of Pkn14, the second member of an STPK cascade that leads to phophorylation of MrpC, which inhibits LonD-dependent proteolytic conversion of MrpC to MrpC2 (Nariya and Inouye 2005; Nariya and Inouye 2006). The net effect of starvation is to boost MrpC and MrpC2 accumulation. Because MrpC2 binds better than MrpC to certain promoter regions (Nariya and Inouye 2006; Ueki and Inouye 2003), only MrpC2 is depicted in Figure 7 as auto-activating *mrpC* and activating *mrpAB*, although MrpC likely contributes to these activities. Likewise, output from the Mrp module in terms of activating target genes *fruA* (Ueki and Inouye 2003), *mazF* (Nariya and Inouye 2008), *dev* (S. Mittal, P. Viswanathan, and L.K., unpublished data), *fmgA* (Mittal and Kroos 2009a), and *fmgBC* (Mittal and Kroos 2009b) is depicted solely from MrpC2. On the other hand, only MrpC has been shown to be an antitoxin that binds to the MazF toxin, inhibiting programmed cell death (Nariya and Inouye 2008); whether MrpC2 binds to MazF remains to be tested.

The FruA Module

If A-signaling provides input into *fruA* transcription via the SasR EBP (Guo et al. 2000) (as depicted in Fig. 7), this is likely an indirect effect because the *fruA* promoter is not similar to others recognized by sigma^{54}RNA polymerase. For example, SasR might provide an A-signal-dependent input into the EBP cascade (at the putative sigma^{54}-dependent *nla6* and/or *nla28* promoters) and/or the Mrp module (at the putative sigma^{54}-dependent *mrpAB* promoter), any of which could affect MrpC2 production, which is essential for *fruA* transcription (Ueki and Inouye 2003). The eukaryotic-like transcription factor CarD (Elias-Arnanz et al. 2008), the putative DNA-binding protein HthA (Nielson et al. 2004), and DevT (Boysen et al. 2002) (in a feedback loop from the *dev* operon) positively regulate *fruA* transcription, but whether these are direct inputs is unknown. Two inputs into the FruA module are proposed to stimulate formation of FruA* (Fig. 7). C-signaling (Ellehauge et al. 1998) and MXAN4899 (Jelsbak et al. 2005) have been proposed to cause phosphorylation of FruA, which has a CheY-like receiver domain but it lacks some residues typically required for phosphorylation, its DNA binding was not changed by small molecule phosphodonors (Mittal and Kroos 2009a), and the cognate HPK has not been found. In any case, C-signaling and MXAN4899 do not change the level of FruA, but somehow change its activity, affecting module output, which includes (1) modulation of the Frizilator (Frz in Fig. 7; see also Fig. 6) to bring about aggregation (which enhances C-signaling and therefore FruA* formation in a positive feedback loop), (2) direct or indirect activation of the *act* operon (Gronewold and Kaiser 2007) (enhancing C-signaling and FruA* formation in a second positive feedback loop), and (3) direct activation of target genes *dev* (S. Mittal, P. Viswanathan, and L.K., unpublished data), *fmgA* (Mittal and Kroos 2009a), *fmgBC* (Mittal and Kroos 2009b), *fdgA* (Ueki and Inouye 2005b), and *dofA* (Ueki and Inouye 2005a), some of which play a role in aggregation or sporulation.

It is worth noting that FruA and MrpC2 bind cooperatively to the promoter regions of *fmgA*, *fmgBC*, and *dev*, directly subjecting transcription of these genes to regulation by both the Mrp and FruA modules. Also, transcription of the *dev* operon, which is important for sporulation, requires a third transcription factor, LadA, and *cis*-regulatory elements spanning >1 kb (Viswanathan et al. 2007a; Viswanathan et al. 2007b); complexity resembling regulation of genes in multicellular eukaryotes.

CONCLUDING REMARKS

Why does *M. xanthus* devote a large amount of energy and protein synthetic capacity to reverse its gliding engines? To enable reversal, both the A and the S engines are always found at both cell poles. However, at each pole only one of the two engines is in working order; the other engine either lacks components that are critical for motor activity (incomplete S engines), or have additional components that block motor activity (inactive A engines). In addition both the A and the S engines are controlled by MglA•GTP via effector proteins, which regulate the timing of engine assembly and distinguish old and currently active engines from new engines that have yet to function. New engines are made by constitutive growth mechanisms, following instructions from the template that indicate which of the two engines should be active at which pole. Studies of *Myxococcus* swarming (Wu et al. 2009) show that reversals enable it to build multicellular structures like the rafts and mounds, which are evident in Figure 5, and to build fruiting bodies, Figure 1.

Many biochemical questions remain to be answered concerning engine synthesis and its relation to cell division. Among them, which molecules constitute the proposed template, and how does the template encode direction? Can the inhibition of FrzF protein by FrzE~P, postulated for the frizilator, be shown with purified proteins? Which *M. xanthus* proteins provide the Wza, Wzx, Wzy, and Wba-like functions for capsular polysaccharide synthesis, and how do those proteins feed several polysaccharide chains into a single enlarged Wza-like pore? What is the sugar sequence of the capsular polysaccharide, of the propulsive slime polysaccharide, and are they identical as proposed? How do the Cgl and Agm proteins prevent the nozzles at the leading pole from synthesizing and secreting slime? How are Mgl effector proteins metabolized? Which protein kinase phosphorylates effector RomR and how is that kinase regulated? What is the regulatory target of effector FrzS and what is its action? Is Tgl the target of FrzS or of another effector? How is PilQ disassembled? If by proteolysis, by which protease?

Many unanswered questions concern the control of gene expression and the regulatory networks shown in Figures 6 and 7: How is FruA activated? Does A-signaling directly activate the *act* operon or does it activate via the EBP cascade module? What are the sensory inputs to the EBPs and what contributions do the nearly one-hundred STPKs make that are found in *Myxococcus*.

ACKNOWLEDGMENTS

MR was supported by a fellowship from the Gene Expression in Development and Disease Initiative at Michigan State University. Work in the laboratory of LK was supported by National Science Foundation grant MCB-0744343 and by the Michigan Agricultural Experiment Station.

REFERENCES

Acuna G, Shi W, Trudeau K, Zusman D. 1995. The *cheA* and *cheY* domains of *Myxococcus xanthus* FrzE function independently in vitro as an autokinase and a phosphate acceptor, respectively. *FEBS Letts* **358:** 31–33.

Astling DP, Lee JY, Zusman DR. 2006. Differential effects of chemoreceptor methylation-domain mutations on swarming and development in the social bacterium *Myxococcus xanthus*. *Mol Microbiol* **59:** 45–55.

Behmlander RM, Dworkin M. 1994. Biochemical and structural analyses of the extracellular matrix fibrils of *Myxococcus xanthus*. *J Bacteriol.* **176:** 6295–6303.

Berleman J, Chumley T, Cheung P, Kirby JR. 2006. Rippling is a predatory behavior in *Myxococcus xanthus*. *J Bacteriol* **188:** 5888–5895.

Bitter W, Koster M, Latijnhouwers M, de Cock H, Tommassen J. 1998. Formation of oligomeric rings by XcpQ and PilQ, which are involved in protein transport across the outer membrane of *Pseudomonas aeruginosa*. *Mol Microbiol* **27:** 209–219.

Blackhart BD, Zusman D. 1985a. Frizzy genes of *Myxococcus xanthus* are involved in control of frequency of reversal of gliding motility. *Proc Natl Acad Sci* **82:** 8767–8770.

Blackhart BD, Zusman DR. 1985b. Cloning and complementation analysis of the frizzy genes of *Myxococcus xanthus*. *Mol Gen Genet* **198:** 243–254.

Boysen A, Ellehauge E, Julien B, Søgaard-Andersen L. 2002. The DevT protein stimulates synthesis of FruA, a signal transduction protein required for fruiting body morphogenesis in *Myxococcus xanthus*. *J Bacteriol* **184:** 1540–1546.

Buck M, Gallegos MT, Studholme DJ, Guo Y, Gralla JD. 2000. The bacterial enhancer-dependent σ^{54} (σN) transcription factor. *J Bacteriol* **182:** 4129–4136.

Bui NK, Gray J, Schwarz H, Schumann P, Blanot D, Vollmer W. 2009. The peptidoglycan sacculus of *Myxococcus xanthus* has unusual structural features and is degraded during glycerol-induced myxospore development. *J Bacteriol* **191:** 494–505.

Burchard RP. 1970. Gliding motility mutants of *Myxococcus xanthus*. *J Bacteriol* **104:** 940–947.

Burchard RP. 1974. Growth of surface colonies of the gliding bacterium *Myxococcus xanthus*. *Arch Microbiol* **96:** 247–254.

Bustamante VH, Martínez-Flores I, Vlamakis HC, Zusman D. 2004. Analsis of the Frz signal transduction system of *Myxococcus xanthus* shows the importance of the conserved C-terminal region of the cytoplasmic chemoreceptor FrzCD in sensing signals. *Mol Microbiol* **53:** 1501–1513.

Börner U, Deutsch A, Reichenbach H, Bär M. 2002. Rippling patterns in aggregates of myxobacteria arise from cell-cell collisions. *Phys Rev Lett* **89:** 078101.

Caberoy NB, Giglio K, Suen G, Garza AG. 2010. A cascade of enhancer binding proteins regulates gene expression during multicellular development in *Myxococcus xanthus*. *Proc Natl Acad Sci* (in press).

Cashel M, Gentry DR, Hernandez VJ, Vinella D. 1996. The stringent response. in *Escherichia coli and Salmonella* (ed. F. Neidhardt et al.), pp. 1458–1496. ASM Press, Washington, D.C.

Chiang P, Habash M, Burrows LL. 2005. Disparate subcellular localization patterns of *Pseudomonas aeruginosa* Type IV pilus ATPases involved in twitching motility. *J Bacteriol* **187:** 829–839.

Collins RF, Davidsen L, Derrick JP, Ford RC, Tonjum T. 2001. Analysis of the PilQ secretin from *Neisseria meningitidis* by transmission electron microscopy reveals a dodecameric quaternary structure. *J Bacteriol* **183:** 3825–3832.

Collins RF, Frye SA, Kitmitto A, Ford RC, Tønjum T, Derrick JP. 2004. Structure of the *Neisseria meningitidis* outer membrane PilQ secretin complex at 12 Å resolution. *J Biol Chem* **279:** 39750–39756.

Craig L, Volkmann N, Arvai AS, Pique ME, Yeager M, Egelman EH, Tainer JA. 2006. Type IV pilus structure by cryo-electron microscopy and crystallography: Implications for pilus assembly and functions. *Mol Cell* **23:** 651–662.

Curtis PD, Shimkets LJ. 2008. Metabolic pathways relevant to predation, signaling and development. In *Myxobacteria, multicellularity and differentiation* (ed. D.E. Whitworth), pp. 241–258. ASM Press, Washington, D.C.

D'Andrea LD, Regan L. 2003. TPR proteins: the versatile helix. *Trends in Biochem Sci* **28:** 655–662.

Diodati M, Gill R, Plamann L, Singer M. 2008. Initiation and early developmental events. In *Myxobacteria, multicellularity and differentiation* (ed. D.E. Whitworth), pp. 43–76. ASM Press, Washington D.C.

Durocher D, Jackson SP. 2002. The FHA domain. *FEBS Letts* **513:** 58–66.

Dworkin M. 1962. Nutritional requirements for vegetative growth of *Myxococcus xanthus*. *J Bacteriol* **84:** 250–257.

Dworkin M. 1999. Fibrils as extracellular appendages of bacteria: Their role in contact-mediated cell-cell interactions in *Myxococcus xanthus*. *BioEssays* **21:** 590–595.

Elias-Arnanz M, Fontes M, Panmanabhan S. 2008. Carotinogenesis in *Myxococcus xanthus*: A complex regulatory network. In *Myxobacteria, multicellularity and differentiation* (ed. D.E. Whitworth), pp. 211–225. ASM Press, Washington, D.C.

Ellehauge E, Norregaard-Madsen M, Søgaard-Andersen L. 1998. The FruA signal transduction protein provides a checkpoint for the temporal coordination of intercellular signals in *Myxococcus xanthus* development. *Mol Microbiol* **30:** 807–813.

Evans K, Hobley L, Lambert C, Sockett RE. 2008. Bdellovibrio: Lone hunter cousin of the pack hunting myxobacteria. In *Myxobacteria, multicellularity and differentiation* (ed. D.E. Whitworth), pp. 351–362. ASM Press, Washington, D.C.

Fontes M, Kaiser D. 1999. *Myxococcus* cells respond to elastic forces in their substrate. *Proc Nat Acad Sci* **96:** 8052–8057.

Frye SA, Assalkhou R, Collins RF, Ford RC, Petersson C, Derrick JP, Tonjum T. 2006. Topology of the outer-membrane secretin PilQ from *Neisseria meningitidis*. *Microbiology* **152:** 3751–3764.

Garcia-Heras F, Panmanabhan S, Murillo F, Elias-Arnanz M. 2009. Functional equivalence of HMGA- and histone H1-like domains in a bacterial transcription factor. *Proc Natl Acad Sci* **106:** 13546–13551.

Goldman BS, Nierman WC, Kaiser D, Slater SC, Durkin AS, Eisen JA, Ronning CM, Barbazuk WB, Blanchard M, Field C, et al. 2006. Evolution of sensory complexity recorded in a myxobacterial genome. *Proc Natl Acad Sci* **103:** 15200–15205.

Gronewold TMA, Kaiser D. 2001. The *act* operon controls the level and time of C-signal production for *Myxococcus xanthus* development. *Mol Microbiol* **40:** 744–756.

Gronewold TMA, Kaiser D. 2007. Mutations of the *act* promoter in *Myxococcus xanthus*. *J Bacteriol* **189:** 1836–1844.

Guo D, Wu Y, Kaplan HB. 2000. Identification and characterization of genes required for early *Myxococcus xanthus* developmental gene expression. *J Bacteriol* **182:** 4564–4571.

Hagen DC, Bretscher AP, Kaiser D. 1978. Synergism between morphogenetic mutants of *Myxococcus xanthus*. *Dev Biol* **64:** 284–296.

Hansen JK, Forest KT. 2006. Type IV pilin structures: Insights on shared architecture, fiber assembly, receptor binding and type II secretion. *J Mol Microbiol Biotechnol* **11:** 192–207.

Hartzell P, Kaiser D. 1991a. Function of MglA, a 22-kilodalton protein essential for gliding in *Myxococcus xanthus*. *J Bacteriol* **173:** 7615–7624.

Hartzell P, Kaiser D. 1991b. Upstream gene of the *mgl* operon controls the level of *mglA* protein in *Myxococcus xanthus*. *J Bacteriol* **173:** 7625–7635.

Hartzell P, Shi W, Youdarian P. 2008. Gliding motility of *Myxococcus xanthus*. In *Myxobacteria, multicellularity and differentiation* (ed. D.E. Whitworth), pp. 103–132. ASM Press, Washington, D.C.

Higgs PI, Jagadeesan S, Mann P, Zusman DR. 2008. EspA, an orphan hybrid histidine protein kinase, regulates the timing of expression of key developmental proteins of *Myxococcus xanthus*. *J Bacteriol* **190:** 4416–4426.

Hodgkin J, Kaiser D. 1977. Cell-to-cell stimulation of movement in nonmotile mutants of *Myxococcus*. *Proc Natl Acad Sci* **74:** 2938–2942.

Hodgkin J, Kaiser D. 1979a. Genetics of Gliding Motility in *Myxococcus xanthus* (Myxobacterales): Genes Controlling Movement of Single Cells. *Mol gen Genet* **171:** 167–176.

Hodgkin J, Kaiser D. 1979b. Genetics of Gliding Motility in *Myxococcus xanthus* (Myxobacterales): Two Gene Systems Control Movement. *Mol gen Genet* **171:** 177–191.

Holbrook SR, Kim S-H. 1989. Molecular model of the G protein a subunit based on the crystal structure of the HRAS protein. *Proc Natl Acad Sci* **86:** 1751–1755.

Igoshin O. 2004. "Modeling pattern formation in myxobacteria". PhD thesis in Physics. University of California, Berkeley.

Igoshin O, Goldbetter A, Kaiser D, Oster G. 2004. A biochemical oscillator explains the developmental progression of myxobacteria. *Proc Natl Acad Sci* **101:** 15760–15765.

Igoshin O, Mogilner A, Welch R, Kaiser D, Oster G. 2001. Pattern formation and traveling waves in myxobacteria: Theory and modeling. *Proc Nat Acad Sci* **98:** 14913–14918.

Inouye M, Inouye S, Zusman D. 1979. Gene expression during development of *Myxococcus xanthus*: Pattern of protein synthesis. *Devel Biol* **68:** 579–591.

Inouye S, Nariya H, Munoz-Dorado J. 2008. Protein Ser/Thr kinases and phosphatases in *Myxococcus xanthus*. In *Myxobacteria, multicellularity and differentiation* (ed. D.E. Whitworth), pp. 191–210. ASM Press, Washington, D.C.

Jelsbak L, Kaiser D. 2005. Regulating pilin expression reveals a threshold for type IV pilus assembly in *Myxococcus xanthus*. *J Bacteriol* **187:** 2105–2112.

Jelsbak L, Givskov M, Kaiser D. 2005. Enhancer-binding proteins with a forkhead-associated domain and the sigma54 regulon in *Myxococcus xanthus* fruiting body development. *Proc Natl Acad Sci* **102:** 3010–3015.

Jelsbak L, Søgaard-Andersen L. 2002. Pattern formation by a cell-surface associated morphogen in *Myxococcus xanthus*. *Proc Nat Acad Sci* **99:** 2032–2037.

Kaiser AD. 1979. Social gliding is correlated with the presence of pili in *Myxococcus xanthus*. *Proc Natl Acad Sci* **76:** 5952–5956.

Kaiser AD, Crosby C. 1983. Cell movement and its coordination in swarms of *Myxococcus xanthus*. *Cell Motility* **3:** 227–245.

Kaiser D. 2007. Bacterial swarming, a re-examination of cell movement patterns. *Current Biology* **17:** R561–R570.

Kaiser D. 2008a. Reversing *Myxococcus xanthus* polarity. In *Myxobacteria, multicellularity and differentiation* (ed. D.E. Whitworth), pp. 93–102. ASM Press, Washington, D.C.

Kaiser D. 2008b. *Myxococcus* - from single cell polarity to complex multicellular patterns *Annu Rev Genetics* **42:** 109–130.

Kaiser D. 2009. Are there lateral as well as polar engines for A motile gliding in myxobacteria? *J Bacteriol* **191:** 5336–5341.

Kaiser D, Welch R. 2004. Dynamics of fruiting body morphogenesis. *J Bacteriol* **186:** 919–927.

Kaplan HB, Plamann L. 1996. A *Myxococcus xanthus* cell density-sensing system required for multicellular development. *FEMS Microbiol Letters* **139:** 89–95.

Kearns DB, Shimkets LJ. 2001. Lipid chemotaxis and signal transduction in *Myxococcus xanthus*. *Trends in Microbiol* **9:** 126–129.

Kim SK, Kaiser D. 1990a. Cell alignment required in differentiation of *Myxococcus xanthus*. *Science* **249:** 926–928.

Kim SK, Kaiser D. 1990b. Purification and properties of *Myxococcus xanthus* C-factor, an intercellular signaling protein. *Proc Natl Acad Sci* **87:** 3635–3639.

Kim SK, Kaiser D. 1991. C-factor has distinct aggregation and sporulation thresholds during Myxococcus development. *J Bacteriol* **173:** 1722–1728.

Kimsey HH, Kaiser D. 1991. Targeted disruption of the *Myxococcus xanthus* orotidine 5′-monophosphate decarboxylase gene: Effects on growth and fruiting-body development. *J Bacteriol* **173:** 6790–6797.

Koonin E, Aravind L. 2001. Dynein light chains of the roadblock/LC7 group belong to an ancient protein superfamily implicated in NTPase regulation. *Current Biology* **10:** R774–R776

Kroos L, Hartzell P, Stephens K, Kaiser D. 1988. A link between cell movement and gene expression argues that motility is required for cell-cell signalling during fruiting body development. *Genes and Devel* **2:** 1677–1685.

Kroos L, Inouye S. 2008. Transcriptional regulatory mechanisms during *Myxococcus xanthus* development. In *Myxobacteria, multicellularity and differentiation* (ed. D.E. Whitworth), pp. 149–168. ASM Press, Washington, D.C.

Kruse T, Lobendanz S, Bertheleson NMS, Søgaard-Andersen L. 2001. C-signal: A cell surface-associated morphogen that induces and coordinates multicellular fruiting body morphogenesis and sporulation in *Myxococcus xanthus*. *Mol Microbiol* **40:** 156–168.

Kuhlwein H, Reichenbach H. 1968. *Swarming and morphogenesis in Myxobacteria* Film C893/1965 Institut fur den Wissensch Film, Gottingen, Germany.

Kuspa A, Kroos L, Kaiser D. 1986. Intercellular signaling is required for developmental gene expression in *Myxococcus xanthus*. *Dev Biol* **117:** 267–276.

Kuspa A, Plamann L, Kaiser D. 1992a. Identification of heat-stable A-factor from *Myxococcus xanthus*. *J Bacteriol* **174:** 7360–7369.

Kuspa A, Plamann L, Kaiser D. 1992b. A-signalling and the cell density requirement for *Myxococcus xanthus* development. *J Bacteriol* **174:** 7360–7369.

Leonardy S, Bulyha I, Søgaard-Andersen L. 2008. Reversing cells and oscillating motility proteins. *Mol Biosyst* **4:** 1009–1014.

Leonardy S, Freymark G, Hebener S, Ellehauge E, Søgaard-Andersen L. 2007. Coupling of protein localization and cell movement by a dynamically localized response regulator in *Myxococcus xanthus. EMBO J.* **26:** 4433–4444.

Li S, Lee BU, Shimkets L. 1992. *csgA* expression entrains *Myxococcus xanthus* development. *Genes and Devel* **6:** 401–410.

Li Y, Sun H, Ma X, Lu A, Lux R, Zusman D, Shi W. 2003. Extracellular polysaccharides mediate pilus retraction during social motility of *Myxococcus xanthus. Proc Nat Acad Sci* **100:** 5443–5448.

Lobedanz S, Søgaard-Andersen L. 2003. Identification of the C-signal, a contact-dependent morphogen coordinating multiple developmental responses in *Myxococcus xanthus. Genes and Devel* **17:** 2151–2161.

Lu A, Cho K, Black WP, Duan X-y, Lux R, Yang Z, Kaplan HB, Zusman D, Shi W. 2005. Exopolysaccharide biosynthesis genes required for social motility in *Myxococcus xanthus. Mol Microbiol* **55:** 206–220.

Maier B, Potter L, So M, Seifert HS, Sheetz MP. 2002. Single pilus motor forces exceed 100 pN. *Proc Natl Acad Sci* **99:** 16012–16017.

Manoil C, Kaiser D. 1980a. Accumulation of guanosine tetraphosphate and guanosine pentaphosphate in *Myxococcus xanthus* during starvation and myxospore formation. *J Bacteriol* **141:** 297–304.

Manoil C, Kaiser D. 1980b. Guanosine pentaphosphate and guanosine tetraphosphate accumulation and induction of *Myxococcus xanthus* fruiting body development. *J Bacteriol* **141:** 305–315.

Mattick JS. 2002. Type IV pili and twitching motility. *Annu Rev Microbiol* **56:** 289–314.

Mauriello E, Nan B, Zusman D. 2009b. AglZ regulates adventurous [A-] motility in *Myxococcus xanthus* through its interaction with the cytoplasmic receptor, FrzCD. *Molecular Microbiol* **72:** 964–977.

Mauriello E, Astling DP, Sliusarenko O, Zusman D. 2009a. Localization of a bacterial cytoplasmic receptor is dynamic and changes with cell-cell contacts. *Proc Natl Acad Sci* **106:** 4852–5857.

McBride M, Kohler T, Zusman D. 1992. Methylation of FrzCD, a methyl-accepting taxis protein of *Myxococcus xanthus* is correlated with factors affecting cell behavior. *J Bacteriol* **174:** 4246–4257.

McCleary WR, Zusman DR. 1990a. FrzE of *Myxococcus xanthus* is homologous to both CheA and CheY of *Salmonella typhimurium. Proc Natl Acad Sci* **87:** 5898–5902.

McCleary WR, Zusman DR. 1990b. Purification and characterization of the *Myxococcus xanthus* FrzE protein shows that it has autophosphorylation activity. *J Bacteriol* **172:** 6661–6668.

McCleary WR, McBride MJ, Zusman DR. 1990. Developmental sensory transduction in *Myxococcus xanthus* involves methylation and demethylation of frzCD. *J Bacteriol* **172:** 4877–4887.

Merz AJ, So M, Sheetz MP. 2000. Pilus retraction powers bacterial twitching motility. *Nature* **407:** 98–102.

Mignot T, Merlie JP, Zusman D. 2005. Regulated pole-to-pole oscillations of a bacterial gliding motility protein. *Science* **310:** 855–857.

Mignot T, Shaevitz J, Hartzell P, Zusman D. 2007. Evidence that focal adhesions power bacterial gliding motility. *Science* **315:** 853–856.

Mittal S, Kroos L. 2009a. A combination of unusual transcription factors binds cooperatively to control *Myxococcus xanthus* developmental gene expression. *Proc Natl Acad Sci* **106:** 1965–1970.

Mittal S, Kroos L. 2009b. Combinatorial regulation by a novel arrangement of FruA and MrpC2 transcription factors during *Myxococcus xanthus* development. *J Bacteriol* **191:** 2753–2763.

Nariya H, Inouye S. 2005. Identification of a protein Ser/Thr kinase cascade that regulates essential transcriptional activators in *Myxococcus xanthus* development. *Mol Microbiol* **58:** 367–379.

Nariya H, Inouye S. 2006. A protein Ser/Thr kinase cascade negatively regulates the DNA-binding activity of MrpC, a smaller form of which may be necessary for *Myxococcus xanthus* development. *Mol Microbiol* **60:** 1205–1217.

Nariya H, Inouye M. 2008. MazF, an mRNA interferase, mediates programmed cell death during multicellular *Myxococcus* development. *Cell* **132:** 55–66.

Nesper J, Hill CMD, Paiment A, Harauz G, Beis K, Naismith JH, Whitfield C. 2003. Translocation of group 1 capsular polysaccharide in Escherichia coli serotype K30. *J Biol Chem* **278:** 49763–49772.

Nielson M, Rasmussen A, Ellehauge E, Treuner-Lange A, Søgaard-Andersen L. 2004. HthA, a putative DNA-binding protein, and HthB are important for fruiting body morphogenesis. *Microbiology* **150:** 2171–2183.

Nudleman E, Kaiser D. 2004. Pulling together with type IV pili. *J Molec Microbiol Biotechnol* **7:** 52–62.

Nudleman E, Wall D, Kaiser D. 2005. Cell-to-cell transfer of bacterial outer-membrane lipoproteins. *Science* **309:** 125–127.

Nudleman E, Wall D, Kaiser D. 2006. Polar assembly of the type IV pilus secretin in *Myxococcus xanthus. Mol Microbiol* **60:** 1205–1217.

Nunn D, Bergman S, Lory S. 1990. Products of three accessory genes, *pilB, pilC,* and *pilD,* are required for biogenesis of *Pseudomonas aeruginosa* pili. *J Bacteriol* **172:** 2911–2919.

Pai E, Kabsch W, Krengel U, Holmes KC, John J, Wittinghofer A. 1989. Structure of the guanine-nucleotide-binding domain of the Ha-*ras* oncogene product p21 in the triphosphate conformation. *Nature* **341:** 209–214.

Paiment A, Hocking J, Whitfield C. 2002. Impact of phosphorylation of specific residues in the tyrosine autokinase, Wzc, on its activity in assembly of group 1 capsules in *Escherichia coli J Bacteriol* **184:** 6437–6447.

Pearce M, Mintseris J, Ferreyra J, Gygi S, Darwin KH. 2008. Ubiquitin-like protein involved in the proteasome pathway of *Mycobacterium tuberculosis Science* **322:** 1104–1107.

Pelicic V. 2008. Type IV pili: e pluribus unum? *Mol Microbiol* **68:** 827–837.

Purcell EM. 1977. Life at low Reynolds number. *Am J Phys* **45:** 3–11.

Pérez J, Castañeda-García A, Jenke-Kodama H, Müller R, Muñoz-Dorado J. 2008. Eukaryotic-like protein kinases in the prokaryotes and the myxobacterial kinome. *Proc Natl Acad Sci* **105:** 15950–15955.

Reichenbach H. 1966. *Myxococcus spp (Myxobacteriales) Schwarmentwicklung und Bildung von Protocysten Film E778* Inst fur den Wissensch. Film, Gottingen, Germany.

Reichenbach H. 1968. *Archangium violaceum (Myxobacteriales) Schwarmentwicklung und bildung von protocysten Film E 777/1965* Institut fur den Wissensch. Film, Gottingen, Germany.

Reichenbach H. 1974. *Chondromyces apiculatus (Myxobacteriales) Schwarmentwicklung und morphogenese Film E 779/1965* Institut fur den Wissensch. Film, Gottingen, Germany.

Reichenbach H. 1984. Myxobacteria: a most peculiar group of social prokaryotes. In *Myxobacteria* (ed E. Rosenberg), pp. 1–50. Springer-Verlag, New York.

Reichenbach H. 1993. Biology of the Myxobacteria: Ecology and Taxonomy. In *Myxobacteria II 404 pps* (ed. M. Dworkin and D. Rosenberg), pp. 13–62. ASM Press, Washington D.C.

Reichenbach H, Dworkin M. 1969. Studies on Stigmatella aurantiaca. *J Gen Microbiol* **58:** 3–14.

Rodriguez AM, Spormann AM. 1999. Genetic and molecular analysis of *cglB*, a gene essential for single-cell gliding in *Myxococcus xanthus. J Bacteriol* **181:** 4381–4390.

Rodriguez-Soto JP, Kaiser D. 1997a. Identification and localization of the *tgl* protein, which is required for *Myxococcus xanthus* social motility. *J Bacteriol* **179:** 4372–4381.

Rodriguez-Soto JP, Kaiser D. 1997b. The *tgl* gene: Social motility and stimulation in *Myxococcus xanthus. J Bacteriol* **179:** 4372–4381.

Rolbetski A, Ammon M, Jakovljevic V, Konovalova A, Søgaard-Andersen L. 2008. Regulated secretion of a protease activity activates intercellular signaling during fruiting body formation in M. xanthus. *Developmental Cell* **15:** 627–634.

Sager B, Kaiser D. 1994. Intercellular C-signaling and the traveling waves of Myxococcus. *Genes Devel* **8:** 2793–2804.

Satyshur K, Worzalla G, Meyer L, Heiniger E, Aukema K, Misic A, Forest KT. 2007. Crystal structures of the pilus retraction motor PilT suggests large domain movements and subunit cooperation drive motility. *Structure* **15:** 363–376.

Schneiker S, Perlova O, Alici A, Altmeyer MO, Bartels D, Bekel T, Beyer S, Blöcker H, Bode E, Bode HB, et al. 2007. Complete sequence of the largest known bacterial genome from the myxobacterium *Sorangium cellulosum. Nature Biotechnology* **25:** 1281–1289.

Semmler ABT, Whitchurch CB, Mattick JS. 1999. A re-examination of twitching motility in *Pseudomonas aeruginosa. Microbiology* **145:** 2863–2873.

Shimkets LJ, Gill RE, Kaiser D. 1983. Developmental cell interactions in *Myxococcus xanthus* and the *spoC* locus. *Proc Natl Acad Sci* **80:** 1406–1410.

Simunovic V, Gherardini FC, Shimkets LJ. 2003. Membrane localization of motility, signaling, and polyketide synthase proteins in *Myxococcus xanthus. J Bacteriol* **185:** 5066–5075.

Singer M, Kaiser D. 1995. Ectopic production of guanosine pent-and tetraphosphate can initiate early developmental gene expression in *Myxococcus xanthus. Genes Dev* **9:** 1633–1644.

Sozinova O, Jang Y, Kaiser D, Alber M. 2006. A Three-Dimensional Model of Myxobacterial Fruiting Body Formation. *Proc Natl Acad Sci* **103:** 17255–17259.

Sozinova O, Jang Y, Kaiser D, Alber MS. 2005. Three-dimensional model of myxobacterial aggregation by contact-mediated interaction. *Proc Natl Acad Sci* **102:** 11308–11312.

Spormann AM, Kaiser D. 1999. Gliding mutants of *Myxococcus xanthus* with high reversal frequencies and small displacements. *J Bacteriol* **181:** 2593–2601.

Sproer C, Reichenbach H, Stackebrandt E. 1999. Correlation between morphological and phylogenetic classification of myxobacteria. *Int J Syst Bacteriol* **49:** 1255–1262.

Stanier RY. 1942. Elasticotaxis in Myxobacteria. *J Bacteriol* **44:** 405–412.

Stevens A, Søgaard-Andersen L. 2005. Making waves: Pattern formation by a cell-surface-associated signal. *Trends in Microbiology* **13:** 249–252.

Stock JB, Surette MG, Levit M, Park P. 1995. Two component signal transduction systems: Structure-function relationships and mechanism of catalysis. In *Two component signal transduction* (ed. J.A. Hoch and T.J. Silhavy), pp. 25–51. ASM Press, Washington, D.C.

Strom MS, Nunn DN, Lory S. 1993. A single bifunctional enzyme, PilD, catalyzes cleavage and N-methylation of proteins belonging to the type IV pilin family. *Proc Natl Acad Sci* **90:** 2404–2408.

Sun H, Shi W. 2001. Genetic studies of *mrp*, a locus essential for cellular aggregation and sporulation of *Myxococcus xanthus. J Bacteriol* **183:** 4786–4795.

Søgaard-Andersen L. 2008. Contact-dependent signaling in *Myxococcus xanthus*: The function of the C-signal in fruiting body morphogenesis. (ed. D.E. Whitworth), pp. 77–91. ASM Press, Washington, D.C.

Thaxter R. 1892. On the Myxobacteriaceae, a new order of Schizomycetes. *Bot Gaz* **17:** 389–406.

Tzeng L-F, Singer M. 2005. DNA replication during sporulation in *Myxococcus xanthus* fruiting bodies. *Proc Natl Acad Sci* **102:** 14428–14433.

Tzeng L-F, Ellis TN, Singer M. 2006. DNA replication during aggregation phase is essential for *Myxococcus xanthus* development. *J Bacteriol* **188:** 2774–2779.

Ueki T, Inouye S. 2003. Identification of an activator protein required for the induction of *fruA*, a gene essential for fruiting body development in *Myxococcus xanthus. Proc Natl Acad Sci* **100:** 8782–8787.

Ueki T, Inouye S. 2005a. Activation of a development-specific gene, *dofA*, by FruA, an essential transcription factor for development of *Myxococcus xanthus. J Bacteriol* **187:** 8504–8506.

Ueki T, Inouye S. 2005b. Identification of a gene involved in polysaccharide export as a transcription target of FruA, an essential factor for *Myxococcus xanthus* development. *J Biol Chem* **280:** 32279–32284.

Viswanathan P, Murphy K, Julien B, Garza AG, Kroos L. 2007a. Regulation of *dev*, an operon that includes genes essential for *Myxococcus xanthus* development and

CRISPR-associated genes and repeats. *J Bacteriol* **189:** 3738–3750.

Viswanathan P, Ueki T, Inouye S, Kroos L. 2007b. Combinatorial regulation of genes essential for *Myxococcus xanthus* development involves an upstream response regulator and a downstream LysR-type regulator. *Proc Natl Acad Sci* **104:** 7969–7974.

Ward MJ, Lew H, Zusman DR. 2000. Social motility in *Myxococcus xanthus* requires FrzS, a novel protein with an extensive coiled-coil domain. *Mol Microbiol* **37:** 1357–1371.

Welch R, Kaiser D. 2001. Cell behavior in traveling wave patterns of myxobacteria. *Proc Natl Acad Sci* **98:** 14907–14912.

White DJ, Hartzell PL. 2000. AglU, a protein required for gliding motility and spore maturation of *Myxococcus xanthus*, is related to WD-repeat proteins. *Mol Microbiol* **36:** 662–678.

Whitfield C, Roberts IS. 1999. Structure, assembly and regulation of expression of capsules in *Escherichia coli*. *Mol Microbiol* **31:** 1357–1371.

Wolgemuth C. 2005. Force and flexibility of flailing myxobacteria. *Biophys J* **89:** 945–950.

Wolgemuth C, Hoiczyk E, Kaiser D, Oster G. 2002. How myxobacteria glide. *Current Biology* **12:** 369–377.

Wu SS, Kaiser D. 1997. Regulation of expression of the *pilA* gene in *Myxococcus xanthus*. *J Bacteriol* **179:** 7748–7758.

Wu SS, Wu J, Kaiser D. 1997. The *Myxococcus xanthus* pilT locus is required for social gliding motility although pili are still produced. *Molec Microbiol* **23:** 109–121.

Wu Y, Kaiser D, Jiang Y, Alber M. 2007. Social interactions in myxobacterial swarming. *PLoS Comp Biol* **31:** e253. doi: 210.1371/journal.pcbi.0030253.

Wu Y, Kaiser D, Jiang Y, Alber M. 2009. Periodic reversal of direction allows Myxobacteria to swarm. *Proc Natl Acad Sci* **106:** 1222–1227.

Wugeditsch T, Paiment A, Hocking J, Drummelsmith J, Forrester C, Whitfield C. 2001. Phosphorylation of Wzc, a tyrosine autokinaase, is essential for assembly of group 1 capsular polysaccharides in Escherichia coli. *J Biol Chem* **276:** 2361–2371.

Yang R, Bartle S, Otto R, Stassinopoulos A, Rogers M, Plamann L, Hartzell P. 2004. AglZ is a filament-forming coiled-coil protein required for adventurous gliding motility of *Myxococcus xanthus*. *J Bacteriol* **186:** 6168–6178.

Youderian P, Burke N, White DJ, Hartzell PL. 2003. Identification of genes required for adventurous gliding motility in *Myxococcus xanthus* with the transposable element mariner. *Mol Microbiol* **49:** 555–570.

Yu R, Kaiser D. 2007. Gliding motility and polarized slime secretion. *Molecular Microbiol* **63:** 454–467.

Cyanobacterial Heterocysts

Krithika Kumar[2], Rodrigo A. Mella-Herrera[1,2], and James W. Golden[1]

[1]Division of Biological Sciences, University of California-San Diego, La Jolla, California 92093

[2]Department of Biology, Texas A&M University, College Station, Texas 77843

Correspondence: jwgolden@ucsd.edu

Many multicellular cyanobacteria produce specialized nitrogen-fixing heterocysts. During diazotrophic growth of the model organism *Anabaena* (*Nostoc*) sp. strain PCC 7120, a regulated developmental pattern of single heterocysts separated by about 10 to 20 photosynthetic vegetative cells is maintained along filaments. Heterocyst structure and metabolic activity function together to accommodate the oxygen-sensitive process of nitrogen fixation. This article focuses on recent research on heterocyst development, including morphogenesis, transport of molecules between cells in a filament, differential gene expression, and pattern formation.

Organisms composed of multiple differentiated cell types can possess structures, functions, and behaviors that are more diverse and efficient than those of unicellular organisms. Among multicellular prokaryotes, heterocyst-forming cyanobacteria offer an excellent model for the study of cellular differentiation and multicellular pattern formation. Cyanobacteria are a large group of Gram-negative prokaryotes that perform oxygenic photosynthesis. They have evolved multiple specialized cell types, including nitrogen-fixing heterocysts, spore-like akinetes, and the cells of motile hormogonia filaments. Of these, the development of heterocysts in the filamentous cyanobacterium *Anabaena* (also *Nostoc*) sp. strain PCC 7120 (hereafter *Anabaena* PCC 7120) has been the best studied. Heterocyst development offers a striking example of cellular differentiation and developmental biology in a very simple form: Filaments are composed of only two cell types and these are arrayed in a one-dimensional pattern similar to beads on a string (Figs. 1 and 2).

Many cyanobacterial species are capable of nitrogen fixation. However, oxygenic photosynthesis and nitrogen fixation are incompatible processes because nitrogenase is inactivated by oxygen. Cyanobacteria mainly use two mechanisms to separate these activities: a biological circadian clock to separate them temporally, and multicellularity and cellular differentiation to separate them spatially. For example, the unicellular *Cyanothece* sp. strain ATCC 51142 stores glycogen during the day and fixes nitrogen at night (Toepel et al. 2008), whereas the filamentous *Trichodesmium erythraeum* IMS101 fixes nitrogen during the day in groups of specialized cells (Sandh et al. 2009). Heterocyst-forming cyanobacteria differentiate highly specialized cells

 doi: 10.1101/cshperspect.a000315
Cite this article as *Cold Spring Harb Perspect Biol* 2010;2:a000315

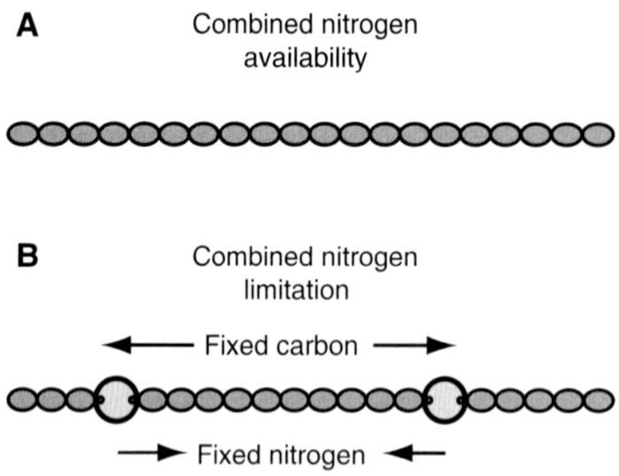

Figure 1. Heterocyst development in *Anabaena* PCC 7120. (*A*) *Anabaena* PCC 7120 grown in medium containing a source of combined nitrogen grows as filaments of photosynthetic vegetative cells. (*B*) In the absence of combined nitrogen, heterocysts differentiate at semiregular intervals, forming a developmental pattern of single heterocysts every 10 to 20 vegetative cells along filaments. Heterocysts are often larger than vegetative cells, have a thicker multilayered envelope, and usually contain cyanophycin granules at their poles adjacent to a vegetative cell.

to provide fixed nitrogen to the vegetative cells in a filament.

In the presence of a source of combined nitrogen such as nitrate or ammonium, *Anabaena* PCC 7120 grows as long filaments containing hundreds of photosynthetic vegetative cells. In the absence of combined nitrogen, it produces heterocysts, which are terminally differentiated nitrogen-fixing cells that form at semiregular intervals between stretches of vegetative cells to produce a multicellular pattern of single heterocysts every ten to twenty vegetative cells along filaments (Figs. 1 and 2). Some heterocyst-forming cyanobacteria show different regulation or display different developmental patterns but these topics are beyond the scope of this article. Heterocyst development involves integration of multiple external and internal signals, communication between the cells in a filament, and temporal and spatial regulation of genes and cellular processes. The study of heterocyst development in *Anabaena* PCC 7120 has proven to be an excellent model for the study of cell fate determination, pattern formation, and differential gene expression during prokaryotic multicellular evelopment. Various aspects of heterocyst development, signaling, and regulation have been the subject of several recent reviews (Meeks and Elhai 2002; Forchhammer 2004; Herrero et al. 2004; Zhang et al. 2006; Aldea et al. 2008; Zhao and Wolk 2008).

Although beyond the scope of this article, it should be noted that cyanobacteria have recently attracted increased attention because of their important roles in environmental carbon and nitrogen fixation (Montoya et al. 2004), and their potential for providing renewable chemicals and biofuels (Dismukes et al. 2008).

CELLULAR DIFFERENTIATION, MULTICELLULARITY, AND TRANSPORT ALLOW SPECIALIZED FUNCTIONS

Cyanobacteria have a Gram-negative cell wall that includes two distinct membranes, the plasma membrane and an outer membrane, and a peptidoglycan layer, which is thicker than in other Gram-negative bacteria, sandwiched between these two membranes (Hoiczyk and Hansel 2000). External to the cell wall is a carbohydrate-enriched glycocalyx that can have different relative amounts of three recognizable layers: a closely associated sheath, a defined capsule, and loosely associated slime. These layers protect the cells from desiccation and presumably from phages and predators. Cyanobacteria contain extensive internal thylakoid membranes (Fig. 2), which are the site of photosynthetic reactions, but these will not be considered further here.

Heterocysts are typically distinguishable from vegetative cells by their somewhat larger and rounder shape, diminished pigmentation, thicker cell envelopes, and usually prominent cyanophycin granules at poles adjacent to vegetative cells (Fig. 2). The additional envelope layers surrounding heterocysts help to protect the enzyme nitrogenase from oxygen (Fay 1992). For details of the heterocyst cell wall and envelope, readers are referred to recent reviews (Awai et al. 2009; Nicolaisen et al. 2009; Pereira et al. 2009). Mature heterocysts provide the microoxic environment required for nitrogen fixation, spatially separating oxygen-evolving photosynthesis in vegetative

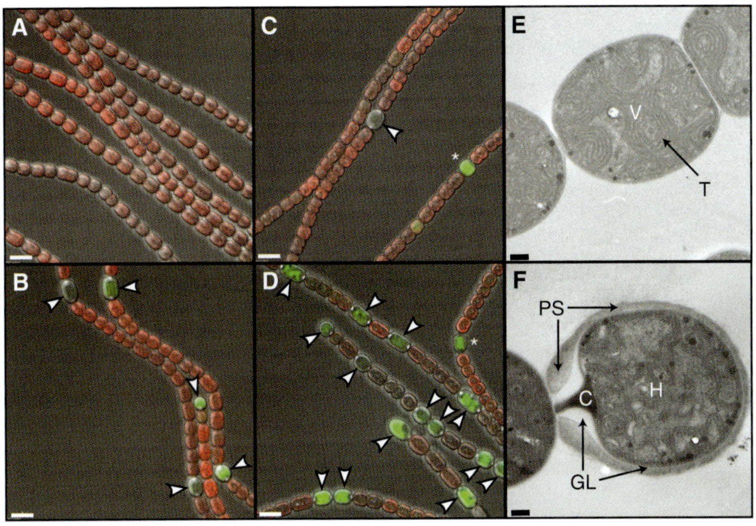

Figure 2. Heterocyst development in *Anabaena* PCC 7120. Filaments of the wild type carrying a *patS-gfp* reporter grown in medium containing nitrate are composed of vegetative cells (*A*), and have undergone heterocyst development 1 d after transfer to medium without combined nitrogen (*B*). A *patS* mutant strain carrying the same *patS-gfp* reporter grown in media containing nitrate contains a small number of heterocysts (*C*), and 1 d after transfer to medium without combined nitrogen shows a higher than normal frequency of heterocysts and an abnormal developmental pattern (*D*). (*A, B, C, D*) Merged DIC (grayscale), autofluorescence of photosynthetic pigments (red), and *patS-gfp* reporter fluorescence (green) microscopic images; arrowheads indicate heterocysts; asterisks indicate proheterocysts; size bar, 5 μm. (*E, F*) Transmission electron micrographs of wild-type vegetative cells (V) and a heterocyst (H) at the end of a filament; T, thylakoid membranes; PS, polysaccharide layer; GL, glycolipid layer; C, polar cyanophycin granule; size bar, 0.2 μm.

cells from nitrogen fixation. Differentiating cells undergo many metabolic and morphological changes (Golden and Yoon 1998). Oxygen-producing photosystem PSII is dismantled during differentiation and heterocysts show an increased rate of respiration (Wolk et al. 1994). Morphological changes include the deposition of two additional envelope layers around the heterocyst: an inner "laminated" layer composed of two heterocyst-specific glycolipids (HGL) and an outer polysaccharide layer (HEP) (Cardemil and Wolk 1979; Cardemil and Wolk 1981; Nicolaisen et al. 2009). The heterocyst envelope polysaccharide layer is sometimes subdivided into a well-defined homogeneous inner layer and an external fibrous layer.

Heterocysts and vegetative cells are mutually interdependent. Because they lack photosystem II and carbon fixation, heterocysts are dependent on vegetative cells for a source of reductant and carbon, which is probably partially supplied as sucrose (Cumino et al. 2007; Marcozzi et al. 2009). In *Anabaena* PCC 7120, vegetative cells must also supply glutamate to heterocysts, which convert it to glutamine and other amino acids (Martin-Figueroa et al. 2000). In return, vegetative cells obtain fixed nitrogen in the form of amino acids from the heterocysts (Meeks and Elhai 2002). The temporal and spatial distributions of fixed carbon and nitrogen were studied using high-resolution nanometer-scale secondary ion mass spectrometry (NanoSIMS), in *Anabaena oscillarioides* (Popa et al. 2007). Newly fixed nitrogen is rapidly exported from heterocysts and distributed to nearby vegetative cells.

The exchange of metabolites and intercellular signals that control the regulated spacing of the heterocysts require movement of molecules between cells along a filament, possibly through a continuous periplasm (Flores et al. 2006). According to this model, molecules exported from one cell would diffuse through the periplasm and then be taken up by other cells along a filament. It was recently shown that GFP expressed from the *patS* promoter, which is expressed at a low basal level in vegetative cells and strongly up-regulated in differentiating heterocysts, and targeted to the periplasm by a cleavable twin-arginine signal peptide, could diffuse through the heterocyst periplasm to nearby vegetative cells (Mariscal et al. 2007). GFP attached to the cytoplasmic membrane was only seen in heterocysts and not in adjacent vegetative cells. However, another group found that GFP targeted to the periplasm of vegetative cells or heterocysts using different cell-type specific promoters (P-*hepA*, P-*patB*, or P-*rbcL*) and the signal sequence of the *Escherichia coli* TorA protein attached to GFP showed no intercellular diffusion from one cell to the next (Zhang et al. 2008). The exported GFP diffused around individual cells but not beyond the cell borders. Although it is unclear why these two research groups came to different conclusions, it may be because of the different signal sequences that were used. The data obtained by Zhang et al. could be the result of a failure to cleave the TorA signal peptide from the GFP reporter, which could result in the GFP being anchored to the membrane or localized to the space on the inner side of the peptidoglycan layer. Electron micrographs of *Anabaena* PCC 7120 intercellular junctions appear to show an intact peptidoglycan layer around each cell and sometimes a distinct "junctional space" between these peptidoglycan layers (Fig. 3) is observed, which could be a barrier for secreted large molecules.

Recent data support the exchange of molecules through intercellular junctions or channels directly connecting the cytoplasm of adjacent cells (Mullineaux et al. 2008). Calcein, a small fluorescent molecule loaded into the cytoplasm of cells, was found to quickly diffuse between cells. Electron micrographs suggest the existence of connections between adjacent cells named microplasmodesmata (Fig. 3) (Giddings and Staehelin 1978; Giddings and Staehelin 1981). The microplasmodesmata may be channels formed by protein oligomers, and FraG (SepJ) has been suggested as a candidate channel forming protein (Flores et al. 2007; Nayar et al. 2007). FraG has a large extracytoplasmic domain that may be involved in spanning the cell wall and bridging the gap between adjacent cells (Flores et al. 2007). Localization studies with a FraG-GFP reporter showed that it is expressed in both cell types and that it is localized at intercellular septa (Flores et al. 2007). *fraG* mRNA levels increase after nitrogen depletion and *fraG* null mutants are unable to differentiate heterocysts completely (Flores et al. 2007; Nayar et al. 2007). The diffusion of calcein required FraG, suggesting that FraG could be the channel forming protein connecting the cytoplasm of two adjacent cells (Mullineaux et al. 2008).

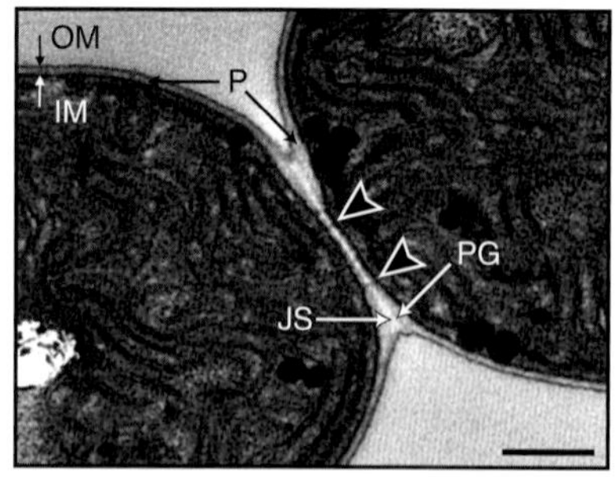

Figure 3. Transmission electron micrograph of the junction between two vegetative cells. Arrowheads indicate microplasmodesmata, which are potential cell-to-cell channels. Note the "junctional space" between the cell wall peptidoglycan layers of the two cells, which may indicate a partial barrier between the periplasmic compartments of adjacent cells. IM, inner membrane; OM, outer membrane; P, periplasm; PG, peptidoglycan cell wall; JS, junctional space. Size bar, 0.5 μm.

GENE EXPRESSION AND MORPHOGENESIS DURING HETEROCYST DEVELOPMENT

The timeline of heterocyst development begins with sensing combined-nitrogen limitation and culminates with nitrogen fixation in the mature heterocyst. Heterocyst development is complete in about 20 hours at 30°C and involves cellular differentiation of selected vegetative cells into heterocysts and less obvious changes in gene expression and metabolism in the remaining vegetative cells. Several studies using DNA microarray methods show global changes in gene expression after nitrogen step-down (Ehira et al. 2003; Sato et al. 2004; Ehira and Ohmori 2006b; Campbell et al. 2007). The process of differentiation is reversible if a source of combined nitrogen or the inhibitory PatS pentapeptide is added within 9–12 hours after nitrogen deprivation, after which point the cells are committed to forming heterocysts (Fig. 4) (Thiel and Pratte 2001; Yoon and Golden 2001).

Initiation and Early Stages of Heterocyst Development

The presence of a source of combined nitrogen such as ammonium or nitrate inhibits the differentiation of heterocysts. In cyanobacteria, 2-oxoglutarate, an intermediate in the Krebs cycle, constitutes the signal for nitrogen deprivation (Laurent et al. 2005; Zhang et al. 2006). The Krebs cycle in cyanobacteria is incomplete because of the lack of 2-oxoglutarate dehydrogenase. As a result, 2-oxoglutarate's main function is to serve as a precursor in a variety of biosynthetic reactions. It is the primary carbon skeleton for incorporation of ammonium and is considered the metabolic junction between carbon and nitrogen balance in cyanobacteria (Muro-Pastor et al. 2001; Vazquez-Bermúdez et al. 2003; Muro-Pastor et al. 2005). Nitrogen-limiting conditions result in an increase in the levels of 2-oxoglutarate. An artificial analog of 2-oxoglutarate, 2,2-difluoropentanoic acid, DFPA, added to medium resulted in heterocyst development even in the presence of ammonium, showing that 2-oxoglutarate plays a key role in controlling heterocyst development (Laurent et al. 2005).

NtcA, a transcriptional regulator belonging to the CRP (cyclic AMP receptor protein) family of proteins, senses 2-oxoglutarate levels (Fig. 5). The protein is conserved in all cyanobacteria and regulates a number of genes involved in carbon and nitrogen metabolism (Herrero et al. 2004; Marcozzi et al. 2009). In *Anabaena* PCC 7120, NtcA is required for the expression of the genes in pathways for ammonium and nitrate assimilation, as well as heterocyst development (Wei et al. 1993; Ramasubramanian et al. 1994). The *ntcA* gene is induced soon after nitrogen deprivation and is autoregulated (Ramasubramanian et al. 1994; Ramasubramanian et al. 1996; Muro-Pastor et al. 2002). *ntcA* mutants are unable to use nitrate as the sole source of nitrogen and are blocked from initiating heterocyst development (Frias et al. 1994; Wei et al. 1994). Genes that are activated by NtcA typically have the consensus binding site TGTA-(N_8)-TACA centered at −41.5 nucleotides upstream of the transcription start point (TSP) (Herrero et al. 2004). The DNA binding activity of NtcA is enhanced in the presence of 2-oxoglutarate, and 2-oxoglutarate is necessary for transcriptional activation by NtcA (Tanigawa et al. 2002; Vazquez-Bermúdez et al. 2002). Additionally, DFPA, the synthetic analogue of 2-oxoglutarate, stimulates DNA binding activity of NtcA in vitro (Laurent et al. 2005; Chen et al. 2006).

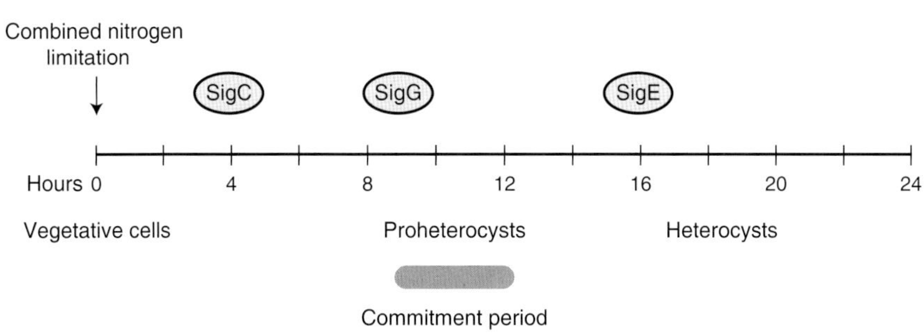

Figure 4. Developmentally regulated σ factors in *Anabaena* PCC 7120. The positions of the σ factor icons mark the times that GFP transcriptional reporters for the σ factor genes are up-regulated during heterocyst development. Filaments are composed of vegetative cells before nitrogen limitation. Immature proheterocysts are observed during approximately the same period of time when cells become committed to complete differentiation. Mature heterocysts are present by 20 h. See text for details.

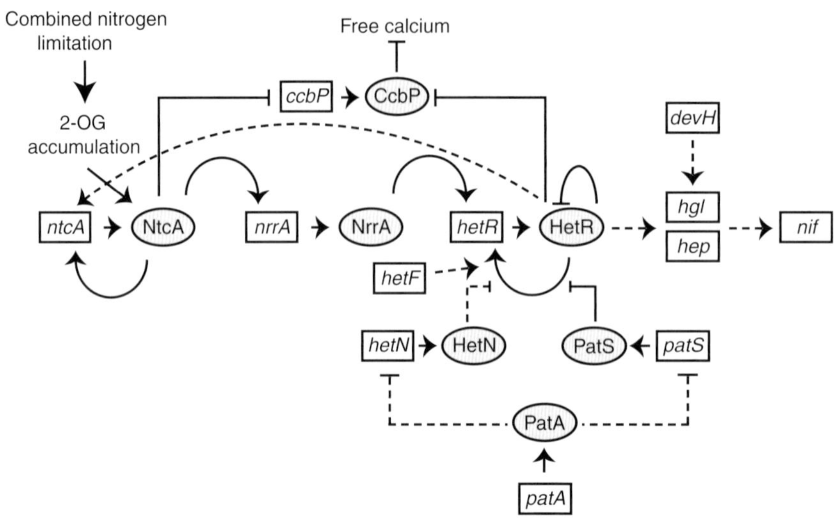

Figure 5. Model of regulatory interactions during heterocyst development. For clarity, the figure shows only selected genes, proteins, and events. Open boxes represent genes and gray ovals represent proteins. Lines ending in arrows and bars indicate positive and negative interactions, respectively. Dashed lines represent indirect and/or unknown interactions or missing steps. Short arrows are between genes and their products. See the text for details.

HetR is a master regulator of heterocyst development and plays a key role in differentiation and pattern formation (Buikema and Haselkorn 1991b). Null mutants of *hetR* fail to produce heterocysts, and overexpression of *hetR* (Buikema and Haselkorn 1991a; Buikema and Haselkorn 2001), and particular point mutants (Khudyakov and Golden 2004) result in increased heterocyst frequency. *hetR* is one of the earliest genes induced in differentiating cells and is positively autoregulated (Fig. 5) (Black et al. 1993; Buikema and Haselkorn 2001). Transcription of *hetR* increases as early as 30 minutes after nitrogen deprivation, and by 3.5 hours, expression is confined to spaced foci arranged in a pattern similar to that of differentiating cells.

HetR protein is a serine type protease that has autoprotease activity and DNA binding activity that requires formation of a HetR homodimer (Zhou et al. 1998; Huang et al. 2004). The heterocyst inhibitory peptide PatS interferes with HetR DNA-binding activity in vitro (Huang et al. 2004). Mutations that affect these activities block heterocyst development at an early stage; however, the role of particular amino acid residues is unclear (Dong et al. 2000; Risser and Callahan 2007). A $hetR_{R223W}$ point mutant is insensitive to the main inhibitory signals of pattern formation, PatS, and HetN (Fig. 5), and produces a conditionally lethal phenotype because of complete differentiation under nitrogen limiting conditions (Khudyakov and Golden 2004).

Expression of *ntcA* and *hetR* show a mutual dependency during heterocyst development (Muro-Pastor et al. 2002). *hetR* is not induced in an *ntcA* mutant and *ntcA* expression is transiently induced in a HetR-dependent manner (Frias et al. 1994; Muro-Pastor et al. 2002). Expression of some genes involved in the developmental process is dependent on both *ntcA* and *hetR*; however, in some cases, the dependency is probably because of the interdependent up-regulation of *hetR* and *ntcA* (Hebbar and Curtis 2000; Fiedler et al. 2001; Valladares et al. 2004). NrrA, a response regulator, has been identified as the regulatory link between NtcA and HetR (Fig. 5) (Ehira and Ohmori 2006a). *nrrA* is transcribed in differentiating cells within 3 hours after nitrogen deprivation and is directly dependent on NtcA (Ehira and Ohmori 2006b; Muro-Pastor et al. 2006). Earlier work on the transposon reporter strain TLN14, now known to be a fusion to the *nrrA* gene, showed rapid induction 1 hour after nitrogen stepdown (Cai and Wolk 1997). An *nrrA* mutant strain shows a delay in heterocyst development caused by a delay in accumulation of HetR, and extra copies of *nrrA* result in increased expression of *hetR*, and thus, increased heterocyst frequency (Ehira and Ohmori 2006a; Ehira and Ohmori 2006b).

In *Anabaena* PCC 7120, the calcium-binding luminescent protein aequorin was used to detect an increase in intracellular calcium levels after nitrogen deprivation (Torrecilla et al. 2004). The Ca^{++} reporter obelin was used to show a 10-fold higher Ca^{++} concentration in heterocysts compared with vegetative cells at 4 hours after nitrogen deprivation (Zhao et al. 2005). The increase in Ca^{++} concentrations is because of the decreased expression of *ccbP*, which encodes a calcium sequestering protein, CcbP (Fig. 5) (Zhao et al. 2005). *ccbP* message is down-regulated in differentiating cells and absent in mature heterocysts. Inactivation of *ccbP* causes a multiple-contiguous-heterocyst (Mch) phenotype and overexpression inhibits heterocyst development. It has been hypothesized that a regulatory pathway consisting of HetR, CcbP, and NtcA controls intracellular free calcium; HetR specifically degraded CcbP in a calcium-dependent manner, and *ccbP* down-regulation required 2-oxoglutarate-dependent binding of NtcA to its promoter (Shi et al. 2006). The increase in Ca^{++} in differentiating cells is thought to be important for HetR's Ca^{++}-dependent serine protease and/or other Ca^{++}-dependent proteolytic activities.

HetF influences heterocyst development by a positive effect on *hetR* expression (Fig. 5) (Wong and Meeks 2001; Risser and Callahan 2008). In mutants of *hetF*, expression of *hetR* was not localized to heterocysts and initiation of heterocyst development was not seen (Wong and Meeks 2001). Overexpression of *hetF* produces a multiple-contiguous-heterocyst phenotype, but only in the absence of combined nitrogen (Wong and Meeks 2001). Like HetR, HetF is likely to be a protease (Risser and Callahan 2008). An elegant mosaic analysis was used to show that HetF is required specifically in cells that differentiate (Risser and Callahan 2008). However, it is also thought that HetF plays a role in restricting *hetR* expression and the accumulation of HetR protein to differentiating cells (Wong and Meeks 2001).

Like *hetF*, *patA* also influences heterocyst development via a positive effect on *hetR* expression, but its effect is related to pattern formation (Fig. 5) (Liang et al. 1992; Risser and Callahan 2008). The *patA* gene encodes a response regulator similar to the CheY protein, which functions as a phosphorylation-activated switch (Liang et al. 1992). *patA* mutants form heterocysts almost exclusively at the ends of filaments. This mutant phenotype is maintained even when *hetR* is overexpressed in a *patA* mutant background, suggesting that *patA* acts downstream of *hetR* (Liang et al. 1992; Buikema and Haselkorn 2001). It is possible that PatA influences heterocyst development by attenuating the negative effects of the main inhibitory signals of heterocyst pattern formation, PatS and HetN (Orozco et al. 2006).

The *hetC* gene, which encodes a member of the family of ATP-binding cassette type exporters, is required for an early step in the differentiation of heterocysts as observed by a P_{hetC}-*gfp* reporter, which showed an increase in expression in proheterocysts and heterocysts (Khudyakov and Wolk 1997; Muro-Pastor et al. 1999). A *hetC* mutant carrying a P_{hetR}-*gfp* reporter shows a pattern of weakly fluorescent cells that are blocked from further stages of development (Xu and Wolk 2001; Wang and Xu 2005). It is possible that the *hetC* mutant fails to complete an early step in morphogenesis of the envelope that then triggers a developmental checkpoint that prevents further differentiation.

Two novel genes, *hetL* and *asr1734*, have been shown to be involved in regulating heterocyst development, but their exact roles and biochemical functions remain unclear. HetL is a pentapeptide-repeat protein composed almost entirely of 40 tandem pentapeptide repeats forming 10 complete coils (Ni et al. 2009). The *hetL* gene was isolated in a genetic screen designed to identify genes involved in PatS signaling (Liu and Golden 2002). Overexpression of *hetL* in a *patS*-overexpression strain allows heterocysts to form by bypassing the PatS inhibitory signal. Overexpression of *hetL* in the wild type produces a multiple-contiguous-heterocyst phenotype. *hetL* overexpression even induces partial heterocyst development in an *ntcA*-null mutant. A *hetL*-null mutant shows normal heterocyst development and diazotrophic growth, indicating that *hetL* plays a nonessential role in heterocyst development.

The *asr1734* gene is found in only heterocyst-forming cyanobacteria and a $P_{asr1734}$-*gfp* reporter showed localized expression in proheterocysts and heterocysts after nitrogen stepdown. An *asr1734* knockout mutant shows elevated levels of *ntcA* mRNA and forms 15% heterocysts and a weak Mch phenotype in media without fixed nitrogen. Overexpression of *asr1734* inhibits heterocyst development in the wild type and in two genetic backgrounds that stimulate heterocyst formation, a *patS* null and a $hetR_{R223W}$ mutant, suggesting that Asr1734 acts downstream of PatS and HetR (Wu et al. 2007).

Cyclic-di-GMP signaling appears to be involved in heterocyst development. The *all2874* gene, which encodes a diguanylate cyclase, is required for normal heterocyst development during growth under conditions of high light intensity (Neunuebel and Golden 2008). An *all2874* mutant shows a significant reduction in heterocyst frequency and reduced vegetative cell size. An *all2874* mutant strain carrying a P_{patS}-*gfp* transcriptional reporter did not show normal up-regulation of the reporter, indicating that the decrease in heterocyst frequency is because of an early block in differentiation.

Heterocyst Cellular Differentiation Produces an Environment for Nitrogen Fixation

The middle and later stages of heterocyst development are distinguished by structural and physiological changes. These changes begin with morphogenesis of the heterocyst envelope by the deposition of an outer polysaccharide layer and an inner glycolipid layer, which decrease the entry of oxygen into the heterocyst (Fig. 2) (Fay 1992). Mutants that lack the envelope polysaccharide or the glycolipid layer are unable to grow diazotrophically in the presence of air (Wolk et al. 1988; Wolk 1996; Fan et al. 2005; Huang et al. 2005; Nicolaisen et al. 2009).

Deposition of the external polysaccharide layer is one of the earliest morphological changes during heterocyst differentiation. DevR and HepK, which comprise a two component regulatory system, are involved in biosynthesis of the polysaccharide layer (Zhou and Wolk 2003). The *hep* genes *hepA*, *hepB*, and *hepC*, and a cluster of genes present around *hepA* are required for the deposition of the polysaccharide layer (Zhu et al. 1998; Huang et al. 2005; Wang et al. 2007).

The heterocyst glycolipid layer is assembled beneath the polysaccharide layer and is composed of fatty alcohols glycosidically linked to sugar residues. The *hglB*, *hglC*, *hglD*, and *hglE* genes along with a cluster of nearby genes are required for the synthesis of these glycolipids (Campbell et al. 1997; Fan et al. 2005). DevH, a trans-acting regulatory protein, is required for the formation of the glycolipid layer, either by directly regulating the expression of the genes or indirectly through other gene products (Fig. 5) (Ramirez et al. 2005). The *hglK* gene is required for the localization of the glycolipids and may be directly involved in their deposition (Black et al. 1995). ORF *all5341*, named *hglT*, is predicted to encode a glycosyl transferase and is required for the formation of the glycolipid layer (Awai and Wolk 2007).

Differentiation and maturation of heterocysts is dependent on DevBCA and HgdD, which are thought to be a glycolipid exporter and outer membrane efflux tunnel, respectively (Fiedler et al. 1998; Moslavac et al. 2007a). The autolysin HcwA, which is presumably involved in cell wall remodeling, is required for heterocyst maturation (Zhu et al. 2001). The combination of autolysin activity and turgor pressure is presumably responsible for the slightly larger and rounder shape of heterocysts. The regulatory genes *hepK*, *hepN*, *henR*, and *hepS* are also required for normal heterocyst maturation (Lechno-Yossef et al. 2006).

Recent proteomic studies of the heterocyst and vegetative cell wall have identified proteins in the inner and outer membranes as a step toward understanding the dynamics of the cell wall proteome during cell differentiation

(Moslavac et al. 2005; Moslavac et al. 2007b). Analysis of proteins of the outer membrane of vegetative cells and of heterocysts revealed a high similarity, suggesting that the relative protein concentrations within the outer membrane may vary between the two cell types but not the composition (Nicolaisen et al. 2009).

The cell-wall-related proteins can be categorized as proteins involved in (1) signal transduction, (2) synthesis, and (3) transport. Among the signal transduction proteins in *Anabaena* PCC 7120, histidine kinases and protein phosphatases have recently been shown to act downstream of NtcA and have been suggested to be involved in heterocyst cell wall formation (Wang et al. 2002; Cheng et al. 2006). In *Anabaena* PCC 7120, 131 genes encode putative histidine kinases, 52 genes code for serine/threonine and tyrosine kinases, and 13 genes code for a group of histidine and Ser/Thr hybrid (HSTKI) proteins (Ohmori et al. 2001; Phalip et al. 2001). Two genes, *pkn30* and *pkn44*, that encode HSTKI proteins are involved in regulation of HGL2 synthesis (Shi et al. 2007). A single mutant of either kinase has no obvious phenotype, but a double mutant has an immature HEP layer and is defective in synthesis of the HGL2. PrpJ, a Ser/Thr phosphatase encoded by *all1731*, is involved in the regulation of the HGL1 (Jang et al. 2007).

Among the proteins that define cell wall structure, OmpA, encoded by *all3289* and *alr5049*, acts as a physical linker between the outer membrane and the peptidoglycan layer. The *alr1278* gene encoding OstA/Imp is thought to be involved in remodeling of the lipid content of the outer membrane during heterocyst development (Nicolaisen et al. 2009). OstA/Imp is involved in transport of lipid A to the cell surface (Bos et al. 2004). Expression analysis showed that *alr1278* was up-regulated early following nitrogen deprivation (Ehira and Ohmori 2006b).

The outer membrane of *Anabaena* PCC 7120 includes integral membrane proteins and lipoproteins attached to the outer membrane by amino-terminal lipids (Nicolaisen et al. 2009). These proteins comprise porin-type proteins, proteins that define the cell wall structure, and proteins involved in transport and assembly (Nicolaisen et al. 2009). In *Anabaena* PCC 7120, All4499 and All4550, which show homology with the outer membrane porin OprB of *Pseudomonas aeruginosa*, were identified in membranes of vegetative cells and were the most abundant outer membrane proteins in heterocysts (Moslavac et al. 2005; Moslavac et al. 2007b). Genome-wide expression analysis of genes after nitrogen deprivation showed an increase in expression of *all4499* 8 hours after nitrogen deprivation and a decrease in expression of *all4550* 24 hours after nitrogen depletion. In addition, ORF *alr0834*, classified as encoding a general porin, is up-regulated during the early stages of heterocyst differentiation (Ehira and Ohmori 2006b).

A large number of outer membrane proteins are involved in protein transport and assembly. *all0406* encodes a protein with high similarity to autotransporters, which are Gram-negative extracellular proteins that are characterized by the ability to translocate themselves across the outer membrane (Nicolaisen et al. 2009). HgdD, a TolC like protein, is an outer membrane efflux channel and functions in HGL-layer assembly. The *hgdD* gene is up-regulated during heterocyst development and the protein is essential for HGL-layer formation. A *hgdD* knockout mutant is defective in deposition of the HGL layer and the phenotype is similar to the defect observed for a *devBCA* knockout. These data suggest the formation of a DevBCA-HgdD secretion complex that is essential for the formation of the HGL layer (Moslavac et al. 2007a).

Nitrogen Fixation in Heterocysts

Heterocyst development culminates with nitrogen fixation, which is the ATP-dependent process of reducing atmospheric nitrogen to ammonia by the enzyme nitrogenase. In many heterocyst-forming cyanobacteria, including *Anabaena* PCC 7120, nitrogenase is synthesized only in heterocysts, where it is protected from irreversible inactivation by oxygen (Fay 1992). However, some heterocyst-forming cyanobacteria contain an additional set of *nif* genes that allows nitrogen fixation in vegetative cells under anoxic conditions (Thiel et al. 1995). Nitrogenase is a well-conserved enzyme in all nitrogen-fixing organisms and consists of two components: dinitrogenase (Mo-Fe protein) and dinitrogenase reductase (Fe protein) (Rubio and Ludden 2008). Dinitrogenase reductase serves to transfer electrons from electron donors such as ferredoxin or flavodoxin to dinitrogenase. In *Anabaena* sp. PCC 7120, the *fdxH* gene codes for a heterocyst-specific ferredoxin (Masepohl et al. 1997). In addition to the common molybdenum nitrogenase, some cyanobacteria contain an alternative nitrogenase that uses a vanadium cofactor (Pratte et al. 2006).

In *Anabaena* PCC 7120, the nitrogen-fixation (*nif*) genes are expressed specifically in heterocysts late during development, between 18 and 24 hours after nitrogen deprivation (Elhai and Wolk 1990; Golden et al. 1991). The *nifHDK* operon encodes the molybdenum-containing nitrogenase enzyme complex. Upstream of the *nifHDK* operon is another *nif* operon, *nifB-fdxN-nifS-nifU* (Mulligan and Haselkorn 1989). The *nifVZT* genes form a *nif* gene cluster separated from the main *nif* gene cluster (Stricker et al. 1997). Other cyanobacterial *nif* genes include *nifE*, *nifN*, *nifX*, and *nifW* (Wolk et al. 1994; Thiel et al. 1995). The transcriptional regulators that control the expression of the *nif* genes in cyanobacteria have not been identified.

A heterocyst-specific uptake hydrogenase, encoded by the *hupSL* operon, recovers reductant from H_2 produced by nitrogenase (Tamagnini et al. 2007). In *Nostoc punctiforme*, the upstream region of the *hupSL* operon contains an NtcA binding site and NtcA binds to the *hupSL* promoter in vitro, but deletion of the binding site did not affect the expression of *hupSL* significantly (Holmqvist et al. 2009). This study showed that sequences required for heterocyst-specific expression of the *hupSL* genes are located in a region starting 57 base pairs upstream of the transcription start point and extending 258 base pairs downstream. Whereas NtcA is not required for transcriptional regulation of *hupSL* in *N. punctiforme*, NtcA is required for the induction of *hupSL* in *Anabaena variabilis* after nitrogen step-down, as shown by reduced induction of *hupSL* in an *ntcA* mutant (Weyman et al. 2008). A consensus NtcA-binding site is upstream of the *A. variabilis hupSL* operon, and NtcA binds to this region (Weyman et al. 2008).

Heterocyst development is accompanied by changes in the photosynthetic apparatus and carbon metabolism to provide ATP and low potential reductant for nitrogen fixation (Wolk et al. 1994; Meeks and Elhai 2002). Respiration by cytochrome c oxidase generates ATP for nitrogen fixation and helps create the microoxic environment. Whereas *cox1* encoding mitochondrial type cytochrome c oxidase is expressed in vegetative cells, *cox2* and *cox3* are expressed only in heterocysts and are required for growth during nitrogen fixation (Jones and Haselkorn 2002; Valladares et al. 2003).

In *Anabaena* PCC 7120, three DNA rearrangements take place during the late stages of heterocyst development that affect nitrogen fixation and uptake hydrogenase operons (Golden et al. 1985; Golden et al. 1988; Carrasco et al. 1995; Carrasco and Golden 1995; Golden 1997; Carrasco et al. 2005). The rearrangements are the result of developmentally programmed site-specific recombination between direct repeats that flank the DNA elements that are deleted from the chromosome (Golden 1997). One rearrangement excises an 11-kb element from the *nifD* gene (Golden et al. 1985; Brusca et al. 1989). The excision requires *xisA* located on the 11-kb element (Golden and Wiest 1988; Brusca et al. 1990; Lammers et al. 1990; Henson et al. 2008). A *xisA* null mutant forms heterocysts but is unable to excise the element or grow on media lacking a source of combined nitrogen (Golden and Wiest 1988; Golden et al. 1991). A second rearrangement excises a 55-kb element from the *fdxN* gene (Golden et al. 1988) and requires the *xisF*, *xisH*, and *xisI* genes (Carrasco et al. 1994; Ramaswamy et al. 1997). A third programmed DNA arrangement deletes a 10.5-kb element from the *hupL* gene and requires the *xisC* recombinase gene (Carrasco et al. 1995; Carrasco et al. 2005). All three DNA elements appear to be parasitic DNA sequences that are passed to daughter vegetative cells as they are carried innocuously in genes that are only required in heterocysts. And, all three DNA elements have acquired the ability to excise from the chromosome in terminally differentiated heterocysts so that the genes in which they reside can be expressed. This developmental regulation has apparently evolved independently for each element but the regulatory mechanism is not yet known.

RNA Polymerase σ Factors are Developmentally Regulated

The control of gene expression in bacteria is primarily regulated at the level of transcription initiation. Bacteria contain an RNA polymerase (RNAP) enzyme that requires an additional σ factor for promoter recognition and transcription initiation. In many cases, a bacterium uses specific sigma (σ) factors to activate a set of genes in response to environmental and intracellular signals (Campbell et al. 2008). For example, in *Bacillus subtilis*, temporally and spatially regulated σ factors modulate the expression of genes during starvation-induced sporulation (Kroos 2007; Campbell et al. 2008). It was hypothesized that global changes in gene expression during heterocyst development in *Anabaena* PCC 7120 are regulated by σ factors (Khudyakov and Golden 2001; Aldea et al. 2007). The σ 70 family of σ factors has been divided into four major groups: Group 1 is responsible for transcription of housekeeping genes, group 2 is similar to group 1 but not essential under laboratory growth conditions, group 3 is involved in the expression of genes associated with sporulation, motility, stress response, and heat shock, and group 4 participates in multiple biological processes (Campbell et al. 2008). *Anabaena* PCC 7120 has 12 putative σ factors identified by sequence similarity (Aldea et al. 2007; Yoshimura et al. 2007). Genetic analysis has not identified any σ factor genes that are essential for heterocyst differentiation or nitrogen fixation, suggesting that there is a level of functional redundancy between the σ factors (Khudyakov and Golden 2001).

Genetic analysis using *gfp* as a reporter identified three σ-factor genes, *sigC*, *sigG*, and *sigE*, that are developmentally up-regulated after nitrogen deprivation (Aldea et al. 2007). The expression of *sigC* is up-regulated 4 hours after nitrogen step-down, and by 10 hours, the expression is restricted mostly to single cells in a heterocyst-like pattern (Fig. 4). SigC may be involved in regulating early heterocyst-specific genes (Aldea et al. 2007). The expression of *sigE* (alr4249) is low in vegetative cells grown in nitrate-containing media and is up-regulated in individual cells at about 16 hours after nitrogen depletion, suggesting that SigE may be involved in the expression of late heterocyst-specific genes such as the *nif* and *hup* genes. Expression of a *sigG* reporter is seen in vegetative cells in media containing nitrate but decreases in all cells soon after nitrogen depletion. At around 9 hours after nitrogen step-down, increased *sigG* expression is localized to single cells and by 16 hours, expression is localized to morphologically distinguishable proheterocyst cells; at 24 hours, heterocysts show somewhat decreased reporter expression. These data suggest that SigG is involved in the expression of genes during the middle stages of differentiation such as those involved in morphogenesis and the creation of a microoxic environment, and possibly genes that are necessary for commitment to complete differentiation (Aldea et al. 2007).

HETEROCYST PATTERN FORMATION REQUIRES DYNAMIC SIGNALING

Pattern formation has been observed in many prokaryotic systems. For example, *Myxococcus xanthus* forms a complex three-dimensional pattern when starving bacteria self-organize to form fruiting bodies (Jelsbak and Søgaard-Andersen 2002), and *Pseudomonas aeruginosa* produces biofilms during infection (Klausen et al. 2003). Temporary patterns have been studied in *Escherichia coli* and *Salmonella typhimurium* when these organisms swim in gradients of nutritional chemoattractant (Blair 1995; Budrene and Berg 1995). The developmentally regulated pattern of heterocysts in multicellular cyanobacteria offers a model to understand biological pattern formation, signals that control pattern formation, and downstream regulatory pathways.

A longstanding model of heterocyst pattern formation is that differentiating cells produce an inhibitory signal that diffuses along a filament to form a gradient that inhibits differentiation of neighboring cells in a concentration-dependent manner (Wilcox et al. 1973; Wolk 1996). The *patS* gene in *Anabaena* PCC 7120 is essential for normal pattern formation and is predicted to encode a 13- to 17-amino-acid peptide depending on the start codon used in vivo (Fig. 5) (Yoon and Golden 1998; Yoon and Golden 2001). All filamentous cyanobacteria for which a genome sequence is available contain *hetR* and *patS* genes, and although the *patS* genes encode peptides from 13 to 90 amino acids, they all have an identical five amino acid sequence, RGSGR, at their carboxy-terminal ends (Zhang et al. 2009). This PatS pentapeptide sequence may be the essential element of the predicted diffusible inhibitor controlling, at least in part, heterocyst pattern.

The *patS* gene is transcribed early during heterocyst development. Expression analysis using northern RNA blots and a *patS-lacZ* reporter showed that *patS* is up-regulated by 6 hours after nitrogen deprivation (Yoon and Golden 1998). A *patS-gfp* reporter strain in nitrate-containing media showed low levels of expression in all cells. At 8–10 hours after nitrogen step-down, increased fluorescence was localized to small groups of cells or individual cells (Yoon and Golden 2001). By 12–14 hours after nitrogen step-down, most of the GFP signal was localized in individual cells that were arranged in a pattern similar to that of mature heterocysts, and by 18 hours, the fluorescence was localized almost entirely to heterocysts (also see Fig. 2).

A *patS* null mutant produces heterocysts even in the presence of nitrate, and forms multiple contiguous heterocysts after nitrogen deprivation with up to 30% of the cells differentiating to form heterocysts, whereas extra copies or overexpression of *patS* results in a complete suppression of heterocysts (Yoon and Golden 1998; Yoon and Golden 2001). Mutations in the last five carboxy-terminal residues of PatS result in a loss-of-function phenotype. A synthetic RGSGR pentapeptide, PatS-5, blocks heterocyst differentiation at nanomolar concentrations, whereas a four amino acid synthetic peptide did not inhibit differentiation (Yoon and Golden 1998). Addition of PatS-5 to growth medium of a *patS* null mutant at a concentration that decreases the frequency of heterocysts to the wild-type level does not restore a wild-type heterocyst pattern, but ectopic expression of *patS* from a heterocyst-specific promoter, *hepA*, in a *patS* null mutant background does restore the normal pattern. These data are consistent with PatS functioning as a diffusible inhibitor and that a gradient of PatS is required for the establishment of a normal pattern (Golden and Yoon 2003). However, other somewhat more complicated mechanisms are possible and direct support for this hypothesis would require, for example, the development of methods to detect gradients of small molecules in cyanobacterial filaments.

The PatS signal must move from differentiating cells to neighboring cells to inhibit HetR. However, a *patS5* minigene expressed in heterocysts in a *patS* null background does not produce a normal pattern, indicating that PatS-5 produced by the minigene cannot function as a cell-to-cell signal and may remain in the cytoplasm of the differentiated cell, possibly because of the lack of a domain that is needed for transport to adjacent cells or export to the periplasmic space (Wu et al. 2004). The ability of large peptides containing an internal RGSGR sequence motif to inhibit heterocyst development supports the idea that the receptor for PatS is localized in the cytoplasm (Wu et al. 2004). In vitro studies show that the DNA-binding activity of HetR is inhibited by PatS-5 pentapeptide in a dose-dependent manner, implying that the HetR-PatS ratio is important for controlling heterocyst differentiation and indicating that HetR is a PatS receptor (Huang et al. 2004). Up-regulation of *hetR* expression is abolished when PatS-5 is added to the growth medium (Huang et al. 2004). *hetR* mutant *hetR*$_{R223W}$ is insensitive to the PatS inhibitory signal and overexpression of the *hetR*$_{R223W}$ allele results in a conditionally lethal phenotype after nitrogen deprivation because nearly all cells differentiate (Khudyakov and Golden 2004).

Overexpression of *patS* and *hetR* in a synthetic operon inhibits heterocyst formation, indicating that *patS* acts downstream of *hetR*, which is consistent with PatS inhibiting the activity of HetR (Orozco et al. 2006).

The *hetN* gene encodes a protein similar to ketoacyl reductase. Like *patS*, overexpression of *hetN* gene results in complete suppression of heterocyst development (Fig. 5). A *hetN* null mutant shows a wild-type pattern at 24 hours after nitrogen depletion but forms multiple contiguous heterocysts by 48 hours, suggesting that *hetN* is not necessary for de novo heterocyst pattern, and instead is required for the maintenance of the preexisting heterocyst pattern (Callahan and Buikema 2001). A *patS-hetN* double mutant shows complete differentiation of heterocysts in the absence of combined nitrogen, producing a lethal phenotype. These data indicate that the two inhibitory signal pathways may be independent (Borthakur et al. 2005). Together, *patS*- and *hetN*-mediated inhibitory pathways are the primary mechanism for establishing heterocyst pattern.

However, other factors can influence pattern formation, such as the cell cycle, the physiology of individual cells, signals or nutrients from vegetative cells, and the products of nitrogen fixation (Golden and Yoon 2003; Sakr et al. 2006; Aldea et al. 2008). It is known that single cells from fragmented filaments will not differentiate but instead always divide first, followed by differentiation of one of the daughter cells, and it has been suggested that an activator, possibly 2-oxoglutarate, of differentiation is produced by vegetative cells and serves as a regulator of pattern formation (Zhao and Wolk 2008). There is some evidence for the products of nitrogen fixation supplied from heterocysts contributing to the average spacing between heterocysts along filaments (Yoon and Golden 2001; Aldea et al. 2008). However, data from *A. variabilis*, which can fix nitrogen in vegetative cells, have been interpreted as an argument against the products of nitrogen fixation having a significant effect on heterocyst pattern (Thiel and Pratte 2001). This seemingly simple question has been difficult to answer because mutants in which heterocysts cannot supply nitrogen to the filament cannot grow diazotrophically, which makes determining the maintenance of heterocyst pattern impossible.

In *N. punctiforme*, the novel genes *patU* and *patN* are required to inhibit excess heterocyst differentiation (Meeks et al. 2002). In *Anabaena* PCC 7120, a cluster of genes that influence pattern formation and heterocyst differentiation, *hetZ*, *patU5*, and *patU3*, is conserved among heterocyst- and nonheterocyst-forming filamentous cyanobacteria (Zhang et al. 2007). All three genes are up-regulated after nitrogen deprivation and encode proteins with unknown function. A *hetZ* mutant does not differentiate heterocysts, unlike a *patU3* mutant, which forms multiple contiguous heterocysts. When *patU3* is inactivated in a *patA* mutant background, the *patA* mutant phenotype is abolished, restoring the formation of intercalary heterocysts (Zhang et al. 2007). Although these genes are clearly involved in the regulation of heterocyst development, their biochemical functions remain to be determined.

CONCLUDING REMARKS

Cyanobacteria have evolved several differentiated cell types in addition to heterocysts, including akinetes and the cells of hormogonia, to perform specialized functions. In *N. punctiforme*, vegetative cells have three developmental fates: heterocysts, akinetes, and hormogonial cells (Meeks et al. 2001; Meeks et al. 2002). Akinetes are perennating sporelike cells formed to endure cold and desiccation. Heterocysts and akinetes show pattern formation that must be regulated by cell-to-cell communication between cells within a filament. In the absence of heterocysts, akinetes form at random positions along a filament, whereas the presence of heterocysts influences akinete placing. In some organisms, akinetes differentiate from cells immediately adjacent to heterocysts, and in others chains of akinetes are produced in which the youngest are furthest away from the heterocysts. Hormogonia are typically short filaments of differentiated cells that lack heterocysts and often express gas vesicles and show gliding motility. One role of hormogonia is to disperse and inhabit new environments (Tandeau de Marsac 1994; Meeks and Elhai 2002). Although our knowledge of these additional forms of cyanobacterial development are limited compared with our understanding of heterocyst development, it is evident that cyanobacteria are capable of striking and complex multicellular development that involves the integration of environmental, extracellular, and intracellular signals to control cellular differentiation and morphogenesis. The study of multicellular developmental biology in cyanobacteria, as well as other bacterial developmental systems, has uprooted the common misconception of bacterial organisms as simple autonomous single cells.

ACKNOWLEDGMENTS

We thank anonymous reviewers for their comments and suggestions that helped improve this article. Preparation of this article was partially supported by award 0925126 from the National Science Foundation to J.W.G.

REFERENCES

Aldea MR, Kumar K, Golden JW. 2008. Heterocyst development and pattern formation. In *Chemical communication among bacteria* (ed. W.W. C. and B.L. Bassler), pp. 75–90. ASM Press, Washington, DC.

Aldea MR, Mella-Herrera RA, Golden JW. 2007. σ factor genes *sigC*, *sigE*, and *sigG* are upregulated in heterocysts of the cyanobacterium *Anabaena* sp. strain PCC 7120. *J Bacteriol* **189:** 8392–8396.

Awai K, Wolk CP. 2007. Identification of the glycosyl transferase required for synthesis of the principal glycolipid characteristic of heterocysts of *Anabaena* sp. strain PCC 7120. *FEMS Microbiol Lett* **266:** 98–102.

Awai K, Lechno-Yossef S, Wolk CP. 2009. Heterocyst Envelope Glycolipids. In *Lipids in photosynthesis: essential and regulatory functions* (ed. H. Wada and N. Murata), pp. 179–202. Springer, New York.

Black K, Buikema WJ, Haselkorn R. 1995. The *hglK* gene is required for localization of heterocyst-specific glycolipids in the cyanobacterium *Anabaena* sp. strain PCC 7120. *J Bacteriol* **177:** 6440–6448.

Black TA, Cai Y, Wolk CP. 1993. Spatial expression and autoregulation of *hetR*, a gene involved in the control of heterocyst development in *Anabaena*. *Mol Microbiol* **9:** 77–84.

Blair DF. 1995. How bacteria sense and swim. *Annu Rev Microbiol* **49:** 489–522.

Borthakur PB, Orozco CC, Young-Robbins SS, Haselkorn R, Callahan SM. 2005. Inactivation of *patS* and *hetN* causes lethal levels of heterocyst differentiation in the filamentous cyanobacterium *Anabaena* sp. PCC 7120. *Mol Microbiol* **57:** 111–123.

Bos MP, Tefsen B, Geurtsen J, Tommassen J. 2004. Identification of an outer membrane protein required for the transport of lipopolysaccharide to the bacterial cell surface. *Proc Natl Acad Sci* **101:** 9417–9422.

Brusca JS, Chastain CJ, Golden JW. 1990. Expression of the *Anabaena* sp. strain PCC 7120 *xisA* gene from a heterologous promoter results in excision of the *nifD* element. *J Bacteriol* **172:** 3925–3931.

Brusca JS, Hale MA, Carrasco CD, Golden JW. 1989. Excision of an 11-kilobase-pair DNA element from within the *nifD* gene in *Anabaena variabilis* heterocysts. *J Bacteriol* **171:** 4138–4145.

Budrene EO, Berg HC. 1995. Dynamics of formation of symmetrical patterns by chemotactic bacteria. *Nature* **376:** 49–53.

Buikema WJ, Haselkorn R. 1991a. Characterization of a gene controlling heterocyst differentiation in the cyanobacterium *Anabaena* 7120. *Genes Dev* **5:** 321–330.

Buikema WJ, Haselkorn R. 1991b. Isolation and complementation of nitrogen fixation mutants of the cyanobacterium *Anabaena* sp. strain PCC 7120. *J Bacteriol* **173:** 1879–1885.

Buikema WJ, Haselkorn R. 2001. Expression of the *Anabaena hetR* gene from a copper-regulated promoter leads to heterocyst differentiation under repressing conditions. *Proc Natl Acad Sci* **98:** 2729–2734.

Cai Y, Wolk CP. 1997. Nitrogen deprivation of Anabaena sp. strain PCC 7120 elicits rapid activation of a gene cluster that is essential for uptake and utilization of nitrate. *J Bacteriol* **179:** 258–266.

Callahan SM, Buikema WJ. 2001. The role of HetN in maintenance of the heterocyst pattern in *Anabaena* sp. PCC 7120. *Mol Microbiol* **40:** 941–950.

Campbell EL, Cohen MF, Meeks JC. 1997. A polyketide-synthase-like gene is involved in the synthesis of heterocyst glycolipids in *Nostoc punctiforme* strain ATCC 29133. *Arch Microbiol* **167:** 251–258.

Campbell EA, Westblade LF, Darst SA. 2008. Regulation of bacterial RNA polymerase σ factor activity: A structural perspective. *Curr Opin Microbiol* **11:** 121–127.

Campbell EL, Summers ML, Christman H, Martin ME, Meeks JC. 2007. Global gene expression patterns of *Nostoc punctiforme* in steady-state dinitrogen-grown heterocyst-containing cultures and at single time points during the differentiation of akinetes and hormogonia. *J Bacteriol* **189:** 5247–5256.

Cardemil L, Wolk CP. 1979. The polysaccharides from heterocyst and spore envelopes of a blue-green alga. Structure of the basic repeating unit. *J Biol Chem* **254:** 736–741.

Cardemil L, Wolk CP. 1981. Isolated heterocysts of *Anabaena variabilis* synthesize envelope polysaccharide. *Biochim Biophys Acta* **674:** 265–276.

Carrasco CD, Golden JW. 1995. Two heterocyst-specific DNA rearrangements of *nif* operons in *Anabaena cylindrica* and *Nostoc* sp. strain Mac. *Microbiology* **141:** 2479–2487.

Carrasco CD, Buettner JA, Golden JW. 1995. Programmed DNA rearrangement of a cyanobacterial *hupL* gene in heterocysts. *Proc Natl Acad Sci* **92:** 791–795.

Carrasco CD, Holliday SD, Hansel A, Lindblad P, Golden JW. 2005. Heterocyst-specific excision of the *Anabaena* sp. strain PCC 7120 *hupL* element requires *xisC*. *J Bacteriol* **187:** 6031–6038.

Carrasco CD, Ramaswamy KS, Ramasubramanian TS, Golden JW. 1994. *Anabaena xisF* gene encodes a developmentally regulated site-specific recombinase. *Genes Dev* **8:** 74–83.

Chen H, Laurent S, Bédu S, Ziarelli F, Chen HL, Cheng Y, Zhang CC, Peng L. 2006. Studying the signaling role of 2-oxoglutaric acid using analogs that mimic the ketone and ketal forms of 2-oxoglutaric acid. *Chem Biol* **13:** 849–856.

Cheng Y, Li JH, Shi L, Wang L, Latifi A, Zhang CC. 2006. A pair of iron-responsive genes encoding protein kinases with a Ser/Thr kinase domain and a His kinase domain are regulated by NtcA in the cyanobacterium *Anabaena* sp. strain PCC 7120. *J Bacteriol* **188:** 4822–4829.

Cumino AC, Marcozzi C, Barreiro R, Salerno GL. 2007. Carbon cycling in *Anabaena* sp. PCC 7120. Sucrose synthesis in the heterocysts and possible role in nitrogen fixation. *Plant Physiol* **143:** 1385–1397.

Dismukes GC, Carrieri D, Bennette N, Ananyev GM, Posewitz MC. 2008. Aquatic phototrophs: Efficient alternatives to land-based crops for biofuels. *Curr Opin Biotechnol* **19:** 235–240.

Dong Y, Huang X, Wu XY, Zhao J. 2000. Identification of the active site of HetR protease and its requirement for heterocyst differentiation in the cyanobacterium *Anabaena* sp. strain PCC 7120. *J Bacteriol* **182:** 1575–1579.

Ehira S, Ohmori M. 2006a. NrrA directly regulates the expression of *hetR* during heterocyst differentiation in the cyanobacterium *Anabaena* sp. strain PCC 7120. *J Bacteriol* **188:** 8520–8525.

Ehira S, Ohmori M. 2006b. NrrA, a nitrogen-responsive response regulator facilitates heterocyst development in the cyanobacterium *Anabaena* sp. strain PCC 7120. *Mol Microbiol* **59:** 1692–1703.

Ehira S, Ohmori M, Sato N. 2003. Genome-wide expression analysis of the responses to nitrogen deprivation in the heterocyst-forming cyanobacterium *Anabaena* sp. strain PCC 7120. *DNA Res* **10:** 97–113.

Elhai J, Wolk CP. 1990. Developmental regulation and spatial pattern of expression of the structural genes for nitrogenase in the cyanobacterium *Anabaena*. *EMBO J* **9:** 3379–3388.

Fan Q, Huang G, Lechno-Yossef S, Wolk CP, Kaneko T, Tabata S. 2005. Clustered genes required for synthesis and deposition of envelope glycolipids in *Anabaena* sp. strain PCC 7120. *Mol Microbiol* **58:** 227–243.

Fay P. 1992. Oxygen relations of nitrogen fixation in cyanobacteria. *Microbiol Rev* **56:** 340–373.

Fiedler G, Arnold M, Hannus S, Maldener I. 1998. The DevBCA exporter is essential for envelope formation in heterocysts of the cyanobacterium *Anabaena* sp. strain PCC 7120. *Mol Microbiol* **27:** 1193–1202.

Fiedler G, Muro-Pastor AM, Flores E, Maldener I. 2001. NtcA-dependent expression of the *devBCA* operon, encoding a heterocyst-specific ATP-binding cassette transporter in *Anabaena* spp. *J Bacteriol* **183:** 3795–3799.

Flores E, Herrero A, Wolk CP, Maldener I. 2006. Is the periplasm continuous in filamentous multicellular cyanobacteria? *Trends Microbiol* **14:** 439–443.

Flores E, Pernil R, Muro-Pastor AM, Mariscal V, Maldener I, Lechno-Yossef S, Fan Q, Wolk CP, Herrero A. 2007. Septum-localized protein required for filament integrity and diazotrophy in the heterocyst-forming cyanobacterium *Anabaena* sp. strain PCC 7120. *J Bacteriol* **189:** 3884–3890.

Forchhammer K. 2004. Global carbon/nitrogen control by PII signal transduction in cyanobacteria: From signals to targets. *FEMS Microbiol Rev* **28:** 319–333.

Frias JE, Flores E, Herrero A. 1994. Requirement of the regulatory protein NtcA for the expression of nitrogen assimilation and heterocyst development genes in the cyanobacterium *Anabaena* sp. PCC 7120. *Mol Microbiol* **14:** 823–832.

Giddings TH, Staehelin LA. 1978. Plasma membrane architecture of *Anabaena cylindrica*: occurrence of microplasmodesmata and changes associated with heterocyst development and the cell cycle. *Eur J Cell Biol* **16:** 235–249.

Giddings TH, Staehelin LA. 1981. Observation of microplasmodesmata in both heterocyst-forming and non-heterocyst forming filamentous cyanobacteria by freeze-fracture electron microscopy. *Arch Microbiol* **129:** 295–298.

Golden JW. 1997. Programmed DNA rearrangements in cyanobacteria. In *Bacterial genomes: Physical structure and analysis* (ed. F.J. de Bruijn, J.R. Lupski and G. Weinstock), pp. 162–173. Chapman and Hall, New York.

Golden JW, Wiest DR. 1988. Genome rearrangement and nitrogen fixation in *Anabaena* blocked by inactivation of *xisA* gene. *Science* **242:** 1421–1423.

Golden JW, Yoon HS. 1998. Heterocyst formation in *Anabaena*. *Curr Opin Microbiol* **1:** 623–629.

Golden JW, Yoon HS. 2003. Heterocyst development in *Anabaena*. *Curr Opin Microbiol* **6:** 557–563.

Golden JW, Robinson SJ, Haselkorn R. 1985. Rearrangement of nitrogen fixation genes during heterocyst differentiation in the cyanobacterium *Anabaena*. *Nature* **314:** 419–423.

Golden JW, Whorff LL, Wiest DR. 1991. Independent regulation of *nifHDK* operon transcription and DNA rearrangement during heterocyst differentiation in the cyanobacterium *Anabaena* sp. strain PCC 7120. *J Bacteriol* **173:** 7098–7105.

Golden JW, Carrasco CD, Mulligan ME, Schneider GJ, Haselkorn R. 1988. Deletion of a 55-kilobase-pair DNA element from the chromosome during heterocyst differentiation of *Anabaena* sp. strain PCC 7120. *J Bacteriol* **170:** 5034–5041.

Hebbar PB, Curtis SE. 2000. Characterization of *devH*, a gene encoding a putative DNA binding protein required for heterocyst function in *Anabaena* sp. strain PCC 7120. *J Bacteriol* **182:** 3572–3581.

Henson BJ, Pennington LE, Watson LE, Barnum SR. 2008. Excision of the *nifD* element in the heterocystous cyanobacteria. *Arch Microbiol* **189:** 357–366.

Herrero A, Muro-Pastor AM, Valladares A, Flores E. 2004. Cellular differentiation and the NtcA transcription factor in filamentous cyanobacteria. *FEMS Microbiol Rev* **28:** 469–487.

Hoiczyk E, Hansel A. 2000. Cyanobacterial cell walls: News from an unusual prokaryotic envelope. *J Bacteriol* **182:** 1191–1199.

Holmqvist M, Stensjö K, Oliveira P, Lindberg P, Lindblad P. 2009. Characterization of the *hupSL* promoter activity in *Nostoc punctiforme* ATCC 29133. *BMC Microbiol* **9:** 54.

Huang X, Dong Y, Zhao J. 2004. HetR homodimer is a DNA-binding protein required for heterocyst differentiation, and the DNA-binding activity is inhibited by PatS. *Proc Natl Acad Sci* **101:** 4848–4853.

Huang G, Fan Q, Lechno-Yossef S, Wojciuch E, Wolk CP, Kaneko T, Tabata S. 2005. Clustered genes required for the synthesis of heterocyst envelope polysaccharide in *Anabaena* sp. strain PCC 7120. *J Bacteriol* **187:** 1114–1123.

Jang J, Wang L, Jeanjean R, Zhang CC. 2007. PrpJ, a PP2C-type protein phosphatase located on the plasma membrane, is involved in heterocyst maturation in the cyanobacterium *Anabaena* sp. PCC 7120. *Mol Microbiol* **64:** 347–358.

Jelsbak L, Søgaard-Andersen L. 2002. Pattern formation by a cell surface-associated morphogen in *Myxococcus xanthus*. *Proc Natl Acad Sci* **99:** 2032–2037.

Jones KM, Haselkorn R. 2002. Newly identified cytochrome c oxidase operon in the nitrogen-fixing cyanobacterium *Anabaena* sp. strain PCC 7120 specifically induced in heterocysts. *J Bacteriol* **184:** 2491–2499.

Khudyakov IY, Golden JW. 2001. Identification and inactivation of three group 2 σ factor genes in *Anabaena* sp. strain PCC 7120. *J Bacteriol* **183:** 6667–6675.

Khudyakov IY, Golden JW. 2004. Different functions of HetR, a master regulator of heterocyst differentiation in *Anabaena* sp. PCC 7120 can be separated by mutation. *Proc Natl Acad Sci* **101:** 16040–16045.

Khudyakov I, Wolk CP. 1997. *hetC*, a gene coding for a protein similar to bacterial ABC protein exporters, is involved in early regulation of heterocyst differentiation in *Anabaena* sp. strain PCC 7120. *J Bacteriol* **179:** 6971–6978.

Klausen M, Heydorn A, Ragas P, Lambertsen L, Aaes-Jørgensen A, Molin S, Tolker-Nielsen T. 2003. Biofilm formation by *Pseudomonas aeruginosa* wild type, flagella and type IV pili mutants. *Mol Microbiol* **48:** 1511–1524.

Kroos L. 2007. The *Bacillus* and *Myxococcus* developmental networks and their transcriptional regulators. *Annu Rev Genet* **41:** 13–39.

Lammers PJ, McLaughlin S, Papin S, Trujillo-Provencio C, Ryncarz AJ 2nd. 1990. Developmental rearrangement of cyanobacterial *nif* genes: Nucleotide sequence, open reading frames, and cytochrome P-450 homology of the *Anabaena* sp. strain PCC 7120 *nifD* element. *J Bacteriol* **172:** 6981–6990.

Laurent S, Chen H, Bedu S, Ziarelli F, Peng L, Zhang CC. 2005. Nonmetabolizable analogue of 2-oxoglutarate elicits heterocyst differentiation under repressive conditions in *Anabaena* sp. PCC 7120. *Proc Natl Acad Sci* **102:** 9907–9912.

Lechno-Yossef S, Fan Q, Ehira S, Sato N, Wolk CP. 2006. Mutations in four regulatory genes have interrelated effects on heterocyst maturation in *Anabaena* sp. strain PCC 7120. *J Bacteriol* **188:** 7387–7395.

Liang J, Scappino L, Haselkorn R. 1992. The *patA* gene product, which contains a region similar to CheY of *Escherichia coli*, controls heterocyst pattern formation in the cyanobacterium *Anabaena* 7120. *Proc Natl Acad Sci* **89:** 5655–5659.

Liu D, Golden JW. 2002. *hetL* overexpression stimulates heterocyst formation in *Anabaena* sp. strain PCC 7120. *J Bacteriol* **184:** 6873–6881.

Marcozzi C, Cumino AC, Salerno GL. 2009. Role of NtcA, a cyanobacterial global nitrogen regulator, in the regulation of sucrose metabolism gene expression in *Anabaena* sp. PCC 7120. *Arch Microbiol* **191:** 255–263.

Mariscal V, Herrero A, Flores E. 2007. Continuous periplasm in a filamentous, heterocyst-forming cyanobacterium. *Mol Microbiol* **65:** 1139–1145.

Martin-Figueroa E, Navarro F, Florencio FJ. 2000. The GS-GOGAT pathway is not operative in the heterocysts. Cloning and expression of *glsF* gene from the cyanobacterium *Anabaena* sp. PCC 7120. *FEBS Lett* **476:** 282–286.

Masepohl B, Schölisch K, Görlitz K, Kutzki C, Böhme H. 1997. The heterocyst-specific *fdxH* gene product of the cyanobacterium *Anabaena* sp. PCC 7120 is important but not essential for nitrogen fixation. *Mol Gen Genet* **253:** 770–776.

Meeks JC, Elhai J. 2002. Regulation of cellular differentiation in filamentous cyanobacteria in free-living and plant-associated symbiotic growth states. *Microbiol Mol Biol Rev* **66:** 94–121.

Meeks JC, Campbell EL, Summers ML, Wong FC. 2002. Cellular differentiation in the cyanobacterium *Nostoc punctiforme*. *Arch Microbiol* **178:** 395–403.

Meeks JC, Elhai J, Thiel T, Potts M, Larimer F, Lamerdin J, Predki P, Atlas R. 2001. An overview of the genome of *Nostoc punctiforme*, a multicellular, symbiotic cyanobacterium. *Photosyn Res* **70:** 85–106.

Montoya JP, Holl CM, Zehr JP, Hansen A, Villareal TA, Capone DG. 2004. High rates of N2 fixation by unicellular diazotrophs in the oligotrophic Pacific Ocean. *Nature* **430:** 1027–1032.

Moslavac S, Bredemeier R, Mirus O, Granvogl B, Eichacker LA, Schleiff E. 2005. Proteomic analysis of the outer membrane of *Anabaena* sp. strain PCC 7120. *J Proteome Res* **4:** 1330–1338.

Moslavac S, Nicolaisen K, Mirus O, Al Dehni F, Pernil R, Flores E, Maldener I, Schleiff E. 2007a. A TolC-like protein is required for heterocyst development in *Anabaena* sp. strain PCC 7120. *J Bacteriol* **189:** 7887–7895.

Moslavac S, Reisinger V, Berg M, Mirus O, Vosyka O, Ploscher M, Flores E, Eichacker LA, Schleiff E. 2007b. The proteome of the heterocyst cell wall in *Anabaena* sp. PCC 7120. *Biol Chem* **388:** 823–829.

Mulligan ME, Haselkorn R. 1989. Nitrogen fixation (*nif*) genes of the cyanobacterium *Anabaena* species strain PCC 7120. The *nifB-fdxN-nifS-nifU* operon. *J Biol Chem* **264:** 19200–19207.

Mullineaux CW, Mariscal V, Nenninger A, Khanum H, Herrero A, Flores E, Adams DG. 2008. Mechanism of intercellular molecular exchange in heterocyst-forming cyanobacteria. *EMBO J* **27:** 1299–1308.

Muro-Pastor AM, Olmedo-Verd E, Flores E. 2006. All4312, an NtcA-regulated two-component response regulator in *Anabaena* sp. strain PCC 7120. *FEMS Microbiol Lett* **256:** 171–177.

Muro-Pastor AM, Valladares A, Flores E, Herrero A. 1999. The *hetC* gene is a direct target of the NtcA transcriptional regulator in cyanobacterial heterocyst development. *J Bacteriol* **181:** 6664–6669.

Muro-Pastor MI, Reyes JC, Florencio FJ. 2001. Cyanobacteria perceive nitrogen status by sensing intracellular 2-oxoglutarate levels. *J Biol Chem* **276:** 38320–38328.

Muro-Pastor MI, Reyes JC, Florencio FJ. 2005. Ammonium assimilation in cyanobacteria. *Photosynth Res* **83:** 135–150.

Muro-Pastor AM, Valladares A, Flores E, Herrero A. 2002. Mutual dependence of the expression of the cell differentiation regulatory protein HetR and the global nitrogen regulator NtcA during heterocyst development. *Mol Microbiol* **44:** 1377–1385.

Nayar AS, Yamaura H, Rajagopalan R, Risser DD, Callahan SM. 2007. FraG is necessary for filament integrity and heterocyst maturation in the cyanobacterium *Anabaena* sp. strain PCC 7120. *Microbiology* **153:** 601–607.

Neunuebel MR, Golden JW. 2008. The *Anabaena* sp. strain PCC 7120 gene all2874 encodes a diguanylate cyclase and is required for normal heterocyst development under high-light growth conditions. *J Bacteriol* **190:** 6829–6836.

Ni S, Sheldrick GM, Benning MM, Kennedy MA. 2009. The 2Å resolution crystal structure of HetL, a pentapeptide repeat protein involved in regulation of heterocyst differentiation in the cyanobacterium *Nostoc* sp. strain PCC 7120. *J Struct Biol* **165:** 47–52.

Nicolaisen K, Hahn A, Schleiff E. 2009. The cell wall in heterocyst formation by *Anabaena* sp. PCC 7120. *J Basic Microbiol* **49:** 5–24.

Ohmori M, Ikeuchi M, Sato N, Wolk P, Kaneko T, Ogawa T, Kanehisa M, Goto S, Kawashima S, Okamoto S, et al. 2001. Characterization of genes encoding multi-domain proteins in the genome of the filamentous nitrogen-fixing cyanobacterium *Anabaena* sp. strain PCC 7120. *DNA Res* **8:** 271–284.

Orozco CC, Risser DD, Callahan SM. 2006. Epistasis analysis of four genes from *Anabaena* sp. strain PCC 7120 suggests a connection between PatA and PatS in heterocyst pattern formation. *J Bacteriol* **188:** 1808–1816.

Pereira S, Zille A, Micheletti E, Moradas-Ferreira P, De Philippis R, Tamagnini P. 2009. Complexity of cyanobacterial exopolysaccharides: Composition, structures, inducing factors and putative genes involved in their biosynthesis and assembly. *FEMS Microbiol Rev* **33:** 917–941.

Phalip V, Li JH, Zhang CC. 2001. HstK, a cyanobacterial protein with both a serine/threonine kinase domain and a histidine kinase domain: Implication for the mechanism of signal transduction. *Biochem J* **360:** 639–644.

Popa R, Weber PK, Pett-Ridge J, Finzi JA, Fallon SJ, Hutcheon ID, Nealson KH, Capone DG. 2007. Carbon and nitrogen fixation and metabolite exchange in and between individual cells of *Anabaena oscillarioides*. *ISME J* **1:** 354–360.

Pratte BS, Eplin K, Thiel T. 2006. Cross-functionality of nitrogenase components NifH1 and VnfH in *Anabaena variabilis*. *J Bacteriol* **188:** 5806–5811.

Ramasubramanian TS, Wei TF, Golden JW. 1994. Two *Anabaena* sp. strain PCC 7120 DNA-binding factors interact with vegetative cell- and heterocyst-specific genes. *J Bacteriol* **176:** 1214–1223.

Ramasubramanian TS, Wei TF, Oldham AK, Golden JW. 1996. Transcription of the *Anabaena* sp. strain PCC 7120 *ntcA* gene: Multiple transcripts and NtcA binding. *J Bacteriol* **178:** 922–926.

Ramaswamy KS, Carrasco CD, Fatma T, Golden JW. 1997. Cell-type specificity of the *Anabaena fdxN*-element rearrangement requires *xisH* and *xisI*. *Mol Microbiol* **23:** 1241–1249.

Ramirez ME, Hebbar PB, Zhou R, Wolk CP, Curtis SE. 2005. *Anabaena* sp. strain PCC 7120 gene *devH* is required for synthesis of the heterocyst glycolipid layer. *J Bacteriol* **187:** 2326–2331.

Risser DD, Callahan SM. 2007. Mutagenesis of *hetR* reveals amino acids necessary for HetR function in the heterocystous cyanobacterium *Anabaena* sp. strain PCC 7120. *J Bacteriol* **189:** 2460–2467.

Risser DD, Callahan SM. 2008. HetF and PatA control levels of HetR in *Anabaena* sp. strain PCC 7120. *J Bacteriol* **190:** 7645–7654.

Rubio LM, Ludden PW. 2008. Biosynthesis of the iron-molybdenum cofactor of nitrogenase. *Annu Rev Microbiol* **62:** 93–111.

Sakr S, Jeanjean R, Zhang CC, Arcondeguy T. 2006. Inhibition of cell division suppresses heterocyst development in *Anabaena* sp. strain PCC 7120. *J Bacteriol* **188:** 1396–1404.

Sandh G, El-Shehawy R, Diez B, Bergman B. 2009. Temporal separation of cell division and diazotrophy in the marine diazotrophic cyanobacterium *Trichodesmium erythraeum* IMS101. *FEMS Microbiol Lett* **295:** 281–288.

Sato N, Ohmori M, Ikeuchi M, Tashiro K, Wolk CP, Kaneko T, Okada K, Tsuzuki M, Ehira S, Katoh H, et al. 2004. Use of segment-based microarray in the analysis of global gene expression in response to various environmental stresses in the cyanobacterium *Anabaena* sp. PCC 7120. *J Gen Appl Microbiol* **50:** 1–8.

Shi L, Li JH, Cheng Y, Wang L, Chen WL, Zhang CC. 2007. Two genes encoding protein kinases of the HstK family are involved in synthesis of the minor heterocyst-specific glycolipid in the cyanobacterium *Anabaena* sp. strain PCC 7120. *J Bacteriol* **189:** 5075–5081.

Shi Y, Zhao W, Zhang W, Ye Z, Zhao J. 2006. Regulation of intracellular free calcium concentration during heterocyst differentiation by HetR and NtcA in *Anabaena* sp. PCC 7120. *Proc Natl Acad Sci* **103:** 11334–11339.

Stricker O, Masepohl B, Klipp W, Böhme H. 1997. Identification and characterization of the *nifV-nifZ-nifT* gene region from the filamentous cyanobacterium *Anabaena* sp. strain PCC 7120. *J Bacteriol* **179:** 2930–2937.

Tamagnini P, Leitão E, Oliveira P, Ferreira D, Pinto F, Harris DJ, Heidorn T, Lindblad P. 2007. Cyanobacterial hydrogenases: Diversity, regulation and applications. *FEMS Microbiol Rev* **31:** 692–720.

Tandeau de Marsac N. 1994. Differentiation of hormogonia and relationships with other biological processes. In *The molecular biology of cyanobacteria* (ed. D.A. Bryant), pp. 825–840. Kluwer Academic Publishers, Dordrecht.

Tanigawa R, Shirokane M, Maeda Si S, Omata T, Tanaka K, Takahashi H. 2002. Transcriptional activation of NtcA-dependent promoters of *Synechococcus* sp. PCC 7942 by 2-oxoglutarate in vitro. *Proc Natl Acad Sci* **99:** 4251–4255.

Thiel T, Pratte B. 2001. Effect on heterocyst differentiation of nitrogen fixation in vegetative cells of the cyanobacterium *Anabaena variabilis* ATCC 29413. *J Bacteriol* **183:** 280–286.

Thiel T, Lyons EM, Erker JC, Ernst A. 1995. A second nitrogenase in vegetative cells of a heterocyst-forming cyanobacterium. *Proc Natl Acad Sci* **92:** 9358–9362.

Toepel J, Welsh E, Summerfield TC, Pakrasi HB, Sherman LA. 2008. Differential transcriptional analysis of the cyanobacterium *Cyanothece* sp. strain ATCC 51142 during light-dark and continuous-light growth. *J Bacteriol* **190:** 3904–3913.

Torrecilla I, Leganés F, Bonilla I, Fernández-Piñas F. 2004. A calcium signal is involved in heterocyst differentiation in the cyanobacterium *Anabaena* sp. PCC7120. *Microbiology* **150:** 3731–3739.

Valladares A, Herrero A, Pils D, Schmetterer G, Flores E. 2003. Cytochrome c oxidase genes required for nitrogenase activity and diazotrophic growth in *Anabaena* sp. PCC 7120. *Mol Microbiol* **47:** 1239–1249.

Valladares A, Muro-Pastor AM, Herrero A, Flores E. 2004. The NtcA-dependent P1 promoter is utilized for *glnA* expression in N_2-fixing heterocysts of *Anabaena* sp. strain PCC 7120. *J Bacteriol* **186:** 7337–7343.

Vazquez-Bermúdez MF, Herrero A, Flores E. 2002. 2-Oxoglutarate increases the binding affinity of the NtcA (nitrogen control) transcription factor for the *Synechococcus glnA* promoter. *FEBS Lett* **512:** 71–74.

Vazquez-Bermúdez MF, Herrero A, Flores E. 2003. Carbon supply and 2-oxoglutarate effects on expression of nitrate reductase and nitrogen-regulated genes in *Synechococcus* sp. strain PCC 7942. *FEMS Microbiol Lett* **221:** 155–159.

Wang Y, Xu X. 2005. Regulation by *hetC* of genes required for heterocyst differentiation and cell division in *Anabaena* sp. strain PCC 7120. *J Bacteriol* **187:** 8489–8493.

Wang L, Sun YP, Chen WL, Li JH, Zhang CC. 2002. Genomic analysis of protein kinases, protein phosphatases and two-component regulatory systems of the cyanobacterium *Anabaena* sp. strain PCC 7120. *FEMS Microbiol Lett* **217:** 155–165.

Wang Y, Lechno-Yossef S, Gong Y, Fan Q, Wolk CP, Xu X. 2007. Predicted glycosyl transferase genes located outside the HEP island are required for formation of heterocyst envelope polysaccharide in *Anabaena* sp. strain PCC 7120. *J Bacteriol* **189:** 5372–5378.

Wei TF, Ramasubramanian TS, Golden JW. 1994. *Anabaena* sp. strain PCC 7120 *ntcA* gene required for growth on nitrate and heterocyst development. *J Bacteriol* **176:** 4473–4482.

Wei TF, Ramasubramanian TS, Pu F, Golden JW. 1993. *Anabaena* sp. strain PCC 7120 *bifA* gene encoding a sequence-specific DNA-binding protein cloned by *in vivo* transcriptional interference selection. *J Bacteriol* **175:** 4025–4035.

Weyman PD, Pratte B, Thiel T. 2008. Transcription of *hupSL* in *Anabaena variabilis* ATCC 29413 is regulated by NtcA and not by hydrogen. *Appl Environ Microbiol* **74:** 2103–2110.

Wilcox M, Mitchison GJ, Smith RJ. 1973. Pattern formation in the blue-green alga, *Anabaena*. I. Basic mechanisms. *J Cell Sci* **12:** 707–723.

Wolk CP. 1996. Heterocyst formation. *Annu Rev Genet* **30:** 59–78.

Wolk CP, Ernst A, Elhai J. 1994. Heterocyst metabolism and development. In *The molecular biology of cyanobacteria* (ed. D.A. Bryant), pp. 769–863. Kluwer Academic Publishers, Dordrecht.

Wolk CP, Cai Y, Cardemil L, Flores E, Hohn B, Murry M, Schmetterer G, Schrautemeier B, Wilson R. 1988. Isolation and complementation of mutants of *Anabaena* sp. strain PCC 7120 unable to grow aerobically on dinitrogen. *J Bacteriol* **170:** 1239–1244.

Wong FC, Meeks JC. 2001. The *hetF* gene product is essential to heterocyst differentiation and affects HetR function in the cyanobacterium *Nostoc punctiforme*. *J Bacteriol* **183:** 2654–2661.

Wu X, Lee DW, Mella RA, Golden JW. 2007. The *Anabaena* sp. strain PCC 7120 asr1734 gene encodes a negative regulator of heterocyst development. *Mol Microbiol* **64:** 782–794.

Wu X, Liu D, Lee MH, Golden JW. 2004. *patS* minigenes inhibit heterocyst development of *Anabaena* sp. strain PCC 7120. *J Bacteriol* **186:** 6422–6429.

Xu X, Wolk CP. 2001. Role for *hetC* in the transition to a nondividing state during heterocyst differentiation in *Anabaena* sp. *J Bacteriol* **183:** 393–396.

Yoon HS, Golden JW. 1998. Heterocyst pattern formation controlled by a diffusible peptide. *Science* **282:** 935–938.

Yoon HS, Golden JW. 2001. PatS and products of nitrogen fixation control heterocyst pattern. *J Bacteriol* **183:** 2605–2613.

Yoshimura H, Okamoto S, Tsumuraya Y, Ohmori M. 2007. Group 3 σ factor gene, *sigJ*, a key regulator of desiccation tolerance, regulates the synthesis of extracellular polysaccharide in cyanobacterium *Anabaena* sp. strain PCC 7120. *DNA Res* **14:** 13–24.

Zhang JY, Chen WL, Zhang CC. 2009. *hetR* and *patS*, two genes necessary for heterocyst pattern formation, are widespread in filamentous nonheterocyst-forming cyanobacteria. *Microbiology* **155:** 1418–1426.

Zhang LC, Chen YF, Chen WL, Zhang CC. 2008. Existence of periplasmic barriers preventing green fluorescent protein diffusion from cell to cell in the cyanobacterium *Anabaena* sp. strain PCC 7120. *Mol Microbiol* **70:** 814–823.

Zhang W, Du Y, Khudyakov I, Fan Q, Gao H, Ning D, Wolk CP, Xu X. 2007. A gene cluster that regulates both heterocyst differentiation and pattern formation in *Anabaena* sp. strain PCC 7120. *Mol Microbiol* **66:** 1429–1443.

Zhang CC, Laurent S, Sakr S, Peng L, Bédu S. 2006. Heterocyst differentiation and pattern formation in cyanobacteria: A chorus of signals. *Mol Microbiol* **59:** 367–375.

Zhao J, Wolk CP. 2008. Developmental Biology of Heterocysts, 2006. In *Myxobacteria: Multicellularity and differentiation* (ed. D.E. Whitworth), pp. 397–418. ASM Press, Washington, DC.

Zhou R, Wolk CP. 2003. A two-component system mediates developmental regulation of biosynthesis of a heterocyst polysaccharide. *J Biol Chem* **278:** 19939–19946.

Zhao Y, Shi Y, Zhao W, Huang X, Wang D, Brown N, Brand J, Zhao J. 2005. CcbP, a calcium-binding protein from *Anabaena* sp. PCC 7120, provides evidence that calcium ions regulate heterocyst differentiation. *Proc Natl Acad Sci* **102:** 5744–5748.

Zhou R, Wei X, Jiang N, Li H, Dong Y, Hsi KL, Zhao J. 1998. Evidence that HetR protein is an unusual serine-type protease. *Proc Natl Acad Sci* **95:** 4959–4963.

Zhu J, Kong R, Wolk CP. 1998. Regulation of *hepA* of *Anabaena* sp. strain PCC 7120 by elements 5' from the gene and by *hepK*. *J Bacteriol* **180:** 4233–4242.

Zhu J, Jager K, Black T, Zarka K, Koksharova O, Wolk CP. 2001. HcwA, an autolysin, is required for heterocyst maturation in *Anabaena* sp. strain PCC 7120. *J Bacteriol* **183:** 6841–6851.

Biofilms

Daniel López, Hera Vlamakis, and Roberto Kolter

Department of Microbiology and Molecular Genetics, Harvard Medical School, Boston, Massachusetts 02115

Correspondence: rkolter@hms.harvard.edu

The ability to form biofilms is a universal attribute of bacteria. Biofilms are multicellular communities held together by a self-produced extracellular matrix. The mechanisms that different bacteria employ to form biofilms vary, frequently depending on environmental conditions and specific strain attributes. In this review, we emphasize four well-studied model systems to give an overview of how several organisms form biofilms: *Escherichia coli, Pseudomonas aeruginosa, Bacillus subtilis,* and *Staphylococcus aureus.* Using these bacteria as examples, we discuss the key features of biofilms as well as mechanisms by which extracellular signals trigger biofilm formation.

Bacteria are able to grow adhered to almost every surface, forming architecturally complex communities termed biofilms. In biofilms, cells grow in multicellular aggregates that are encased in an extracellular matrix produced by the bacteria themselves (Branda et al. 2005; Hall-Stoodley and Stoodley 2009). Biofilms impact humans in many ways as they can form in natural, medical, and industrial settings. For instance, formation of biofilms on medical devices, such as catheters or implants often results in difficult-to-treat chronic infections (Hall-Stoodley et al. 2004; Donlan 2008; Hatt and Rather 2008). Moreover, infections have been associated with biofilm formation on human surfaces such as teeth, skin, and the urinary tract (Hatt and Rather 2008). However, biofilms on human surfaces are not always detrimental. For example, dental plaque biofilms comprise dozens of species and the community composition frequently determines the presence or absence of disease. In dental plaque, there is a progression of colonization and the presence of beneficial species antagonizes colonization by detrimental organisms (Kreth et al. 2008). But biofilms form everywhere. For example, biofilms form on the hulls of ships and inside pipes where they cause severe problems (de Carvalho 2007). On the other hand, in many natural settings, biofilm formation often allows mutualistic symbioses. For instance, *Actinobacteria* often grow on ants, allowing the ants to maintain pathogen-free fungal gardens (Currie 2001; Danhorn and Fuqua 2007). Given the vast potential benefits and detriments that biofilms can confer, it is essential that we understand how bacteria thrive in these communities.

There are numerous benefits that a bacterial community might obtain from the formation of biofilms. Biofilms confer resistance to many antimicrobials, protection from protozoan

 doi: 10.1101/cshperspect.a000398
Cite this article as *Cold Spring Harb Perspect Biol* 2010;2:a000398

grazing, and protection against host defenses (Mah and O'Toole 2001; Matz and Kjelleberg 2005; Anderson and O'Toole 2008). One possible reason for the increased resistance to environmental stresses observed in biofilm cells appears to be the increase in the portion of persister cells within the biofilm (Lewis 2005). Despite being genetically identical to the rest of the population, persister cells are resistant to many antibiotics and are nondividing. Persister cells have been proposed to be protected from the action of antibiotics because they express toxin–antitoxin systems where the target of the antibiotics is blocked by the toxin modules (Lewis 2005). In addition to an increase in persisters, the presence of an extracellular matrix protects constituent cells from external aggressions. Extracellular matrices also act as a diffusion barrier to small molecules (Anderson and O'Toole 2008; Hall-Stoodley and Stoodley 2009). Related to this, in biofilms the diffusion of nutrients, vitamins, or cofactors is slower resulting in a bacterial community in which some of cells are metabolically inactive. Furthermore, the rate of bacterial growth is influenced by the fact that cells within a biofilm are confined to a limited space (Stewart and Franklin 2008). This condition is similar to the stationary phase created in laboratory conditions. Hence, biofilm formation in a way represents the natural stationary phase of bacterial growth. During stationary phase, bacteria profoundly change their physiology by increasing production of secondary metabolites such as antibiotics, pigments, and other small-molecules (Martin and Liras 1989). These secondary metabolites also function as signaling molecules to initiate the process of biofilm formation or to inhibit biofilm formation by other organisms that inhabit the same habitat (Lopez and Kolter 2009). In this article, we review the metabolic processes that characterize biofilm formation for a handful of well-studied bacterial organisms: *Pseudomonas aeruginosa*, *Escherichia coli*, *Staphylococcus aureus*, and *Bacillus subtilis*. In addition, we address the function of secondary metabolites and their role as signaling molecules during biofilm formation.

STRUCTURAL COMPONENTS OF BIOFILMS

The molecular mechanisms that regulate biofilm formation vary greatly among different species, and even vary between different strains of the same species. However, some features are recognized as general attributes of biofilm formation (Monds and O'Toole 2009). For instance, all biofilms contain an extracellular matrix that holds cells together. This matrix is often composed of a polysaccharide biopolymer along with other components such as proteins or DNA (Branda et al. 2005). The nature of the matrix exopolysaccharide greatly varies depending on growth conditions, medium, and substrates.

P. aeruginosa is a gram-negative pathogen that makes biofilms by producing three distinct exopolysaccharides: alginate, PEL, and PSL. The importance and contribution of each exopolysaccharide to the matrix varies depending on the strain studied (Ryder et al. 2007; Tart and Wozniak 2008). For example, alginate is a produced by mucoid strains of *P. aeruginosa* that are often isolated from lungs of cystic fibrosis patients. The *pel* gene cluster, encoding a glucose-rich polymer termed PEL, is found in most of the strains analyzed to date, yet its expression strongly varies among strains (Branda et al. 2005). The reference strain PA14 used in many laboratories harbors a partial deletion of the *psl* locus, which prevents the PSL mannose-rich polysaccharide from being made (Friedman and Kolter 2004).

The soil-dwelling Gram-positive bacterium *B. subtilis* is also studied as a model organism for biofilm formation. Different *B. subtilis* strains are able to secrete two distinct polymers: the polysaccharide EPS and poly-δ-glutamate (PGA). Both of these molecules have been described to participate in the process of biofilm formation (Stanley and Lazazzera 2005; Branda et al. 2006). Yet, they contribute differently depending on the strain and conditions studied. For example, in colony biofilms the undomesticated strain NCIB3610 requires exopolysaccharide EPS for biofilm formation (Fig. 1). However, no colony biofilm defect is observed in a mutant strain lacking the ability

Figure 1. Colony morphology of *B. subtilis* strain 3610 wild type and matrix mutant (*eps*). Top view of cells after 3 d of growth on 1.5% agar MSgg medium. Bar is 5 mm.

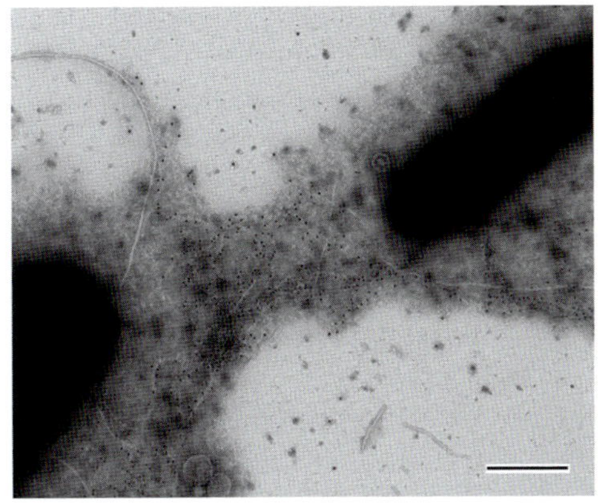

Figure 2. Electron micrograph of *B. subtilis* strain 3610 immunogold labeled with anti-TasA antibody (black dots). Bar is 0.5 μm. Image courtesy of Diego Romero.

to produce PGA (Branda et al. 2006). Instead, cells that overproduced PGA formed structureless, mucoid colonies. Another undomesticated strain of *B. subtilis*, RO-FF-1 naturally produces PGA and forms mucoid colonies. PGA production is important for surface-adhered biofilm formation in both RO-FF-1 and the laboratory strain JH642 (Stanley and Lazazzera 2005). In contrast, the strain NCIB3610 is unable to form robust surface-adhered biofilms (Branda et al. 2006).

Another bacterial model used to study biofilm formation is the Gram-positive pathogen *S. aureus*. Most strains of *S. aureus* use a polymer of *N*-acetyl glucosamine (PNAG) also referred to as polysaccharide intercellular adhesin (PIA), to form biofilms (O'Gara 2007). The *ica* operon encodes the machinery that synthesizes this polymer, yet not all strains carry this operon. Even in some of those strains that carry the *ica* operon, deletion of the operon does not impair their ability to make biofilm via an *ica*-independent pathway (O'Gara 2007; Otto 2008). This alternative mechanism relies on the ability of *S. aureus* to express a variety of adhesin proteins that allow cells to attach and colonize a large number of different surfaces (Lasa and Penades 2006).

As alluded to above, the extracellular matrix of biofilms also harbors adhesive proteins. For instance, *S. aureus* matrix harbors biofilm-associated proteins (termed Bap) that are required for biofilm formation (Lasa and Penades 2006; Latasa et al. 2006). These proteins are found anchored to the cell wall of *S. aureus* and serve to hold cells together within the biofilm, probably by interacting with other proteins on the surface of neighboring cells. In certain strains, the expression of Bap proteins eliminates the requirement for exopolysaccharides for biofilm formation (Cucarella et al. 2004). Unlike the multitude of Bap proteins found in *S. aureus* biofilms, *B. subtilis* expresses a single major protein associated with the extracellular matrix termed TasA. Mutants deficient in TasA fail to form biofilms despite the fact that they can still produce exopolysaccharide (Branda et al. 2006). TasA has recently been shown to form extracellular filaments that have amyloid-like properties and is thought to play a structural or architectural role in the extracellular matrix (Fig. 2) (Romero et al. 2010). TasA is not the first amyloid-like protein to be implicated in biofilm formation. In *E. coli*, the curli protein also forms amyloid filaments and is critical for biofilm formation (Chapman et al. 2002; Barnhart and Chapman 2006). Other proteinaceous structures important for biofilm formation are pili and fimbriae. These cell appendages are used to adhere cells to each other or to different surfaces. *E. coli* produces Type I fimbriae that are required for adherence to mannose-containing receptors. These fimbriae are important for biofilm formation on plastic surfaces as well as on host cells during urinary tract infections (Pratt and Kolter 1998;

Wright et al. 2007). *P. aeruginosa* also has many surface proteins that contribute to biofilm formation. For example, mutants unable to produce type IV pili or the CupA fimbriae were defective in surface-adhered biofilms (O'Toole and Kolter 1998; Vallet et al. 2001; D'Argenio et al. 2002). In addition, there are other matrix-associated lectin-binding proteins that recognize and bind carbohydrate moieties. These facilitate cell–matrix or cell–cell interactions within the biofilm. *P. aeruginosa*, has two lectin-binding proteins involved in biofilm formation (LecA and LecB) (Tielker et al. 2005; Diggle et al. 2006).

In addition to the exopolysaccharides and proteins, extracellular DNA (eDNA) also provides structural integrity to the biofilm. Biofilm matrix in *P. aeruginosa* contains significant amounts of eDNA. The addition of DNase to cultures inhibits biofilm formation and dissolves mature biofilms (Whitchurch et al. 2002). *S. aureus* biofilms also have eDNA in the matrix and it functions to provide stability to the biofilms. It is thought that this eDNA is caused by cell lysis and subsequent release of genomic DNA (Rice et al. 2007).

The participation of multiple types of molecules such as polysaccharides, DNA, and proteins in the formation of extracellular matrix makes it impossible to present a single summary of biofilm matrix. Similarly, although it is tempting to suggest matrix is a biofilm feature that could be targeted for biofilm control, the variation between matrices of different strains and species makes it impossible to find a unifying attribute. Furthermore, as described earlier, depending on the conditions, different components of the matrix become more or less important for the integrity of the biofilm.

CELL HETEROGENEITY WITHIN BIOFILMS

Although single-species biofilms can arise from a single cell and should therefore be clonal populations, they are often composed of phenotypically distinct subpopulations. Within biofilms specialized cell types often arise because of differences in gene expression but not in gene composition (Fux et al. 2005; An and Parsek 2007; Spormann 2008; Stewart and Franklin 2008). Cell differentiation in bacterial communities depends on the extracellular conditions to which cells are exposed. The formation of gradients of nutrients, oxygen or electron acceptors throughout the biofilm creates microenvironments to which cells respond by altering their gene expression (Spormann 2008; Stewart and Franklin 2008). For instance, in *P. aeruginosa* oxygen only penetrates the outer regions of the biofilm. When the enzymatic activity of the oxygen-dependent enzyme alkaline phosphatase was measured in cross sections of biofilms the activity of the enzyme was correlated with the cells located in areas more exposed to the oxygen (Xu et al. 1998). Similarly, *S. aureus* biofilms displayed an aerobic zone closer to the surface of the biofilm. When the metabolically active zones of the biofilm were identified by localizing the areas where DNA and protein synthesis occured, two distinct strata were observed. One correlated with the area exposed to oxygen and a second was at the base of the colony at the agar surface closer to the nutrients. This indicates that in *S. aureus* about two-thirds of the biofilm was metabolically inactive (Rani et al. 2007).

The physiological state of cells within a biofilm can also be monitored by analyzing cell-type specific gene expression for each defined subpopulation of cells. This technique can be applied only when each cell differentiation pathway is well understood at the molecular level, as is the case for the model organism *B. subtilis*. This organism sporulates during starvation, forming metabolically inactive spores resistant to many environmental stresses (Piggot and Hilbert 2004). Using *lacZ* or GFP transcriptional fusions to sporulation-specific genes, sporulating cells were observed to preferentially localize in the aerial structures that form on colony biofilms (Branda et al. 2001; Veening et al. 2006). The localization of spores to the apical region of the aerial structures was confirmed at the single-cell level by thin sectioning frozen colony biofilms of cells that had cell-type specific promoters fused to fluorescent proteins. This technique was applied to the localization of other subpopulation of cells such as matrix producers or motile cells, cells that express flagella that allow these cells to swim. Sporulation, matrix production, and motility were shown to occur in distinct subpopulations within the biofilm (Fig. 3) (Vlamakis et al. 2008). Furthermore, the percentage and localization of each cell type was dynamic. Early stages of biofilm formation show an abundance of motile cells whereas as the biofilm matured, many of the motile cells differentiated into matrix producing cells. At later time points the subpopulation of sporulating cells arose primarily from the matrix producing subpopulation (Vlamakis et al. 2008). Strains harboring fluorescent reporters have also been used to detect the presence of surfactin-producing cells, the signaling molecule that triggers the differentiation of matrix producers (Lopez et al. 2009c). As we discuss later, this unidirectional signaling represents a remarkable event in the field of bacterial cell differentiation.

SIGNALING IN BIOFILM FORMATION

Because the formation of a biofilm can be considered a mechanism to protect the bacterial community from external insults, it seems reasonable that specific extracellular cues regulate activation of the metabolic pathways that lead to biofilm formation. These external cues come from diverse sources. Signals can be produced and secreted by the bacterial community itself, in which case the molecules are termed autoinducers. Autoinducers accumulate extracellularly and the concentration of autoinducer can be correlated with population density. At high concentrations, autoinducers trigger signal transduction cascades that lead to multicellular responses in the bacterial population. This mechanism of cell–cell communication in bacteria (termed quorum sensing) controls a large number of developmental processes included those related to biofilm formation (Camilli and Bassler 2006).

QUORUM-SENSING MOLECULES THAT INDUCE BIOFILM FORMATION

In *P. aeruginosa*, along with many other Gram-negative organisms, quorum-sensing systems respond to a class of autoinducer termed acyl homoserine lactones (AHLs). *P. aeruginosa* possesses two AHL quorum-sensing systems: *las* and *rhl*. Each system has its own AHL synthase (LasI and RhlI) and its own transcriptional regulator (LasR and RhlR). The AHL signals produced by the synthases are *N*-(3-oxododecanoyl)-HSL and *N*-butyryl-HSL,

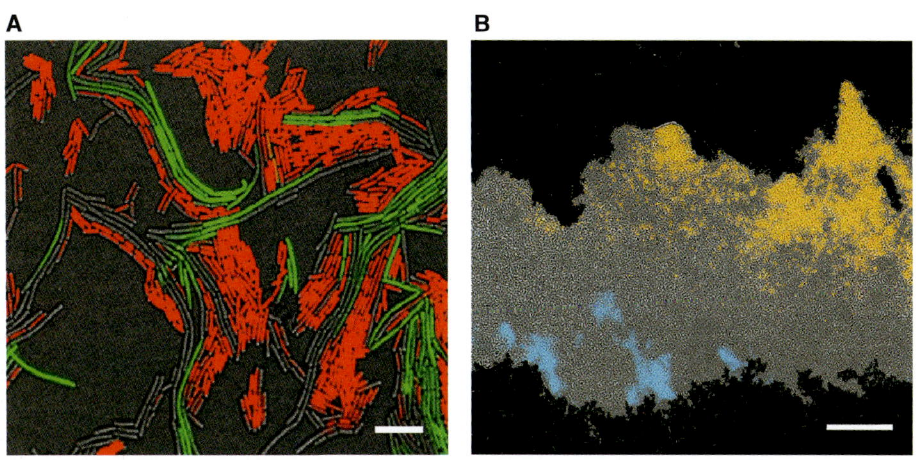

Figure 3. Heterogeneity in *B. subtilis* biofilms. (*A*) Top view of cells at the onset of colony development. Overlay of fluorescence images with DIC (gray), motile (red), and matrix-producing (green) cells. Bar 5 μm. (*B*) Thin-sectioned three-day-old biofilm. Agar is at the bottom and the center of the colony is on the right. Overlay of fluorescence images with DIC (gray), motile (blue), and sporulating (orange) cells. Bar 50 μm.

respectively (de Kievit 2009). Work by many groups has found that depending on the strain and experimental conditions, the importance of these two systems in biofilm formation varies (Hentzer et al. 2004; de Kievit 2009). In strain PAO1 both Las and Rhl systems were important for extracellular DNA release, which plays a role in biofilm matrix and structure (Allesen-Holm et al. 2006). In the strain PA14 the Las system is essential for biofilm architecture probably through control of the production of the PEL exopolysaccharide (Sauer et al. 2002; Sakuragi and Kolter 2007; Yang et al. 2007).

In Gram-positive organisms, the autoinducers are often peptides and these are detected outside the cell. To be detected extracellularly, autoinducer molecules are generally sensed by membrane-associated sensor kinases, which activate cognate response regulators by phosphorylation. That, in turn, activates the expression of the target genes (Novick and Geisinger 2008). In *S. aureus*, the autoinducer molecule is a peptide (AIP) derived from the product of the *agrD* gene. This peptide is processed to yield a cyclic peptide containing a thiolactone ring. AIP is secreted and detected by AgrC, which activates the regulator AgrA. AgrA positively regulates the transcription of genes including those that code for several extracellular proteases involved in the dispersal of the biofilm (Balaban and Novick 1995; Ji et al. 1995; Yarwood et al. 2004; O'Gara 2007). Thus, in the case of *S. aureus*, quorum-sensing negatively regulates biofilm formation (Boles and Horswill 2008). Biofilm formation in *S. aureus* involves several sequential stages such as initial attachment, cell-to-cell adhesion, maturation, and final detachment (Otto 2004). Adhesion to a surface is the initial step to transition from planktonic cells to biofilm formation in *S. aureus*. This step is favored only when the *agr* quorum-sensing systems is inhibited. Once cells successfully attach to a surface, bacteria accumulate forming a complex architecture, which involves the production of exopolysaccharide (known as PNAG or PIA). Eventually small clusters of cells detach from the mature biofilm; a step that is important for the dispersal of the community (Yarwood et al. 2004; Yao and Strauch 2005). Mutants lacking the *agr* gene form thicker biofilms than wild type. This is not attributed to cell growth or death, but rather to the inability of cells to detach from the mature biofilm (Vuong et al. 2000).

For *B. subtilis*, the production and secretion of the quorum-sensing molecule surfactin is important for biofilm formation (Lopez et al. 2009a). Aside from its surfactant properties, surfactin causes potassium leakage from the cytoplasm. Potassium leakage is sensed by a membrane-associated sensor kinase, KinC, to specifically trigger the expression of the genes involved in extracellular matrix production (Lopez et al. 2009a). Because of the nature of the stimulus, various small-molecules were identified that induce matrix production via KinC. These differ largely in their molecular structure. What is conserved is their ability to induce potassium leakage by making ion-selective pores in the membrane of *B. subtilis*. Among these molecules are the macrolide polyenes nystatin and amphotericin as well as the peptide antibiotics gramidicin and valinomycin, all of which are produced by soil-dwelling bacteria. This particular mechanism of quorum-sensing that recognizes the mode of action of the signaling molecule rather than its structure permits *B. subtilis* to sense and respond to a diverse number of signals. These signals are not only to self-produced molecules but also natural products that are secreted by other soil-dwelling organisms (Lopez et al. 2009a).

Despite the fact that surfactin is a secreted molecule that should be able to interact with all of the cells within the population, not all of the cells respond to the presence of the molecule. This heterogeneity in response can be explained by the mechanism of gene activation (Lopez et al. 2009a). Once surfactin activates the sensor kinase KinC, that leads to activation of the kinase's cognate regulator, Spo0A. Spo0A is a transcriptional regulator whose activity depends on the level of phosphorylated protein within the cell. Matrix gene expression is triggered only when Spo0A~P accumulates above certain levels (Fujita et al. 2005). This mechanism restricts the expression of matrix-related genes to the subpopulation of cells that accumulates the required level of Spo0A~P, triggering a cascade of events resulting in a bimodal population where some cells produce matrix and others do not (Chai et al. 2008).

Interestingly, the subpopulation of surfactin producers is different from the subpopulation of cells (matrix producers) that respond to surfactin (Lopez et al. 2009c). The *srfAA-AD* operon is responsible for surfactin production and is directly controlled by another regulator termed ComA. Activation of ComA by phosphorylation (ComA~P) is driven by another quorum-sensing system that senses the presence of the extracellular pheromone ComX via the membrane-associated sensor kinase ComP (Magnuson et al. 1994; Nakano 1991). This system, in which a primary signal controls the production of secondary system, might serve as a timing mechanism to regulate the activation of diverse metabolic pathways sequentially during the course of development.

Surfactin acts as a uni-directional signal in this particular quorum-sensing system, in which one subpopulation of cells produces the molecule, whereas another population of cells responds to it and produces matrix (Lopez et al. 2009c). This mechanism adds more complexity to the concept of "quorum sensing" or, as we have referred previously "autoinduction," where all cells are physiologically similar thus, able to produce the signal and respond to it (Fuqua et al. 1994; Camilli and Bassler 2006). In the typical quorum-sensing scenario, signaling would be referred as autocrine. The uni-directional signaling recently described in *B. subtilis* is the first example of paracrine signaling in bacteria where the subpopulation producing the signal does not respond to it (Lopez et al. 2009c).

MOLECULES THAT INDUCE BIOFILM FORMATION INDEPENDENT OF QUORUM SENSING

In addition to quorum-sensing molecules, a diversity of other signals trigger biofilm formation. These include secondary metabolites such as antibiotics, pigments, and siderophores. At subinhibitory concentrations many antibiotics function not to kill cells, but rather as signals that trigger changes in gene expression (Yim et al. 2007). Subinhibitory concentrations of the antibiotic imipenem induced expression of the polysaccharide alginate in *P. aeruginosa* biofilms (Bagge et al. 2004). Hence, imipenem-exposed biofilms were thicker and covered more of the substratum than nontreated biofilms. Similarly, subinhibitory concentrations of the aminoglycoside antibiotic tobramycin induced biofilm formation in *P. aeruginosa* and *E. coli* (Hoffman et al. 2005). The mechanism of action of this signaling process is not understood.

In addition to sensing the presence of antibiotics produced by other organisms, *P. aeruginosa* responds to small-molecules that it produces. For example, the redox-active pigments termed phenazines have been described to have antibiotic activity or function as virulence factors in eukaryotic hosts (Price-Whelan et al. 2006). Within biofilms, the phenazine pyocyanin functions in extracellular electron transfer to generate energy for growth. Having a small, diffusible molecule to shuttle electrons in a biofilm where the diffusion solubility may be limited is beneficial for the community (Hernandez and Newman 2001). Phenazines in *P. aeruginosa* also function as signaling molecules in biofilm formation, as a mutant unable to produce phenazines produced dramatically more wrinkled colony morphology than a wild-type strain (Dietrich et al. 2008). This difference was because of the induction of SoxR-regulated genes in response to phenazines; both phenazine overproducing strains and *soxR* mutant strains formed flat, featureless colonies.

In the case of *S. aureus*, the activation of the quorum-sensing system inhibits biofilm formation. Thus, small molecules that inhibit quorum sensing also favor biofilm formation. This was recently described for the furanones, which are natural products derived from marine algae (de Nys et al. 2006). These small molecules are able to inhibit the quorum-sensing systems of many Gram-negative bacteria (Wu et al. 2004). Additionally, the molecules were tested for inhibition of the quorum-sensing system in *Staphylococci*. At high concentrations, furanones had bactericidal effect on *S. epidermidis* and *S. aureus* Interestingly, *S. aureus* treated with subinhibitory concentrations of the marine furanones resulted in an inhibition of the quorum-sensing system coupled with an increase in the ability of *S. aureus* to make biofilms (Kuehl et al. 2009). Similar results were previously observed with subinhibitory concentrations of other well-known antimicrobials such as tetracycline or quinupristin-dalfopristin. Cultures of *Staphylococcus epidermidis* treated with subinhibitory concentrations of these molecules enhance the expression of genes responsible for exopolysaccharide production. Also, a weaker induction was observed when treated with sub-inhibitory concentrations of the antibiotic erythromycin. The mechanism underlying this effect is not well understood yet (Rachid et al. 2000).

Other small-molecules induce biofilm formation in *B. subtilis* independent of quorum sensing. Specific molecules with antibiotic properties trigger differentiation of cells into the subpopulation of matrix producers. This occurs though the Spo0A genetic pathway that triggers differentiation into matrix producers. In addition to regulating matrix gene expression, Spo0A~P also triggers a second differentiation pathway called cannibalism (Gonzalez-Pastor et al. 2003; Ellermeier et al. 2006; Claverys and Havarstein 2007). Cells expressing cannibalism genes produce, and are resistant to, two toxins: Skf and Sdp. As described earlier, different gene sets are regulated by different levels of Spo0A phosphate, thus only the cells in the population that have achieved high enough levels of Spo0A~P are able to express the cannibalism toxins and resistance machinery (Fujita et al. 2005). This leaves a sensitive portion of the population that has not achieved high enough levels of Spo0A~P. It is this sensitive portion of the population that lyses once some cells express *skf* and *sdp* genes. Dead siblings serve as food for the community to overcome nutritional limitation, and this delays the onset of sporulation. This could benefit the community because spore development is energy intensive and, once committed, cells may not exit this state for prolonged periods. Thus, by sacrificing a portion of the population, *B. subtilis* can delay the entry into sporulation for as long as possible (Gonzalez-Pastor et al. 2003; Ellermeier et al. 2006).

Both matrix production and cannibalism are triggered by the same genetic cascade, and these two traits are indeed expressed concomitantly in the same subpopulation of cells (Lopez et al. 2009b). Consistent with this, the signaling molecule surfactin, which is responsible for differentiation of the subpopulation of matrix producers, also triggers cannibal toxin production. When the cannibalism toxins are secreted to the extracellular space, only the subpopulation of matrix producers are favored to grow, because it is the only subpopulation that expresses the immunity machinery to the action of the cannibalism toxins. This gives that subpopulation the advantage in that they can use the nutrients released by their killed siblings causing the matrix/cannibal cells to increase relative to other cell types. With the increase in the relative number of matrix-producing cells, these communities are able to produce more extracellular matrix and form stronger biofilms. This behavior constitutes a mechanism to eliminate cell types that might no longer be required for the development of the community to promote the growth of other subpopulations such as matrix producers (Lopez et al. 2009b).

Other antimicrobial peptides could mimic the effect of the cannibalism toxins. This is because of the ability of the cannibalism resistance machinery to work nonspecifically for several similar molecules (Butcher and Helmann 2006). One such molecule is nisin, a peptide antibitiotic putatively similar to the Skf toxin. The presence of nisin therefore promotes the subpopulation of matrix producers in *B. subtilis* communities much like the cannibalism toxins (Fig. 4) (Lopez et al. 2009b). This ability to sense small-molecules produced by diverse soil microorganisms suggests a broad mechanism that *B. subtilis* uses to respond to surrounding bacterial communities by altering its development.

Figure 4. Effect of the antimicrobial nisin on *B. subtilis* biofilm morphology. Cells closer to the disk containing nisin are more wrinkled due to the presence of more matrix-producing cells. Bar is 3 mm. (Reprinted, with permission, from Lopez et al 2009b [© Wiley].)

SUMMARY

As presented, the majority of microbes are able to develop multicellular biofilm communities. These communities are composed of subpopulations of different cell types that provide additional benefits to the organisms. There are numerous differences among the mechanisms that induce biofilm formation in different species. Even considering only four of the best-known bacterial models (*E. coli*, *P. aeruginosa*, *S. aureus*, and *B. subtilis*), the discrepancies between the processes involved in the formation of biofilms among these examples are broad. A vast array of exopolysaccharides, secreted proteins and cell-surface adhesins contribute to the structural integrity of biofilms. Small molecules such as homoserine lactones, antibiotics, and other secondary metabolites play a critical role in the development and maintenance of biofilm communities.

REFERENCES

Allesen-Holm M, Barken KB, Yang L, Klausen M, Webb JS, Kjelleberg S, Molin S, Givskov M, Tolker-Nielsen T. 2006. A characterization of DNA release in *Pseudomonas aeruginosa* cultures and biofilms. *Mol Microbiol* **59:** 1114–1128.

An D, Parsek MR. 2007. The promise and peril of transcriptional profiling in biofilm communities. *Curr Opin Microbiol* **10:** 292–296.

Anderson GG, O'Toole GA. 2008. Innate and induced resistance mechanisms of bacterial biofilms. in *Bacterial Biofilms* (ed. Romeo T.), pp. 85–105. Springer, Heidelberg.

Bagge N, Schuster M, Hentzer M, Ciofu O, Givskov M, Greenberg EP, Hoiby N. 2004. *Pseudomonas aeruginosa* biofilms exposed to imipenem exhibit changes in global gene expression and β-lactamase and alginate production. *Antimicrob Agents Chemother* **48:** 1175–1187.

Balaban N, Novick RP. 1995. Autocrine regulation of toxin synthesis by *Staphylococcus aureus*. *Proc Natl Acad Sci* **92:** 1619–1623.

Barnhart MM, Chapman MR. 2006. Curli biogenesis and function. *Annu Rev Microbiol* **60:** 131–147.

Boles BR, Horswill AR. 2008. Agr-mediated dispersal of *Staphylococcus aureus* biofilms. *PLoS Pathog* **4:** e1000052.

Branda SS, Chu F, Kearns DB, Losick R, Kolter R. 2006. A major protein component of the *Bacillus subtilis* biofilm matrix. *Mol Microbiol* **59:** 1229–1238.

Branda SS, Gonzalez-Pastor JE, Ben-Yehuda S, Losick R, Kolter R. 2001. Fruiting body formation by *Bacillus subtilis*. *Proc Natl Acad Sci* **98:** 11621–11626.

Branda SS, Vik S, Friedman L, Kolter R. 2005. Biofilms: The matrix revisited. *Trends Microbiol* **13:** 20–26.

Butcher BG, Helmann JD. 2006. Identification of *Bacillus subtilis* σ-dependent genes that provide intrinsic resistance to antimicrobial compounds produced by Bacilli. *Mol Microbiol* **60:** 765–782.

Camilli A, Bassler BL. 2006. Bacterial small-molecule signaling pathways. *Science* **311:** 1113–1116.

Chai Y, Chu F, Kolter R, Losick R. 2008. Bistability and biofilm formation in *Bacillus subtilis*. *Mol Microbiol* **67:** 254–263.

Chapman MR, Robinson LS, Pinkner JS, Roth R, Heuser J, Hammar M, Normark S, Hultgren SJ. 2002. Role of *Escherichia coli* curli operons in directing amyloid fiber formation. *Science* **295:** 851–855.

Claverys JP, Havarstein LS. 2007. Cannibalism and fratricide: Mechanisms and raisons d'etre. *Nat Rev Microbiol* **5:** 219–229.

Cucarella C, Tormo MA, Ubeda C, Trotonda MP, Monzon M, Peris C, Amorena B, Lasa I, Penades JR. 2004. Role of biofilm-associated protein bap in the pathogenesis of bovine *Staphylococcus aureus*. *Infect Immun* **72:** 2177–2185.

Currie CR. 2001. A community of ants, fungi, and bacteria: A multilateral approach to studying symbiosis. *Annu Rev Microbiol* **55:** 357–380.

D'Argenio DA, Calfee MW, Rainey PB, Pesci EC. 2002. Autolysis and autoaggregation in *Pseudomonas aeruginosa* colony morphology mutants. *J Bacteriol* **184:** 6481–6489.

Danhorn T, Fuqua C. 2007. Biofilm formation by plant-associated bacteria. *Annu Rev Microbiol* **61:** 401–422.

de Carvalho CC. 2007. Biofilms: Recent developments on an old battle. *Recent Pat Biotechnol* **1:** 49–57.

de Kievit TR. 2009. Quorum sensing in *Pseudomonas aeruginosa* biofilms. *Environ Microbiol* **11:** 279–288.

de Nys R, Givskov M, Kumar N, Kjelleberg S, Steinberg PD. 2006. Furanones. *Prog Mol Subcell Biol* **42:** 55–86.

Dietrich LE, Teal TK, Price-Whelan A, Newman DK. 2008. Redox-active antibiotics control gene expression and community behavior in divergent bacteria. *Science* **321:** 1203–1206.

Diggle SP, Stacey RE, Dodd C, Camara M, Williams P, Winzer K. 2006. The galactophilic lectin, LecA, contributes to biofilm development in *Pseudomonas aeruginosa*. *Environ Microbiol* **8:** 1095–1104.

Donlan RM. 2008. Biofilms on central venous catheters: Is eradication possible? *Curr Top Microbiol Immunol* **322:** 133–161.

Ellermeier CD, Hobbs EC, Gonzalez-Pastor JE, Losick R. 2006. A three-protein signaling pathway governing immunity to a bacterial cannibalism toxin. *Cell* **124:** 549–559.

Friedman L, Kolter R. 2004. Two genetic loci produce distinct carbohydrate-rich structural components of the *Pseudomonas aeruginosa* biofilm matrix. *J Bacteriol* **186:** 4457–4465.

Fujita M, Gonzalez-Pastor JE, Losick R. 2005. High- and low-threshold genes in the Spo0A regulon of *Bacillus subtilis*. *J Bacteriol* **187:** 1357–1368.

Fuqua WC, Winans SC, Greenberg EP. 1994. Quorum sensing in bacteria: The LuxR-LuxI family of cell density-responsive transcriptional regulators. *J Bacteriol* **176:** 269–275.

Fux CA, Costerton JW, Stewart PS, Stoodley P. 2005. Survival strategies of infectious biofilms. *Trends Microbiol* **13:** 34–40.

Gonzalez-Pastor JE, Hobbs EC, Losick R. 2003. Cannibalism by sporulating bacteria. *Science* **301:** 510–513.

Hall-Stoodley L, Stoodley P. 2009. Evolving concepts in biofilm infections. *Cell Microbiol* **11:** 1034–1043.

Hall-Stoodley L, Costerton JW, Stoodley P. 2004. Bacterial biofilms: From the natural environment to infectious diseases. *Nat Rev Microbiol* **2:** 95–108.

Hatt JK, Rather PN. 2008. Role of bacterial biofilms in urinary tract infections. in *Bactarial Biofilms* (ed. Romeo T.), 163–192. Springer, Heidelberg.

Hentzer M, Givskov M, Eberl L. 2004. Quorum sensing in biofilms: Gossip in slime city. in *Microbial Biofilms* (ed. O'Toole M.G.a.G.), pp. 118–140. ASM Press, Washington DC.

Hernandez ME, Newman DK. 2001. Extracellular electron transfer. *Cell Mol Life Sci* **58:** 1562–1571.

Hoffman LR, D'Argenio DA, MacCoss MJ, Zhang Z, Jones RA, Miller SI. 2005. Aminoglycoside antibiotics induce bacterial biofilm formation. *Nature* **436:** 1171–1175.

Ji G, Beavis RC, Novick RP. 1995. Cell density control of staphylococcal virulence mediated by an octapeptide pheromone. *Proc Natl Acad Sci* **92:** 12055–12059.

Kreth J, Zhang Y, Herzberg MC. 2008. Streptococcal antagonism in oral biofilms: *Streptococcus sanguinis* and *Streptococcus gordonii* interference with *Streptococcus mutans*. *J Bacteriol* **190:** 4632–4640.

Kuehl R, Al-Bataineh S, Gordon O, Luginbuehl R, Otto M, Textor M, Landmann R. 2009. Furanone enhances biofilm of staphylococci at subinhibitory concentrations by luxS repression. *Antimicrob Agents Chemother* **53:** 4157–4166.

Lasa I, Penades JR. 2006. Bap: A family of surface proteins involved in biofilm formation. *Res Microbiol* **157:** 99–107.

Latasa C, Solano C, Penades JR, Lasa I. 2006. Biofilm-associated proteins. *C R Biol* **329:** 849–857.

Lewis K. 2005. Persister cells and the riddle of biofilm survival. *Biochemistry (Mosc)* **70:** 267–274.

Lopez D, Kolter R. 2009. Extracellular signals that define distinct and coexisting cell fates in *Bacillus subtilis*. *FEMS Microbiol Rev.*

Lopez D, Fischbach MA, Chu F, Losick R, Kolter R. 2009a. Structurally diverse natural products that cause potassium leakage trigger multicellularity in *Bacillus subtilis*. *Proc Natl Acad Sci* **106:** 280–285.

Lopez D, Vlamakis H, Losick R, Kolter R. 2009b. Cannibalism enhances biofilm development in *Bacillus subtilis*. *Mol Microbiol* **74:** 609–618.

Lopez D, Vlamakis H, Losick R, Kolter R. 2009c. Paracrine signaling in a bacterium. *Genes Dev* **23:** 1631–1638.

Magnuson R, Solomon J, Grossman AD. 1994. Biochemical and genetic characterization of a competence pheromone from *B. subtilis*. *Cell* **77:** 207–216.

Mah TF, O'Toole GA. 2001. Mechanisms of biofilm resistance to antimicrobial agents. *Trends Microbiol* **9:** 34–39.

Martin MF, Liras P. 1989. Organization and expression of genes involved in the biosynthesis of antibiotics and other secondary metabolites. *Annu Rev Microbiol* **43:** 173–206.

Matz C, Kjelleberg S. 2005. Off the hook–how bacteria survive protozoan grazing. *Trends Microbiol* **13:** 302–307.

Monds RD, O'Toole GA. 2009. The developmental model of microbial biofilms: Ten years of a paradigm up for review. *Trends Microbiol* **17:** 73–87.

Nakano MM, Xia LA, Zuber P. 1991. Transcription initiation region of the srfA operon, which is controlled by the comP-comA signal transduction system in *Bacillus subtilis*. *J Bacteriol* **173:** 5487–5493.

Novick RP, Geisinger E. 2008. Quorum sensing in staphylococci. *Annu Rev Genet* **42:** 541–564.

O'Gara JP. 2007. ica and beyond: Biofilm mechanisms and regulation in *Staphylococcus epidermidis* and *Staphylococcus aureus*. *FEMS Microbiol Lett* **270:** 179–188.

O'Toole GA, Kolter R. 1998. Flagellar and twitching motility are necessary for *Pseudomonas aeruginosa* biofilm development. *Mol Microbiol* **30:** 295–304.

Otto M. 2004. Quorum-sensing control in Staphylococci – a target for antimicrobial drug therapy? *FEMS Microbiol Lett* **241:** 135–141.

Otto M. 2008. Staphylococcal biofilms. in *Bacterial Biofilms* (ed. Romeo T.), pp. 207–228. Springer, Heidelberg.

Piggot PJ, Hilbert DW. 2004. Sporulation of *Bacillus subtilis*. *Curr Opin Microbiol* **7:** 579–586.

Pratt LA, Kolter R. 1998. Genetic analysis of Escherichia coli biofilm formation: Roles of flagella, motility, chemotaxis and type I pili. *Mol Microbiol* **30:** 285–293.

Price-Whelan A, Dietrich LE, Newman DK. 2006. Rethinking 'secondary' metabolism: Physiological roles for phenazine antibiotics. *Nat Chem Biol* **2:** 71–78.

Rachid S, Ohlsen K, Witte W, Hacker J, Ziebuhr W. 2000. Effect of subinhibitory antibiotic concentrations on polysaccharide intercellular adhesin expression in biofilm-forming *Staphylococcus epidermidis. Antimicrob Agents Chemother* **44:** 3357–3363.

Rani SA, Pitts B, Beyenal H, Veluchamy RA, Lewandowski Z, Davison WM, Buckingham-Meyer K, Stewart PS. 2007. Spatial patterns of DNA replication, protein synthesis, and oxygen concentration within bacterial biofilms reveal diverse physiological states. *J Bacteriol* **189:** 4223–4233.

Rice KC, Mann EE, Endres JL, Weiss EC, Cassat JE, Smeltzer MS, Bayles KW. 2007. The cidA murein hydrolase regulator contributes to DNA release and biofilm development in Staphylococcus aureus. *Proc Natl Acad Sci* **104:** 8113–8118.

Romero D, Aguilar C, Losick R, Kolter R. 2010. Amyloid fibers provide structural integrity to *Bacillus subtilis* BIOFILMS. *Proc Natl Acad Sci* **107:** 2230–2234.

Ryder C, Byrd M, Wozniak DJ. 2007. Role of polysaccharides in *Pseudomonas aeruginosa* biofilm development. *Curr Opin Microbiol* **10:** 644–648.

Sakuragi Y, Kolter R. 2007. Quorum-sensing regulation of the biofilm matrix genes (pel) of *Pseudomonas aeruginosa. J Bacteriol* **189:** 5383–5386.

Sauer K, Camper AK, Ehrlich GD, Costerton JW, Davies DG. 2002. *Pseudomonas aeruginosa* displays multiple phenotypes during development as a biofilm. *J Bacteriol* **184:** 1140–1154.

Spormann AM. 2008. Physiology of microbes in biofilms. in *Bacterial Biofilms* (ed. Romeo T.), pp. 17–36. Springer, Heidelberg.

Stanley NR, Lazazzera BA. 2005. Defining the genetic differences between wild and domestic strains of *Bacillus subtilis* that affect poly-gamma-dl-glutamic acid production and biofilm formation. *Mol Microbiol* **57:** 1143–1158.

Stewart PS, Franklin MJ. 2008. Physiological heterogeneity in biofilms. *Nat Rev Microbiol* **6:** 199–210.

Tart AH, Wozniak DJ. 2008. Shifting paradigms in *Pseudomonas aeruginosa* biofilm research. in *Bacterial Biofilms* (ed. Romeo T.), pp. 193–206. Springer, Heidelberg.

Tielker D, Hacker S, Loris R, Strathmann M, Wingender J, Wilhelm S, Rosenau F, Jaeger KE. 2005. *Pseudomonas aeruginosa* lectin LecB is located in the outer membrane and is involved in biofilm formation. *Microbiology* **151:** 1313–1323.

Vallet I, Olson JW, Lory S, Lazdunski A, Filloux A. 2001. The chaperone/usher pathways of *Pseudomonas aeruginosa*: Identification of fimbrial gene clusters (cup) and their involvement in biofilm formation. *Proc Natl Acad Sci* **98:** 6911–6916.

Veening JW, Kuipers OP, Brul S, Hellingwerf KJ, Kort R. 2006. Effects of phosphorelay perturbations on architecture, sporulation, and spore resistance in biofilms of *Bacillus subtilis. J Bacteriol* **188:** 3099–3109.

Vlamakis H, Aguilar C, Losick R, Kolter R. 2008. Control of cell fate by the formation of an architecturally complex bacterial community. *Genes Dev* **22:** 945–953.

Vuong C, Saenz HL, Gotz F, Otto M. 2000. Impact of the agr quorum-sensing system on adherence to polystyrene in *Staphylococcus aureus. J Infect Dis* **182:** 1688–1693.

Whitchurch CB, Tolker-Nielsen T, Ragas PC, Mattick JS. 2002. Extracellular DNA required for bacterial biofilm formation. *Science* **295:** 1487.

Wright KJ, Seed PC, Hultgren SJ. 2007. Development of intracellular bacterial communities of uropathogenic *Escherichia coli* depends on type 1 pili. *Cell Microbiol* **9:** 2230–2241.

Wu H, Song Z, Hentzer M, Andersen JB, Molin S, Givskov M, Hoiby N. 2004. Synthetic furanones inhibit quorum-sensing and enhance bacterial clearance in *Pseudomonas aeruginosa* lung infection in mice. *J Antimicrob Chemother* **53:** 1054–1061.

Xu KD, Stewart PS, Xia F, Huang CT, McFeters GA. 1998. Spatial physiological heterogeneity in *Pseudomonas aeruginosa* biofilm is determined by oxygen availability. *Appl Environ Microbiol* **64:** 4035–4039.

Yang L, Barken KB, Skindersoe ME, Christensen AB, Givskov M, Tolker-Nielsen T. 2007. Effects of iron on DNA release and biofilm development by *Pseudomonas aeruginosa. Microbiology* **153:** 1318–1328.

Yao F, Strauch MA. 2005. Independent and interchangeable multimerization domains of the AbrB, Abh, and SpoVT global regulatory proteins. *J Bacteriol* **187:** 6354–6362.

Yarwood JM, Bartels DJ, Volper EM, Greenberg EP. 2004. Quorum sensing in *Staphylococcus aureus* biofilms. *J Bacteriol* **186:** 1838–1850.

Yim G, Wang HH, Davies J. 2007. Antibiotics as signalling molecules. *Philos Trans R Soc Lond B Biol Sci* **362:** 1195–1200.

Electron Cryotomography

Elitza I. Tocheva[1], Zhuo Li[1,2], and Grant J. Jensen[1,2]

[1]Division of Biology, California Institute of Technology, Pasadena, California 91125

[2]Howard Hughes Medical Institute, California Institute of Technology, Pasadena, California 91125

Correspondence: jensen@caltech.edu

Electron cryotomography (ECT) is an emerging technology that allows thin samples such as macromolecular complexes and small bacterial cells to be imaged in 3-D in a nearly native state to "molecular" (~4 nm) resolution. As such, ECT is beginning to deliver long-awaited insight into the positions and structures of cytoskeletal filaments, cell wall elements, motility machines, chemoreceptor arrays, internal compartments, and other ultrastructures. This article describes the technique and summarizes its contributions to bacterial cell biology. For comparable recent reviews, see (Subramaniam 2005; Jensen and Briegel 2007; Murphy and Jensen 2007; Li and Jensen 2009). For reviews on the history, technical details, and broader application of electron tomography in general, see for example (Subramaniam and Milne 2004; Lucić et al. 2005; Leis et al. 2008; Midgley and Dunin-Borkowski 2009).

INTRODUCTION—THE STORY OF FtsZ

ECT can produce three-dimensional (3-D) reconstructions of intact cells in near-native states to "molecular" resolution (~4 nm), and has thus begun providing unprecedented views into the ultrastructure of bacterial cells. The exciting potential of ECT as compared to other light and electron microscopy techniques is powerfully illustrated by the history of FtsZ.

In 1980 Lutkenhaus et al. identified a temperature-sensitive, filament-forming *Escherichia coli* mutant unable to undergo cell division but unaffected in its ability to replicate and segregate DNA (Lutkenhaus et al. 1980). The gene responsible for the observed phenotype was identified and named *ftsZ* (for *f*ilamentous *t*emperature *s*ensitive *Z*). A decade later, Bi and Lutkenhaus showed through immuno-electron microscopy that FtsZ protein localized near the constriction site of dividing cells (Bi and Lutkenhaus 1991). This seminal study was performed using what we would now call "traditional" electron microscopy (EM): exponentially growing *E. coli* cells were chemically fixed (2% glutaraldehyde), pelleted and washed in buffer, dehydrated in a gradient of ethanol (from 30% to 100%), stained with uranyl acetate, resin embedded, cured at 60°C, sectioned, exposed to an FtsZ-specific antibody, and finally reacted with a colloidal-gold-labeled secondary antibody. Although longitudinal cell sections like the one shown in Figure 1A showed that FtsZ localized to the mid-plane of dividing cells, unfortunately no detailed ultrastructure was visible, probably because the fixation,

Cite this article as *Cold Spring Harb Perspect Biol* 2010;2:a003442

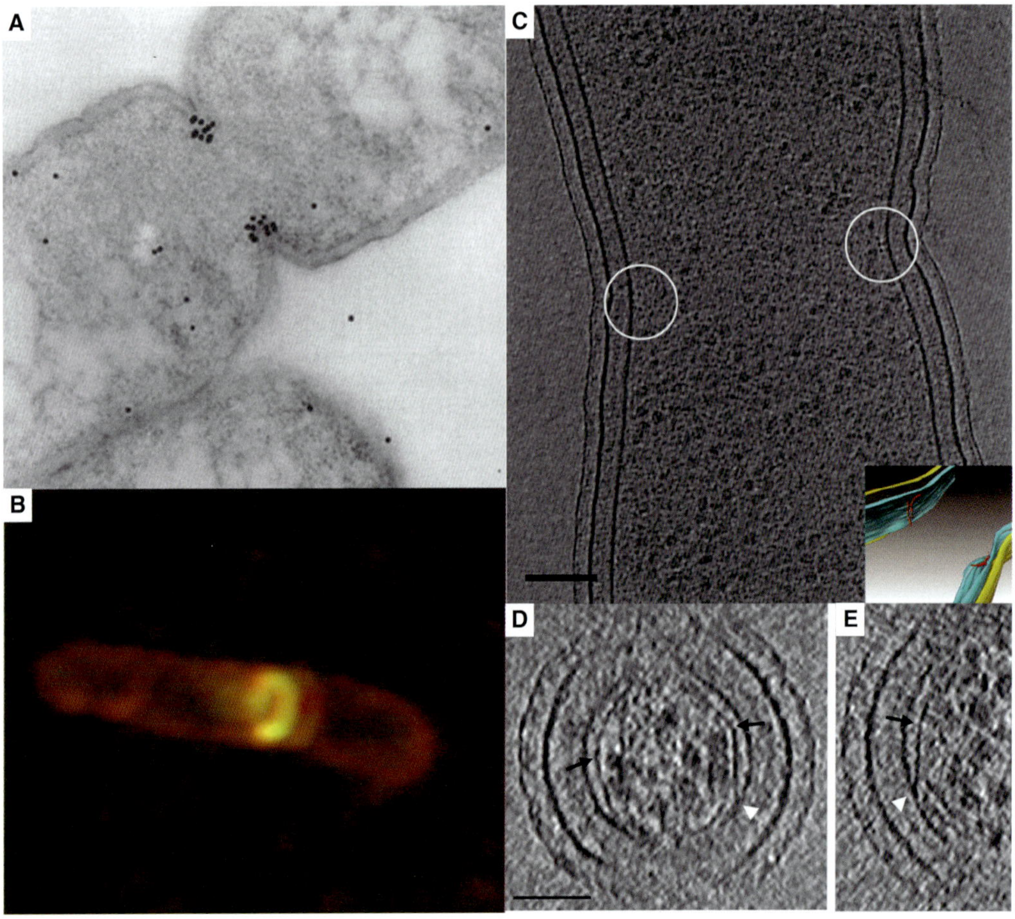

Figure 1. Microscopy of FtsZ. (*A*) Immunolocalization of FtsZ in a dividing *Escherichia coli* cell. The location of FtsZ is indicated by the black dots, which are gold-labeled antibodies. Adapted from (Lutkenhaus 1993) with permission from Blackwell Publishing ltd. (*B*) Fluorescent light microscopy image of *E. coli* cells expressing low levels of FtsZ-GFP showing the formation of a new FtsZ ring (gold) in a dividing cell (outlined in red). Reprinted from (Margolin 1998), © 1998, with permission from Elsevier Ltd. (*C*) An 8-nm thick cryo-tomographic slice through a dividing *Caulobacter crescentus* cell showing FtsZ filaments in cross-section (small dark dots near the center of the circles next to the membrane). Scale bar 100 nm. The inset shows the 3-D segmentation of the same cell. The inner membrane, outer membrane and the FtsZ arclike filaments are colored in blue, yellow and red, respectively. (*D*) and (*E*) 6.7-nm thick tomographic slices containing examples of straight and curved (white arrow) segments of FtsZ filaments. Abrupt "kinks" were sometimes seen (black arrows) as well as direct connections of straight filaments to the membrane (white arrowheads). Scale bar 50 nm. Fig. 1*C*–*E* adapted from (Li et al. 2007) with permission from Nature Publishing Group.

dehydration, and other harsh preparative steps destroyed it. Thus although FtsZ was proposed to form a ringlike structure (the "Z-ring"), no further mechanistic clues could be gained.

Just a few years later, the cloning and recombinant expression of the green fluorescent protein (GFP) from *Aequorea* sp. revolutionized cell biology, allowing specific proteins of interest to be tagged and imaged in living cells by fluorescence light microscopy (fLM) (reviewed by Tsien 1998). In 1996, Ma et al. fused FtsZ with GFP and confirmed that it formed a dynamic ring at the midplane of dividing cells (Ma et al. 1996) (Fig. 1B) (similar results were also obtained by immunofluorescence (Addinall and Lutkenhaus 1996; Levin and Losick 1996)). Later, sophisticated fluorescence recovery after photobleaching (FRAP) experiments showed that FtsZ monomers entered and left the ring with a half-time of ~9–30 s (Stricker et al. 2002; Anderson et al. 2004). Despite fLM's ability to reveal dynamics and deliver 3-D information (through deconvolution and confocal techniques), the mechanism of constriction remained unclear. The resolution of light microscopy was insufficient to show, for instance, whether FtsZ monomers were actually forming filaments or just localizing to the midcell near the membrane. Crystal structures showing FtsZ's clear homology to eukaryotic tubulin and in vitro polymerization studies argued strongly that it did form filaments (Löwe and Amos 1998). Among many hypotheses, one, for instance, was that in analogy to eukaryotic cytokinesis, perhaps FtsZ formed a complete ring or spiral and an unknown protein drove adjacent protofilaments in opposite directions, generating a constriction force (Bramhill 1997). The FRAP data, however, seemed incompatible with this idea. Moreover, although not a problem in this specific case (of localizing FtsZ), it is worth noting that GFP and its relatives can cause proteins to mislocalize. A recent comprehensive tagging screen showed, for instance, that less than one-third of proteins that show a particular localization pattern when tagged at the amino terminus show the same localization when the tag is moved to the carboxyl terminus (most are delocalized whereas a small percent localize differently) (Werner et al. 2009)! Although immunofluorescence can be used to confirm particular patterns (of untagged proteins), this again introduces the problems of fixation.

It was in this context that the first ECT images of bacterial cell division were obtained. Images of dividing *Caulobacter crescentus* cells showed that FtsZ forms short, separated, arc-like filaments rather than a complete ring or spiral (Fig. 1C) (Li et al. 2007). In one cell there were, for instance, nine filaments on one side of the cell and none on the other. Individual 4-nm-wide protofilaments were irregularly spaced (rather than bundled) and connected to the inner membrane by additional protein densities. Perhaps most intriguingly, some FtsZ filaments were curved whereas others were straight (Figs. 1D,E). Because like its eukaryotic tubulin homologs, FtsZ appears to favor curved and straight conformations in vitro when bound to GDP and GTP, respectively (Lu et al. 2000), these new structural details strongly supported a model first proposed by Erickson that FtsZ itself generates the force needed to constrict the cell through nucleotide-hydrolysis-driven conformational changes in short, independently acting protofilaments (Erickson 1997).

As illustrated by these imaging studies on FtsZ, the main advantages of ECT over other microscopy techniques are the preservation of the sample in a near-native state (that preserves filaments, for instance) and the resolution to visualize even small details directly (like filament curvature and connectivity), in 3-D. How this is accomplished is described next.

TECHNICAL AND PROCEDURAL DETAILS

Sample Preparation

One of the main challenges in biological electron microscopy is preserving the sample in as native state as possible within the high vacuum of the microscope. As mentioned earlier, this has traditionally been done through chemical fixation, dehydration, and plastic-embedment. To eliminate the artifacts produced by these harsh treatments, Dubochet and his colleagues developed the method of "plunge-freezing" (Dubochet et al. 1983). A few microliters of an aqueous sample such as a cell culture or a solution of purified protein is applied to the surface of a thin, perforated layer of carbon supported by a standard EM grid. This can be done by pipet or by simply dipping the EM grid into the sample. When working with adherent cells, it can be advantageous to add EM grids to the cell culture for hours or days so the cells grow naturally on and attach to the carbon surface (Seybert et al. 2006). In these cases gold grids are usually used in place of the more toxic but cheaper copper grids. To facilitate subsequent image alignment, 5- or 10-nm colloidal gold clusters are also typically added to the sample. These can be either dried onto the grid before the sample is added, mixed into the sample itself, or both.

Next, to reduce the sample to just a thin film (~0.5-μm thick or less), the grid is blotted by filter-paper pads on either one side or both sides simultaneously for a second or two. The blotted grid is then rapidly plunged into liquid ethane or, more conveniently, into an ethane-propane mixture (Tivol et al. 2008). This freezes the samples so quickly (>10^6 K/s) (Dubochet et al. 1988) that the water molecules stop diffusing before they can form the energetically favorable hydrogen bond networks of crystalline ice, retaining instead a disordered, liquid-water-like arrangement referred to as "vitreous" ice (Angell 2004). Dissolved macromolecules, their complexes, and even entire cells are thereby preserved in a near-native, "frozen-hydrated" state (sufficiently native in fact to continue living if later thawed (Erk et al. 1998). If the plunging process is not rapid enough, ice crystals form and denature the sample. Fortunately, denatured samples are immediately recognized in the EM, as ice crystals have a pronounced and unambiguous appearance. Automatic plunge-freezers with multiple adjustable parameters have dramatically improved the reproducibility of plunge-freezing; one commonly used, commercially available machine is called the "Vitrobot" (Frederik and Hubert 2005). For a more detailed description of ECT sample preparation with the Vitrobot, see (Iancu et al. 2006a). Once the sample is frozen across a grid, it is ready for imaging in the EM.

Tilt-Series Acquisition and Fundamental Limitations

The word "tomography" means imaging by sections or sectioning. The most familiar use of the word is the medical "CT," or "computed tomography" scan, wherein X-ray projection images through a subject are recorded from a number of directions and then merged to produce a 3-D anatomical model. Similarly, in electron tomography, a "tilt-series" of projection images are recorded of a single object like a bacterial cell as it is incrementally tilted around one, and sometimes two axes, and these images are then merged to produce a 3-D "reconstruction" or "tomogram" (Fig. 2). The basic workflow is that a grid is inserted into the EM, a target is chosen and centered under the electron beam, a projection image is recorded, the sample is rotated (tilted) a degree or two, another projection image is recorded, and the cycle of rotation and imaging is repeated as far as useful images can be obtained (until the sample becomes prohibitively thick or the grid or grid holder begins to block the beam, usually

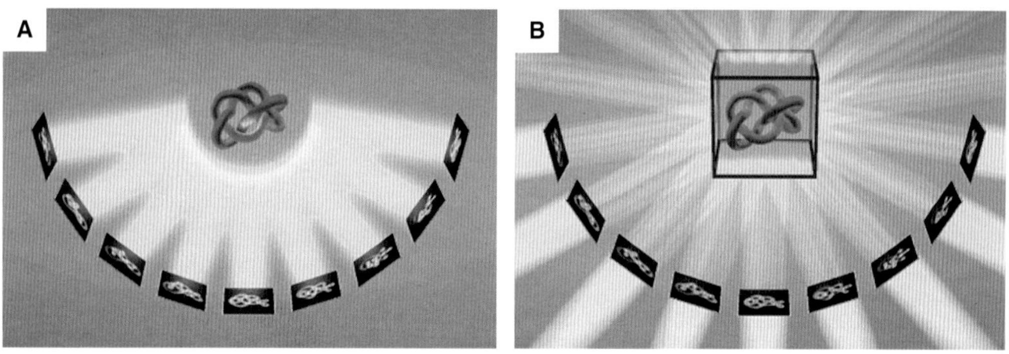

Figure 2. Principles of data acquisition and reconstruction in ECT. (*A*) A set of 2-D projection images is recorded while tilting the object around one axis. (*B*) The back-projection method of 3-D reconstruction in ECT: For each projection, a back-projection body is calculated, and the sum of all the back-projection bodies represents the density distribution of the original object. Reprinted, with permission from (Baumeister et al. 1999).

~65°). Images of the inverse tilt angles (i.e., 0° to −65°) are recorded similarly, or alternatively, the tilt-series can begin at one extreme tilt angle (like 65°) and proceed through the untilted position to the opposite extreme (i.e., −65°). For a movie showing an example tilt-series of a frozen-hydrated bacterial cell, see Supp. Movie S2 of (Komeili et al. 2006).

Because the resolution of modern EMs is subatomic (Batson et al. 2002), and there is no limit to how finely the tilt range can be sampled, one might hope that this procedure would deliver atomic-resolution images of cells. Unfortunately, for frozen-hydrated biological materials, radiation damage prohibits this. As the imaging electrons pass through the sample, they can remain unscattered, scatter elastically, scatter inelastically, or suffer multiple scattering events. Although image contrast (the information content) is produced by the interference of the unscattered and the elastically scattered electrons, the inelastically scattered electrons gradually destroy the sample. Inelastic scattering events break covalent bonds, deposit heat, and more rarely even knock atomic nuclei out of place. Because for every useful elastic scattering event there are approximately 3 damaging inelastic scattering events (Henderson 1995), as more and more electrons are used to build up an image, sample damage accumulates. The originally sharp edges of macromolecular structures degrade and eventually "bubbles" of (presumably) radiolytic fragments appear and catastrophically disrupt the structure (Comolli and Downing 2005; Iancu et al. 2006b; Wright et al. 2006). Thus the most fundamentally limiting factor in ECT is the total number of electrons that can be used to record images before the sample is destroyed. Depending on how much damage is tolerated, in practice total doses of ~1–200 electrons/Å^2 are typically used, spread across the tilt-series. As a result, individual atoms or even protein secondary structures are not resolved, but the position of domains and the rough morphology of complexes can be, as was described for FtsZ earlier.

After damaging the sample, inelastically scattered electrons emerge with lower energies (and thus longer wavelengths), and therefore focus differently as they pass through the electron lenses. To prevent inelastically scattered electrons from blurring the image, energy filters remove them. Because the mean free path (i.e., the average distance between scattering events) in aqueous materials of the 300 kV electrons that are typically used in ECT is only ~350 nm (Grimm et al. 1996), about half of the electrons sent through even a very slender bacterium are inelastically scattered and then removed by the energy filter. As the sample is tilted, the path length the electrons must traverse through the sample increases as the secant of the tilt angle (becoming twice as thick, for instance, at 60° tilt), and more and more electrons are lost to inelastic scattering. To build up meaningful images, higher doses can be used, but at the cost of increased radiation damage. Moreover, in addition to requiring higher doses per image, thicker samples in general also require more images to achieve the same resolution because the tilt range has to be sampled more finely (Derosier and Klug 1968). Thus a second fundamental limitation in ECT is that the interpretability (clarity and resolution) of reconstructions degrades with sample thickness, with 0.5 μm being a useful practical limit.

Traditionally, thicker samples such as cells and tissues have always been sectioned. This is straightforward for plastic-embedded materials, but methods to "cryo-section" frozen-hydrated samples have only recently met success (Al-Amoudi et al. 2004). Larger samples (such as suspensions of large bacterial cells, biofilms, or even tissues) are first frozen under high pressure (2100 atmospheres) to prevent ice crystal formation (McDonald and Auer 2006), and then the frozen blocks are cryo-sectioned. Although the procedure is technically challenging and fraught with artifacts such as knife-marks, crevasses, and compression, nevertheless it is beginning to provide much-needed images of larger samples in near-native states, as illustrated in some of the examples presented later.

Because with even the thinnest samples, useful images at tilt angles higher than ~65–70° cannot usually be collected, there is a "wedge" of information (the tilt angles surrounding 90°) that remains unmeasured. As a result, the resolution of the 3-D reconstruction in the direction parallel to the electron beam is significantly worse than the resolution perpendicular. In simple visual terms, this causes spherical objects to appear somewhat ellipsoidal (smeared in the direction of the beam), and continuous objects such as filaments and membranes are more visible in some orientations than in others. This is why in "xz" or "yz" tomographic slices (such as Fig. 1D), the membranes do not appear to connect around the "top" and "bottom" of the cell. Although the missing wedge may be reduced to a missing "pyramid" by rotating the grid 90° and collecting a second, orthogonal tilt-series (a so-called "dual-axis" data set), this procedure is more than twice as time consuming, the dose that can be used per image is halved, and alignment errors between the tilt-series can erode the benefit (Nickell et al. 2003; Iancu et al. 2005). Thus a third fundamental limitation in ECT is the anisotropic resolution caused by tilt limitations (the "missing wedge").

In practice, the acquisition of even one tilt-series is complicated. Once the sample is plunge-frozen, it must be kept at temperatures below the vitrification point (~120 K) throughout storage, loading, and data collection to prevent the water from crystallizing (which would denature the specimen). For this purpose, special cryo-stages have been developed for transmission electron microscopes that are cooled by liquid nitrogen or helium. Unfortunately, no stage is perfectly eucentric and these cryo-stages are particularly susceptible to thermal drift. As a result, as the sample is rotated, it is displaced laterally and vertically within the column. Thus to kept the target centered under the beam and at a constant focus, before each image is taken the beam must be electronically shifted, the image has to be correspondingly unshifted to center it back on the CCD camera, and the strength of the objective lens has to be adjusted. To find the magnitude and direction of the shifts and focus changes that should be applied before each image is taken, without exposing the target to any unnecessary dose, the movements can either be "predicted" by modeling trends from previous images in the tilt-series (Zheng et al. 2004) or measured indirectly by deflecting the beam to an off-target, "tracking" position and recording reference images. Fortunately there are several software packages that perform these tedious and intricate operations automatically now, including SerialEM (Mastronarde 2005), Leginon (Suloway et al. 2009), UCSF Tomo (Zheng et al. 2007), the TOM toolbox (Nickell et al. 2005), and others.

The ideal instrument for ECT is a fully automatable intermediate-voltage (~300 kV) EM with a highly stable and eucentric cryo-stage, an energy filter, and a large format and low-noise CCD. Unfortunately because of their extreme cost, there are still only a few such instruments being used for ECT in the world today. The natural contrast of unstained biological macromolecules suspended in vitreous ice in such microscopes is extremely low, because it arises from the small difference in their densities. Because the ultimate resolution limitation is radiation damage, imaging parameters such as defocus, magnification, and tilt increment are therefore carefully chosen to maximize contrast and reduce noise without becoming even more limiting (than radiation damage). The defocus is set high enough (~4–16 μm) to generate strong phase contrast, but not so high that the highest spatial frequencies of interest are inverted by the contrast transfer function (Fernández et al. 2006). Magnifications are set high enough that the pixel size of the reconstruction (typically 0.5–1.0 nm) is one-third or less of the desired resolution, but not further, so that the dose per pixel and field of view can remain as large as possible. The tilt increment is set fine enough to capture the highest spatial frequency of interest, but no smaller so the total dose per image can be maximized.

3-D Reconstruction and Interpretation

As mentioned earlier, because no goniometer is perfect, specimens move laterally and vertically within the column throughout the tilt-series. The images must therefore be precisely aligned before a 3-D reconstruction can be calculated. As an additional challenge, because of the physics of electron optics, changes in height/focus within the column cause images to rotate and show subtly different magnifications. Further, although the tilt angle of each image is approximately known, the actual angles reached must be determined more accurately. Sophisticated software has therefore been written to refine estimates of the translations, rotation, magnification, tilt axis, and tilt angle of each image in the tilt-series (Mastronarde 2008). The colloidal gold beads typically added to the samples provide precise fiducial markers to facilitate this process.

Once the images are aligned, 3-D reconstructions can be calculated with a variety of algorithms. The most intuitive is "back-projection," in which a reconstruction is built up by "smearing" the densities in each image back through space in the opposite direction they were projected (Fig. 2B) (Crowther et al. 1970b). To understand this reconstruction in Fourier space, the key principle is that the 2-D Fourier transform of a projection image is a central slice of the 3-D Fourier transform of the object (the "Projection theorem") (Crowther et al. 1970a). Thus the 3-D Fourier transform of the sample can be "filled" with the transforms of the 2-D images, and then re-sampled onto a regular (for instance Cartesian) coordinate system and inverse transformed to produce a real-space reconstruction (Lee et al. 2008). Various software packages have been written to perform these calculations, including IMOD (Mastronarde 2008), the TOM toolbox (Nickell et al. 2005), and RAPTOR (Amat et al. 2008). Once reconstructed, tomograms can be "denoised" to improve image contrast and enhance interpretability (Frangakis and Hegerl 2001; Narasimha et al. 2008) and/or "segmented" to allow specific features to be visualized in isolation or as surfaces (Pruggnaller et al. 2008).

Tomograms are rich in detail. The layers of the cell envelope and large ultrastructures such as flagella, magnetosomes, storage granules, carboxysomes, and gas vesicles are readily recognizable through their unique and known characteristics, but more sophisticated methods must be used to identify others within the crowded environment of the cell (Grunewald et al. 2003). In the case of chemoreceptor arrays, for instance, bundles of parallel filament-like densities extending into the cytoplasm from the inner membrane were immediately suspected to be chemoreceptor arrays because they matched expectations for their gross architecture and dimensions. To confirm this assignment, however, Weis et al. chemically fixed cells and immunolabeled components of the chemoreceptor arrays (Weis et al. 2003). This worked well because the distinctive "zippered" and "micellar" structures of overexpressed receptors persisted through the required chemical fixation, and so could still be recognized in the immunolabeled sections. Unfortunately this is not always true, as described earlier in the case of FtsZ, where the labels revealed the general location of FtsZ but the filaments themselves were no longer visible.

A potentially more general strategy for identifying structures is correlated fLM and ECT. The basic idea is to immobilize cells expressing a fluorescent fusion protein on an EM grid, record fLM images marking the localization of the fusion protein in each cell, then move the sample into an EM and record high-resolution cryo-tomograms of the same cells. For this purpose light microscope cryo-stages have recently been developed (Sartori et al. 2007; Schwartz et al. 2007), but because the sample is kept frozen, oil immersion lenses cannot be used and unfortunately the resolution is too poor to localize fusion proteins within small bacterial cells (unpublished observations). Briegel et al. developed an alternative protocol, immobilizing cells on an EM grid with poly-L-lysine before inspection at room temperature in an oil-immersion, high-resolution fLM (Briegel et al. 2008). These cells remained in place when they were removed from the fLM and plunge-frozen, so matching cryo-tomograms were obtainable. The perfect correlation between the location of (mCherry-fused) chemoreceptor foci in the fLM images and the presence of the putative chemoreceptor array structure in the cryo-tomograms confirmed their identity. Unfortunately, to preserve the cells' integrity and stop internal rearrangements, they had to be lightly fixed before they were adhered to the grid surface with poly-L-lysine, so this method has not worked for other structures like certain cytoskeletal filaments that are destroyed by light fixation. Ideally, methods to label proteins with genetically fusable tags that are directly visible in cryo-tomograms of native cells are needed. To this end, several groups are trying to develop heavy metal cluster tags based on metallothionein or other metal-binding proteins (Nishino et al. 2007; Diestra et al. 2008).

In the absence of specific tags, another strategy that has been employed to identify objects in tomograms is manipulating their abundance. In the case of FtsZ, for example, the filaments at the constriction site were identified by comparing tomograms of wild-type cells with cells overexpressing FtsZ, where there were many more filaments showing the same localization patterns and structural characteristics, and cells overexpressing a hyper-stable mutant of FtsZ, in which hundreds of again similar but much longer filaments were seen. No similar filaments were seen in cells depleted for FtsZ.

Objects in tomograms can also sometimes be identified by their structural "signatures." ParM filaments within cryo-sections of *E. coli* cells were identified this way, for instance (Salje et al. 2008). Power spectra of the filaments showing their longitudinal repeat and interfilament packing distances matched spacings known from crystal structures and images of ParM filament bundles in vitro. Sophisticated software has been written to identify particular isolated complexes by their morphological similarity to a known template (Bohm et al. 2000). In this way the positions and orientations of the ribosomes in *Spiroplasma melliferum*, a very slender bacterium ($\sim$0.3 μm diameter) were estimated (Ortiz et al. 2006). When a large number of identical or nearly identical objects are present, higher signal-to-noise ratios can be obtained by averaging subtomograms (Schmid and Booth 2008).

In summary, ECT produces 3-D images of intact cells in near-native states to "molecular resolution," but the sample must be thin ($<$ 0.5 μm), the interpretability of the resulting tomograms is limited by radiation damage, the resolution is anisotropic because of tilt limitations, the procedure is complicated and requires expensive electron cryo-microscopes, and identifying structures of interest in the tomograms can be challenging. Nevertheless in just the last few years since it has become practical, ECT has already provided a wealth of new details about bacterial ultrastructures like the cytoskeleton, cell envelope, surface appendages, chemoreceptor arrays, and other large structures. A number of examples are summarized later.

FIRST CONTRIBUTIONS OF ECT TO BACTERIAL CELL BIOLOGY

The Bacterial Cytoskeleton

For decades it was thought that bacteria lacked cytoskeletons. Homolog of all three major eukaryotic cytoskeletal protein families (actin, tubulin, and intermediate filaments) were then identified in bacteria by their sequence, structural, and biochemical similarities (Michie and Löwe 2006), and were further found by fLM to localize in elongated patterns consistent with a filamentous structure (Pogliano 2008). As illustrated by the FtsZ example in the Introduction, ECT has now both confirmed by *direct visualization* that at least some of these do form filaments in vivo and provided vital new clues to their structures and functions.

The first published report of the direct visualization of a cytoskeletal filament in a bacterium by ECT described the species *Spiroplasma melliferum* (Fig. 3A) (Kürner et al. 2005). Two ribbons of parallel 11-nm filaments were seen underneath the cytoplasmic membrane flanking a central ribbon of 4-nm filaments. Although the peripheral filaments were identified as the Fib protein, the central ribbon was speculated to be MreB. This "three-ribbon" model corresponds to a total ribbon width of 146 nm. More recently, however, results by Trachtenberg et al. reported that the filament bundle was continuous, uniform, and only 70 nm wide and proposed a "single-ribbon" model (Trachtenberg et al. 2008). The cause of these differences is unclear.

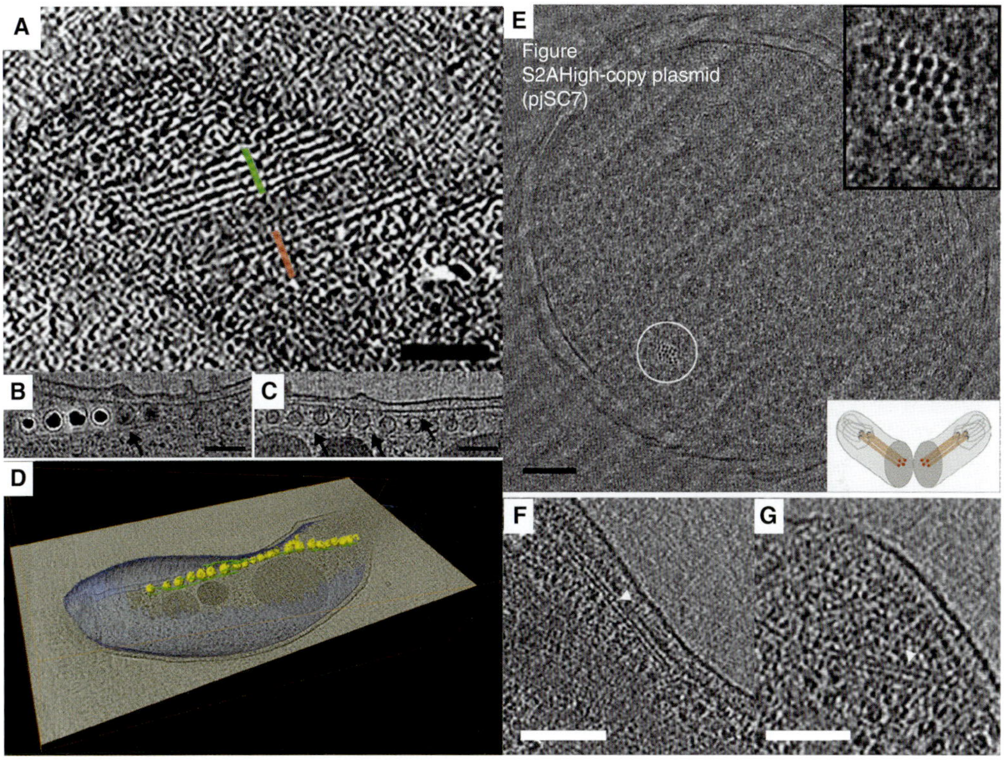

Figure 3. Examples of bacterial cytoskeletal filaments. (*A*) Tomographic slice through a *Spiroplasma melliferum* cell. The two outer ribbons are composed of five thick filaments with a spacing of ~11 nm (highlighted in green and red). Adapted from (Kürner et al. 2005) with permission from the American Association for the Advancement of Science. (*B*) Tomographic slice showing the magnetosome chains in *Magnetospirillum magneticum* sp. AMB-1 flanked by long MamK filaments (black arrows). The magnetite crystals appear as dense black balls. (*C*) Tomographic view of a magnetosome chain grown in the absence of iron. (*D*) 3-D segmentation of a *M. magneticum* cell showing the magnetosomes (yellow) and their associated filaments (green) with respect to the whole cell (blue), superimposed on a tomographic slice through the same cell (gray level image). Fig. 3B–D adapted from (Komeili et al. 2006) with permission from the American Association for the Advancement of Science. (*E*) Projection image through a vitreous cryo-section of an *E. coli* cell carrying the high-copy ParMRC plasmid pJSC7. Small bundles of filaments appear as black dots (circled). Upper inset: enlarged view of bundle, lower inset: Schematic representation of the plane of imaging which is perpendicular to the filaments to increase contrast and reveal their packing. Adapted from (Salje et al. 2008) with permission from the American Association for the Advancement of Science. (*F*) and (*G*) Tomographic slices through *Bdellovibrio bacteriovorus* cells showing cytoskeletal filaments (white arrowhead) oriented parallel to the plane of the plasma membrane. Adapted from (Borgnia et al. 2008) with permission from the American Society for Microbiology. Scale bar for Fig. 3E–G is 100 nm.

ECT led to the discovery of a novel bacterial actin homolog, MamK, in magnetotactic bacteria. Magnetotactic bacteria can navigate in geomagnetic fields with the help of unique organelle-like compartments called magnetosomes. Magnetosomes are magnetite (Fe_3O_4) or gregite (Fe_3S_4) crystals partially enclosed by a lipid bilayer, and are found in linear chains of 15 or more within cells (Bazylinski and Frankel 2004). Because traditional thin-section EM images often showed the magnetosomes near but not connected to the cytoplasmic membrane, it was thought that they were free-standing, fully-membrane-enclosed bacterial organelles. When frozen-hydrated *Magnetospirillum magneticum* sp. AMB-1 cells were imaged in 3-D by ECT, however, it became clear that the magnetosomes were instead just invaginations of the inner membrane (Komeili et al. 2006). A network of cytoskeletal filaments was seen flanking the magnetosome chain (Fig. 3B–D). One gene, *mamK*, from the *mam* operon involved in magnetosome genesis and function was identified as a homolog of the bacterial actin MreB. ECT reconstructions of cells with an in-frame deletion of *mamK* showed no filaments, and the magnetosome chains were disordered. When the mutant cells were complemented with a plasmid containing the *mamK* gene, the filaments reappeared and the magnetosomes were once again aligned. Concurrently, Scheffel et al. imaged the strain *Magnetospirillum gryphiswaldense* sp. MSR-1 and revealed similar filamentous structures associated with magnetosome chains (Scheffel et al. 2006). Surprisingly, in cells missing the *mamJ* gene, which encodes a small acidic protein, filaments were still present but the magnetosomes did not appear to be connected to them and no longer showed a linear arrangement. Together, these studies led to the model that MamJ links individual magnetosomes to MamK filaments. Importantly, these studies also showed that one of the functions of bacterial cytoskeletal filaments is to organize internal organelle-like compartments, just as they do in eukaryotic cells.

As mentioned briefly earlier, ParM has also been visualized in vivo by ECT. ParM is a bacterial actin homolog involved in low-copy R1 plasmid segregation. Because *E. coli* cells are too thick for direct imaging after plunge-freezing, Salje et al. cryo-sectioned cells overexpressing ParM and saw filament bundles (Fig. 3E) (Salje et al. 2008). The bundles were identified as ParM by noting that their computed diffraction patterns matched the diffraction patterns of ParM filaments assembled in vitro. In a wild-type (low copy R1 plasmid number) strain, the authors observed bundles of three to five intracellular ParM filaments localized within the periphery of the nucleoid, strongly supporting the model that one filament is needed to separate each R1 plasmid pair.

A number of other filaments have been described that still need to be identified. Four different types of cytoskeletal bundles, for instance, were observed in *Caulobacter crescentus* (Briegel et al. 2006). A different filament bundle was seen in *Bdellovibrio bacteriovorus* (Fig. 3F,G) (Borgnia et al. 2008). Taken together these examples show that filamentous proteins are both general and abundant in bacteria. Not surprisingly, the list of candidate filamentous proteins is rapidly expanding and now includes, for instance, actin homologs MreB, ParM, FtsA, MamK, AlfA, and others; tubulin homologs FtsZ, BtubA/B, TubZ-Ba, TubZ-Bt, and others; intermediate filaments CreS, CfpA, Scc, and AglZ, and the recently identified ParA/MinD superfamily (Pogliano 2008). ECT will likely be used to study the structure and function of all of these and others yet to be identified.

The Bacterial Cell Envelope

Although traditional thin-section EM images revealed the basic architecture of Gram-positive and negative bacterial envelopes, ECT and simple projection imaging of vitreous cryo-sections has now revealed key new details free of the potential complications of fixation artifacts. The mycobacterial cell wall has been intensively studied for decades, for instance, because its unique composition makes *Mycobacterium tuberculosis* remarkably resistant to antibiotics (Dover et al. 2004). Because the cell wall is comprised of an unusual peptidoglycan-arabinogalactan polymer with covalently bound mycolic acids and pore-forming proteins, it was thought to form an asymmetrical bilayer. In two independent studies, Hoffmann et al. and Zuber et al. applied ECT to vitreous cryo-sections of *Mycobacterium smegmatis*, *Mycobacterium bovis* BCG, and *Corynebacterium glutamicum* (Hoffmann et al. 2008; Zuber et al. 2008). The images showed that the outermost layer of the envelope is indeed a symmetric lipid bilayer (Fig. 4A). By imaging mycolic acid-deficient *C. glutamicum*, both groups also showed that mycolic acids are constituents of the outer membrane. Their models differed, however, in

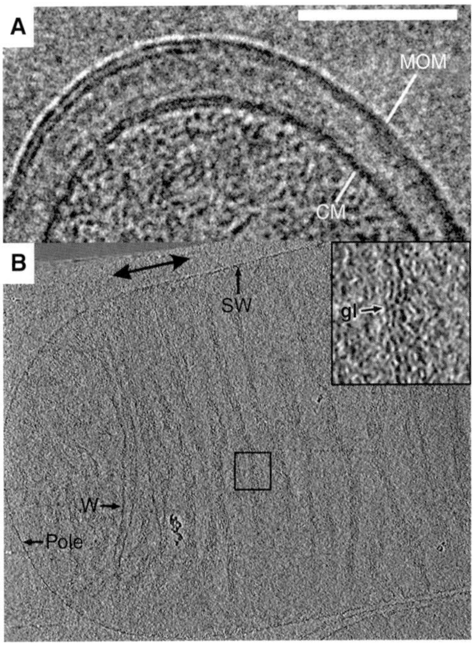

Figure 4. The bacterial cell envelope. (*A*) Projection image from a cryo-section of *Mycobacterium smegmatis*. The cytoplasmic and mycobacterial outer membranes are marked as CM and MOM, respectively and the bilayer structure of both is discernible. Scale bar 100 nm. Adapted from (Hoffmann et al. 2008) with permission. © 2008 National Academy of Sciences, USA. (*B*) 10-nm thick tomographic slice through an *E. coli* strain XL-10 sacculus. Abbreviations: gl, glycan strand; SW, side wall; w, wrinkle. The double-headed arrow denotes the saccular polar axis. Inset shows four-fold enlarged view of the boxed region. Adapted from (Gan et al. 2008) with permission. © 2008 National Academy of Sciences, USA.

whether the meromycolic chains were folded or unfolded.

The mechanical strength of the bacterial cell envelope arises from a mesh of peptide-crosslinked glycan strands. Although the chemical composition and subunit structure of this peptidoglycan layer has been known for decades, its higher-order architecture had remained unclear. Two fundamentally different models had been suggested, namely the "layered" and "scaffold" models (Vollmer and Holtje 2004). By imaging intact sacculi purified from two Gram-negative bacteria, *E. coli* (Fig. 4B) and *C. crescentus*, Gan et al. showed that the peptidoglycan is just a single layer thick and that individual glycan strands are oriented in the plane of the sacculus perpendicular to the long axis of the cell approximately 5–8 nm apart (Gan et al. 2008). This observation ruled out the scaffold model and instead established a "disordered, circumferential, layered" model.

Motility and Surface Appendages

ECT has also provided insight into several different forms of bacterial motility. The cryo-tomograms of *S. melliferum* mentioned earlier led to a model of its motility based on differential contraction of the filament ribbons (Kürner et al. 2005). The gliding bacterium *Flavobacterium johnsoniae* was shown to possess tufts of ~5-nm-wide external filaments emanating from the inner surface of the outer membrane (Fig. 5A) (Liu et al. 2007). These filaments were absent in a nonmotile *gldF* mutant cell but were restored in the same mutant complemented with plasmid-encoded GldF, a component of a putative ATP-binding cassette transporter. Cryo-tomograms of *Mycoplasma pneumoniae* revealed the molecular architecture of its "attachment organelle" and suggested that it was in fact a multijointed, flexible motility machine (Henderson and Jensen 2006; Seybert et al. 2006). ECT of the pathogenic spirochete *Treponema denticola* revealed cytoplasmic filaments and platelike structures possibly anchoring those filaments to the inner membrane (Izard et al. 2008). The cytoplasmic filaments, earlier suggested to be composed of CfpA (a unique protein to spirochetes) (Izard et al. 2001), were visible directly beneath the flagella in the cytoplasm, suggesting a role in motility. Cryo-tomograms of the spirochete *Treponema primitia* revealed novel structures such as bowls, arches, fibrils, and two layers of peptidoglycan (Fig. 5B) (Murphy et al. 2008), strongly supporting the "rolling cylinder" model of motility originally proposed by Berg (Berg 1976). Charon et al. imaged the Lyme disease spirochete, *Borrelia burgdorferi*, and revealed that the periplasmic flagella adopt a

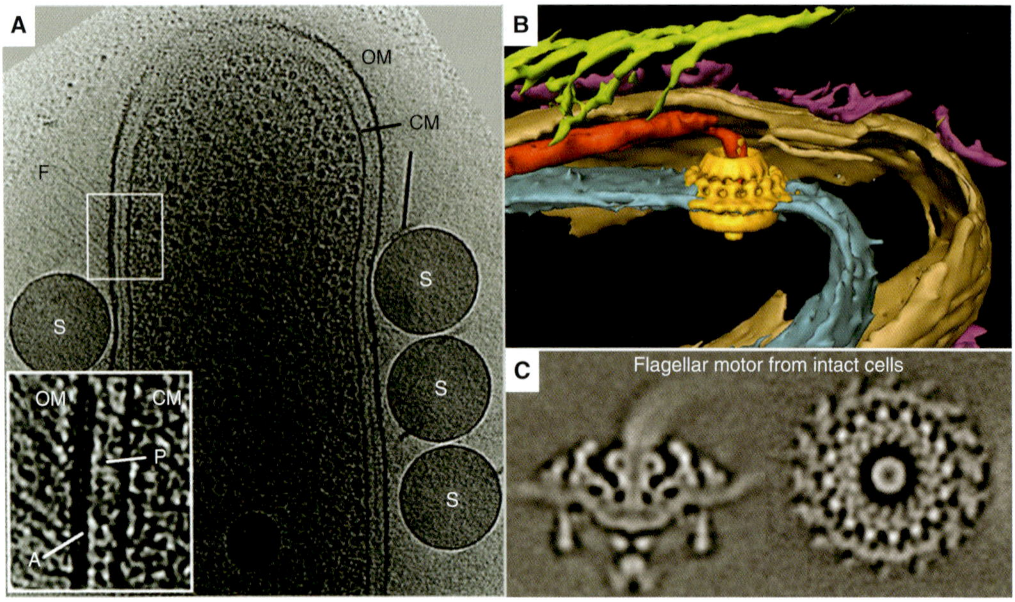

Figure 5. Motility and surface appendages. (*A*) 3-nm thick tomographic slice through a wild-type *Flavobacterium johnsoniae* cell. Features arising from the cytoplasmic membrane (CM), outer membrane (OM), peptidoglycan (P), cell surface filaments (F), and added latex spheres (S) are labeled. The inset shows an expanded view of the periplasmic region at a location where filaments are observed. The densities arising from the outer membrane, cytoplasmic membrane, peptidoglycan layer, and patch (*A*) at the base of the outer membrane can be seen. Scale bar 300 nm. Adapted from (Liu et al. 2007) with permission from the American Society for Microbiology. (*B*) Manually segmented 3-D view of a *Treponema primitia* cell shows surface hooks (green), bowls (purple), inner membrane (blue), outer membrane (brown), and the averaged flagellar motor (yellow) embedded in the position of the motor. Adapted from (Murphy and Jensen 2007). © 2007 BioTechniques. Used by Permission. (*C*) Side (left) and top (right) views of the averaged flagellar motor from *Borrelia burgdorferi*, without rotational symmetry imposed. Adapted from (Liu et al. 2009) with permission from the American Society for Microbiology.

flat-ribbon configuration (Charon et al. 2008) instead of the previously reported stacked bundle observed by traditional-EM (Motaleb et al. 2000).

At the heart of swimming motility is the flagellar motor. This quintessential molecular machine is a structural biological marvel, but because the torque-generating stator is embedded in the inner membrane and binds to the peptidoglycan, isolating intact motors for structure determination has not been possible. The first structure of a complete flagellar motor, at ~7 nm resolution, was therefore obtained by averaging 20 individual motors from cryo-tomograms of *T. primitia* (Fig. 5B) (Murphy et al. 2006). The stator showed 16-fold symmetry and possessed direct contacts to the rotor, C ring, and a novel P-ring-like structure. This work illustrated how some important macromolecular complexes may not be purifiable, and will therefore have to be studied in situ. ECT is a technology that can do that, albeit at low resolution. More recently Liu et al. reached ~3.5 nm resolution by averaging over a thousand flagellar motors from cryo-tomograms of *B. burgdorferi* (Fig. 5C) (Liu et al. 2009). By not imposing rotational symmetry, asymmetric aspects of the stator structure and rotor-stator interactions were revealed including curvature in the stator itself and flexibility in the rotor-stator interaction. It was proposed that flexibility in the motor structure may be needed to maintain connectivity among the flagellar proteins during rotation.

Chemotaxis

The chemotactic system in bacteria senses attractants and repellants in the environment. The main structural units of the system (the methyl-accepting chemotaxis proteins, or MCPs) are arranged in arrays and are usually located at the cell pole. After numerous biochemical, structural, and modeling studies (Kim et al. 1999; Shimizu et al. 2000; Studdert and Parkinson 2005), averaging cryo-tomograms of the serine Tsr receptor overexpressed in *E. coli* gave rise to the first full-length 3-D structure of an MCP (Fig. 6A,B) (Khursigara et al. 2008b). The result fit well with the known atomic models of the various domains. Structures were obtained both in the presence and absence of attractant, revealing conformational changes in the HAMP domain of the receptor likely responsible for signal conversion. The averaged structures of the two states showed a "trimer-of-dimers" interaction, confirming the functional unit suggested by previous crystallographic studies (Kim et al. 2002).

The extraordinary sensitivity, gain, and signal integration showed by chemoreceptor systems is thought to be made possible by the grouping of chemoreceptors into organized arrays (Hazelbauer et al. 2008). Several competing models of these interactions have been proposed over the years based on crystal structures (Kim et al. 2002), pulsed electron spin relaxation (Park et al. 2006), cross-linking (Studdert and Parkinson 2005), cryo-EM reconstructions (Wolanin et al. 2006), and tomograms of cells overexpressing chemoreceptors (Zhang et al. 2004). ECT of *C. crescentus* cells by two groups then showed that chemoreceptors are arranged

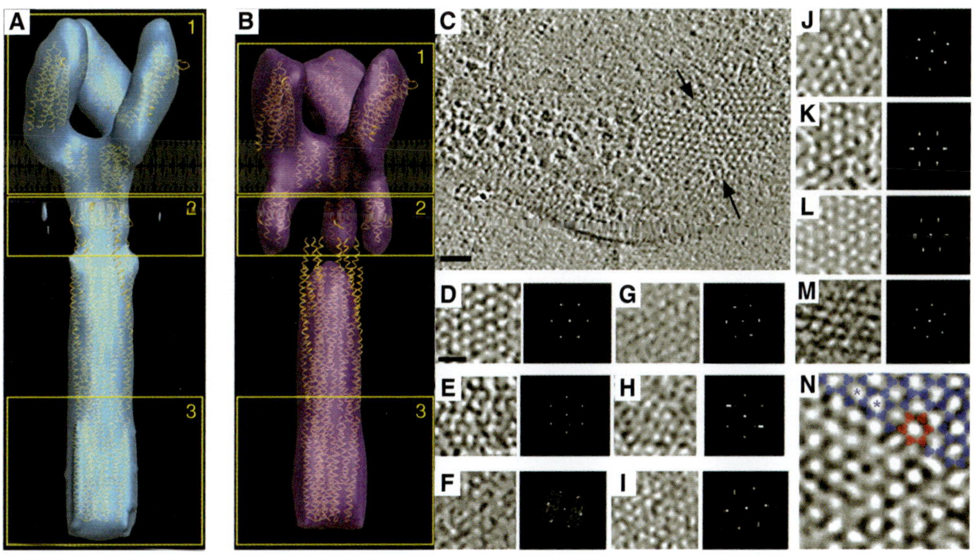

Figure 6. Chemotaxis. (*A*) Compact and (*B*) Expanded averaged conformations of the Tsr_{QEQE} receptor. The structural coordinates corresponding to chemoreceptor models are loosely fit to a single trimer of receptor dimers. Adapted from (Khursigara et al. 2008b) with permission. © 2008 National Academy of Sciences, USA. (*C*) Hexagonal arrangement of receptors in different bacteria. Top view of a chemoreceptor array (between black arrows) in *Thermotoga maritima.* A subregion of the hexagonally ordered lattice and its corresponding power spectrum showing the ~12 nm periodicity are enlarged in the inset. Scale bar 50 nm. (*D–M*) Top views of receptor arrays in other organisms. (*D*) *T. maritima*; (*E*) *A. longum*; (*F*) *C. jejuni*; (*G*) *H. hepaticus*; (*H*) *M. magneticum*; (*I*) *H. neapolitanus*; (*J*) *R. sphaeroides*; (*K*) *E. coli*; (*L*) *V. cholerae*; (*M*) *T. primitia.* Scale bar 25 nm, power spectra enlarged. (*N*) Trimer of dimers (blue) fit into the vertices of the hexagonal lattice in a chemoreceptor array. Six trimers of dimers (red) enclose one hexagon. The spacing from the center of one hexagon to another (blue asterisks) is 12 nm. Adapted from (Briegel et al. 2009) with permission. © 2009 National Academy of Sciences, USA.

in a hexagonal array with a center-to-center spacing of 12 nm (Briegel et al. 2008; Khursigara et al. 2008a). Docking crystal structures of chemoreceptor "trimer-of-dimers" at the vertices of this hexagonal pattern yielded a compelling fit. ECT of 13 different species has now shown that this same 12-nm hexagonal architecture is conserved across a broad array of bacterial phyla, and is thus probably universally conserved (Fig. 6C–G) (Briegel et al. 2009).

Metabolism

Bacteria were once thought to be simple bags of enzymes. It is now understood, however, that many reactions are spatially organized and localized (Norris et al. 2007). Because ECT can produce molecular resolution 3-D images of unique assemblies of macromolecules, it will be an essential tool to understand such spatial regulation. One of the most stunning examples of spatial organization is the carboxysome, a proteinaceous microcompartment that sequesters the enzyme ribulose 1,5-bisphosphate carboxylase/oxygenase (RuBisCO) (Yeates et al. 2008). In cyanobacteria and many chemoautotrophic bacteria, the encapsulated RuBisCO catalyzes the first step of carbon fixation in the Calvin-Benson-Bassham cycle. It is thought that by sequestering RuBisCO near carbonic anhydrases, carboxysomes increase the local concentration of substrate and enzyme and inhibit side reactions. Schmid et al. first reported the structure of carboxysomes by ECT from *Halothiobacillus neapolitanus* and showed that they are regular icosahedra of different sizes (Fig. 7A) (Schmid et al. 2006). The authors averaged subclasses of particles and suggested that the different sizes might be a result of different packing arrangements of the shell proteins. Iancu et al. imaged carboxysomes from *Synechococcus* strain WH8102, showing that they too were regular icosahedra, but suggested that the different sizes arose from different "T-numbers" (Iancu et al. 2007). These authors further showed through simulation that simple packing forces could explain the concentric shells of RuBisCO seen in both studies.

Because the presumed function of carboxysomes is to sequester and organize specific reactions within the cell, however, their positions and relationships to the rest of the cell must also be understood. ECT has been used by two groups now to image carboxysomes within their cellular context (Ting et al. 2007; Iancu et al. 2010). ECT of intact *H. neapolitanus* cells by Iancu et al. showed that carboxysomes in vivo are clustered around and have extensive proteinaceous contacts with polyphosphate storage granules. This work also showed that carboxysomes in vivo are much more structurally heterogeneous than purified preparations and frequently contain internal storage granules that are lost during purification. Finally, ECT of cells provided images of assembling carboxysomes, giving insight into processes and sorting phenomena that are unlikely to ever be reproduced in vitro.

ECT has also been used to elucidate the organization of bacterial photosynthetic membranes (Fig. 7B). Ting et al. applied ECT to frozen-hydrated cells and vitreous sections and showed that closely related strains of cyanobacteria from the genus *Prochlorococcus* differ in their cell size, cell-wall structure, carboxysomes size, and photosynthetic lamellae structure (Ting et al. 2007). Studies on another photosynthetic bacterium, *Rhodopseudomonas viridis* (now *Blastochloris viridis*), revealed tunnel-like structures connecting the tightly packed photosynthetic membranes to the inner membrane (Konorty et al. 2008). A tightly packed two-dimensional hexagonal lattice of proteins within the membranes prompted questions about whether electron carriers such as quinones could diffuse freely. To address that issue, Konorty et al. then used traditional-EM, ECT and spectroscopy to identify structural and functional differences between cells grown in the presence or absence of light. The authors show that increased molecular order is a general trend in cells characterized by low metabolic activity likely as a result of slower quinone diffusion and fewer bc_1 complexes (Konorty et al. 2009).

Although individual enzyme structures can be obtained by X-ray crystallography, NMR spectroscopy, electron crystallography, and/or

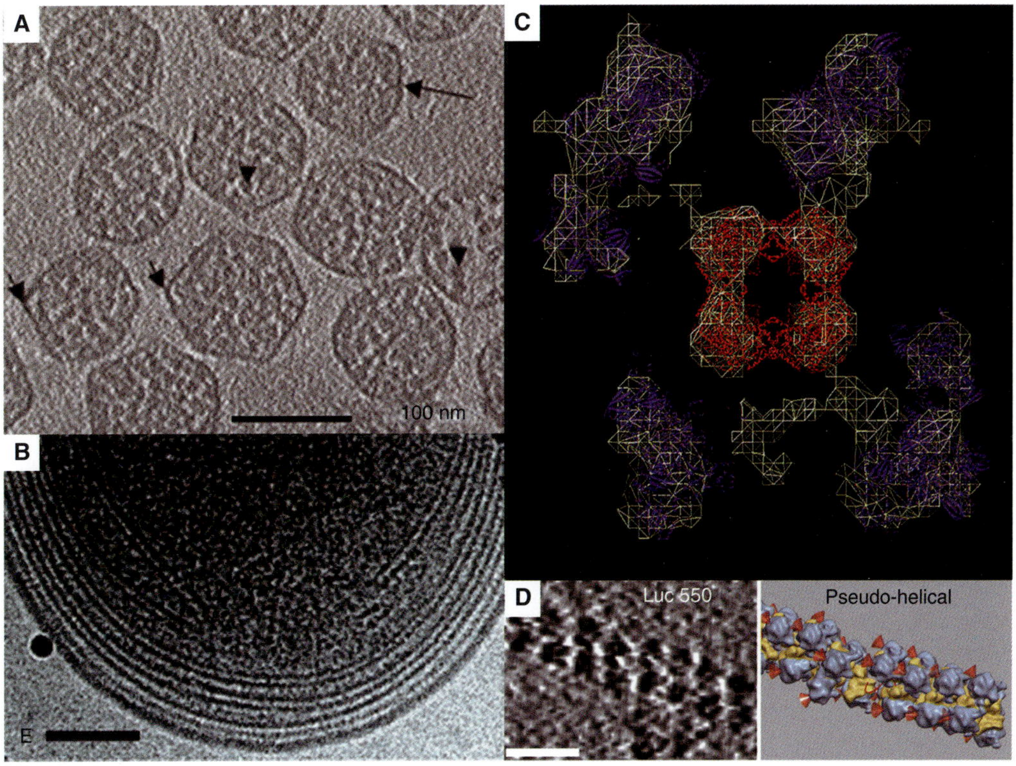

Figure 7. Metabolic macromolecular complexes. (*A*) 4-nm tomographic slice through a field of isolated carboxysomes from *H. neapolitanus*. Long arrow: thin shell surrounding the carboxysome. Shorter arrows: fivefold vertices of several particles. Arrowheads: RuBisCO molecules. Scale bar 100 nm. Adapted from (Schmid et al. 2006) with permission from Elsevier Ltd. (*B*) Projection image of the intracytoplasmic membranes at the pole of a *Prochlorococcus* MIT9313 cell. The black particles are colloidal gold. Scale bar 100 nm. Adapted from (Ting et al. 2007) with permission from the American Society for Microbiology. (*C*) Tomographic slice through the reconstruction of a multimeric PDMC complex. The wire surface encloses the part of the volume of the complete complex that appears in this slice. Crystal structures of the central E2 core (red) and the peripheral E1 or E3 domains (blue) are shown as one of several potential fits. Note that although the crystal structure of the symmetrically arranged E2 core could be unambiguously docked, the resolution was insufficient to distinguish the E1 from E3 subunits or their precise orientations. Adapted from (Murphy and Jensen 2005) with permission from Elsevier Ltd. (*D*) Tomographic slice (*left*) through a translationally stalled polysome in vitro. Isosurfaces (*right*) are represented in color (large ribosomal subunit in blue and small subunit in yellow). Red cones point to the peptide exit tunnel on each 50S subunit. Scale bar 50 nm. Adapted from (Brandt et al. 2009) with permission from Elsevier Ltd.

cryo-EM-based single-particle analysis, all of these techniques require large numbers of identical copies of the biological specimen. Because many enzymes function as parts of heterogeneous complexes that are too dynamic to present a single uniform conformation, techniques like ECT that can reveal the structure of unique objects such as these, albeit at much lower resolution, are essential. ECT was used, for example, to determine the quaternary structures of the *E. coli* pyruvate and 2-oxoglutarate dehydrogenases (PDHC and OGDHC, respectively) to ~5.5 nm resolution (Fig. 7C) (Murphy and Jensen 2005). PDHC and OGDHC are multimeric enzymes composed of three protein subunits (E1, E2, and E3). Atomic models of all three subunits (de Kok et al. 1998) and single particle reconstructions of the E1E2 complex (Milne et al. 2002) were available, but the architecture of the intact complex remained unclear.

Murphy et al. showed that the E1 and E3 subunits are flexibly tethered to the cubic E2 core. Protein complexes as small as 80 kDa were clearly visible in the reconstructions, suggesting that ECT might prove helpful in elucidating the quaternary structure and conformational flexibility of many large complexes. Concurrent tomographic study of the E2E3 complex in vitro further supported the notion that the distribution of E1 and E3 molecules around the E2 core is random (Milne et al. 2006).

Brandt et al. used a template-matching approach to identify 70S ribosomes in actively translating *E. coli* cell lysates (Brandt et al. 2009). Even though various crystal structures of ribosomal complexes have been solved, little is known about the spatial arrangement of ribosomes in the cell during protein synthesis. Ribosomes in lysates were found to be densely packed around an mRNA template and showed preferred orientations (either staggered or helical) (Fig. 7D). Although the mRNA template was found to be on the inside of the ribosome cluster, the tRNAs and newly synthesized proteins were entering and exiting from the outside, minimizing interaction between newly synthesized proteins and lowering the probability of aggregate formation.

CONCLUSION AND FUTURE DIRECTIONS

As can be seen from these examples, ECT has become a powerful tool to investigate bacterial ultrastructure. Improvements in both the quality and number of cryo-tomograms being produced should be expected. Beginning with the sample, further improvements in methods for cryo-sectioning will make ECT of serial cryosections routine for larger cells and even biofilms. The development of genetically encodable, electron-dense tags (Mercogliano and DeRosier 2007; Nishino et al. 2007; Diestra et al. 2008), improvements in the technologies for correlating fluorescence and cryo-EM images (Lučić et al. 2007; Schwartz et al. 2007; Briegel et al. 2008), and increasingly sophisticated computational tools will allow specific molecules to be located and identified. Instrumentally, the development of direct electron detectors, phase plates, and aberration correctors are expected to dramatically improve image quality in just the next few years. As additional institutions acquire the expensive microscopes needed and more practitioners are trained, the number of ECT studies will also obviously increase. Lastly, automated data collection (Suloway et al. 2009) and processing (Amat et al. 2008), coupled with the development of databases for the management and distribution of the large datasets produced by ECT (Martone et al. 2003) (Ding, J.H. and Jensen, G.J., *in preparation*), will make it possible to collect and compare thousands of tomograms. Together with complementary advances in genetics, biochemistry, structural genomics, light microscopy, and computational modeling, it therefore seems that elucidating the structure, function, and location of every macromolecule in key model bacterial cells is a plausible goal for our generation (Morris and Jensen 2008).

REFERENCES

Addinall SG, Lutkenhaus J. 1996. FtsA is localized to the septum in an FtsZ-dependent manner. *J Bacteriol* **178:** 7167–7172.

Al-Amoudi A, Chang JJ, Leforestier A, McDowall A, Salamin LM, Norlén LP, Richter K, Blanc NS, Studer D, Dubochet J. 2004. Cryo-electron microscopy of vitreous sections. *EMBO J* **23:** 3583–3588.

Amat F, Moussavi F, Comolli LR, Elidan G, Downing KH, Horowitz M. 2008. Markov random field based automatic image alignment for electron tomography. *J Struct Biol* **161:** 260–275.

Anderson DE, Gueiros-Filho FJ, Erickson HP. 2004. Assembly dynamics of FtsZ rings in *Bacillus subtilis* and *Escherichia coli* and effects of FtsZ-regulating proteins. *J Bacteriol* **186:** 5775–5781.

Angell CA. 2004. Amorphous water. *Annu Rev Phys Chem* **55:** 559–583.

Batson PE, Dellby N, Krivanek OL. 2002. Sub-angstrom resolution using aberration corrected electron optics. *Nature* **418:** 617–620.

Baumeister W, Grimm R, Walz J. 1999. Electron tomography of molecules and cells. *Trends Cell Biol* **9:** 81–85.

Bazylinski DA, Frankel RB. 2004. Magnetosome formation in prokaryotes. *Nat Rev Microbiol* **2:** 217–230.

Berg HC. 1976. How spirochetes may swim. *J Theor Biol* **56:** 269–273.

Bi EF, Lutkenhaus J. 1991. FtsZ ring structure associated with division in *Escherichia coli*. *Nature* **354:** 161–164.

Bohm J, Frangakis AS, Hegerl R, Nickell S, Typke D, Baumeister W. 2000. Toward detecting and identifying macromolecules in a cellular context: Template matching applied to electron tomograms. *Proc Natl Acad Sci* **97:** 14245–14250.

Borgnia MJ, Subramaniam S, Milne JL. 2008. Three-dimensional imaging of the highly bent architecture of *Bdellovibrio bacteriovorus* by using cryo-electron tomography. *J Bacteriol* **190:** 2588–2596.

Bramhill D. 1997. Bacterial cell division. *Annu Rev Cell Dev Biol* **13:** 395–424.

Brandt F, Etchells SA, Ortiz JO, Elcock AH, Hartl FU, Baumeister W. 2009. The native 3D organization of bacterial polysomes. *Cell* **136:** 261–271.

Briegel A, Dias DP, Li Z, Jensen RB, Frangakis AS, Jensen GJ. 2006. Multiple large filament bundles observed in *Caulobacter crescentus* by electron cryotomography. *Mol Microbiol* **62:** 5–14.

Briegel A, Ding HJ, Li Z, Werner J, Gitai Z, Dias DP, Jensen RB, Jensen GJ. 2008. Location and architecture of the *Caulobacter crescentus* chemoreceptor array. *Mol Microbiol* **69:** 30–41.

Briegel A, Ortega DR, Tocheva EI, Wuichet K, Li Z, Chen S, Muller A, Iancu CV, Murphy GE, Dobro MJ et al. 2009. Universal architecture of bacterial chemoreceptor arrays. *Proc Natl Acad Sci* **106:** 17181–17186.

Charon NW, Goldstein SF, Marko M, Hsieh C, Gebhardt LL, Motaleb MA, Wolgemuth CW, Limberger RJ, Rowe N. 2008. The flat ribbon configuration of the periplasmic flagella of *Borrelia burgdorferi* and its relationship to motility and morphology. *J Bacteriol* **191:** 600–607.

Comolli LR, Downing KH. 2005. Dose tolerance at helium and nitrogen temperatures for whole cell electron tomography. *J Struct Biol* **152:** 149–156.

Crowther RA, DeRosier DJ, Klug A. 1970b. The reconstruction of a three-dimensional structure from projections and its application to electron microscopy. *Proc Roy Soc Lond A* **317:** 319–340.

Crowther RA, Amos LA, Finch JT, De Rosier DJ, Klug A. 1970a. Three dimensional reconstructions of spherical viruses by fourier synthesis from electron micrographs. *Nature* **226:** 421–425.

de Kok A, Hengeveld AF, Martin A, Westphal AH. 1998. The pyruvate dehydrogenase multi-enzyme complex from Gram-negative bacteria. *Biochim Biophys Acta* **1385:** 353–366.

Derosier DJ, Klug A. 1968. Reconstruction of 3 dimensional structures from electron micrographs. *Nature* **217:** 130.

Diestra E, Fontana J, Guichard P, Marco S, Risco C. 2008. Visualization of proteins in intact cells with a clonable tag for electron microscopy. *J Struct Biol* **165:** 157–168.

Dover LG, Cerdeño-Tárrage AM, Pallen MJ, Parkhill J, Besra GS. 2004. Comparative cell wall core biosynthesis in the mycolated pathogens, *Mycobacterium tuberculosis* and *Corynebacterium diphtheriae*. *FEMS Microbiol Rev* **28:** 225–250.

Dubochet J, Adrian M, Chang JJ, Homo JC, Lepault J, McDowall AW, Schultz P. 1988. Cryo-electron microscopy of vitrified specimens. *Q Rev Biophys* **21:** 129–228.

Dubochet J, McDowall AW, Menge B, Schmid EN, Lickfeld KG. 1983. Electron microscopy of frozen-hydrated bacteria. *J Bacteriol* **155:** 381–390.

Erickson HP. 1997. FtsZ, a tubulin homologue in prokaryote cell division. *Trends Cell Biol* **7:** 362–367.

Erk I, Nicolas G, Caroff A, Lepault J. 1998. Electron microscopy of frozen biological objects: A study using cryosectioning and cryosubstitution. *J Microsc* **189:** 236–248.

Fernández JJ, Li S, Crowther RA. 2006. CTF determination and correction in electron cryotomography. *Ultramicroscopy* **106:** 587–596.

Frangakis AS, Hegerl R. 2001. Noise reduction in electron tomographic reconstructions using nonlinear anisotropic diffusion. *J Struct Biol* **135:** 239–250.

Frederik PM, Hubert DH. 2005. Cryoelectron microscopy of liposomes. *Methods Enzymol* **391:** 431–448.

Gan L, Chen S, Jensen GJ. 2008. Molecular organization of Gram-negative peptidoglycan. *Proc Natl Acad Sci* **105:** 18953–18957.

Grimm R, Typke D, Barmann M, Baumeister W. 1996. Determination of the inelastic mean free path in ice by examination of tilted vesicles and automated most probable loss imaging. *Ultramicroscopy* **63:** 169–179.

Grunewald K, Medalia O, Gross A, Steven AC, Baumeister W. 2003. Prospects of electron cryotomography to visualize macromolecular complexes inside cellular compartments: Implications of crowding. *Biophys Chem* **100:** 577–591.

Hazelbauer GL, Falke JJ, Parkinson JS. 2008. Bacterial chemoreceptors: High-performance signaling in networked arrays. *Trends Biochem Sci* **33:** 9–19.

Henderson R. 1995. The potential and limitations of neutrons, electrons and X-rays for atomic resolution microscopy of unstained biological molecules. *Q Rev Biophys* **28:** 171–193.

Henderson GP, Jensen GJ. 2006. Three-dimensional structure of *Mycoplasma pneumoniae*'s attachment organelle and a model for its role in gliding motility. *Mol Microbiol* **60:** 376–385.

Hoffmann C, Leis A, Niederweis M, Plitzko JM, Engelhardt H. 2008. Disclosure of the mycobacterial outer membrane: Cryo-electron tomography and vitreous sections reveal the lipid bilayer structure. *Proc Natl Acad Sci* **105:** 3963–3967.

Iancu CV, Ding HJ, Morris DM, Dias DP, Gonzales AD, Martino A, Jensen GJ. 2007. The structure of isolated *Synechococcus* strain WH8102 carboxysomes as revealed by electron cryotomography. *J Mol Biol* **372:** 764–773.

Iancu CV, Morris DM, Dou Z, Heinhorst S, Cannon GC, Jensen GJ. 2010. Organization, Structure, and Assembly of alpha-Carboxysomes Determined by Electron Cryotomography of Intact Cells. *J Mol Biol* **396:** 105–117.

Iancu CV, Tivol WF, Schooler JB, Dias DP, Henderson GP, Murphy GE, Wright ER, Li Z, Yu Z, Briegel A et al. 2006a. Electron cryotomography sample preparation using the Vitrobot. *Nat Protocols* **1:** 2813–2819.

Iancu CV, Wright ER, Benjamin J, Tivol WF, Dias DP, Murphy GE, Morrison RC, Heymann JB, Jensen GJ. 2005. A "flip-flop" rotation stage for routine dual-axis electron cryotomography. *J Struct Biol* **151:** 288–297.

Iancu CV, Wright ER, Heymann JB, Jensen GJ. 2006b. A comparison of liquid nitrogen and liquid helium as cryogens for electron cryotomography. *J Struct Biol* **153:** 231–240.

Izard J, Samsonoff WA, Limberger RJ. 2001. Cytoplasmic filament-deficient mutant of *Treponema denticola* has pleiotropic defects. *J Bacteriol* **183:** 1078–1084.

Izard J, Hsieh CE, Limberger RJ, Mannella CA, Marko M. 2008. Native cellular architecture of *Treponema denticola* revealed by cryo-electron tomography. *J Strct Biol* **163:** 10–17.

Jensen GJ, Briegel A. 2007. How electron cryotomography is opening a new window onto prokaryotic ultrastructure. *Curr Opin Struct Biol* **17:** 260–267.

Khursigara CM, Wu X, Subramaniam S. 2008a. Chemoreceptors in *Caulobacter crescentus*: Trimers of receptor dimers in a partially ordered hexagonally packed array. *J Bacteriol* **190:** 6805–6810.

Khursigara CM, Wu X, Zhang P, Lefman J, Subramaniam S. 2008b. Role of HAMP domains in chemotaxis signaling by bacterial chemoreceptors. *Proc Natl Acad Sci* **105:** 16555–16560.

Kim KK, Yokota H, Kim SH. 1999. Four-helical-bundle structure of the cytoplasmic domain of a serine chemotaxis receptor. *Nature* **400:** 787–792.

Kim SH, Wang W, Kim KK. 2002. Dynamic and clustering model of bacterial chemotaxis receptors: Structural basis for signaling and high sensitivity. *Proc Natl Acad Sci* **99:** 11611–11615.

Komeili A, Li Z, Newman DK, Jensen GJ. 2006. Magnetosomes are cell membrane invaginations organized by the actin-like protein MamK. *Science* **311:** 242–245.

Konorty M, Brumfeld V, Vermeglio A, Kahana N, Medalia O, Minsky A. 2009. Photosynthetic system in Blastochloris viridis revisited. *Biochemistry* **48:** 4753–4761.

Konorty M, Kahana N, Linaroudis A, Minsky A, Medalia O. 2008. Structural analysis of photosynthetic membranes by cryo-electron tomography of intact *Rhodopseudomonas viridis* cells. *J Struct Biol* **161:** 393–400.

Kürner J, Frangakis A, Baumeister W. 2005. Cryo-electron tomography reveals the cytoskeletal structure of *Spiroplasma melliferum*. *Science* **307:** 436–438.

Lee E, Fahimian BP, Iancu CV, Suloway C, Murphy GE, Wright ER, Castano-Diez D, Jensen GJ, Miao J. 2008. Radiation dose reduction and image enhancement in biological imaging through equally-sloped tomography. *J Struct Biol* **164:** 221–227.

Leis A, Rockel B, Andrees L, Baumeister W. 2008. Visualizing cells at the nanoscale. *Trends Biochem Sci* **34:** 60–70.

Levin PA, Losick R. 1996. Transcription factor Spo0A switches the localization of the cell division protein FtsZ from a medial to a bipolar pattern in *Bacillus subtilis*. *Genes Development* **10:** 478–488.

Li Z, Jensen GJ. 2009. Electron cryotomography: A new view into microbial ultrastructure. *Curr Opin Microbiol* **12:** 333–340.

Li Z, Trimble MJ, Brun YV, Jensen GJ. 2007. The structure of FtsZ filaments in vivo suggests a force-generating role in cell division. *EMBO J* **26:** 4694–4708.

Liu J, Lin T, Botkin DJ, McCrum E, Winkler H, Norris SJ. 2009. Intact flagellar motor of Borrelia burgdorferi revealed by cryo-electron tomography: Evidence for stator ring curvature and rotor/C-ring assembly flexion. *J Bacteriol* **191:** 5026–5036.

Liu J, McBride MJ, Subramaniam S. 2007. Cell surface filaments of the gliding bacterium *Flavobacterium johnsoniae* revealed by cryo-electron tomography. *J Bacteriol* **189:** 7503–7506.

Löwe J, Amos LA. 1998. Crystal structure of the bacterial cell-division protein FtsZ. *Nature* **391:** 203–206.

Lu C, Reedy M, Erickson HP. 2000. Straight and curved conformations of FtsZ are regulated by GTP hydrolysis. *J Bacteriol* **182:** 164–170.

Lucić V, Förster F, Baumeister W. 2005. Structural studies by electron tomography: from cells to molecules. *Annu Rev Biochem* **74:** 833–865.

Lucić V, Kossel AH, Yang T, Bonhoeffer T, Baumeister W, Sartori A. 2007. Multiscale imaging of neurons grown in culture: from light microscopy to cryo-electron tomography. *J Struct Biol* **160:** 146–156.

Lutkenhaus J. 1993. FtsZ ring in bacterial cytokinesis. *Mol Microbiol* **9:** 403–409.

Lutkenhaus JF, Wolf-Watz H, Donachie WD. 1980. Organization of genes in the ftsA-envA region of the *Escherichia coli* genetic map and identification of a new fts locus (ftsZ). *J Bacteriol* **142:** 615–620.

Ma X, Ehrhardt DW, Margolin W. 1996. Colocalization of cell division proteins FtsZ and FtsA to cytoskeletal structures in living *Escherichia coli* cells by using green fluorescent protein. *Proc Natl Acad Sci* **93:** 12998–13003.

Margolin W. 1998. A green light for the bacterial cytoskeleton. *Trends Microbiol* **6:** 233–238.

Martone ME, Zhang S, Gupta A, Qian X, He H, Price DL, Wong M, Santini S, Ellisman MH. 2003. The cell-centered database: A database for multiscale structural and protein localization data from light and electron microscopy. *Neuroinformatics* **1:** 379–395.

Mastronarde DN. 2005. Automated electron microscope tomography using robust prediction of specimen movements. *J Struct Biol* **152:** 36–51.

Mastronarde DN. 2008. Correction for non-perpendicularity of beam and tilt axis in tomographic reconstructions with the IMOD package. *J Microsc* **230:** 212–217.

McDonald KL, Auer M. 2006. High-pressure freezing, cellular tomography, and structural cell biology. *BioTechniques* **41:** 137–143.

Mercogliano C, DeRosier D. 2007. Concatenated metallothionein as a clonable gold label for electron microscopy. *J Struct Biol* **160:** 70–82.

Michie KA, Löwe J. 2006. Dynamic filaments of the bacterial cytoskeleton. *Annu Rev Biochem* **75:** 467–492.

Midgley PA, Dunin-Borkowski RE. 2009. Electron tomography and holography in materials science. *Nat Mater* **8:** 271–280.

Milne JL, Shi D, Rosenthal PB, Sunshine JS, Domingo GJ, Wu X, Brooks BR, Perham RN, Henderson R, Subramaniam S. 2002. Molecular architecture and mechanism of an icosahedral pyruvate dehydrogenase complex: A multifunctional catalytic machine. *EMBO J* **21:** 5587–5598.

Milne JL, Wu X, Borgnia MJ, Lengyel JS, Brooks BR, Shi D, Perham RN, Subramaniam S. 2006. Molecular structure of a 9-MDa icosahedral pyruvate dehydrogenase subcomplex containing the E2 and E3 enzymes using cryoelectron microscopy. *J Biol Chem* **281:** 4364–4370.

Morris DM, Jensen GJ. 2008. Toward a biomechanical understanding of whole bacterial cells. *Annu Rev Biochem* **77:** 583–613.

Motaleb MA, Corum L, Bono JL, Elias AF, Rosa P, Samuels DS, Charon NW. 2000. *Borrelia burgdorferi* periplasmic flagella have both skeletal and motility functions. *Proc Natl Acad Sci* **97:** 10899–10904.

Murphy GE, Jensen GJ. 2005. Electron cryotomography of the *E. coli* pyruvate and 2-oxoglutarate dehydrogenase complexes. *Structure* **13:** 1765–1773.

Murphy GE, Jensen GJ. 2007. Electron cryotomography. *BioTechniques* **43:** 413–421.

Murphy GE, Leadbetter JR, Jensen GJ. 2006. In situ structure of the complete *Treponema primitia* flagellar motor. *Nature* **442:** 1062–1064.

Murphy GE, Matson EG, Leadbetter JR, Berg HC, Jensen GJ. 2008. Novel ultrastructures of *Treponema primitia* and their implications for motility. *Mol Microbiol* **67:** 1184–1195.

Narasimha R, Aganj I, Bennett AE, Borgnia MJ, Zabransky D, Sapiro G, McLaughlin SW, Milne JL, Subramaniam S. 2008. Evaluation of denoising algorithms for biological electron tomography. *J Struct Biol* **164:** 7–17.

Nickell S, Forster F, Linaroudis A, Net WD, Beck F, Hegerl R, Baumeister W, Plitzko JM. 2005. TOM software toolbox: Acquisition and analysis for electron tomography. *J Struct Biol* **149:** 227–234.

Nickell S, Hegerl R, Baumeister W, Rachel R. 2003. Pyrodictium cannulae enter the periplasmic space but do not enter the cytoplasm, as revealed by cryo-electron tomography. *J Struct Biol* **141:** 34–42.

Nishino Y, Yasunaga T, Miyazawa A. 2007. A genetically encoded metallothionein tag enabling efficient protein detection by electron microscopy. *J Electron Microsc* **56:** 93–101.

Norris V, Den Blaauwen T, Cabin-Flaman A, Doi R, Harshey R, Janniere L, Jimenez-Sanchez A, Jin D, Levin P, Mileykovskaya E et al. 2007. Functional Taxonomy of Bacterial Hyperstructures. *Microbiol Mol Biol R* **71:** 230–253.

Ortiz JO, Förster F, Kürner J, Linaroudis AA, Baumeister W. 2006. Mapping 70S ribosomes in intact cells by cryoelectron tomography and pattern recognition. *J Struct Biol* **156:** 334–341.

Park SY, Borbat PP, Gonzalez-Bonet G, Bhatnagar J, Pollard AM, Freed JH, Bilwes AM, Crane BR. 2006. Reconstruction of the chemotaxis receptor-kinase assembly. *Nat Struct Mol Biol* **13:** 400–407.

Pogliano J. 2008. The bacterial cytoskeleton. *Curr Opin Cell Biol* **20:** 19–27.

Pruggnaller S, Mayr M, Frangakis A. 2008. A visualization and segmentation toolbox for electron microscopy. *J Struct Biol* **164:** 161–165.

Salje J, Zuber B, Löwe J. 2008. Electron Cryomicroscopy of *E. coli* Reveals Filament Bundles Involved in Plasmid DNA Segregation. *Science* **323:** 509–512.

Sartori A, Gatz R, Beck F, Rigort A, Baumeister W, Plitzko JM. 2007. Correlative microscopy: Bridging the gap between fluorescence light microscopy and cryo-electron tomography. *J Struct Biol* **160:** 135–145.

Scheffel A, Gruska M, Faivre D, Linaroudis A, Plitzko JM, Schüler D. 2006. An acidic protein aligns magnetosomes along a filamentous structure in magnetotactic bacteria. *Nature* **440:** 110–114.

Schmid MF, Booth CR. 2008. Methods for aligning and for averaging 3D volumes with missing data. *J Struct Biol* **161:** 243–248.

Schmid MF, Paredes AM, Khant HA, Soyer F, Aldrich HC, Chiu W, Shively JM. 2006. Structure of *Halothiobacillus neapolitanus* carboxysomes by cryo-electron tomography. *J Mol Biol* **364:** 526–535.

Schwartz CL, Sarbash VI, Ataullakhanov FI, McIntosh JR, Nicastro D. 2007. Cryo-fluorescence microscopy facilitates correlations between light and cryo-electron microscopy and reduces the rate of photobleaching. *J Microsc* **227:** 98–109.

Seybert A, Herrmann R, Frangakis A. 2006. Structural analysis of *Mycoplasma pneumoniae* by cryo-electron tomography. *J Struct Biol* **156:** 342–354.

Shimizu TS, Le Novère N, Levin MD, Beavil AJ, Sutton BJ, Bray D. 2000. Molecular model of a lattice of signalling proteins involved in bacterial chemotaxis. *Nat Cell Biol* **2:** 792–796.

Stricker J, Maddox P, Salmon ED, Erickson HP. 2002. Rapid assembly dynamics of the *Escherichia coli* FtsZ-ring demonstrated by fluorescence recovery after photobleaching. *Proc Natl Acad Sci U S A* **99:** 3171–3175.

Studdert CA, Parkinson JS. 2005. Insights into the organization and dynamics of bacterial chemoreceptor clusters through in vivo crosslinking studies. *Proc Natl Acad Sci* **102:** 15623–15628.

Subramaniam S. 2005. Bridging the imaging gap: Visualizing subcellular architecture with electron tomography. *Curr Opin Struct Biol* **8:** 316–322.

Subramaniam S, Milne JL. 2004. Three-dimensional electron microscopy at molecular resolution. *Annu Rev Bioph Biom* **33:** 141–155.

Suloway C, Shi J, Cheng A, Pulokas J, Carragher B, Potter CS, Zheng SQ, Agard DA, Jensen GJ. 2009. Fully automated, sequential tilt-series acquisition with Leginon. *J Struct Biol.*

Ting CS, Hsieh C, Sundararaman S, Mannella C, Marko M. 2007. Cryo-electron tomography reveals the comparative three-dimensional architecture of Prochlorococcus, a globally important marine cyanobacterium. *J Bacteriol* **189:** 4485–4493.

Tivol WF, Briegel A, Jensen GJ. 2008. An improved cryogen for plunge freezing. *Microsc Microanal* **14:** 375–379.

Trachtenberg S, Dorward LM, Speransky VV, Jaffe H, Andrews SB, Leapman RD. 2008. Structure of the cytoskeleton of Spiroplasma melliferum BC3 and its interactions with the cell membrane. *J Mol Biol* **378:** 776–787.

Tsien RY. 1998. The green fluorescent protein. *Annu Rev Biochem* **67:** 509–544.

Vollmer W, Holtje JV. 2004. The architecture of the murein (peptidoglycan) in gram-negative bacteria: Vertical scaffold or horizontal layer(s)? *J Bacteriol* **186:** 5978–5987.

Weis RM, Hirai T, Chalah A, Kessel M, Peters PJ, Subramaniam S. 2003. Electron microscopic analysis of membrane assemblies formed by the bacterial chemotaxis receptor Tsr. *J Bacteriol* **185:** 3636–3643.

Werner JN, Chen EY, Guberman JM, Zippilli AR, Irgon JJ, Gitai Z. 2009. Quantitative genome-scale analysis of protein localization in an asymmetric bacterium. *Proc Natl Acad Sci* **106:** 7858–7863.

Wolanin PM, Baker MD, Francis NR, Thomas DR, DeRosier DJ, Stock JB. 2006. Self-assembly of receptor/signaling complexes in bacterial chemotaxis. *Proc Natl Acad Sci* **103:** 14313–14318.

Wright ER, Iancu CV, Tivol WF, Jensen GJ. 2006. Observations on the behavior of vitreous ice at ~82 and ~12 K. *J Struct Biol* **153:** 241–252.

Yeates TO, Kerfeld CA, Heinhorst S, Cannon GC, Shively JM. 2008. Protein-based organelles in bacteria: Carboxysomes and related microcompartments. *Nat Rev Micro* **6:** 681–691.

Zhang P, Bos E, Heymann J, Gnaegi H, Kessel M, Peters PJ, Subramaniam S. 2004. Direct visualization of receptor arrays in frozen-hydrated sections and plunge-frozen specimens of *E. coli* engineered to overproduce the chemotaxis receptor Tsr. *J Microsc* **216:** 76–83.

Zheng QS, Braunfeld MB, Sedat JW, Agard DA. 2004. An improved strategy for automated electron microscopic tomography. *J Struct Biol* **147:** 91–101.

Zheng SQ, Keszthelyi B, Branlund E, Lyle JM, Braunfeld MB, Sedat JW, Agard DA. 2007. UCSF tomography: An integrated software suite for real-time electron microscopic tomographic data collection, alignment, and reconstruction. *J Struct Biol* **157:** 138–147.

Zuber B, Chami M, Houssin C, Dubochet J, Griffiths G, Daffe M. 2008. Direct visualization of the outer membrane of mycobacteria and corynebacteria in their native state. *J Bacteriol* **190:** 5672–5680.

Single-Molecule and Superresolution Imaging in Live Bacteria Cells

Julie S. Biteen[1,2] and W.E. Moerner[1]

[1]Department of Chemistry, Stanford University, Stanford, California 94305-5080

[2]Department of Chemistry, University of Michigan, Ann Arbor, Michigan 48109-1055

Correspondence: wmoerner@stanford.edu

Single-molecule imaging enables biophysical measurements devoid of ensemble averaging, gives enhanced spatial resolution beyond the diffraction limit, and permits superresolution reconstructions. Here, single-molecule and superresolution imaging are applied to the study of proteins in live *Caulobacter crescentus* cells to illustrate the power of these methods in bacterial imaging. Based on these techniques, the diffusion coefficient and dynamics of the histidine protein kinase PleC, the localization behavior of the polar protein PopZ, and the treadmilling behavior and protein superstructure of the structural protein MreB are investigated with sub-40-nm spatial resolution, all in live cells.

Since its advent 20 years ago, single-molecule fluorescence imaging has given rise to a host of exciting experiments (Ambrose and Moerner 1991). Beyond enabling fundamental investigations of the physics of emissive molecules, one main advantage of this technique is its use in biologically relevant, live-cell experiments. Optical fluorescence microscopy is an important instrument for cell biology, as light can be used to noninvasively probe a sample with relatively small perturbation of the specimen, enabling dynamical observation of the motions of internal structures in living cells. Single-molecule epifluorescence microscopy extends these capabilities by achieving nanometer-scale resolution, taking advantage of the fact that one can precisely characterize the point spread function (PSF) of a microscope, allowing the center of a distribution, and thus the exact position of an emitter, to be localized with accuracy much better than the diffraction limit itself. This localization accuracy improves beyond the diffraction limit roughly as one over the square root of the number of detected photons (Thompson et al. 2002). Detecting 100 photons from a single, isolated molecule can therefore improve the resolution of an optical measurement from the ~250-nm diffraction limit down to 25 nm.

Single-molecule imaging has been used in the investigation of a number of live-cell samples. In 2000, the lateral heterogeneity of the plasma membrane was investigated by tracing the motion of single dye-labeled lipids in native human airway smooth muscle (HASM) cells (Schütz et al. 2000), and epidermal growth factor (EGF) receptor signaling was explored with a fluorescent protein fusion and a labeled ligand

 doi: 10.1101/cshperspect.a000448
Cite this article as *Cold Spring Harb Perspect Biol* 2010;2:a000448

(Sako et al. 2000). Single fluorophore-labeled molecules have subsequently been used in many ways (Moerner 2003), for instance to investigate the effect of varying cholesterol concentration on the mobility of proteins in the plasma membrane of Chinese hamster ovary (CHO) cells (Vrljic et al. 2002; Vrljic et al. 2005) and to explore the real-time dynamic behavior of cell-penetrating-peptide (CPP) molecular transporters on the plasma membrane of CHO cells (Lee et al. 2008). Furthermore, in 2001, Harms et al. characterized the emission of fluorescent proteins in biocompatible environments and noted that the yellow fluorescent protein EYFP was well-suited to single-molecule imaging in cells (Harms et al. 2001). Such fluorescent proteins can be genetically encoded as tags for native proteins in cells; these fusions have been used in many live-cell single-molecule experiments.

More recently, single-molecule epifluorescence microscopy has been used to probe the inner workings of live bacteria. The small size of prokaryotic cells makes the optical diffraction limit particularly noticeable, which has stimulated the push toward superlocalization and superresolution to overcome this obstacle. As a result, the nascent field of bacterial structural biology has benefited greatly from single-molecule investigations of proteins in live cells. The overall shapes of such cells can be seen in a standard light microscope, but those interested in probing subcellular details, such as protein structure and localization, have typically had to resort to in vitro characterization combined with extrapolation to the cellular environment, as well as to indirect methods such as biochemical assays. Although cryo-electron microscopy can provide extremely high spatial resolution, fixation or plunge-freezing is essential, and methods for identifying specific proteins out of many are still lacking. As a consequence, bacterial cell biology is an area of study ripe for investigation with direct, noninvasive optical methods of probing position, coupling and structure, with resolution below the standard diffraction limit.

Several groups have extended single-molecule imaging techniques to live bacterial samples. In 2004, single PleC proteins were visualized in *Caulobacter crescentus* cells (Deich et al. 2004), and the behavior of this system is described in more detail later. More recently, Xie and coauthors have used single-molecule fluorescence techniques to study DNA-binding proteins, mRNA, and membrane proteins to provide much insight into the mechanisms of bacterial gene expression; these efforts have been documented in a recent review (Xie et al. 2008). As well, Conley et al. used covalently linked Cy3-Cy5-thiol switchable fluorophores to illuminate the stalks of *C. crescentus* cells with high resolution (Conley et al. 2008). In this article, we focus on the application of single-molecule imaging and single-molecule-based superresolution imaging to investigate the localization, movement, and structure of three important proteins, PleC, PopZ, and MreB, in live *C. crescentus* cells.

METHODS FOR LIVE-CELL SINGLE-MOLECULE IMAGING

Sample Preparation

Single-molecule fluorescence imaging of proteins in living cells can be accomplished via a genetically encoded fusion of the protein of interest to a photostable fluorescent protein (FP) emitting at long enough wavelengths to avoid cellular autofluorescence. However, single-molecule imaging requires a low copy number of fluorophores in the cell at any one time, so that the ~250-nm diameter image spots from single molecules do not overlap. This low concentration can be achieved in several ways. First, the protein-FP fusion can be integrated into the chromosome in lieu of the native protein copy. In such strains, the initial FP concentration is generally too high for single-molecule imaging, so the emitting FP population needs to be reduced via photobleaching before measurements are made (Deich et al. 2004). Second, the protein-FP fusion can be integrated into the chromosome under the control of an inducible promoter while maintaining the untagged wild-type protein in the strain. This has the advantage of allowing the experimenter to control the level of tagged protein, while maintaining a significant background of the native protein to minimize any physiological effects of the tagging (Kim et al. 2006; Bowman et al. 2008). Finally, a large number of protein-FP fusions can be expressed in the cell, and the number of emitting FPs in any imaging frame can be subject to optical control, for instance by bleaching all of the FPs and then photo-reactivating a sparse subpopulation (Biteen et al. 2008).

To resolve the emission from single FPs above background, cell and sample autofluorescence must be minimized. In *Caulobacter crescentus*, this is accomplished by growing cells in minimal M2G medium for >24 hours, and by avoiding blue and green GFP mutants. The experiments discussed later all use the enhanced yellow fluorescent protein EYFP (S65G, V68L, S72A, T203Y), which can be excited at 514 nm. The high resolution provided by single-molecule imaging also requires the cells to lie still on a substrate. This can be accomplished by preparing a pad of 1.5%–2% agarose in M2G onto which the cells are deposited for imaging. The sample of cells on agarose must also contain fiduciary markers that can quantify stage drift over the imaging time. Fiducials can be any bright, nonbleaching emitters, for example quantum dots (Kim et al. 2006) or fluorescent beads (Biteen et al. 2008).

Imaging

Single-molecule imaging of bacterial cells can be accomplished in a standard widefield epifluorescence microscopy configuration, and the general imaging considerations have been described in detail (Moerner and Fromm 2003). Confocal or total-internal-reflection methods are often not necessary because the entire cell itself is mostly within the depth of focus of a high numerical aperture (NA) microscope (~300–700 nm). Because the photon emission rate from a single molecule is typically 10 orders of magnitude smaller than the number of photons per second irradiating each pixel of the detector, appropriate filtering is necessary to reject scattered laser light. To optimize the number of detected photons, the single-molecule fluorescence microscope should incorporate a high-NA oil-immersion objective, sharp and carefully chosen filters, and a sensitive detector capable of detecting single photons. In modern experiments, this detector is generally an electron-multiplying charge-coupled detector (EMCCD) (Kim et al. 2006; Biteen et al. 2008; Bowman et al. 2008), though single-molecule imaging has also been accomplished in live cells with an intensified CCD (Deich et al. 2004).

For imaging of EYFP-protein fusions, the sample is excited with a 514-nm Ar^+ ion or solid state laser. The excitation beam is converted from linear to circular polarization with a $\lambda/4$ waveplate to avoid the preferential detection of certain chromophore orientations (which often cannot be directly correlated with the orientation of the protein under investigation because of the FP fusion). In the detection pathway, a 525-nm long-pass dichroic mirror and a 530-nm long-pass filter are used to reject scattered laser light with an optical density of 10^8–10^{10}. In experiments in which photoswitching is useful, photoreactivation of EYFP to control the emitting fluorophore density is performed with a 407-nm laser. This is also circularly polarized, and is coupled into the excitation pathway with a dichroic mirror. For example, superresolution imaging based on single-molecule fluorescence and photoreactivation is accomplished by capturing different sparse subsets of emitters in each imaging frame. On photobleaching of a sparse subset, a new subset is generated by photoreactivation and the imaging cycle continues. The final image is generated based on the sum of many single-molecule imaging frames (Betzig et al. 2006; Hess et al. 2006; Rust et al. 2006; Biteen et al. 2008).

EMCCD cameras can operate at rates as fast as ~50 frames/second for a fairly large field, and even faster for subimages. The integration time is chosen based on the experimental system. A longer integration time increases the signal-to-noise when detecting quasi-stationary or slow-moving FPs. On the other hand, a shorter integration time is needed to resolve freely or quickly diffusing FPs. Typical integration times used in *C. crescentus* are 15.4–32.2 ms for the quickly diffusing unpolymerized proteins PopZ and MreB (Kim et al. 2006; Bowman et al. 2008), 100 ms for the slowly diffusing mobile protein PleC (Deich et al. 2004), and 100 ms for MreB monomers incorporated into filaments (Kim et al. 2006; Biteen et al. 2008).

Data Processing

Single molecules of the protein-FP fusion are identified by several criteria, including digital (one-frame) photobleaching, appropriate number of detected photons, and a diffraction-limited emission pattern (point-spread function, PSF). The position of a nanometer-sized isolated emitter is at the center of the PSF; this center can be identified by eye as the brightest pixel in the emission spot (Deich et al. 2004; Bowman et al. 2008), or the molecule can be more accurately localized by fitting the PSF to a 2D, symmetric Gaussian function (Kim et al. 2006; Biteen et al. 2008). High-density super-resolution images based on multiple cycles of single-molecule imaging frames require the fitting software to be automated (Biteen et al. 2008), and it is essential to carefully analyze the data in terms of numbers of photons detected and localization precision to represent the position determinations in an unbiased manner.

Absolute protein positions and direction of movement are determined relative to fiducials and to white-light or dark-field images of the *C. crescentus* cells. The midpoint and endpoints of each cell are observable, the stalked end is recognized by its stalk, and the cell cycle stage can be identified from the cell morphology and the presence or absence of a stalk. It would also be possible to use differential interference contrast (DIC) images, but the changes in microscope optics compared with those for single-molecule fluorescence imaging would have to be carefully calibrated. The techniques described in this article measure images in only two dimensions, but 2D measurements of diffusion on the surface of the bacterial cell membrane can be corrected appropriately by simulating 3D diffusional movement along an appropriately curved membrane (Deich et al. 2004; Kim et al. 2006).

LOCALIZATION MECHANISM OF THE HISTIDINE PROTEIN KINASE PleC

C. crescentus is characterized by an asymmetric cell division process. Each round of dimorphic *C. crescentus* cell division gives rise to two distinct daughter cells: A motile, nonreplicating swarmer cell with a polar flagellum and a replicating stalked cell with an adhesive stalk. The localization of the histidine kinase PleC regulates polar organelle formation, motility, and asymmetric cell division by localizing to the flagellar pole at specific points of the cell cycle. PleC is localized to the flagellar pole of swarmer cells. The swarmer cell must differentiate into a stalked cell, shedding its flagellum and building a stalk, to replicate and divide. During this differentiation, PleC becomes delocalized. As the stalked cell progresses into a predivisional cell, a new flagellum is formed opposite the stalk, and PleC is localized to the new flagellar pole (Wheeler and Shapiro 1999). Three different mechanisms have been postulated for the crucial process of PleC localization: free diffusion of PleC through the inner membrane followed by capture and immobilization at a binding site (Rudner et al. 2002; Shapiro et al. 2002), directed insertion of PleC into the inner membrane at the flagellar pole at the time of translation (Steinhauer et al. 1999; Robbins et al. 2001), and active transport of PleC to the pole by a motor protein. Deich et al. used single-molecule microscopy to distinguish between these putative mechanisms by measuring the diffusion coefficients and movement directionality of PleC molecules in *C. crescentus* cells as a function of cell type and position within the cell (Deich et al. 2004).

Strain EJ148 of *C. crescentus* was created with PleC-EYFP fusions replacing the wild-type PleC on the chromosome and expressed under control of the PleC promoter. The EYFP tags were exposed to 514-nm illumination until all but a sparse subset remained in the emissive state. These isolated single-molecule emitters were imaged and their positions recorded as a function of time to measure the movement of PleC molecules in the cells. Figure 1 shows representative fluorescence images of this system.

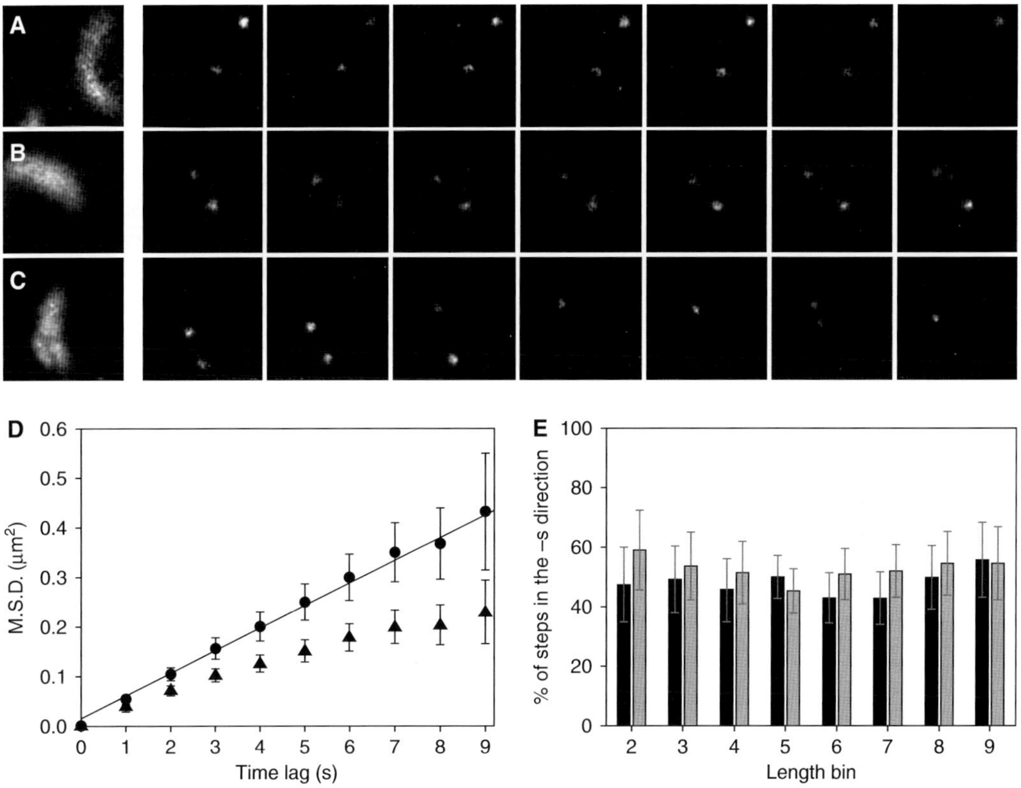

Figure 1. Dynamics of single PleC histidine kinase molecules. (*A–C*) *Leftmost* image: Darkfield image of the cell under investigation. Columns 2–7: Consecutive fluorescence images of single PleC-EYFP molecules spaced by different time intervals: (*A*) images acquired every 100 ms, (*B–C*) images acquired every 1 s. (*D*) Measured mean-square displacement (triangles) and geometry-corrected mean-square displacement (circles) versus time lag. The corrected data slope of 0.049 $\mu m^2/s$ corresponds to a diffusion coefficient, $D = 12 \pm 2\ \mu m^2/s$. (*E*) Direction of motion in stalked cells without localized PleC (black bars) and swarmer and predivisional cells with localized PleC (gray bars), in which $-s$ indicates motion toward the flagellar pole. Each bin represents one tenth of the cell length, with the polar bins excluded. (Reprinted, with permission, from Deich et al. 2004 [National Academy of Sciences].)

The first image in each of the series A–C is a darkfield image illustrating the position, size, and orientation of the cell under investigation. The other images in each series represent successive 100-ms imaging frames of the same cells in which each punctate spot is the fluorescence image of a single PleC-EYFP molecule. Figure 1A shows images acquired uninterruptedly (every 100 ms). Because of the relatively slow dynamics of the membrane-bound PleC, such continuous imaging bleached the molecules before significant motion could be observed. Indeed, the lower molecule in Figure 1A is bleached during the last frame, after only 600 ms of imaging. Time-lapse imaging was therefore used to increase the information attained before photobleaching. Figures 1B and C show two different cells in which a 100-ms imaging frame is acquired every 1 second. In this way, more movement is observed before photobleaching.

The single-molecule investigation of PleC in *C. crescentus* showed several key aspects of PleC dynamics. Bulk microscopy indicates the presence of a bright fixed PleC locus at the flagellar pole of swarmer and predivisional cells, but by resolving PleC-EYFP on a single-molecule level, a population of mobile PleC molecules could be identified in these cells. Mobile PleC molecules were also seen in the stalked cells, which show no evidence of localization in bulk microscopy. The trajectories of 400 such mobile molecules were further investigated by tracking the PleC-EYFP molecules, and the measured mean-square displacement (MSD) of the molecules at different time lags is shown in the triangles in Figure 1D. Because the observed motion is a 2D projection of 3D motion along a membrane surface, the actual distance moved by the molecule is larger than the observed movement. By simulating diffusion on a cell surface, modeled as a cylinder with spherical ends, the measured MSD is converted to a geometry-corrected MSD (circles in Fig. 1D). The geometry-corrected MSD is linear with the time lag, Δt, consistent with Brownian diffusion with a diffusion coefficient, $D = \mathrm{MSD}/(4\Delta t)$ of 12 $\pm 2\ \mu\mathrm{m}^2/\mathrm{s}$.

A normalized coordinate system was created in which the variable s ($0 < s < 1$) indicates the position of a molecule along the long axis of a *C. crescentus* cell. The position $s = 0$ represents the flagellar pole, and $s = 1$ the stalked pole. The geometry-corrected coefficient D for the mobile PleC-EYFP molecules were classified based on the cell type in which they were imaged (stalked, swarmer, or predivisional) and their position in the cell along the s axis. No measurable difference in D was observed for the different cell types. The geometry-corrected coefficient D was also not found to vary significantly as a function of s, except for a small reduction in D at the cell flagellar pole of swarmer and predivisional cells, perhaps attributable to immobilization at the binding site during the course of observation. Finally, Figure 1E shows the direction of PleC-EYFP motion along the s axis, in which $-s$ indicates motion toward the flagellar pole, as a function of position and cell type. Here, stalked cells, which do not have localized PleC, are in black, and swarmer and predivisional cells, which have localized PleC in addition to the mobile PleC under consideration, are in gray, and the cell is divided into ten equal bins, so length bin 2 includes data for molecules starting with $0.1 < s < 0.2$. All molecules were equally likely to move in the $+s$ or $-s$ direction, regardless of cell type or PleC position.

The motion of PleC in the membrane appears to be governed by Brownian diffusion, and is the same at all times in the cell cycle and all positions in the cell. In addition, the absence of directed motion argues against the existence of an active transport mechanism. Overall, the results of single-molecule investigations of PleC in *C. crescentus* were consistent with the diffusion and capture model of PleC localization at the flagellar pole.

DIFFUSION DYNAMICS OF THE POLAR PROTEIN PopZ

The ability of single-molecule microscopy to track individual intracellular proteins was subsequently applied to the proline-rich protein PopZ in live *C. crescentus* cells. PopZ is required to anchor the *C. crescentus* chromosome origins to the cell poles via the DNA-binding protein ParB, an anchoring that is critical for the control of temporal and spatial localization in the cell cycle. PopZ is localized to the flagellar end of nonreplicating swarmer cells. On differentiation into a stalked cell, a second concentration of PopZ begins to accumulate at the opposite pole (distal from the stalk) with an intensity that increases as the cell cycle progresses. The mechanism of PopZ localization was investigated on a single-molecule level (Bowman et al. 2008).

The merodiploid strain GB175 of *C. crescentus* was created with an unperturbed *popZ* locus and *popZ-eyfp* under control of the vanillate promoter at the *vanA* chromosomal locus. Leakage of the vanillate promoter ensured the expression of 1–10 copies of PopZ-EYFP per cell in a background of untagged PopZ molecules. These isolated single-molecule emitters were imaged and their positions recorded as a function of time, and two subpopulations of PopZ were identified: pole-localized molecules that remained fixed within the measurement localization accuracy (60 nm), and mobile molecules that were resolved with fast (32.2-ms) imaging frames. Figures 2A and D show the motion of two molecules tracked in one cell under investigation. In Figure 2A, the tracks are overlaid on a transmitted white-light image of the cell with the cell outline in black, whereas in Figure 2D, the longitudinal positions of the molecules are recorded as a function of imaging time. In Figures 2A and D, the red traces show the position of a polar-localized molecule, and the green traces show the trajectory of a mobile molecule that explores a large portion of the cell interior. Figures 2B–F show the trajectories of 12 fixed (red) and mobile (green) PopZ-EYFP molecules in five cells.

The mobile PopZ-EYFP molecules showed random, diffusive movement in the cytoplasm during most of the time that they were tracked, though on some occasions, they remained fixed at a pole (e.g., Fig. 2D at 3–4 s). This movement is consistent with a diffusion and

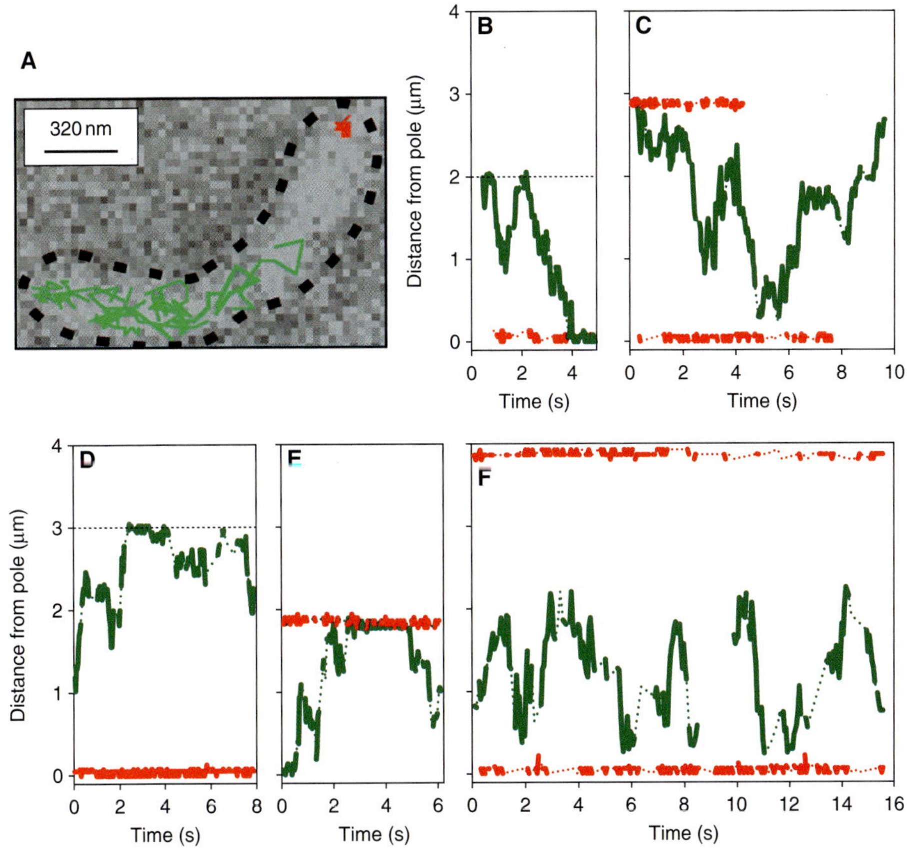

Figure 2. Visualization of single PopZ-YFP molecules in live cells. (*A*) Positions of two PopZ-EYFP molecules within a cell, with colored lines tracking the distance moved between 32.2-ms frames. One molecule (red) remains localized to the pole, and the other (green) has increased mobility. The tracks are overlaid on a white-light image of the cell, which is outlined in black. (*B–F*) Time-dependent behavior of 12 single molecules in five cells. Red and green lines follow stationary and mobile molecules, respectively. The colored lines are dotted during dark (blinking-off) periods. Black horizontal dotted lines in (*B*) and (*D*) mark the positions of the poles opposite to the pole at which stationary molecules are localized. The trajectories in (*D*) correspond to the longitudinal position of the molecules tracked in (*A*) as a function of time. (Reprinted, with permission, from Bowman et al. 2008 [© Elsevier].)

capture mechanism for PopZ localization at the cell pole. The spatial dynamics of single PopZ molecules would not have been observable without single-molecule imaging.

MOVEMENT OF THE STRUCTURAL PROTEIN MreB

The structural protein MreB mediates polarity, chromosome segregation, and cell shape in *C. crescentus* (Gitai et al. 2004; Dye et al. 2005; Gitai et al. 2005). Bulk-level imaging and biochemical studies of the cytoplasmic MreB protein are consistent with the formation of a dynamic superstructure made up of actin-like filaments. This structure provides the machinery for key cellular processes, and the dynamics of MreB molecules were investigated on a single-molecule level (Kim et al. 2006).

The merodiploid strain LS3813 of *C. crescentus* was generated by integrating a single *Pxyl::eyfp-mreB* construct into the *xylX* locus in a *C. crescentus* chromosome already containing a wild-type, unlabeled copy of MreB under its endogenous promoter (Gitai et al. 2004). Cells with only a small number of isolated EYFP-MreB molecules were prepared by incubating the cells in the M2G minimal medium with 0.0006%–0.003% xylose for 4 hours. The cells were imaged with 15.4-ms imaging frames to resolve the tagged MreB molecules. Figure 3A1 shows one such frame, in which three single molecules were captured. The image is smoothed by applying a low-pass filter (3×3 kernels of 0.0625, 0.125, 0.0625, 0.125, 0.25, 0.125, 0.0625, 0.125, and 0.0625) in Figure 3A2 for enhanced visibility. In this cell, two subpopulations of MreB were distinguished from one another: slow-moving molecules and fast-moving molecules (arrowheads and arrow, respectively, in Fig. 3A1). The trajectory of the fast-moving molecule in Figure 3A1, obtained by fitting the fluorescence image in every frame to a 2D Gaussian function, is plotted in Figure 3A3. This molecule diffused rapidly and explores a large portion of the cell. 450 successive frames were summed to form the fluorescence image in Figure 3A4. After this 7-second integration time, the fluorescence from the two slow-moving molecules was still evident, but emission from the fast-moving molecule was smeared out over many pixels and did not appear. The white line in Figure 3A shows the outline of the *C. crescentus* cell.

Because the dynamics of these molecules were unchanged in the presence of the MreB-depolymerizing drug A22, the fast-moving EYFP-MreB molecules were ascribed to a free, unpolymerized population. The behavior of the unpolymerized single molecules was further characterized based on 111 trajectories. The observed mean-square displacement (MSD) of the fast-moving molecules is plotted as a function of time lag (Δt) in the open circles of Figure 3B. Based on the first four points, a diffusion coefficient of $D = \text{MSD}/(4\Delta t)$ of 1.11 ± 0.18 $\mu m^2/s$ was extracted. This value of D is smaller than expected for a free, cytoplasmic protein, but is consistent with the motion of a membrane-associated protein. Because the observed motion represents a 2D projection, simulations were again performed to correct for the three-dimensionality of the cell, and the geometry-corrected MSD of the molecules is plotted in the filled circles of Figure 3B. The geometry-corrected MSD is linear with the time lag, Δt, consistent with diffusion along the cell membrane with $D = \text{MSD}/(4\Delta t)$ of 1.75 ± 0.17 $\mu m^2/s$. The velocity autocorrelation function, $C_V(\tau)$, was also calculated for these molecules. As shown in Figure 3C, for the fast-moving molecules, $C_V(\tau)$ dropped to zero at the very first time step, indicating a nondirected random walk.

Further experiments addressed the behavior of the slow-moving MreB. Because slow-moving EYFP-MreB molecules were not observed in the presence of the MreB-depolymerizing drug A22, the slow-moving molecules represent polymerized MreB. The dynamics of 120 single members of this subpopulation were carefully examined. Because the molecules were stationary within the 30-nm resolution of the measurement over the course of the 7-second integration time shown in Figure 3A4, time-lapse imaging was used. Specifically, 100-ms imaging frames were separated by 9.9 seconds. In this way, the slow movement of polymerized MreB molecules

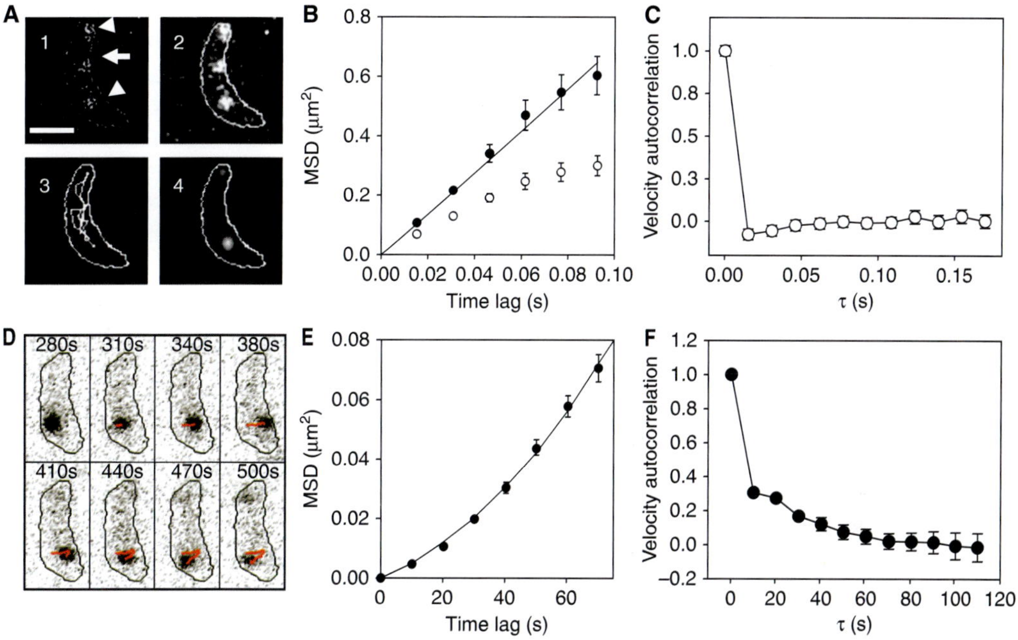

Figure 3. Analysis of motion of EYFP-MreB. (*A*) 15.4-ms integration time fluorescence images of single EYFP-MreB molecules in a *C. crescentus* cell. White line shows the cell outline. (*A*1) 15.4-ms epifluorescence image showing three single EYFP-MreB molecules. The *top* and *bottom* molecules (arrowheads) are stationary on this timescale, the *middle* molecule (arrow) is mobile. (*A*2) Smoothed image of (*A*1) obtained by applying a low-pass filter. (*A*3) Recorded trajectory of the mobile (*middle*) molecule in (*A*1). (*A*4) Summed image of 450 sequential imaging frames. The fluorescence from the two stationary molecules is still evident, whereas the middle molecule does not appear on this 6.93-s integration time scale. (Scale bar, 1 μm.) (*B*) Measured (open circles) and geometry-corrected (filled circles) MSD of fast-moving MreB molecules. The solid line is a linear fit of the corrected data. (C) Velocity autocorrelation of fast-moving molecules; this quantity drops to zero at the very first time lag. (*D*) Inverted-contrast, 100-ms integration time fluorescence images of single slow-moving EYFP-MreB molecules in a *C. crescentus* cell at 30-s intervals. The red arrows illustrate the directional movement of the molecules and the black line shows the cell outline. (*E*) Measured MSD versus time lag for slow-moving MreB molecules. The solid line is a quadratic fit to the data, indicative of directional motion. (*F*) Velocity autocorrelation of slow-moving MreB; this quantity remains positive over at least 80 s. (Reprinted, with permission, from Kim et al. 2006 [National Academy of Sciences].)

could be followed over a longer time period before photobleaching. Figure 3D shows 8 such fluorescence images in reverse contrast, in which a single molecule is tracked for 220 seconds. The molecule moves from left to right, then turns and moves right to left at a downward angle, shown by the red line. The observed MSD of the slow-moving MreB molecules is plotted as a function of Δt in Figure 3E. Here, the MSD was characterized by a quadratic dependence on time lag, as is typical of directed motion. Also consistent with directed motion is the computed velocity autocorrelation function, $C_V(\tau)$, shown in Figure 3F, in which $C_V(\tau)$ remains positive until $\tau = 80$ s.

The slow, directed motion of polymerized MreB was further explored by analyzing the trajectories of each individual molecule in the context of two putative models for the motion of an MreB molecule in an MreB filament: (1) the filaments into which the monomers are incorporated were moving, or (2) the monomers themselves were treadmilling though largely stationary filaments, analogous to the motion of actin. If the motion of polymerized MreB is because of whole-filament movement, then the observation time for the slow-moving single molecules should be limited by the photobleaching time of the EYFP, independent of experiment timescale. In the inset to Figure 4A, the total emission time of EYFP-MreB before photobleaching under continuous emission was measured, with an average on-time of 4.6 seconds. However, the true irradiation time before photobleaching of the molecules imaged with 9.9-second time lapses, shown in Figure 4A, was found to be only 0.8 seconds. Most of the fluorescence from the polymerized MreB molecules therefore disappeared before photobleaching occurred, likely as a result of dissociation from the end of a filament and the onset of fast diffusive motion not resolvable with the 100-ms imaging frames. This result indicates that MreB molecules treadmill through filaments with fixed ends.

Given that the polymerized MreB molecules showed directed motion, a speed value was extracted from each single-molecule trajectory (Fig. 4B). The average speed was 6.0 ± 0.2 nm/s. From its crystal structure, the length of an MreB monomer is 5.4 nm (van den Ent et al. 2001); the average speed therefore corresponded

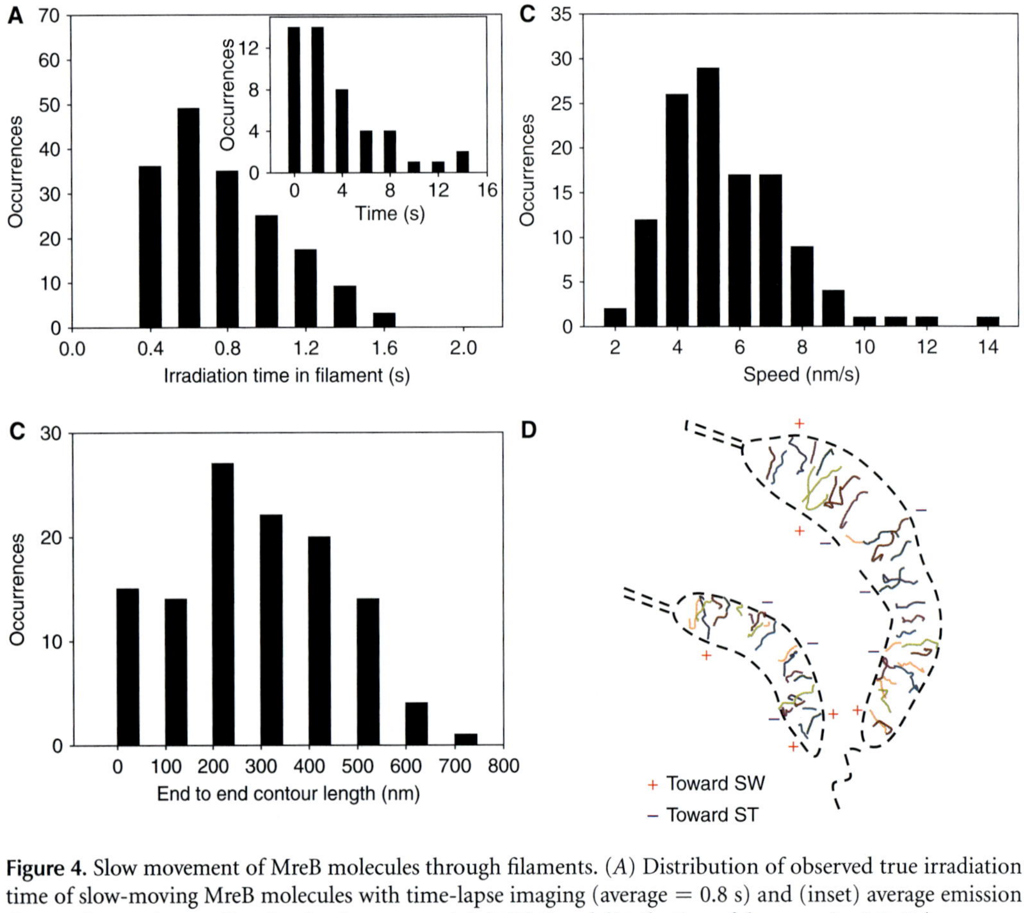

Figure 4. Slow movement of MreB molecules through filaments. (*A*) Distribution of observed true irradiation time of slow-moving MreB molecules with time-lapse imaging (average = 0.8 s) and (inset) average emission time under continuous illumination (average = 4.6 s). (*B*) Speed distribution of slow-moving MreB (average = 6.0 ± 0.2 nm/s). (*C*) Distribution of end-to-end contour lengths measured from the movement of single slow-moving MreB molecules (average = 332 nm). (*D*) Representative trajectories of MreB filaments in a stalked (ST) cell and in a predivisional cell, plotted on normalized cell. Representative global direction assignments are given by "+" (toward the swarmer [SW] pole) or "−" [toward the ST pole]). (Reprinted, with permission, from Kim et al. 2006 [National Academy of Sciences].)

to 1.2 monomer additions per second in steady-state fixed MreB filaments. The average length traveled by a polymerized MreB molecule before dissociation, which corresponds to the filament length, was also measured by considering only MreB molecules that became polymerized after the start of imaging and that dissociated before the end of imaging. The MreB filament lengths of 128 trajectories are shown in Figure 4C, in which the average filament length was 392 $\pm$23 nm—quite small relative to the average cell length of 3.5 μm. A representative subset of these trajectories is plotted on normalized cell shapes in Figure 4D. Of note, most of the trajectories moved perpendicular to the cell long axis. A smaller number of trajectories were oblique. These nonperpendicular trajectories were characterized as moving toward the swarmer pole (+) or toward the stalked pole (−). No preferred orientation was observed.

CELL-CYCLE-DEPENDENT SUPERRESOLUTION STRUCTURE OF MreB

By fitting the point-spread function (PSF) of isolated nanoscale emitters with 30-nm localization accuracy, and by following these superlocalized molecules over time, the previous experiment attained superresolution via single-molecule tracking, as displayed in Figure 4D. This approach was, however, limited by the isolated fluorophore requirement to ~1–3 tagged MreB molecules, and therefore ~1–3 trajectories per cell. To address this, photoactivated localization microscopy (PALM) was used to extend the investigation of EYFP-MreB in *C. crescentus* cells to cells expressing high concentrations of these molecules, and thus to visualize the superstructure formed by a collection of MreB filaments in these cells (Biteen et al. 2008). In particular, bulk visualization of MreB in *C. crescentus* indicates that MreB forms a cell-cycle-dependent structure (Gitai et al. 2004; Dye et al. 2005), and single-molecule-based superresolution imaging was applied to measure this structure.

C. crescentus cells expressing 100–1000 copies of EYFP-MreB were created by incubating the strain LS3813, described above, in the M2G minimal medium with 0.006%–0.075% xylose for 4 hours. The cells were imaged with 100-ms imaging frames to resolve only the slow-moving, polymerized MreB molecules. The initial density of fluorescent tags in these cells was initially too great for isolated, single-molecule imaging, but this could be addressed based on the fact that apparently bleached EYFP molecules can be reactivated with violet light (Dickson et al. 1997). Therefore, in live *C. crescentus* cells, after the bleaching of all emissive EYFP-MreB molecules, a sparse subset could be reactivated with 407-nm irradiation, in which the reactivation intensity of 10^3–10^4 W/cm^2 was chosen such that at most one EYFP molecule was reactivated in each diffraction-limited region. The positions of these isolated molecules were determined by PSF fitting, and the reactivation and imaging process was repeated many times until a superresolution image could be reconstructed from the sum of the localizations. To illustrate the photoreactivation probes, Figures 5A–F show the fluorescence emission from EYFP-MreB molecules in a single cell over time. Each punctate, white spot is the emission from a single molecule, and the fluorescence images (acquired with 100-ms integration times) are superimposed on a reversed-contrast white-light image of the cell. After the initial bleaching step, two single EYFP-MreB molecules are observed in the cell (Fig. 5A). Additional imaging with 514-nm light bleached all the fluorophores (Fig. 5B). A 2-second dose of 407-nm laser illumination returned some EYFP-MreB to an emissive state (Figs. 5C and E), though all fluorophores are apparently bleached prior to the 407-nm pulse (Figs. 5D and F).

The positions of these molecules, calculated from a fit of the fluorescence images to a 2D Gaussian function, were recorded for every such frame, and at the end of a 4-minute acquisition period, these localization events were summed. To form a final superresolution image, each localized molecule was depicted in the reconstruction as a unit-area Gaussian with fixed width equal to the localization accuracy of 30–40 nm. This technique, termed live-cell PALM, enabled the imaging of the EYFP-MreB superstructure in *C. crescentus*. In particular, it

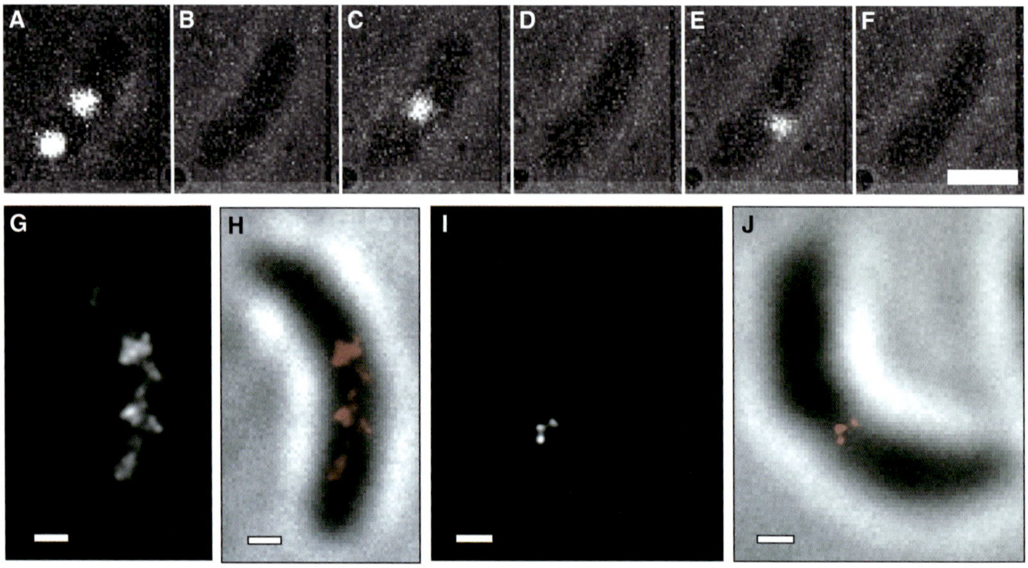

Figure 5. Superresolution imaging of MreB superstructure. (*A–F*) Reactivation of single EYFP-MreB fusions in the same live *C. crescentus* cell using 407 nm light. Fluorescence images of single EYFP-MreB molecules from 514 nm excitation are overlaid on a reversed-contrast white-light image of the cell. (*A*) Initial image showing two isolated emissive EYFP-MreB molecules. (*B*), (*D*), (*F*) After photobleaching by 514-nm irradiation, the cell contains no emissive EYFP-MreB. (*C*), (*E*) Reactivated EYFP-MreB molecules are observed following a short 407-nm reactivation pulse. (Scale bar, 1 µm.) (*G–J*) TL-PALM images of EYFP-MreB in *C. crescentus* cells created by fitting molecule positions in imaging frames such as those in (*A*), (*C*), and (*E*). (*G*), (*H*) Quasi-helical structure in a stalked cell. (*I*), (*J*) Midplane ring in a predivisional cell. Fluorescence PALM images are shown in (*G*) and (*I*). The PALM images in (*H*) and (*J*) are from the same cells as in (*G*) and (*I*), respectively, overlaid on a reversed-contrast white-light image of the cell. (Scale bars, 300 nm). (Reprinted, with permission, from Biteen et al. 2008 [Nature Publishing Group].)

resolved several MreB bands spanning the length of stalked cells, and a tightly-focused midplane ring in predivisional cells. Importantly, understanding superresolution features derived from many image acquisitions requires careful consideration of the emitter photophysics and the dynamics, if any, of the underlying structure. Because, as described above, polymerized MreB molecules treadmill slowly along MreB filaments, time-lapse (TL) imaging, as described above, was incorporated into live-cell PALM to increase the number of localization events given the fact that the maximum labeling concentration in live-cell experiments is limited because of changes in the cell morphology at high concentrations of fusion protein. Specifically, a 900-ms delay was introduced between each 100-ms imaging frame, and the sample was only illuminated with 514-nm light during the short acquisition time. 407-nm reactivation pulses were again applied after photobleaching to generate novel sparse subsets of emissive EYFP-MreB.

Figures 5G–J present the results from TL-PALM imaging of EYFP-MreB in *C. crescentus*. Two distinct MreB superstructures were identified in the cells: a quasi-helical arrangement in a stalked cell (Figs. 5G,H) and a midplane ring in the predivisional cell (Figs. 5I,J). In Figures 5H and J, the false-color PALM reconstruction is superimposed on a reversed-contrast, white-light image of the cell in which the EYFP-MreB was visualized. The images obtained by TL-PALM were more continuous than those acquired without the introduction of dark periods. Furthermore, the use of TL increased the number of localization events (487 in Fig. 5G and 330 in Fig. 5J) to the point in which the Nyquist criterion for 40-nm resolution was satisfied (Shroff et al. 2008), giving rise to a true superresolution reconstruction of the MreB superstructure in *C. crescentus*.

CONCLUSIONS

Single-molecule imaging permits high-resolution measurements of fluorescent emitters in time and space. This noninvasive, nonperturbative technique is quite useful for the study of single protein molecules in live cells. By applying the techniques of single-molecule microscopy to EYFP-tagged proteins in live *C. crescentus* cells, the diffusion dynamics, localization patterns, and structure of three key proteins, PleC, PopZ, and MreB, could be explored. By providing a noninvasive tool for high-resolution imaging that is beginning to approach the spatial resolution of cryo-electron microscopy, single-molecule, superresolution imaging can enable investigations of many more intracellular processes and has become an important biophysical tool.

Because the single-molecule-based imaging techniques described here are based on wide-field microscopy, they can further provide the basis for more sophisticated experiments. Specific challenges that are being addressed include improving the axial resolution of single-molecule microscopes (Huang et al. 2008; Juette et al. 2008; Pavani et al. 2009), incorporating multiple excitation and emission wavelengths (Bates et al. 2007; Shroff et al. 2007), and imaging thick samples (Vaziri et al. 2008).

ACKNOWLEDGMENTS

The authors warmly thank members of the Lucy Shapiro laboratory, specifically Ellen Judd, Harley McAdams, Grant Bowman, and Zemer Gitai, as well as members of the Moerner laboratory, specifically Jason Deich, Marcelle Koenig, So Yeon Kim, Anika Kinkhabwala, Michael Thompson, and Nicole Tselentis, for their contributions to this work. This research was supported in part by DARPA Grant MDA-972-00-1-0032, NSF Grant MCB-0212503, Department of Energy Grant DE-FG02-04ER63777, and NIH Grants P20-HG003638, R01-GM086196, and R01-GM085437.

REFERENCES

Ambrose WP, Moerner WE. 1991. Fluorescence spectroscopy and spectral diffusion of single impurity molecules in a crystal. *Nature* **349:** 225–227.

Bates M, Huang B, Dempsey GT, Zhuang X. 2007. Multicolor super-resolution imaging with photo-switchable fluorescent probes. *Science* **317:** 1749–1753.

Betzig E, Patterson GH, Sougrat R, Lindwasser OW, Olenych S, Bonifacino JS, Davidson MW, Lippincott-Schwartz J, Hess HF. 2006. Imaging intracellular fluorescent proteins at nanometer resolution. *Science* **313:** 1642–1645.

Biteen JS, Thompson MA, Tselentis NK, Bowman GR, Shapiro L, Moerner WE. 2008. Super-resolution imaging in live *Caulobacter crescentus* cells using photoswitchable EYFP. *Nat Meth* **5:** 947–949.

Bowman GR, Comolli LR, Zhu J, Eckart M, Koenig M, Downing KH, Moerner WE, Earnest T, Shapiro L. 2008. A polymeric protein anchors the chromosomal origin/ParB complex at a bacterial cell pole. *Cell* **134:** 945–955.

Conley NR, Biteen JS, Moerner WE. 2008. Cy3-Cy5 covalent heterodimers for single-molecule photoswitching. *J Phys Chem B* **112:** 11878–11880.

Deich J, Judd EM, McAdams HH, Moerner WE. 2004. Visualization of the movement of single histidine kinase molecules in live *Caulobacter* cells. *Proc Nat Acad Sci* **101:** 15921–15926.

Dickson RM, Cubitt AB, Tsien RY, Moerner WE. 1997. On/Off blinking and switching behavior of single green fluorescent protein molecules. *Nature* **388:** 355–358.

Dye NA, Pincus Z, Theriot JA, Shapiro L, Gitai Z. 2005. Two independent spiral structures control cell shape in *Caulobacter. Proc Nat Acad Sci* **102:** 18608–18613.

Gitai Z, Dye N, Shapiro L. 2004. An actin-like gene can determine cell polarity in bacteria. *Proc Nat Acad Sci* **101:** 8643–8648.

Gitai Z, Dye NA, Reisenauer A, Wachi M, Shapiro L. 2005. MreB actin-mediated segregation of a specific region of a bacterial chromosome. *Cell* **120:** 329–341.

Harms GS, Cognet L, Lommerse PHM, Blab GA, Schmidt T. 2001. Autofluorescent proteins in single-molecule research: applications to live cell imaging microscopy. *Biophys J* **80:** 2396–2408.

Hess ST, Girirajan TPK, Mason MD. 2006. Ultra-high resolution imaging by fluorescence photoactivation localization microscopy. *Biophys J* **91:** 4258–4272.

Huang B, Wang W, Bates M, Zhuang X. 2008. Three-dimensional super-resolution imaging by stochastic optical reconstruction microscopy. *Science* **319:** 810–813.

Juette MF, Gould TJ, Lessard MD, Mlodzianoski MJ, Nagpure BS, Bennett BT, Hess ST, Bewersdorf J. 2008. Three-dimensional sub-100 nm resolution fluorescence microscopy of thick samples. *Nat Meth* **5:** 527–529.

Kim SY, Gitai Z, Kinkhabwala A, Shapiro L, Moerner WE. 2006. Single molecules of the bacterial actin MreB undergo directed treadmilling motion in *Caulobacter crescentus. Proc Nat Acad Sci* **103:** 10929–10934.

Lee H-D, Dubikovskaya EA, Hwang H, Semyonov AN, Wang H, Jones LR, Twieg RJ, Moerner WE, Wender PA. 2008. Single-molecule motions of oligoarginine transporter conjugates on the plasma membrane of chinese hamster ovary cells. *J Am Chem Soc* **130:** 9364–9370.

Moerner WE. 2003. Optical measurements of single molecules in cells. *Trends Analyt Chem* **22:** 544–548.

Moerner WE, Fromm DP. 2003. Methods of single-molecule fluorescence spectroscopy and microscopy. *Rev Sci Instrum* **74:** 3597–3619.

Pavani SRP, Thompson MA, Biteen JS, Lord SJ, Liu N, Twieg RJ, Piestun R, Moerner WE. 2009. Three-dimensional, single-molecule fluorescence imaging beyond the diffraction limit by using a double-helix point spread function. *Proc Nat Acad Sci* **106:** 2995–2999.

Robbins JR, Monack D, McCallum SJ, Vegas A, Pham E, Goldberg MB, Theriot JA. 2001. The making of a gradient: IcsA (VirG) polarity in *Shigella flexneri*. *Mol Microbiol* **41:** 861–872.

Rudner DZ, Pan Q, Losick RM. 2002. Evidence that subcellular localization of a bacterial membrane protein is achieved by diffusion and capture. *Proc Nat Acad Sci* **99:** 8701–8706.

Rust MJ, Bates M, Zhuang X. 2006. Sub-diffraction-limit imaging by stochastic optical reconstruction microscopy (STORM). *Nat Meth* **3:** 793–795.

Sako Y, Minoghchi S, Yanagida T. 2000. Single-molecule imaging of EGFR signalling on the surface of living cells. *Nat Cell Biol* **2:** 168–172.

Schütz GJ, Kada G, Pastushenko VP, Schindler H. 2000. Properties of lipid microdomains in a muscle cell membrane visualized by single molecule microscopy. *EMBO J* **19:** 892–901.

Shapiro L, McAdams H, Losick R. 2002. Generating and exploiting polarity in bacteria. *Science* **298:** 1942–1946.

Shroff H, Galbraith CG, Galbraith JA, Betzig E. 2008. Live-cell photoactivated localization microscopy of nanoscale adhesion dynamics. *Nat Meth* **5:** 417–423.

Shroff H, Galbraith CG, Galbraith JA, White H, Gillette J, Olenych S, Davidson MW, Betzig E. 2007. Dual-color superresolution imaging of genetically expressed probes within individual adhesion complexes. *Proc Nat Acad Sci* **104:** 20308–20313.

Steinhauer J, Agha R, Pham T, Varga AW, Goldberg MB. 1999. The unipolar *Shigella* surface protein IcsA is targeted directly to the bacterial old pole: IcsP cleavage of IcsA occurs over the entire bacterial surface. *Mol Microbiol* **32:** 367–377.

Thompson RE, Larson DR, Webb WW. 2002. Precise nanometer localization analysis for individual fluorescent probes. *Biophys J* **82:** 2775–2783.

van den Ent E, Amos LA, Lowe J. 2001. Prokaryotic origin of the actin cytoskelecton. *Nature* **413:** 39–44.

Vaziri A, Tang J, Shroff H, Shank CV. 2008. Multilayer three-dimensional super resolution imaging of thick biological samples. *Proc Nat Acad Sci* **105:** 20221–20226.

Vrljic M, Nishimura SY, Brasselet S, Moerner WE, McConnell HM. 2002. Translational diffusion of individual class II MHC membrane proteins in cells. *Biophys J* **83:** 2681–2692.

Vrljic M, Nishimura SY, Moerner WE, McConnell HM. 2005. Cholesterol depletion suppresses the translational diffusion of class II major histocompatibility complex proteins in the plasma membrane. *Biophys J* **88:** 334–347.

Wheeler RT, Shapiro L. 1999. Differential localization of two histidine kinases controlling bacterial cell differentiation. *Mol Cell* **4:** 683–694.

Xie XS, Choi PJ, Li GW, Lee NK, Lia G. 2008. Single-molecule approach to molecular biology in living bacterial cells. *Annu Rev Biophys* **37:** 417–444.

Automated Quantitative Live Cell Fluorescence Microscopy

Michael Fero[1] and Kit Pogliano[2]

[1]Department of Developmental Biology, Stanford University School of Medicine, Stanford, California 94305
[2]Division of Biological Sciences, University of California San Diego, La Jolla, California 92093-0349
Correspondence: mike.fero@stanford.edu

Advances in microscopy automation and image analysis have given biologists the tools to attempt large scale systems-level experiments on biological systems using microscope image readout. Fluorescence microscopy has become a standard tool for assaying gene function in RNAi knockdown screens and protein localization studies in eukaryotic systems. Similar high throughput studies can be attempted in prokaryotes, though the difficulties surrounding work at the diffraction limit pose challenges, and targeting essential genes in a high throughput way can be difficult. Here we will discuss efforts to make live-cell fluorescent microscopy based experiments using genetically encoded fluorescent reporters an automated, high throughput, and quantitative endeavor amenable to systems-level experiments in bacteria. We emphasize a quantitative data reduction approach, using simulation to help develop biologically relevant cell measurements that completely characterize the cell image. We give an example of how this type of data can be directly exploited by statistical learning algorithms to discover functional pathways.

Over the last decade, advances in microscope automation, fluorescent labeling, sample preparation, image processing, pattern recognition, and statistical learning have ushered in a new era of cell biology experiments based on unbiased genome-wide cell-based assays. In eukaryotic cell biology, these advances have been exploited to investigate the distribution and patterning of labeled subcellular structures and the "location proteome" defined by the distribution and intensity of localized protein reporters (Chen et al. 2006). In the prokaryotic world, the challenges surrounding fluorescence microscopy at the diffraction limit has meant a somewhat slower adoption of highly automated methods and analysis designed for large-scale genetic screens or assays. Nevertheless, large-scale approaches are now being used more routinely during investigations of bacterial systems, from the identification of entire location proteomes (Werner et al. 2009) to the detection of modular transcription and signaling pathways (Christen and Fero 2009). Because of the size and complexity of the eukaryotic cell, generalized pattern recognition and classification approaches are often crucial. In these approaches, conditioned, parameterized, preclassified image data are used by

 doi: 10.1101/cshperspect.a000455
Cite this article as *Cold Spring Harb Perspect Biol* 2010;2:a000455

statistical learning algorithms to build models that capture the differences between classes of images. A properly constructed model can then be used to automatically analyze a previously unclassified experimental data set (Boland et al. 1998; Boland and Murphy 1999; Boland and Murphy 2001; Huang and Murphy 2004; Chen and Murphy 2005; Jones et al. 2009). In prokaryotic cell biology a similar situation exists, with one major difference; the simplicity of the bacterial cell allows us to entertain the idea of a cell parameterization based not only on generalized image properties, but also on specific cell measurements of high biological relevance. For example, instead of measuring hard to interpret properties based on generalized image parameters and a global image transform, such as a Zernike or Fourier transform, one can conceive of measuring a more relevant set of parameters related directly to gene function or to the biology of interest, such the position and quantity of both localized and delocalized fluorescent signals, the length and width of the cell, the degree of pinch at the division plane or sporulation septum, the precise location of the poles or other points of inflection, as well as the location of membrane structures such as pili or stalks and the membrane outline itself. In this way, albeit with a certain degree of additional work, the image can be reduced to a set of high information content parameters. This type of data can be used more efficiently in any type of subsequent analysis, whether direct or via a model-building statistical learning exercise. In addition, reducing the cell to a small set of highly informative parameters allows the option of combining those single-cell measurements into quantitative ensemble-based phenotypes where the entire distribution of measured values is recorded rather than just means and standard deviations. With the addition of genotype information, this data can then be used directly in unsupervised statistical learning approaches such as hierarchical clustering to do quantitative phenotypic profiling (Ohya et al. 2005). In this article we outline the process of doing such an automated analysis on a bacterial system. To keep our discussion firmly rooted in a practical example, we will use a case-study approach based on work performed in the α-proteobacteria *Caulobacter crescentus*. However, most of what we discuss can be applied to other bacteria and single-cell eukaryotes. It should also be noted that the efficacy of a quantitative analysis depends on the constructs used to express fluorescently tagged reporter proteins. Even modest overproduction of some proteins, particularly those involved in signal transduction and cell division, can have deleterious effects on cell viability and on cell shape (Gregory et al. 2008). We will assume that for the purposes of quantitative analyses, tagged protein reporters are fully functional proteins produced from genes in their native chromosomal context.

AUTOMATED LIVE CELL FLUORESCENT MICROSCOPY

Live cell, 100×, oil immersion, fluorescence microscopy presents some problems for a high-throughput system not found in lower resolution systems. First, the geometry is tightly constrained, with a ~2 mm gap between the focal plane and the condenser, and ~0.24 mm spacing between the 100× objective and the focal plane. The cells must be kept alive, which usually means imaging on a 1% agarose + nutrient gel pad and limiting the amount of time the samples are under observation. Fluorescence microscopy usually entails imaging in multiple optical channels; the multiplexed mutant screen example described later used one phase contrast plus three fluorescent channels. This means the optical system has to be stable for several seconds while cubes or filter wheels are automatically switched. The need for optical coupling oil also drives the design of the microscopy plate and choice of inverted versus upright scope geometry.

Data Acquisition and Autofocusing

Acquisition timescales range from tens of milliseconds for optical channel exposure times to seconds required for some mechanical system motions, all the way to the hundreds of minutes to needed to push through thousands of samples. Several issues arise when trying to automatically image multiple live samples. Image drift, the availability of nutrients and oxygen over long time periods, the cells doubling time, and drying out of agarose pads are among the issues that determine how long a sample plate can sit on the microscope and thus how long a data-taking run can last. Experience shows that data-taking runs lasting approximately 1 h work well. If imaging in a 96-well microplate format this implies a throughput of 96 sample-wells per hour. Thus, we have roughly 30 s to: move to the well, coarse focus, fine focus, image capture, and write the data in phase contrast plus several fluorescent channels. Focusing is a large contributor to the time it takes to image a sample well. Because of fluorophore bleaching, using any fluorescent channel to optimize focus is usually unwise. This leads to using an alternative white-light channel such as phase contrast or DIC to set the focus for subsequent fluorescent images. There are several schemes for calculating z-focus quality scores (Groen et al. 1985; Firestone et al. 1991; Price and Gough 1994; Geusebroek et al. 2000) but in our experience we have had good results with a simple autocorrelation function. Images are acquired in a fast loop while changing the z-focus depth. The z-focus quality score is calculated for each z-focus depth. A fast parabolic fit to the points surrounding the maximum of the graph of all z-focus quality scores yields the optimal z-focus depth. This is a very fast way of finding the best focus and works well for multiple closely spaced image fields. However, when the automated imaging system has to move over a macroscopic distance, for example to the next well in a 96- or 384-well microscopy plate, the fine focusing described earlier will not be sufficient for finding a badly out-of-focus image field. Even very small tilts in the plane of the microplate can put an image very far out of focus after translating the objective by several millimeters. To help the autofocusing heuristic an overall calibration or offset is needed to account for any small misalignments between the perpendicular to the sample plane and the optical axis. One method is to use an offset calculated by manually taking focused image data at 3–4 points on the microscopy plate and then fitting a plane to the calibration points. This is performed before automated data taking begins. The *z*-coordinate of the of the fitted calibration plane at the location of the center of each microscopy-plate-well becomes the zero offset for subsequent autofocusing. Alternatively, if available on the microscope setup, one can use laser range finding to calculate the distance to the cover slip. This distance plus the cover slip thickness gives the starting point for subsequent fine focusing. The fine focusing step is implemented with a precision piezo electric objective insert or peizo electric actuated stage insert. An approach using the image data is used because the best focus is only found using the full image as observed through the optical system and CCD camera. Even though the procedure outlined earlier will yield satisfactory results in the fluorescent channels, the correct focal plane in the white-light channel used for autofocusing will not necessarily match the correct focal plane for different fluorescent channels. However, this is a repeatable physical effect and can be compensated for by measuring the focal plane offsets for each fluorescent channel versus each other and versus the optimal phase contrast focal plane. Once measured, this becomes a constant offset when setting the z-focus for that particular fluorescent channel.

Image Stability

Two important considerations for automating data taking and analysis are that one can take a well-focused fluorescent image without resorting to a z-stack, and that repeated frames, such as those found in a time series, are in good alignment. During the analysis, it is possible to correct for out-of-alignment frames, as it is possible to automatically search for the best-focused image out of a z-stack. However, correcting for issues resulting from poor data takes time and is likely to result in an increase in the overall systematic error of a quantitative measurement. An environmentally controlled microscope setup will minimize drift in x–y

or *z*-focus coordinates regardless of vibration, temperature, or humidity changes in the surrounding environment. Image drift over some relevant time scale should be less than the maximum resolution of the optical system. For 100× fluorescent microscopy this means stability against drift at less than a ten nano meters per second. Systems that achieve this level of stability can image in phase contrast plus several fluorescent channels while keeping the optical drift at the subpixel level. Note that it can take several minutes for the microscopy plate to come into thermal equilibrium after being introduced to a controlled environment. Acquiring images too early can lead to significant drift and result in misregistered and out-of-focus images.

Data Model and Storage

For flexibly analyzing large amounts of data the images must be indexed, for example by the optical channel imaged (c), the z-depth of focus in a multiple z-stack set of images (z), a time index for the purposes of time-lapse (t), the selected image field for a particular sample (f), and finally an index for the unique sample (w). Once image data is acquired it needs to be associated with its corresponding metadata and stored. Getting complete metadata associated with the correct image data can be a vexing problem. Image information from the microscope and camera that can be acquired automatically can be easily saved. Ensuring that information provided by the experimenter is properly saved is more difficult. One scheme requires the upload of an experimental metadata file at the start of a data-taking run. However, user friendly tools for properly annotating image data with user-generated experimental metadata after the fact are also important. Regarding computer power and data storage, the capability of software allowing individual investigators to organize, manage, and analyze a large amount of image data is improving so rapidly that any comment made here will be quickly out of date. In our experience, data analysis ambitions tend to scale with increasing computer and data storage capabilities, to the point where dedicated image servers are used for image libraries and a multiple node compute clusters are used for image analysis.

AUTOMATED QUANTITATIVE IMAGE ANALYSIS

Our example of an automated analysis work flow starts with a high-quality phase contrast or DIC image and one or more fluorescent-channel images for each image field as seen in Figure 1. In the first analysis stage, cell finding

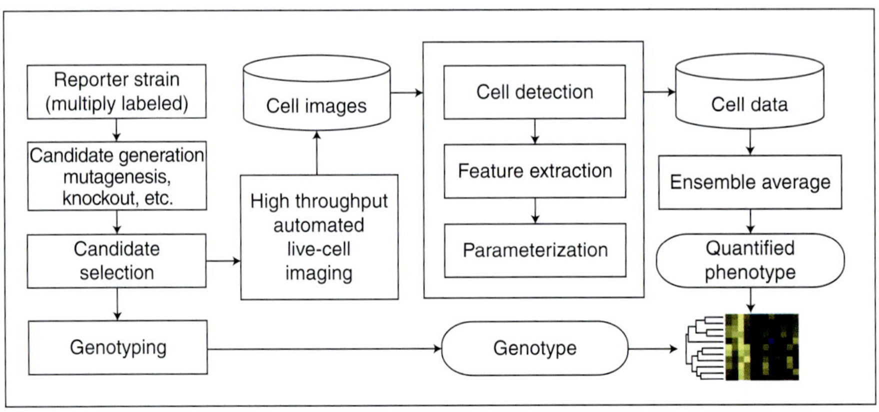

Figure 1. Automated image analysis. Stored images are segmented by individual cell and parameterized using all available image information. Parameterized data is stored to disk for second-pass analysis where cell parameters are converted to biologically relevant metrics that can be used for quantitative phenotyping.

is followed by subcellular feature detection and parameterization at the single-cell level as shown in Figure 2. The data reduction factor at this stage is 10–20-fold depending on the complexity of the image. In the second stage of the analysis, the individual cell measurements are combined into high-statistics ensemble averages and standard deviations. A typical ensemble might consist of all of the cells from all of the images taken for a particular sample well. At this stage ensemble averages are translated in to biologically relevant phenotypes and the results can be tabulated, plotted, or subjected to more advanced classification algorithms. Note that by making individual cell measurements, we have access to the distribution of each measured parameter on an ensemble basis and not just the average and standard deviation. There are observable phenotypes that do not significantly change the average or standard deviation, but do noticeably change the distribution of measured values. For example, high intensity measurements in the tail of a distribution of localized fluorescence signals may disappear in a particular mutant background, while the core distribution remains intact (Christen and Fero 2009). The reason for separating the cell detection and parameterization from the rest of the analysis is that the processing time for the two stages differs markedly. Because cell parameterization can require 100 times more CPU time than ensemble characterization, it is not wise to directly couple the two parts of the analysis. Once the images have been processed and the cells reduced to their data summary form, it is not necessary to repeat this processing step. Instead, the investigator can concentrate on refining the second stage of the analysis, which is where most of the biological content lies.

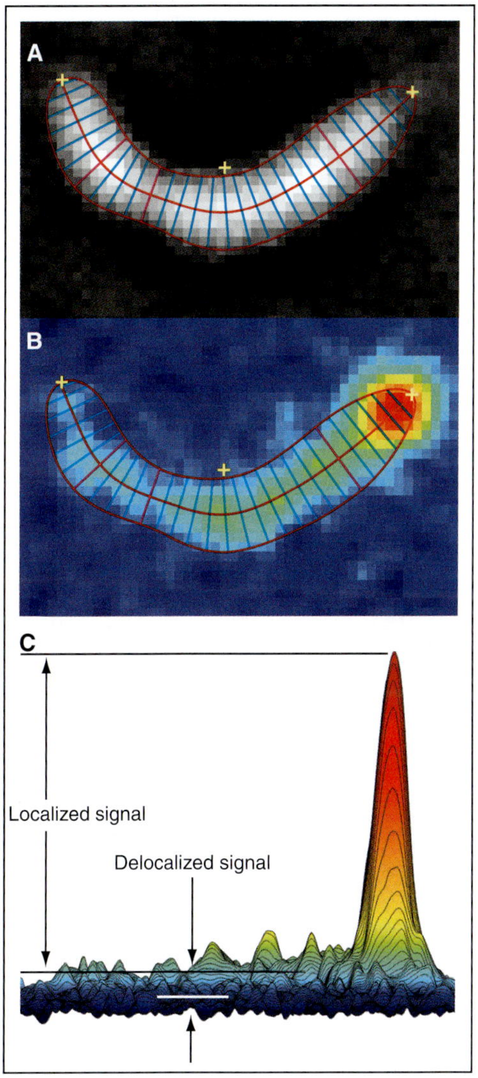

Figure 2. Measurements made on an individual bacterial cell. (*A*) The membrane profile from a subpixel resolution spline fit to a phase-contrast image is combined with, (*B*) subpixel resolution measurements of fluorescent foci and delocalized fluorescent signal to provide, (*C*) a parameterized view of the cell including localized and delocalized fluorescent contributions.

STAGE ONE: CELL LEVEL MEASUREMENTS

Cell Finding and Image Segmentation

Every automated analysis starts with finding cells. Under the controlled experimental conditions of a bacterial fluorescence microscopy experiment, the cell density can be tuned to provide a distribution of cells with a minimal number of touching or overlapping cells. Unlike many eukaryotic cell types, where cell finding and segmentation can be a challenging task (Kasson et al. 2005), cell finding in bacterial systems is a fairly straight forward, usually involving either a thresholding or edge-detection, region-filling approach performed on the phase contrast channel. To compensate for uneven illumination of the phase-contrast channel a simple block-filtering algorithm can be used for background flattening. This step followed by an automatically adjusted thresholding and region-finding algorithm works well to identify cells. It is not crucial at this stage to perfectly identify the cell boundary. Rather, the aim is to properly tag the potential cell locations. After tagging the location of each cell, the original image can be segmented into a number of sub-images, one for each cell, in preparation for detailed examination.

Membrane Profiling

When analyzing the smaller subimages the cell-finding algorithm is repeated, in a more refined way, to get a good measure of the cell profile in two dimensions. The goal, in our example, is a membrane profile representing a section half-way through the cell in the *z*-direction. This can be a difficult step, if care is not taken delicate structures that are marginally visible are lost. In some cases it is possible to assume that the membrane boundary has no sharp discontinuities and a spline-fit (De Boor 1978) can be fit to a threshold boundary yielding a correct subpixel representation of the membrane profile as shown in Figure 2A. Once a membrane profile is available, rod shaped cells can be defined by an internal coordinate scheme defined by the longitudinal midcell line, the poles, and several measures of cell width. Note that here we have taken care to define an internal coordinate system defined by an midcell line that runs from one cell pole to the other, with orthogonal width measures at right angles to the midcell line.

Cell Image Simulation

Vital to the process of developing algorithms that yield results with believable error estimates is the implementation of a simple cell-image simulation, as shown in Figure 3A. With a simulation we know, at the outset, what the "true" (expected) values are for the parameters that

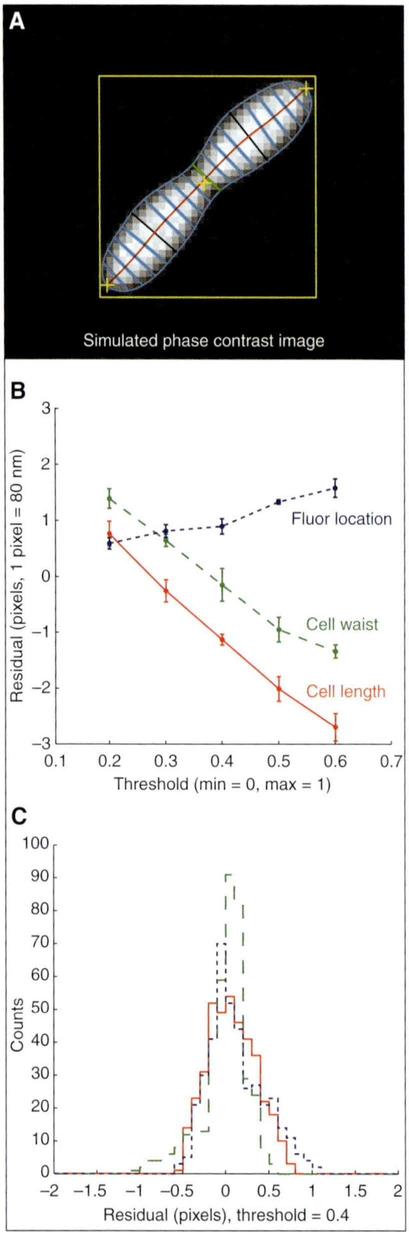

Figure 3. Using cell image simulation when developing algorithms. (*A*) An image simulation of a bacterial cell has been used to test a membrane-fitting algorithm. (*B*) The difference between the predicted and measured length and width of the cell as a function of threshold shows a systematic bias. (*C*) The effect on measuring a fluorescent spot relative to the measured pole of the cell. After correcting for the threshold effect a simulation run of 400 cells shows the resolution of the algorithms, given by the width of the residual histograms as shown.

will eventually be measured by the analysis algorithm. Starting with an idealized cell profile, one can apply the effect of the optical system (in this case the optical point-spread-function plus the finite pixel size of the imaging system), plus an additional dose of simulated noise that emulates the noise seen in the actual images, to create a pseudo-cell image that for our purposes cannot be distinguished from a real cell image. When an analysis algorithm is applied to this image, its output can be compared with the input to the simulation to yield a residual plot of the expected value minus the measured value as a function of the measured parameter, as shown in Figure 3B. As seen in the figure, this analysis has quantified the degree of the negative correlation between a threshold parameter used in the membrane finding algorithm and two measures of cell size. This immediately tells the investigator that, depending on the threshold parameter used, the measurement may have a systematic error of up to four pixels. The measurement only falls into line with the input value to the simulation for fairly low values of the threshold. If these threshold values are too low for reliable cell finding, one option is to use a higher threshold, and then correct the length and width measurements with a factor determined from the simulation. The next step is to run a simple Monte Carlo simulation where numerous simulated cells are generated with varying sizes and amounts of noise. A histogram of the residual values determined from the difference between expected and measured values (with corrections) will produce a Gaussian-like distribution, the width of which can be taken to be a lower bound on the systematic error because of the optical properties of the system and the measurement algorithm itself. Figure 3C shows the result of such a Monte Carlo simulation for 400 randomly produced cells. The residual histograms show that subpixel resolution can be achieved if parameter-dependent algorithms are properly tuned. Without the adjustments determined from cell-image simulation, systematic errors larger than the inherent resolution of the optical system can be unwittingly introduced. Finally, the algorithm can be checked for accuracy with a standard of known range in diameters such as those given by commercially available polystyrene beads (Werner et al. 2009).

Localized and Delocalized Florescent Signals

Properly imaged fluorescent signals from GFP-protein fusions have excellent signal-to-noise characteristics. It is often a simple matter to find the fluorescent maxima and, with criteria based on proximity of the peaks to the structures determined by the membrane profile, assign the fluorescence to the correct cell as shown in Figure 2B. If point-like in origin, the fluorescent signal will look like a well defined two-dimensional Gaussian peak with a width defined by the point spread function of the optical system, as shown in Figure 2C. Knowledge of the underlying source of the fluorescent signal puts powerful constraints on the signal and helps to eliminate spurious pixel noise (with widths much smaller than the PSF) (Szelezniak 2007) and out-of-focus channels or fluorescent contamination or aggregates (with widths much larger than the PSF). Slightly less simple to determine is the level of the localized fluorescent signal, and substantially more difficult to determine is the level or gradient across the cell of nonlocalized signal. When one signal underlies another it is often hard to disentangle them because one may be a background contribution to the other. To make the analysis rigorous we recommend renormalizing the image data, scaling it between zero and one, where a baseline of zero is set to the value of the minimum-intensity pixel and a normalized intensity of one was assigned to the brightest pixel. The scale factors and offsets should be preserved so that after analysis the image can be translated back into the original fluorescence units. Cell image simulation is again useful here for developing a well understood algorithm that will yield a measurement with reasonable systematic errors. To characterize the amplitude of the localized signal one can fit a two-dimensional Gaussian to the data with starting parameters for a nonlinear search algorithm given by (1) starting XY coordinates given by the peak centroid, (2) a starting width determined by the PSF, and (3) a starting amplitude given by the amplitude of the brightest pixel in the fluorescent peak. The output from the fit can be compared with a simulated point like light source and compared as a method for tuning the peak finding algorithm. In this way it is possible to obtain subpixel resolution on the XY coordinate of the center of the peak. The absolute amplitude of the localized signal is more difficult and remains dependent on the amount of background attributed to delocalized, cytoplasmic signal and whether it contributes to the localized signal in an additive way. We have tested various techniques via cell image simulation to account for the delocalized signal and have found one that works reasonably well. The technique involves counting the number of discrete "found" regions inside the cell boundary in the (unfiltered) fluorescent channel of interest, as a continually varying threshold is applied. It is observed that the threshold yielding the maximum number of found regions corresponds well with the median of the cytoplasmic contribution to cell fluorescence.

STAGE TWO: ENSEMBLE LEVEL MEASUREMENTS

Cell Cycle Example

We have divided image analysis into an image processing stage where the individual cells are parameterized, and a second stage where ensemble averages are made and the cells are characterized biologically. We have mentioned the advantages of doing this from a processing point of view, the second stage being much less CPU intensive than the first. Another advantage is procedural; we have left the difficult technical task of reducing the cell image to a set of numbers and can now concentrate on the biology. This stage of the analysis is often repeated and tuned numerous times during the course of an analysis, with input from various collaborators, so it is useful not to have to reanalyze all of the images every time someone has a good idea about what to measure. What is measured at the ensemble level depends on the experiment being performed and is thus often unique. For example, if one were performing a cell cycle time-course experiment on polar protein abundance and localization with synchronized cells, it would be reasonable to plot the peak polar signal averaged over an ensemble of high quality cells for each time point in the cell cycle. An example of this type of analysis is shown in Figure 4 for a *Caulobacter crescentus* cell cycle experiment. The values shown are from ensemble averages over many cells and the error bars reflect the error in those measurements. Although it took several hundred seconds to parameterize all of the imaged cells during the first stage of this analysis, the second stage, involving selection of high quality data, determining the mean of the peak fluorescent signals in the three channels and plotting the data, takes only seconds.

Mutant Screening Example

A large-scale mutant screen provides a more challenging example. In this case it is necessary to process images from thousands of mutants, and then classify them on the basis of several criteria such as polar-localized signal amplitude and location, cytoplasmic-delocalized signal amplitude, localized protein polarity, as well as a host-cell shape parameters. As an example we show in Figure 5 the localized fluorescence amplitude signals from a triply labeled strain of *Caulobacter crescentus* used in a high throughput screen for mutants that mislocalized one or more of the reporter proteins. Each data point is an ensemble average over a number of cells whose peak signal is in the high intensity tail of the peak signal distribution. The data shows the distinct difference between reporter signals resulting from transposon disruptions in the PodJ and TacA genes and the unmutagenized control strains. In Figure 5 we can simultaneously observe three-dimensional phenotypic data, labeled by genotype. However, incorporating more fluorescence based phenotypic information requires analysis tools capable of dealing with higher-dimensional data. Because we know the genotype of every mutant analyzed for each fluorescent

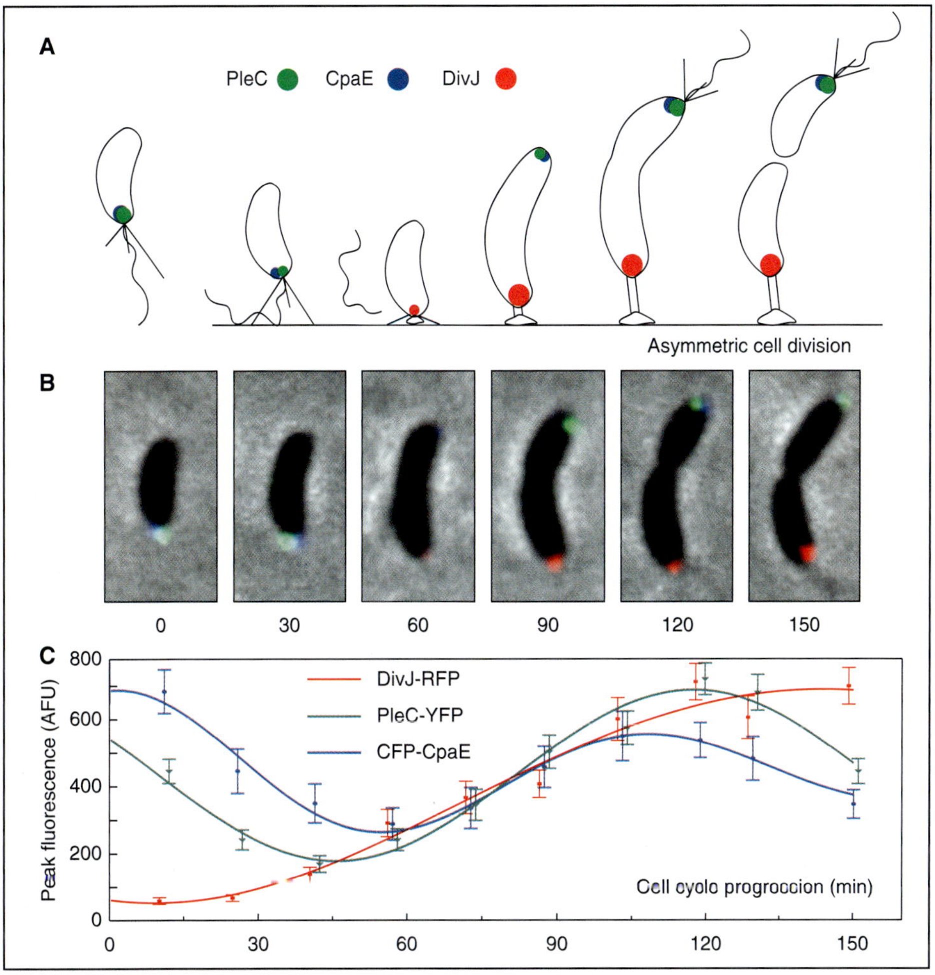

Figure 4. Polar localized protein concentration as a function of cell cycle. (*A*) An example of cell cycle dependent protein localization in *Caulobacter.* (*B*) An example single-cell image shows the correspondence with the visible changes in localized protein concentration (*C*) Automated analysis detects time variation in polar fluorescence intensities of the CpaE-CFP, PleC-YFP, and DivJ-RFP reporter fusions.

phenotype, we can use hierarchical clustering as shown in Figure 6, taking advantage of all the fluorescent data available in the screen. The cluster shown in Figure 6 represents a robust cluster of mutants that emerged from a much larger dendrogram of over 800 mutants. The genes identified in the cluster correspond to the elements of a pathway controlling the relocalization of the sensor histidine kinase DivJ to the stalked pole at the swarmer to stalked transition (Biondi et al. 2006; Radhakrishnan et al. 2008; Christen and Fero 2009). Thus, it is possible to completely automate a fluorescence microscopy based screen that leads to functional gene module identification.

Shape Characterization

The shape of a mutant or non–wild-type cell is often a critical part of a fluorescence microscopy experiment. Although one of the easiest things to pick out by eye, it is surprisingly difficult to parameterize mutant shapes, because it is hard to anticipate all the shapes one might

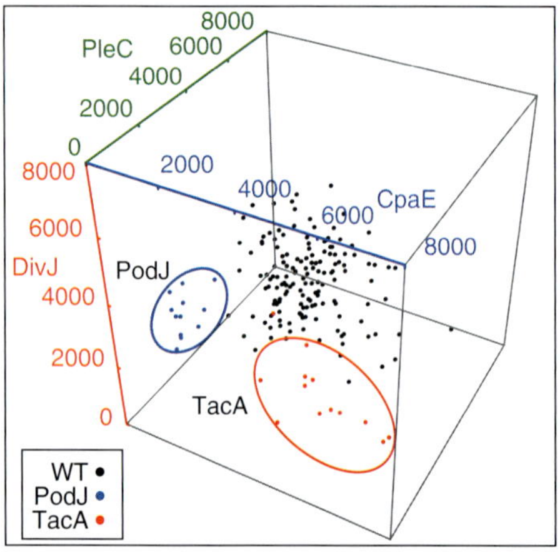

Figure 5. Ensemble measurement of polar localized protein based on individual cell measurements. The polar fluorescence intensities of the CpaE-CFP, PleC-YFP, and DivJ-RFP reporter fusions are plotted for the nonmutagenized control strain (black) as well as gene disruptions in two separate open reading frames coding for the PodJ (blue) and TacA (red) proteins.

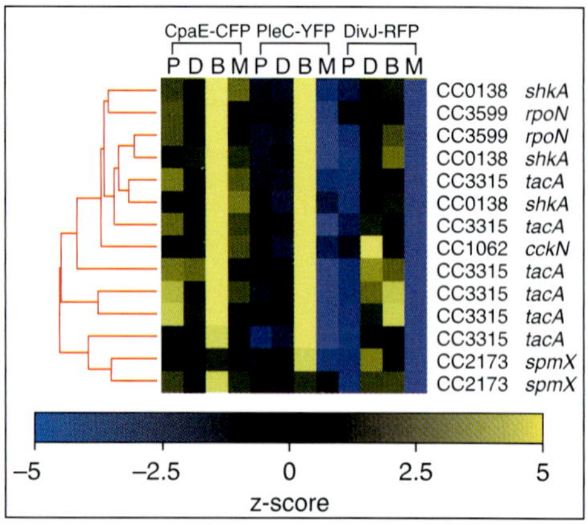

Figure 6. Example of Genotype-Phenotype microscopy based analysis. A collection of automatically determined cell metrics is used to identify a functional cluster indicating a gene module co-ordinating temporal and spatial protein localization. Heat map representation of the calculated z-score values from localized protein abundance (L), delocalized protein (D), the fraction of cells identified as being biopolar (B), and the fraction similarly identified as monopolar (M) are shown for a highly significant gene cluster designated by the dendrogram.

encounter. The membrane parameterization techniques we have described do a reasonably good job following a complex membrane profile but the chances of missing an interesting phenotype increase as the cells begin to differ markedly from what was expected by the algorithm developer. Thus, mutant shape characterization can benefit from more global analysis techniques. In these approaches, the cell is not assumed to have a particular form that yields to parameterization in the way that we have described above. Rather, the cells are interpreted as observations that can be compared either to each other or to a control set and tested for similarity, or to a set of alternative models for classification. This process lends itself to the use of pattern recognition and supervised or unsupervised statistical learning techniques. Supervised techniques require a training data set and produce a model that can then be applied to a test set for purposes of classification. The canonical examples of these approaches are classification and regression trees and various types of neural nets (Hastie et al. 2009). In an unsupervised approach, no assumption about what is expected is made and the algorithm is expected to bootstrap its way to finding unique classes based on the data itself (Heidmann 2005). Image data is not often directly amenable to these approaches and is usually first reduced in complexity by the computation of numerical representations of the image. After a cell has been thus re-interpreted it can be classified via one of the supervised or unsupervised approaches. Good examples of morphology analysis are found in Pincus and Jones (Pincus and Theriot 2007; Jones et al. 2009).

CONCLUDING REMARKS

Well-defined prokaryotic fluorescence microscopy experiments can now be performed in an almost completely automated way. Once a process has been developed, image data acquisition, first and second stage analysis and the presentation of final results can all be done without ever requiring the viewing, converting, or measuring of individual images by the biologist. In making this statement we do not which to trivialize the effort involved in developing automated acquisition and analysis tools. A brief reflection on cost vs. benefit will make it clear to the experimenter whether the upfront investment in automated data taking and analysis is required. We have illustrated a seamless pipeline approach, which requires a minimum amount of user intervention but a maximum of software developer effort. Unless the experiment is very high throughput, most experiments today can be performed quite successfully using nonautomated image acquisition and semiautomated quantitative analysis like those available using stand-alone tools such as ImageJ (Girish and Vijayalakshmi 2004; Collins 2007). However, as the interest in performing high-throughput assays and screens at the genomic or meta-genomic level in prokaryotic systems grows, the value of integrated, automated systems for data acquisition and analysis will become evident.

ACKNOWLEDGMENTS

This work was funded by National Institutes of Health Grant K25 GM070972-01A2 (M.F.).

REFERENCES

Biondi EG, Skerker JM, Arif M, Prasol MS, Perchuk BS, Laub MT. 2006. A phosphorelay system controls stalk biogenesis during cell cycle progression in Caulobacter crescentus. *Mol Microbiol* **59:** 386–401.

Boland MV, Murphy RF. 1999. Automated analysis of patterns in fluorescence-microscope images. *Trends Cell Biol* **9:** 201–202.

Boland MV, Murphy RF. 2001. A neural network classifier capable of recognizing the patterns of all major subcellular structures in fluorescence microscope images of HeLa cells. *Bioinformatics* **17:** 1213–1223.

Boland MV, Markey MK, Murphy RF. 1998. Automated recognition of patterns characteristic of subcellular structures in fluorescence microscopy images. *Cytometry* **33:** 366–375.

Chen X, Murphy RF. 2005. Objective clustering of proteins based on subcellular location patterns. *J Biomed Biotechnol* **2005:** 87–95.

Chen X, Velliste M, Murphy RF. 2006. Automated interpretation of subcellular patterns in fluorescence microscope images for location proteomics. *Cytometry* **69:** 631–640.

Christen B, Fero M, Hillson N, Bowman G, Hong S-H, Shapiro L, McAdams H. 2010. High-throughput identification of protein localization dependency networks. *Proc Natl Acad Sci* **107:** 4681–4686.

Collins TJ. 2007. ImageJ for microscopy. *BioTechniques* **43:** 25–30.

De Boor C. 1978. *A practical guide to splines*. Springer-Verlag, New York.

Firestone L, Cook K, Culp K, Talsania N, Preston K Jr. 1991. Comparison of autofocus methods for automated microscopy. *Cytometry* **12:** 195–206.

Geusebroek JM, Cornelissen F, Smeulders AW, Geerts H. 2000. Robust autofocusing in microscopy. *Cytometry* **39:** 1–9.

Girish V, Vijayalakshmi A. 2004. Affordable image analysis using NIH Image/ImageJ. *Ind J Cancer* **41:** 47.

Gregory JA, Becker EC, Pogliano K. 2008. *Bacillus subtilis* MinC destabilizes FtsZ-rings at new cell poles and contributes to the timing of cell division. *Genes Develop* **22:** 3475–3488.

Groen FC, Young IT, Ligthart G. 1985. A comparison of different focus functions for use in autofocus algorithms. *Cytometry* **6:** 81–91.

Hastie T, Tibshirani R, Friedman JH. 2009. *The elements of statistical learning: Data mining, inference, and prediction.* Springer, New York.

Heidmann G. 2005. Unsupervised image categorization. *Image and Vision Computing* **23:** 861–876.

Huang K, Murphy RF. 2004. From quantitative microscopy to automated image understanding. *J Biomed Optics* **9:** 893–912.

Jones TR, Carpenter AE, Lamprecht MR, Moffat J, Silver SJ, Grenier JK, Castoreno AB, Eggert US, Root DE, Golland P, et al. 2009. Scoring diverse cellular morphologies in image-based screens with iterative feedback and machine learning. *Proc Natl Acad Sci* **106:** 1826–1831.

Kasson PM, Huppa JB, Davis MM, Brunger AT. 2005. A hybrid machine-learning approach for segmentation of protein localization data. *Bioinformatics* **21:** 3778–3786.

Ohya Y, Sese J, Yukawa M, Sano F, Nakatani Y, Saito TL, Saka A, Fukuda T, Ishihara S, Oka S, et al. 2005. High-dimensional and large-scale phenotyping of yeast mutants. *Proc Natl Acad Sci* **102:** 19015–19020.

Pincus Z, Theriot JA. 2007. Comparison of quantitative methods for cell-shape analysis. *J Micros* **227:** 140–156.

Price JH, Gough DA. 1994. Comparison of phase-contrast and fluorescence digital autofocus for scanning microscopy. *Cytometry* **16:** 283–297.

Radhakrishnan SK, Thanbichler M, Viollier PH. 2008. The dynamic interplay between a cell fate determinant and a lysozyme homolog drives the asymmetric division cycle of Caulobacter crescentus. *Genes Dev* **22:** 212–225.

Szelezniak M. 2007. Cluster finder in the HFT readout.

Werner JN, Chen EY, Guberman JM, Zippilli AR, Irgon JJ, Gitai Z. 2009. Quantitative genome-scale analysis of protein localization in an asymmetric bacterium. *Proc Natl Acad Sci* **106:** 7858–7863.

Index